System Analysis
Techniques

System Analysis Techniques

Edited by

J. DANIEL COUGER, D.B.A.
Professor of Computer and Management Science
University of Colorado

and

ROBERT W. KNAPP, Ph.D.
Associate Professor of Business Administration
University of Colorado

JOHN WILEY & SONS

New York   London   Sydney   Toronto

Copyright © 1974, by John Wiley & Sons, Inc.

All rights reserved. Published simultaneously in Canada.

No part of this book may be reproduced by any means, nor transmitted, nor translated into a machine language without the written permission of the publisher.

Library of Congress Cataloging in Publication Data

Couger, J. Daniel, comp.
   System analysis techniques.

   Includes bibliographies.
   1. System analysis. 2. Management information systems. I. Knapp, Robert W., 1935–   joint comp. II. Title.

T57.6.C68   658.4′032   73–14818
ISBN 0-471-17735-0

Printed in the United States of America

10 9 8 7 6 5 4

Dedicated to
Dean WILLIAM H. BAUGHN
who was greatly supportive in our efforts
to implement an information system development
curriculum at the University of Colorado.

PREFACE

The need for this book became apparent during the design of a degree program in Information Systems Development. One of the authors served as a member of the eight-man A.C.M. committee which was provided a $98,000 grant by N.S.F. to develop this degree program. The committee is the business counterpart of the committee that designed the Computer Science program, Curriculum 68, widely adopted in the U.S. and overseas. With the way paved by Curriculum 68, the degree program in Information Systems Development has been well received in U.S. colleges and universities.

Two of the courses in the information system curriculum are the course in system analysis and the course in system design. The system *analysis* course is concerned with analyzing existing systems, determining managerial information requirements, and evaluating the benefits to be derived through computerization of management systems. The system *design* course is concerned with the design of new systems and development of system specifications for programming. The system analysis phase produces the logical design for the new system; the system design phase produces the physical design.

One of the primary tasks of the A.C.M. curriculum committee was preparation of a bibliography to support the courses. The committee was not able to locate a book that covered the techniques to be taught in the system analysis course, particularly those techniques which use the computer as an aid in system analysis. However, a number of good journal articles and papers describe these techniques. These publications have been assembled for this readings book.

It is not the typical readings book, however. The typical readings book serves as a supplementary textbook. Additional materials are provided to enable our book to serve as a textbook for a system analysis course. An instructor's manual provides questions and problems, and their solutions, plus a recommended approach to coverage of the material. The instructor's manual also provides a bibliography of cases which can be used to give students practical experience in application of system analysis techniques.

PREFACE

The authors would like to give special recognition to the A.C.M. committee members who spent many hours in designing the information system curriculum: Daniel Teichroew (Chairman), Russell Armstrong, Robert Ashenhurst, Robert Benjamin, Gordon Davis, Martin Greenberger, John Lubin, James McKenney, Howard Morgan, Frederic Tonge. We also appreciate the efforts of our fine typists, Marilyn Mann and Lois Morey.

Colorado Springs, Colorado

J. Daniel Couger
Robert W. Knapp

CONTENTS

General Introduction 1

PART I
The System Perspective

Introduction to
PART I 5

The History and Status of General Systems Theory 9
LUDWIG VON BERTALANFFY

Towards a System of Systems Concepts 27
RUSSELL L. ACKOFF

PART II
Techniques for Analyzing Systems

Introduction to
SECTION I 41
Overview of the
Four Generations
of System Analysis

Evolution of Business Systems Analysis Techniques 43
J. DANIEL COUGER

CONTENTS

Introduction to
SECTION II   83

Second Generation
System Analysis Techniques

MIS Development Procedures   87
ROBERT G. MURDICK

Study Organization Plan Documentation Techniques   94
IBM

Flowcharting with the ANSI Standard: A Tutorial   128
NED CHAPIN

Decision Tables — What, Why and How   162
RAYMOND M. FERGUS

Gridcharting   180
W. HARTMAN, H. MATTES, and A. PROEME

Business Information Systems Analysis and Design   185
HONEYWELL

ADS: A Technique in Systems Documentation   190
HUGH J. LYNCH

Introduction to
SECTION III   205

Third Generation
System Analysis Techniques

The Languages Used in Decision Tables   209
S. POLLACK, H. HICKS, and W. HARRISON

The Time Automated Grid System (TAG)   220
IBM

An Information Algebra CODASYL DEVELOPMENT COMMITTEE	234
Abstract Formulation of Data Processing Problems JOHN W. YOUNG, JR., and HENRY K. KENT	259
SYSTEMATICS — A Non-Programming Language for Designing and Specifying Commercial Systems for Computers C. B. B. GRINDLEY	275
The Use of Decision Table Within SYSTEMATICS C. B. B. GRINDLEY	282
Some Approaches to the Theory of Information Systems BÖRJE LANGEFORS	292
Problem Statement Languages in MIS DANIEL TEICHROEW	310
Beyond Programming: Practical Steps Toward the Automation of D.P. System Creation JOHN RHODES	328
Problem Statement Analysis: Requirements for the Problem Statement Analyzer (PSA) DANIEL TEICHROEW	336
A Methodology for the Design and Optimization of Information Processing Systems J. F. NUNAMAKER, JR.	359
Introduction to SECTION IV Fourth Generation System Analysis Techniques	377
Automation of System Building DANIEL TEICHROEW and HASAN SAYANI	379

PART III
Cost/Effectiveness Analysis

Introduction to **SECTION I**	393
The Theory Behind Cost/Effectiveness Analysis	
Cost/Benefit Analysis of Information Systems JAMES EMERY	395
Economic Evaluation of Management Information Systems D. F. BOYD and H. S. KRASNOW	426
The Cost and Effectiveness of Computer Systems WILLIAM SHARPE	446
Introduction to **SECTION II**	471
Application of Cost/Effectiveness Analysis Techniques	
Value and Cost of Information R. H. GREGORY and R. L. VAN HORN	473
How to Analyze Computer Project Costs LOUIS FRIED	490
Estimating the Cost of System Implementation LOUIS FRIED	498

System Analysis Techniques

GENERAL INTRODUCTION

Practitioners in almost all disciplines debate the question of their profession as an art or a science. System analysts are no exception; however, the debate is less heated than it was ten years ago. In the early years of the profession few techniques were developed for analyzing potential computer applications. The practice of the profession was indeed an art. However, a body of techniques has evolved for analyzing and designing computer applications. Today the practice of the profession is clearly a science.

This book demonstrates the scientific aspects of the profession. The first part is designed to provide an overall systems perspective. The second part covers system analysis techniques. The third part covers techniques for cost/effectiveness analysis.

The major portion of this book consists of papers describing techniques for use in Phases I and II of the System Development Cycle, which cover documentation of the existing system and analysis of the system to establish requirements for an improved system, that is, development of the logical design. Phase III, development of the physical design, is concerned with the organization of files and with the devices to be used in processing the system. Determination of system cost/effectiveness is an iterative process which provides an interface for Phases II and III. Therefore a section on techniques for determining cost/effectiveness is included in this book. While many books have been published on the subject of Phase III, physical design, few are available on Phases I and II. This gap is enigmatic because the literature is rich with papers describing techniques for system analysis.

The papers selected for Part I provide a background on general systems theory and system concepts. The papers selected for Part II describe the important techniques for determining if a system should be computerized. The first paper describes the development of system analysis techniques and categorizes these techniques into four generations

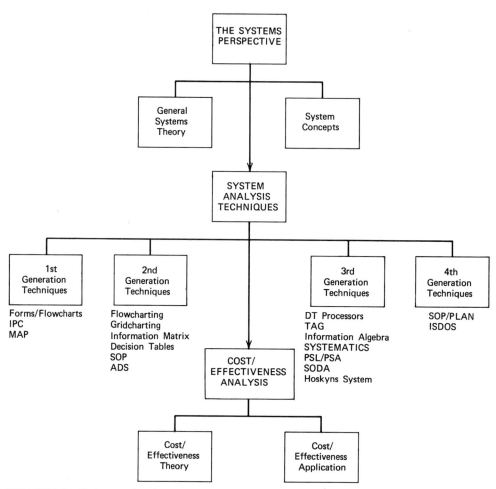

FIGURE 1 Schematic diagram of contents of the book.

of evolution. The remaining papers in Part II describe individual techniques, including manual techniques and computer-aided techniques. Part III provides papers on both the theory and application of cost/effectiveness analysis.

Figure 1 shows the techniques covered and their relationships. The generation of evolution is also highlighted by the illustration.

PART I
The System Perspective

Introduction to PART I

The System Perspective

Aristotle's statement, "The whole is more than the sum of its parts," was the initial impetus to system analysis. A similar concept today is "system synergism:" the manager/computer interactive system is more effective than either of these two elements operating separately.

Following this line of reasoning, we find that subsystems designed independently will not produce optimum benefits for the organization. The system analyst will be in a better position to avoid suboptimization if he understands systems theory. Many systems designed prior to the 1970's were suboptimal because independent systems were designed for interdependent activities.

The two papers in Part I not only provide system perspective but also, taken together, produce a synergistic effect. Such a result occurs through both the quality and the structure of the material. The papers provide a basis for moving through a hierarchic path from the general to applied systems theory.

The notion of general systems theory was first formulated by von Bertalanffy, orally in the 1930's, and in various publications after World War II:

> There exist models, principles, and laws that apply to generalized systems or their subclasses irrespective of their particular kind, the nature of the component elements, and the relations or 'forces' between them. We postulate a new discipline called *General System Theory.* General System Theory is a logico-mathematical field whose task is the formulation and derivation of those general principles that are applicable to 'systems' in general. In this way, exact formulations of terms such as wholeness and sum, differentiation, progressive mechanization, centralization, hierarchic order, finality and

equifinality, etc., become possible, terms which occur in all sciences dealing with 'systems' and imply their logical homology.*

"The History and Status of General Systems Theory" shows why von Bertalanffy remained a stalwart in the field. He was able to expand his theories as a result of his and others' research during the ensuing years. We received permission to include his paper just a month before his death, in June, 1972.

In his section on "trends" von Bertalanffy succinctly explains modern systems theory by categorizing the three principal thrusts of the field:

(1) *Mathematic systems theory:* "mathematics is the exact language permitting rigorous deductions and confirmation (or refusal) of theory";
(2) *System technology:* "the vast realm of techniques, models, mathematical approaches, and so forth, summarized as systems engineering";
(3) *Systems philosophy;* "a new philosophy of nature, . . . contrasting the 'blind laws of nature' of the mechanistic world view and the world process as a Shakespearean tale told by an idiot, with an organismic outlook of the 'world as a great organization'."

Even though the concept of a system is ancient and has been known for many centuries, the theory remained "philosophical," according to von Bertalanffy, and did not become a science because "mathematical techniques were lacking and the problems required a new epistemology." Mathematical systems theory has become extensive; and a variety of approaches have developed "differing in emphasis, focus of interest, mathematical techniques and other respects." The fact that system theories differ is "not an embarrassment or the result of confusion, but rather a healthy development in a new and growing field . . . often encountered in mathematics and science from the geometrical or analytic description of a curve to the equivalence of classical thermodynamics. . . ."

Despite the elegance of von Bertalanffy, it is Ackoff who provides the practical basis for the study of systems, by his definitions of key concepts and terms. He believes, "A scientific field can arise only on the base of a system of concepts. Systems science is not an exception. Systems thinking, if anything, should be carried out systematically!"

*Reprinted in L. von Bertalanffy, *General Systems Theory. Foundations, Development, Applications.* New York: George Braziller, 1968.

Ackoff provides another description of system synergism: "The systems approach to problems focuses on systems taken as a whole, not on their parts separately. Such an approach is concerned with total-system performance even when a change in only one or a few of its parts is contemplated because there are some properties of systems that can only be treated adequately from a holistic point of view."

As a teacher, Ackoff feels "a great need to provide my students with a conceptual framework that will assist them in absorbing and synthesizing of insights. . . ." In meeting this objective Ackoff provides the system perspective necessary for proper selection and use of techniques from the array available to practitioners in this field.

THE HISTORY AND STATUS OF GENERAL SYSTEMS THEORY

LUDWIG VON BERTALANFFY

1.1. HISTORICAL PRELUDE

In order to evaluate the modern "systems approach," it is advisable to look at the systems idea not as an ephemeral fashion or recent technique, but in the context of the history of ideas. (For an introduction and a survey of the field see [15], with an extensive bibliography and Suggestions for Further Reading in the various topics of general systems theory.) In a certain sense it can be said that the notion of system is as old as European philosophy. If we try to define the cental motif in the birth of philosophical-scientific thinking with the Ionian pre-Socratics of the sixth century B.C., one way to spell it out would be as follows. Man in early culture, and even primitives of today, experience themselves as being "thrown" into a hostile world, governed by chaotic and incomprehensible demonic forces which, at best, may be propitiated or influenced by way of magical practices. Philosophy and its descendant, science, was born when the early Greeks learned to consider or find, in the experienced world, an order or *kosmos* which was intelligible and hence controllable by thought and rational action.

One formulation of this cosmic order was the Aristotelian world view with its holistic and teleological notions. Aristotle's statement, "The whole is more than the sum of its parts," is a definition of the basic system problem which is still valid. Aristotelian teleology was eliminated in the later development of Western science, but the problems contained in it, such as the order and goal-directedness of living systems, were negated and bypassed rather than solved. Hence the basic system problem is still not obsolete.

A more detailed investigation would enumerate a long array of thinkers

SOURCE: Von Bertalanffy, L., "The History and Status of General Systems Theory," *Trends in General Systems Theory,* Edited by G. J. Klir, Copyright © 1972 by John Wiley & Sons, Inc., pp. 21-38.

who, in one way or another, contributed notions to what nowadays we call systems theory. If we speak of hierarchic order, we use a term introduced by the Christian mystic, Dionysius the Aeropagite, although he was speculating about the choirs of angels and the organism of the Church. Nicholas of Cusa [5], that profound thinker of the fifteenth century, linking Medieval mysticism with the first beginnings of modern science, introduced the notion of the *coincidentia oppositorum,* the opposition or indeed fight among the parts within a whole which nevertheless forms a unity of higher order. Leibniz's hierarchy of monads looks quite like that of modern systems; his *mathesis universalis* presages an expanded mathematics which is not limited to quantitative or numerical expressions and is able to formalize all conceptual thinking. Hegel and Marx emphasized the dialectic structure of thought and of the universe it produces: the deep insight that no proposition can exhaust reality but only approaches its coincidence of opposites by the dialectic process of thesis, antithesis, and synthesis. Gustav Fechner, known as the author of the psychophysical law, elaborated, in the way of the nature philosophers of the nineteenth century, supraindividual organizations of higher order than the usual objects of observation—for example, life communities and the entire earth, thus romantically anticipating the ecosystems of modern parlance. Incidentally, the present writer wrote a doctoral thesis on this topic in 1925.

Even such a rapid and superficial survey as the preceding one tends to show that the problems with which we are nowadays concerned under the term "system" were not "born yesterday" out of current questions of mathematics, science, and technology. Rather, they are a contemporary expression of perennial problems which have been recognized for centuries and discussed in the language available at the time.

One way to circumscribe the Scientific Revolution of the sixteenth-seventeenth centuries is to say that it replaced the descriptive-metaphysical conception of the universe epitomized in Aristotle's doctrine by the mathematical-positivistic or Galilean conception. That is, the vision of the world as a teleological cosmos was replaced by the description of events in causal, mathematical laws.

We say "replaced," not "eliminated," for the Aristotelian dictum of the whole that is more than its parts still remained. We must strongly emphasize that order or organization of a whole or system, transcending its parts when these are considered in isolation, is nothing metaphysical, not an anthropomorphic superstition or a philosophical speculation; it is a fact of observation encountered whenever we look at a living organism, a social group, or even an atom.

Science, however, was not well prepared to deal with this problem. The second maxim of Descartes' *Discours de la Méthode* was "to break down every problem into as many separate simple elements as might be possible." This, similarly formulated by Galileo as the "resolutive" method, was the conceptual "paradigm" [35] of science from its foundation to modern laboratory work: that is, to resolve and reduce complex phenomena into elementary parts and processes.

This method worked admirably well insofar as observed events were apt to be split into isolable causal chains, that is, relations between two or a few variables. It was at the root of the enormous success of physics and the consequent technology. But questions of many-variable problems always remained. This was the case even in the three-body problem of mechanics; the situation was aggravated when the organization of the living organism or even of the atom, beyond the simplest proton-electron system of hydrogen, was concerned.

Two principal ideas were advanced in order to deal with the problem of order or organization. One was the comparison with man-made machines; the other was to conceive of order as a product of chance. The first was epitomized by Descartes' *bête machine,* later expanded to the *homme machine* of Lamettrie. The other is expressed by the Darwinian idea of natural selection. Again, both ideas were highly successful. The theory of the living organism as a machine in its various disguises—from a mechanical machine or clockwork in the early explanations of the iatrophysicists of the seventeenth century, to later conceptions of the organism as a caloric, chemodynamic, cellular, and cybernetic machine [13]—provided explanations of biological phenomena from the gross level of the physiology of organs down to the submicroscopic structures and enzymatic processes in the cell. Similarly, organismic order as a product of random events embraced an enormous number of facts under the title of "synthetic theory of evolution," including molecular genetics and biology.

Notwithstanding the singular success achieved in the explanation of ever more and finer life processes, basic questions remained unanswered. Descartes' "animal machine" was a fair enough principle to explain the admirable order of processes found in the living organism. But then, according to Descartes, the "machine" had God for its creator. The evolution of machines by events at random rather appears to be self-contradictory. Wristwatches or nylon stockings are not as a rule found in nature as products of chance processes, and certainly the mitochondrial "machines" of enzymatic organization in even the simplest cell or nucleoprotein molecules are incomparably more complex than a watch or the simple polymers which form synthetic fibers. "Survival of the fittest" (or "differential reproduction" in modern terminology) seems to lead to a circuitous argument. Self-maintaining systems must exist before they can enter into competition which leaves systems with higher selective value or differential reproduction predominant. That self-maintenance, however, is the explicandum; it is not provided by the ordinary laws of physics. Rather, the second law of thermodynamics prescribes that ordered systems in which irreversible processes take place tend toward most probable states and hence toward destruction of existing order and ultimate decay [16].

Thus neovitalistic currents, represented by Driesch, Bergson, and others, reappeared around the turn of the present century, advancing quite legitimate arguments which were based essentially on the limits of possible regulations in a "machine," of evolution by random events, and on the goal-directedness of action. They were able, however, to refer only to the old Aristotelian "entelechy" under new names and descriptions, that is, a supernatural, organizing principle or "factor."

Thus the "fight on the concept of organism in the first decades of the twentieth century," as Woodger [56] nicely put it, indicated increasing doubts regarding the "paradigm" of classical science, that is, the explanation of complex phenomena in terms of isolable elements. This was expressed in the question of "organization" found in every living system; in the question whether "random mutations *cum* natural selection provide all the answers to the phenomena of evolution" [32] and thus of the organization of living things; and in the question of goal-directedness, which may be denied but in some way or other still raises its ugly head.

These problems were in no way limited to biology. Psychology, in gestalt theory, similarly and even earlier posed the question that psychological wholes (e.g., perceived *gestalten*) are not resolvable into elementary units such as punctual sensations and excitations in the retina. At the same time sociology [49, 50] came to the conclusion that physicalistic theories, modeled according to the Newtonian paradigm or the like, were unsatisfactory. Even the atom appeared as a minute "organism" to Whitehead.

1.2. FOUNDATIONS OF GENERAL SYSTEMS THEORY

In the late 1920's von Bertalanffy wrote:

> Since the fundamental character of the living thing is its organization, the customary investigation of the single parts and processes cannot provide a complete explanation of the vital phenomena. This investigation gives us no information about the coordination of parts and processes. Thus the chief task of biology must be to discover the laws of biological systems (at all levels of organization). We believe that the attempts to find a foundation for theoretical biology point at a fundamental change in the world picture. This view, considered as a method of investigation, we shall call *"organismic biology"* and, as an attempt at an explanation, *"the system theory of the organism"* [7, pp. 64 ff., 190, 46, condensed].

Recognized "as something new in biological literature" [43], the organismic program became widely accepted. This was the germ of what later became known as general systems theory. If the term "organism" in the above statements is replaced by other "organized entities," such as social groups, personality, or technological devices, this is the program of systems theory.

The Aristotelian dictum of the whole being more than its parts, which was neglected by the mechanistic conception, on the one hand, and which led to a vitalistic demonology, on the other, has a simple and even trivial answer— trivial, that is, in principle, but posing innumerable problems in its elaboration:

> The properties and modes of action of higher levels are not explicable by the summation of the properties and modes of action of their components *taken in isolation*. If, however, we know the *ensemble* of the components and the *relations existing between them,* then the higher levels are derivable from the components [16, p. 148].

Many (including recent) discussions of the Aristotelian paradox and of reductionism have added nothing to these statements: in order to understand an organized whole we must know both the parts and the relations between them.

This, however, defines the trouble. For "normal" science in Thomas Kuhn's sense, that is, science as conventionally practiced, was little adapted to deal with "relations" in systems. As Weaver [51] said in a well-known statement, classical science was concerned with one-way causality or relations between two variables, such as the attraction of the sun and a planet, but even the three-body problem of mechanics (and the corresponding problems in atomic physics) permits no closed solution by analytical methods of classical mechanics. Also, there were descriptions of "unorganized complexity" in terms of statistics whose paradigm is the second law of thermodynamics. However, increasing with the progress of observation and experiment, there loomed the problem of "organized complexity," that is, of interrelations among many but not infinitely many components.

Here is the reason why, even though the problems of "system" were ancient and had been known for many centuries, they remained "philosophical" and did not become a "science." This was so because mathematical techniques were lacking and the problems required a new epistemology; the whole force of "classical" science and its success over the centuries militated against any change in the fundamental paradigm of one-way causality and resolution into elementary units.

The quest for a new "gestalt mathematics" was repeatedly raised a considerable time ago, in which not the notion of quantity but rather that of relations, that is, of form and order, would be fundamental [10, p. 159f.]. However, this demand became realizable only with new developments.

The notion of general systems theory was first formulated by von Bertalanffy, orally in the 1930's and in various publications after World War II:

> There exist models, principles, and laws that apply to generalized systems or their subclasses irrespective of their particular kind, the nature of the component elements, and the relations or "forces" between them. We postulate a new discipline called *General System Theory*. General System Theory is a logico-mathematical field whose task is the formulation and derivation of those general principles that are applicable to "systems" in general. In this way, exact formulations of terms such as wholeness and sum, differentiation, progressive mechanization, centralization, hierarchic order, finality and equifinality, etc., become possible, terms which occur in all sciences dealing with "systems" and imply their logical homology (von Bertalanffy, 1947, 1955; reprinted in [15, pp. 32, 253]).

The proposal of general systems theory had precursors as well as independent simultaneous promoters. Köhler came near to generalizing gestalt theory into general systems theory [33]. Although Lotka did not use the term "general system theory," his discussion of systems of simultaneous differential equations [39] remained basic for subsequent "dynamical" system theory. Volterra's equations [21], originally developed for the competition of species, are applicable to generalized kinetics and dynamics. Ashby, in his early work [1], independently used the same system equations as von Bertalanffy employed, although deriving different consequences.

Von Bertalanffy outlined "dynamical" system theory and gave mathematical descriptions of system properties (such as wholeness, sum, growth, competition, allometry, mechanization, centralization, finality, and equifinality), derived from the system description by simultaneous differential equations. Being a practicing biologist, he was particularly interested in developing the theory of "open systems," that is, systems exchanging matter with environment as every "living" system does. Such theory did not then exist in physical chemistry. The theory of open systems stands in manifold relationships with chemical kinetics in its biological, theoretical, and technological aspects, and with the thermodynamics of irreversible processes, and provides explanations for many special problems in biochemistry, physiology, general biology, and related areas. It is correct to say that, apart from control theory and the application of feedback models, the theory of *Fliessgleichgewicht* and open systems [8, 12] is the part of general systems theory most widely applied in physical chemistry, biophysics, simulation of biological processes, physiology, pharmacodynamics, and so forth [15]. The forecast also proved to be correct that the basic areas of physiology, that is, metabolism, excitation, and morphogenesis (more specifically, the theory of regulation, cell permeability, growth, sensory excitation, electrical stimulation, center function, etc.), would "fuse into an integrated theoretical field under the guidance of the concept of open system" [6, Vol. II, pp. 49 ff.; also 15, p. 137 f.].

The intuitive choice of the open system as a general system model was a correct one. Not only from the physical viewpoint is the "open system" the more general case (because closed systems can always be obtained from open ones to equating transport variables to zero); it also is the general case mathematically because the system of simultaneous differential equations (equations of motion) used for description in dynamical system theory is the general form from which the description of closed systems derives by the introduction of additional constraints (e.g., conservation of mass in a closed chemical system) (cf. [46], p. 80 f.).

At first the project was considered to be fantastic. A well-known ecologist, for example, was "hushed into awed silence" by the preposterous claim that general system theory constituted a new realm of science [24], not foreseeing that it would become a legitimate field and the subject of university instruction within some fifteen years.

Many objections were raised against its feasibility and legitimacy [17]. It was not understood then that the exploration of the properties, models, and laws of "systems" is not a hunt for superficial analogies, but rather poses basic and difficult problems which are partly still unsolved [10, p. 200 f.].

According to the program, "system laws" manifest themselves as analogies or "logical homologies" of laws that are formally identical but pertain to quite different phenomena or even appear in different disciplines. This was shown by von Bertalanffy in examples which were chosen as intentionally simple illustrations, but the same principle applies to more sophisticated cases, such as the following:

> It is a striking fact that biological systems as diverse as the central nervous system, and the biochemical regulatory network in cells should be strictly analogous. . . . It is all the more remarkable when it is realized that this particular analogy between different systems at different levels of biological organization is but one member of a large class of such analogies [45].

It appeared that a number of researchers, working independently and in different fields, had arrived at similar conclusions. For example, Boulding wrote to the present author:

> I seem to have come to much the same conclusions as you have reached, though approaching it from the direction of economics and the social sciences rather than from biology — that there is a body of what I have been calling "general empirical theory," or "general system theory" in your excellent terminology, which is of wide applicability in many different disciplines [15, p. 14; cf. 18].

This spreading interest led to the foundation of the Society for General Systems Research (initially named the Society for the Advancement of General System Theory), an affiliate of the American Association for the Advancement of Science. The formation of numerous local groups, the task group on "General Systems Theory and Psychiatry" in the American Psychiatric Association, and many similar working groups, both in the United States and in Europe, followed, as well as various meetings and publications. The program of the Society formulated in 1954 may be quoted because it remains valid as a research program in general systems theory:

> Major functions are to: (1) investigate the isomorphy of concepts, laws, and models in various fields, and to help in useful transfers from one field to another; (2) encourage the development of adequate theoretical models in the fields which lack them; (3) minimize the duplication of the theoretical effort in different fields; (4) promote the unity of science through improving communication among specialists.

In the meantime a different development had taken place. Starting from the development of self-directing missiles, automation, and computer technology, and inspired by Wiener's work, the cybernetic movement became ever more influential. Although the starting point (technology versus basic science, especially biology) and the basic model (feedback circuit versus dynamic system of interactions) were different, there was a communality of interest in problems of organization and teleological behavior. Cybernetics, too, challenged the "mechanistic" conception that the universe was based on the "operation of anonymous particles at random" and emphasized "the search for new approaches, for new and more comprehensive concepts, and for methods capable of deaing with the large wholes of organisms and personalities" [25]. Although it is incorrect to describe modern systems theory as "springing out of the last war effort" [19] — in fact, it had roots quite different from military hardware and related technological developments — cybernetics and related approaches were independent developments which showed many parallelisms with general system theory.

1.3. TRENDS IN GENERAL SYSTEMS THEORY

This brief historical survey cannot attempt to review the many recent developments in general systems theory and the systems approach. For a critical discussion of the various approaches see [30, pp. 97 ff.], and [27, Book II].

With the increasing expansion of systems thinking and studies, the definition of general systems theory came under renewed scrutiny. Some indication as to its meaning and scope may therefore be pertinent. The term "general system theory," was introduced by the present author, deliberately, in a catholic sense. One may, of course, limit it to its "technical" meaning in the sense of mathematical theory (as is frequently done), but this appears inadvisable because there are many "system" problems asking for "theory" which is not presently available in mathematical terms. So the name "general systems theory" may be used broadly, in a way similar to our speaking of the "theory of evolution," which comprises about everything ranging from fossil digging and anatomy to the mathematical theory of selection; or "behavior theory," which extends from bird watching to sophisticated neurophysiological theories. It is the introduction of a new paradigm that matters.

(a) Systems Science; Mathematical Systems Theory

Broadly speaking, three main aspects of general systems theory can be indicated which are not separable in content but are distinguishable in intention. The first may be circumscribed as *systems science,* that is, scientific exploration and theory of "systems" in the various sciences (e.g., physics, biology, psychology, social sciences), with general systems theory being the doctrine of principles applying to all (or defined subclasses of) systems.

Entities of an essentially new sort are entering the sphere of scientific thought. Classical science in its various disciplines, such as chemistry, biology, psychology, or the social sciences, tried to isolate the elements of the observed universes—chemical compounds and enzymes, cells, elementary sensations, freely competing individuals, or whatever else may be the case—in the expectation that by putting them together again, conceptually or experimentally, the whole or system—cell, mind, society—would result and would be intelligible. We have learned, however, that for an understanding not only the elements but their interrelations as well are required—say, the interplay of enzymes in a cell, the interactions of many conscious and unconscious processes in the personality, the structure and dynamics of social systems, and so forth. Such problems appear even in physics, for example, in the interaction of many generalized "forces" and "fluxes" (irreversible thermodynamics; cf. Onsager reciprocal relations), or in the development of nuclear physics, which "requires much experimental work, as well as the development of additional powerful methods for the handling of systems with many, but not infinitely many, particles" [23]. This requires, first, the exploration of the many systems in our observed universe in their own right and specificities. Second, it turns out that there are general aspects, correspondences, and isomorphisms common to "systems." This is the domain of *general systems theory.* Indeed, such parallelisms or isomorphisms appear (sometimes surprisingly) in otherwise totally different "systems."

General systems theory, then, consists of the scientific exploration of "wholes" and "wholeness" which, not so long ago, were considered to be metaphysical notions transcending the boundaries of science. Novel concepts, models, and mathematical fields have developed to deal with them. At the same time, the interdisciplinary nature of concepts, models, and principles applying to "systems" provides a possible approach toward the unification of science.

The goal obviously is to develop general systems theory in mathematical terms (a "logico-mathematical field," as this author wrote in an early statement) because mathematics is the exact language permitting rigorous deductions and confirmation (or refusal) of theory. Mathematical systems theory has become an extensive and rapidly growing field. "System" being a new "paradigm" (in the sense of Thomas Kuhn), contrasting to the predominant, elementalistic approach and conceptions, it is not surprising that a variety of approaches have developed, differing in emphasis, focus of interest, mathematical techniques, and other respects. These elucidate different aspects, properties, and principles of what is comprised under the term "system," and thus serve different purposes of theoretical or practical nature. The fact that "system theories" by various authors look rather different is, therefore, not an embarrassment or the result of confusion, but rather a healthy new and growing field, and indicates presumably necessary and complementary aspects of the problem. The existence of different descriptions is nothing extraordinary and is often encountered in mathematics and science, from the geometrical or analytical description of a curve to the equivalence of classical thermodynamics and statistical mechanics to that of wave mechanics and particle physics. Different and partly opposing approaches should, however, tend toward further integration, in the sense that one is a special case within another, or that they can be shown to be equivalent or complementary. Such developments are, in fact, taking place.

System-theoretical approaches include general system theory (in the narrower sense), cybernetics, theory of automata, control theory, information theory, set, graph and network theory, relational mathematics, game and decision theory, computerization and simulation, and so forth. The somewhat loose term "approaches" is used deliberately because the list contains rather different things, for example, models (such as those of open system, feedback, logical automaton), mathematical techniques (e.g., theory of differential equations, computer methods, set, graph theory), and newly formed concepts or parameters (information, rational game, decision, etc.). These approaches concur, however, in that, in one way or the other, they relate to "system problems," that is, problems of interrelations within a superordinate "whole." Of course, these are not isolated but frequently overlap, and the same problem can be treated mathematically in different ways. Certain typical ways of describing "systems" can be indicated; their elaboration is due, on the one hand, to theoretical problems of "systems" as such and in relation to other disciplines, and, on the other hand, to problems of the technology of control and communication.

No mathematical development or comprehensive review can be given here. The following remarks, however, may convey some intuitive understanding of the various approaches and the way in which they relate to each other.

It is generally agreed that "system" is a *model* of general nature, that is, a conceptual analog of certain rather universal traits of observed entities. The use of models or analog constructs is the general procedure of science (and even of everyday cognition), as it is also the principle of analog simulation by computer. The difference from conventional disciplines is not essential but lies rather in the degree of generality (or abstraction): "system" refers to very general characteristics partaken by a large class of entities conventionally treated in different disciplines. Hence the interdisciplinary nature of general systems theory; at the same time, its statements pertain to formal or structural commonalities abstracting from the "nature of elements and forces in the system" with which the special sciences (and explanations in these) are concerned. In other words, system-theoretical arguments pertain to, and have predictive value, inasmuch as such general structures are concerned. Such "explanation in principle" may have considerable predictive value; for specific explanation, introduction of the special system conditions is naturally required.

A system may be defined as a set of elements in interrelation among themselves and with the environment. This can be expressed mathematically in different ways. Several typical ways of system description can be indicated.

One approach or group of investigations may, somewhat loosely, be circumscribed as *axiomatic,* inasmuch as the focus of interest is a rigorous definition of system and the derivation, by modern methods of mathematics and logic, of its implications. Among other examples are the system descriptions by Mesarovic [41], Maccia and Maccia [40], Beier and Laue [4] (set theory), Ashby [2] (state-determined systems), and Klir [30] (UC = set of all couplings between the elements and the elements and environment; ST = set of all states and all transitions between states).

Dynamical system theory is concerned with the changes of systems in time. There are two principal ways of description: internal and external [47].

Internal description or "classical" system theory (foundations in [9], [11], and [15, pp. 54 ff.]; comprehensive presentation in [46]; an excellent introduction into dynamical system theory and the theory of open systems, following the line of the present author, in [3]) defines a system by a set of n measures, called state variables. Analytically, their change in time is typically expressed by a set of n simultaneous, first-order differential equations:

$$\frac{dQ_n}{dt} = f_i(Q_1, Q_2, \ldots, Q_n). \quad (1.1)$$

These are called dynamical equations or equations of motion. The set of differential equations permits a formal expression of system properties, such as wholeness and sum, stability, mechanization, growth, competition, final and equifinal behavior and others [9, 11, 15]. The behavior of the system is described by the theory of differential equations (ordinary, first-order, if the definition of the system by Eq. 1.1 is accepted), which is a well-known and highly developed field of mathematics. However, as was mentioned previously, system considerations pose quite definite problems. For example, the theory of stability has developed only recently in conjunction with problems of control (and system): the Liapunov functions date from 1892 (in Russian; 1907 in French); but their significance was recognized only recently, especially through the work of mathematicians of the U.S.S.R.

Geometrically, the change of the system is expressed by the trajectories that the state variables traverse in the state space, that is, the n-dimensional space of possible location of these variables. Three types of behavior may be distinguished and defined as follows:

1. A trajectory is called *asymptotically stable* if all trajectories sufficiently close to it at $t = t_0$ approach it asymptotically when $t \to \infty$.
2. A trajectory is called *neutrally stable* if all trajectories sufficiently close to it at $t = 0$ remain close to it for all later time but do not necessarily approach it asymptotically.
3. A trajectory is called *unstable* if the trajectories close to it at $t = 0$ do not remain close to it as $t \to \infty$.

These correspond to solutions approaching a time-independent state (equilibrium, steady state), periodic solutions, and divergent solutions, respectively.

A time-independent state,

$$f_i(Q_1, \ldots, Q_n) = 0, \quad (1.2)$$

can be considered as a trajectory degenerated into a single point. Then, readily visualizable in two-dimensional projection, the trajectories may converge toward a stable node represented by the equilibrium point, may approach it as a stable focus in damped oscillations, or may cycle around it in undamped oscillations (stable solutions). Or else, they may diverge from an unstable node, wander away from an unstable focus in oscillations, or from a saddle point (unstable solutions).

A central notion of dynamical theory is that of *stability*, that is, the response of a system to perturbation. The concept of stability originates in mechanics (a rigid body is in stable equilibrium if it returns to its original position after a sufficiently small displacement; a motion is stable if insensitive to small perturbations), and is generalized to the "motions" of state variables of system. This question is related to that of the existence of equilibrium states. Stability can be analyzed, therefore, by explicit solution of the differential equations describing the system (so-called indirect method, based essentially on discussion of the eigenwerte λ_i of Eq. 1.1). In the case of nonlinear systems, these equations have to be linearized by development into Taylor series and retention of the first term. Linearization, however, pertains only to stability in the vicinity of equilibrium. But stability arguments without actual solution of the differential equations (direct method) and for nonlinear systems are possible by introduction of so-called *Liapunov functions*; these are essentially generalized energy functions, the sign of which indicates whether or not an equilibrium is asymptotically stable [28, 36].

Here the relation of dynamical system theory to control theory becomes apparent; control means essentially that a system which is not asymptotically stable is made so by incorporating a controller, counteracting the motion of the system away from the stable state. For this reason the theory of stability in internal description or dynamical system theory converges with the theory of (linear) control or feedback systems in external description (see below).

Description by ordinary differential equations (Eq. 1.1) abstracts from variations of the state variables in space which would be expressed by partial differential equations. Such field equations are, however, more difficult to handle. Ways of overcoming this difficulty are to assume complete "stirring," so that distribution is homogeneous within the volume considered; or to assume the existence of compartments to which homogeneous distribution applies, and which are connected by suitable interactions (compartment theory) [44].

In *external description,* the system is considered as a "black box"; its relations to the environment and other systems are presented graphically in block and flow diagrams. The system description is given in terms of inputs and outputs (*Klemmenverhalten* in German terminology); its general form is transfer functions relating input and output. Typically, these are assumed to be linear and are represented by discrete sets of values (cf. yes-no decisions in information theory, Turing machine). This is the language of control technology; external description, typically, is given in terms of communication (exchange of information between system and environment and within the system) and control of the system's function with respect to environment (feedback), to use Wiener's definition of cybernetics.

As mentioned, internal and external descriptions largely coincide with descriptions by continuous or discrete functions. These are two "languages" adapted to their respective purposes. Empirically, there is an obvious contrast between regulations due to the free interplay of forces within a dynamical system, and regulations due to constraints imposed by structural feedback mechanisms [15], for example, the "dynamic" regulations in a chemical system or in the network of reactions in a cell on the one hand, and control by mechanisms such as a thermostat or homeostatic nervous circuit on the other. Formally, however, the two "languages" are related and in certain cases demonstrably translatable. For example, an input-output function can (under certain conditions) be developed as a linear nth-order differential equation, and the terms of the latter can be considered as (formal) "state variables"; while their physical meaning remains indefinite, formal "translation" from one language into the other is possible.

In certain cases—for example, the two-factor theory of nerve excitation (in terms of "excitatory and inhibitory factors" or "substances") and network theory (McCulloch nets of "neurons")—description in dynamical system theory by continuous functions and description in automata theory by digital analogs can be shown to be equivalent [45]. Similarly, predator-prey systems, usually described dynamically by Volterra equations, can also be expressed in terms of cybernetic feedback circuits [55]. These are two-variable systems. Whether a similar "translation" can be effectuated in many-variables systems remains (in the present writer's opinion) to be seen.

Internal description is essentially "structural"; that is, it tries to describe the systems' behavior in terms of state variables and their interdependence. External description is "functional"; the system's behavior is described in terms of its interaction with the environment.

As this sketchy survey shows, considerable progress has been made in mathematical systems theory since the program was enunciated and inaugurated some 25 years ago. A variety of approaches, which, however, are connected with each other, have been developed.

Today mathematical system theory is a rapidly growing field; but it is natural that basic problems, such as those of hierarchic order [53], are approached only slowly and presumably will need novel ideas and theories. "Verbal" descriptions and models (e.g., [20], [31], [42], [52]) are not expendable. Problems must be intuitively "seen" and recognized before they can be formalized mathematically. Otherwise, mathematical formalism may impede rather than expedite the exploration of very "real" problems.

A strong system-theoretical movement has developed in psychiatry, largely through the efforts of Gray [26]. The same is true of the behavioral sciences [20] and also of certain areas in which such a development was quite unexpected, at least by the present writer—for example, theoretical geography [29]. Sociology was stated as being essentially "a science of social systems" [14]; not foreseen was, for instance, the close parallelism of general system theory with French structuralism (e.g., Piaget, Levy-Strauss; cf. [37]) and the influence exerted on American functionalism in sociology ([22]: see especially pp. 2, 96, 141).

(b) Systems Technology

The second realm of general systems theory is *systems technology,* that is, the problems arising in modern technology and society, including both "hardware" (control technology, automation, computerization, etc.) and "software" (application of system concepts and theory in social, ecological, economical, etc., problems). We can only allude to the vast realm of techniques, models, mathematical approaches, and so forth, under similar denominations, in order to place it into the perspective of the present study.

Modern technology and society have become so complex that the traditional branches of technology are no longer sufficient; approaches of a holistic or systems, and generalist and interdisciplinary, nature became necessary. This is true in many ways. Modern engineering includes fields such as circuit theory, cybernetics as the study of communication and control (Wiener [54]), and computer techniques for handling "systems" of a complexity unamenable to classical methods of mathematics. Systems of many levels ask for scientific control: ecosystems, the disturbance of which results in pressing problems like pollution; formal organizations like bureaucracies, educational institutions, or armies; socioeconomic systems, with their grave problems of international relations, politics, and deterrence. Irrespective of the questions of how far scientific understanding (contrasted to the admission of irrationality of cultural and historical events) is possible, and to what extent scientific control is feasible or even desirable, there can be no dispute that these are essentially "system" problems, that is, problems involving interrelations of a great number of "variables." The same applies to narrower objectives in industry, commerce, and armament.

The technological demands have led to novel conceptions and disciplines, some displaying great originality and introducing new basic notions such as control and information theory, game, decision theory, the theory of circuits, of queuing, and others. Again it transpired that concepts and models (such as feedback, information, control, stability, circuits) which originated in certain specified fields of technology have a much broader significance, are of an interdisciplinary nature, and are independent of their special realizations, as exemplified by isomorphic feedback models in mechanical, hydrodynamic, electrical, biological, and other systems. Similarly, developments originating in pure and in applied science converge, as in dynamical system theory and control theory. Again, there is a spectrum ranging from highly sophisticated mathematical theory to computer simulation to more or less informal discussion of system problems.

(c) Systems Philosophy

Third, there is the realm of *systems philosophy* [38], that is, the reorientation of thought and world view following the introduction of "system" as a new scientific paradigm (in contrast to the analytic, mechanistic, linear-causal paradigm of classical science). Like every scientific theory of broader scope, general systems theory has its "metascientific" or philosophical aspects. The concept of "system" constitutes a new "paradigm," in Thomas Kuhn's phrase, or a new "philosophy of nature," in the present writer's [14] words, contrasting the "blind laws of nature" of the mechanistic world view and the world process as a Shakespearean tale told by an idiot, with an organismic outlook of the "world as a great organization."

First, we must find out the "nature of the beast": what is meant by "system," and how systems are realized at the various levels of the world of observation. This is *systems analogy*.

What is to be defined and described as system is not a question with an obvious or trivial answer. It will be readily agreed that a galaxy, a dog, a cell, and an atom are "systems." But in what sense and what respects can we speak of an animal or a human society, personality, language, mathematics, and so forth as "systems"?

We may first distinguish *real systems,* that is, entities perceived in or inferred from observation and existing independently of an observer. On the other hand, there are *conceptual systems,* such as logic or mathematics, which essentially are symbolic constructs (but also including, e.g., music); with *abstracted systems* (science) [42] as a subclass, that is, conceptual systems corresponding with reality. However, the distinction is by no means as sharp as it would appear.

Apart from philosophical interpretation (which would take us into the question of metaphysical realism, idealism, phenomenalism, etc.), we would consider as "objects" (which partly are "real systems") entities given by perception because they are discrete in space and time. We do not doubt that a pebble, a table, an automobile, an animal, or a star (and in a somewhat different sense an atom, a molecule, and a planetary system) are "real" and existent independently of observation. Perception, however, is not a reliable guide. Following it, we "see" the sun revolving around the earth, and certainly do not see that a solid piece of matter like a stone "really" is mostly empty space with minute centers of energy dispersed in astronomical distances. The spatial boundaries of even what appears to be an obvious object or "thing" actually are indistinct. From a crystal consisting of molecules, valences stick out, as it were, into the surrounding space; the spatial boundaries of a cell or an organism are equally vague because it maintains itself in a flow of molecules entering and leaving, and it is difficult to tell just what belongs to the "living system" and what does not. Ultimately all boundaries are dynamic rather than spatial.

Hence an object (and in particular a system) is definable only by its cohesion in a broad sense, that is, the interactions of the component elements. In this sense an ecosystem or social system is just as "real" as an individual plant, animal, or human being, and indeed problems like pollution, as a disturbance of the ecosystem, or social problems strikingly demonstrate their "reality." Interactions (or, more generally, interrelations), however, are never directly seen or perceived; they are conceptual constructs. The same is true even of the objects of our everyday world, which by no means are simply "given" as sense data or simple perceptions but also are constructs based on innate or learned categories, the concordance of different senses, previous experience, learning processes, naming (i.e. symbolic processes), etc. all of which largely determine what we actually "see" or perceive [cf. 34]. Thus the distinction between "real" objects and systems as given in observation and "conceptual" constructs and systems cannot be drawn in any common-sense way.

These are profound problems which can only be indicated in this context. The question for general systems theory is what statements can be made regarding material systems, informational systems, conceptual systems, and other types—questions which are far from being satisfactorily answered at the present time.

This leads to *systems epistemology*. As is apparent from the preceding, this is profoundly different from the epistemology of logical positivism or empiricism, even though it shares the same scientific attitude. The epistemology (and metaphysics) of logical positivism was determined by the ideas of physicalism, atomism, and the "camera theory" of knowledge. These, in view of present-day knowledge, are obsolete. As against physicalism and reductionism, the problems and modes of thought occurring in the biological, behavioral and social sciences require equal consideration, and simple "reduction" to the elementary particles and conventional laws of physics does not appear feasible. Compared to the analytical procedure of classical science, with resolution into component elements and one-way or linear causality as the basic category, the investigation of organized wholes of many variables requires new categories of interaction, transaction, organization, teleology, and so forth, with many problems arising for epistemology, mathematical models and techniques. Furthermore, perception is not a reflection of "real things" (whatever their metaphysical status), and knowledge not a simple approximation to "truth" or "reality." It is an interaction between knower and known, and thus dependent on a multiplicity of factors of a biological, psychological, cultural, and linguistic nature. Physics itself teaches that there are no ultimate entities like corpuscles or waves existing independently of the observer. This leads to a "perspective" philosophy in which physics, although its achievements in its own and related fields are fully acknowledged, is not a monopolistic way of knowledge. As opposed to reductionism and theories declaring that reality is "nothing but" (a heap of physical particles, genes, reflexes, drives, or whatever the case may be), we see science, as one of the "perspectives" that man, with his biological, cultural, and linguistic endowment and bondage, has created to deal with the universe into which he is "thrown," or rather to which he is adapted owing to evolution and history.

The third part of systems philosophy is concerned with the relations of man and his world, or what is termed *values* in philosophical parlance. If reality is a hierarchy of organized wholes, the image of man will be different from what it is in a world of physical particles governed by chance events as the ultimate and only "true" reality. Rather, the world of symbols, values, social entities, and cultures is something very "real"; and its embeddedness in a cosmic order of hierarchies tends to bridge the gulf between C. P. Snow's "two cultures" of science and the humanities, technology and history, natural and social sciences, or in whatever way the antithesis is formulated.

This humanistic concern of general systems theory, as this writer understands it, marks a difference to mechanistically oriented system theorists speaking solely in terms of mathematics, feedback, and technology and so giving rise to the fear that systems theory is indeed the ultimate step toward the mechanization and devaluation of man and toward technocratic society. While understanding and emphasizing the role of mathematics and of pure and applied science, this writer does not see that the humanistic aspects can be evaded unless general systems theory is limited to a restricted and fractional vision.

Thus there is indeed a great and perhaps puzzling multiplicity of approaches and trends in general systems theory. This is understandably uncomfortable to him who wants a neat formalism, to the textbook writer and the dogmatist. It is, however, quite natural in the history of ideas and of science, and particularly in the beginning of a new development. Different models and theories may be apt to render different aspects and so are complementary. On the other hand, future developments will undoubtedly lead to further unification.

General systems theory is, as emphasized, a model of certain general aspects of reality. But it is also a way of seeing things which were previously overlooked or bypassed, and in this sense is a methodological maxim. And like every scientific theory of broader compass, it is connected with, and tries to give its answer to, perennial problems of philosophy.

REFERENCES

1. **Ashby, W. R.,** "Effect of Controls on Stability." *Nature* (London), Vol. 155, No. 3933, pp. 242–243, February 1945.
2. **Ashby, W. R.,** *An Introduction to Cybernetics,* 3rd ed. New York: Wiley, 1958.
3. **Beier, W.,** *Biophysik,* 3rd ed. Jena: Fischer, 1968. English translation in preparation.
4. **Beier, W.,** and **W. Laue,** "On the Mathematical Formulation of Open Systems and Their Steady States." In *Unity Through Diversity. Festschrift in Honor of Ludwig von Bertalanffy,* W. Gray and N. Rizzo (Editors), Book II: *General and Open Systems.* New York: Gordon and Breach, 1971.
5. **Bertalanffy, L. von,** *Nicholaus von Kues.* München: G. Müller, 1928.
6. **Bertalanffy, L. von,** *Theoretische Biologie,* Vols. I, II. Berlin: Borntraeger, 1932, 1942; 2nd ed., Bern: Francke, 1951.
7. **Bertalanffy, L. von,** *Modern Theories of Development.* Translated by J. H. Woodger. Oxford: Oxford University Press, 1934; New York: Harper Torchbooks, 1962. German original: *Kritische Theorie der Formbildung.* Berlin: Borntraeger, 1928.
8. **Bertalanffy, L. von,** "Der Organimus als physikalisches System betrachtet." *Die Naturwissenschaften,* Vol. 28, pp. 521–531, 1940. Reprinted in *General System Theory. Foundations, Development, Applications.* New York: George Braziller, 1968.
9. **Bertalanffy, L. von,** "Zu einer allgemeinen Systemlehre." *Blätter für Deutsche Philosophie* Vol. 18, 1945; extract in *Biologia Generalis,* Vol. 19, pp. 114–129, 1949. Reprinted in *General System Theory. Foundations, Development, Applications.* New York: George Braziller, 1968.
10. **Bertalanffy, L. von,** *Problems of Life. An Evaluation of Modern Biological Thought.* London: Watts & Co.; New York: Wiley, 1952; Harper Torchbooks, 1960. German original: *Das biologische Weltbild.* Bern: Francke, 1949.
11. **Bertalanffy, L. von,** "An Outline of General System Theory." *British Journal of the Philosophy of Science,* Vol. 1, pp. 134–164, 1950. Reprinted in *General System Theory. Foundations, Development, Applications.* New York: George Braziller, 1968.
12. **Bertalanffy, L. von,** *Biophysik des Fliessgleichgewichts.* Translated by W. H. Westphal. Braunschweig: Vieweg, 1953. Revised edition with W. Beier and R. Laue in preparation.
13. **Bertalanffy, L. von,** "Zur Geschichte theoretischer Modelle in der Biologie." *Studium Generale,* Vol. 8, pp. 290–298, 1965.
14. **Bertalanffy, L. von,** *Robots, Men and Minds.* New York: George Braziller, 1967.
15. **Bertalanffy, L. von,** *General System Theory. Foundations, Development, Applications.* New York: George Braziller, 1968.
16. **Bertalanffy, L. von,** "Chance or Law." In *Beyond Reductionism,* A. Koestler and J. R. Smythies (Editors). New York: Hutchinson, 1969.
17. **Bertalanffy, L. von,** C. G. Hempel, R. E. Bass, and H. Jonas, "General System Theory: A New Approach to Unity of Science." *Human Biology,* Vol. 23, pp. 302–361, 1951.
18. **Boulding, K. E.,** *The Image.* Ann Arbor: University of Michigan Press, 1965.
19. **Buckley, W.,** *Sociology and Modern Systems Theory.* Englewood Cliffs, N.J.: Prentice-Hall, 1967.
20. **Buckley, W.** (Editor), *Modern Systems Research for the Behavioral Scientist. A Sourcebook.* Chicago: Aldine Publishing Co., 1968.
21. **D'Ancona, V.,** *Der Kampf ums Dasein.* In *Abhandlungen zur exakten Biologie,* L. von Bertalanffy (Editor). Berlin: Borntraeger, 1939. English translation: *The Struggle for Existence.* Leiden: E. J. Brill, 1954.

22. **Demerath, N. J., III,** and **R. A. Peterson** (Editors), *System, Change and Conflict. A Reader on Contemporary Sociological Theory and the Debate over Functionalism.* New York: Free Press, 1967.
23. **De-Shalit, A.,** "Remarks on Nuclear Structure," *Science,* Vol. 153, pp. 1063–1067, 1966.
24. **Egler, F. E.,** "Bertalanffian Organismicism," *Ecology,* Vol. 34, pp. 443–446, 1953.
25. **Frank, L. K., G. E. Hutchinson, W. K. Livingstone, W. S. McCulloch,** and **N. Wiener,** "Teleological Mechanisms." *Annals of the New York Academy of Sciences,* Vol. 50, Art. 4, pp. 187–278, October 1948.
26. **Gray, W., F. D. Duhl,** and **N. D. Rizzo** (Editors), *General Systems Theory and Psychiatry.* Boston: Little Brown, 1968.
27. **Gray, W.,** and **N. Rizzo** (Editors), *Unity Through Diversity. Festschrift in Honor of Ludwig von Bertalanffy.* Espec. Book II: *General and Open Systems,* and Book IV: *General Systems in the Behavioral Sciences.* New York: Gordon and Breach, 1971.
28. **Hahn, W.,** *Theory and Application of Liapunov's Direct Method.* Englewood Cliffs, N.J.: Prentice-Hall, 1963.
29. **Harvey, D.,** *Explanation in Geography.* London: Arnold, 1969; New York: St. Martin's Press, 1969.
30. **Klir, G. J.,** *An Approach to General Systems Theory.* New York: Van Nostrand Reinhold, 1969.
31. **Koestler, A.,** "The Tree and the Candle." In *Unity Through Diversity,* W. Gray and N. Rizzo (Editors), Book II: *General and Open Systems.* New York: Gordon and Breach, 1971.
32. **Koestler, A.,** and **J. R. Smythies** (Editors), *Beyond Reductionism.* New York: Hutchinson, 1969.
33. **Köhler, W.,** "Zum Problem der Regulation." *Roux's Archiv.,* Vol. 112, 1927.
34. **Kraft, V.,** *Die Grundlagen der Erkenntnis und der Moral.* Berlin: Duncker and Humblot, 1968.
35. **Kuhn, T.,** *The Structure of Scientific Revolutions.* Chicago: University of Chicago Press, 1962.
36. **La-Salle, J.,** and **S. Lefschetz,** *Stability by Liapunov's Direct Method.* New York: Academic Press, 1961.
37. **Laszlo, E.,** "Systems and Structures—Toward Bio-social Anthropology." In *Unity Through Diversity,* W. Gray and N. Rizzo (Editors), Book IV: *General Systems in the Behavioral Sciences.* New York: Gordon and Breach, 1971.
38. **Laszlo, E.,** *Introduction to Systems Philosophy.* New York: Gordon and Breach, 1971.
39. **Lotka, A. J.,** *Elements of Physical Biology* (1925). New York: Dover, 1956.
40. **Maccia, E. S.,** and **G. S. Maccia,** *Development of Educational Theory Derived from Three Educational Theory Models,* Project 5–0638. Columbus, Ohio: The Ohio State Research Foundation, 1966.
41. **Mesarovic, M. D.,** "Foundations for a General Systems Theory." In *Views of General Systems Theory,* M. D. Mesarović (Editor), pp. 1–24. New York: Wiley, 1964.
42. **Miller, J. G.,** "Living Systems: Basic Concepts." In *General Systems Theory and Psychiatry,* W. Gray, F. D. Duhl, and N. D. Rizzo (Editors). Boston: Little, Brown, 1968.
43. **Needham, J.,** "Review of 'Theoretische Biologie,' Vol. I, by L. von Bertalanffy." *Nature* (London), Vol. 132, 1933.
44. **Rescigno, A.,** and **G. Segre,** *Drug and Tracer Kinetics.* Waltham, Mass.: Blaisdell, 1966.
45. **Rosen, R.,** "Two-Factor Models, Neural Nets and Biochemical Automata." *Journal of Theoretical Biology,* Vol. 15, pp. 282–297, 1967.
46. **Rosen, R.,** *Dynamical System Theory in Biology,* Vol. I: *Stability Theory and Its Applications.* New York: Wiley, 1970.
47. **Rosen, R.,** "A Survey of Dynamical Descriptions of System Activity." In *Unity Through Diversity,* W. Gray and N. Rizzo (Editors), Book II: *General*

47. ...*and Open Systems.* New York: Gordon and Breach, 1971.
48. **Schwarz, H.,** *Einführung in die moderne Systemtheorie.* Braunschweig: Vieweg, 1969.
49. **Sorokin, P. A.,** *Contemporary Sociological Theories* (1928). New York: Harper Torchbooks, 1964.
50. **Sorokin, P. A.,** *Sociological Theories of Today.* New York: Harper & Row, 1966.
51. **Weaver, W.,** "Science and Complexity." *American Scientist,* Vol. 36, pp. 536–544, 1948.
52. **Weiss, P. A.,** "Life, Order and Understanding." *The Graduate Journal,* Vol. III, Supplement, University of Texas, 1970.
53. **Whyte, L. L., A. G. Wilson,** and **D. Wilson** (Editors), *Hierarchical Structures.* New York: American Elsevier, 1969.
54. **Wiener, N.,** *Cybernetics,* New York: Wiley, 1948.
55. **Wilbert, H.,** "Feind-Beute-Systeme in kybernetischer Sicht." *Oecologia* (Berlin), Vol. 5, pp. 347–373, 1970.
56. **Woodger, J. H.,** *Biological Principles.* London: Routledge and Kegan Paul, 1929; New York: Humanities Press, 1967.

TOWARDS A SYSTEM OF SYSTEMS CONCEPTS

RUSSELL L. ACKOFF

INTRODUCTION

The concept *system* has come to play a critical role in contemporary science.[1] This preoccupation of scientists in general is reflected among Management Scientists in particular for whom the *systems approach* to problems is fundamental and for whom *organizations*, a special type of system, are the principal subject of study.

The systems approach to problems focuses on systems taken as a whole, not on their parts taken separately. Such an approach is concerned with total-system performance even when a change in only one or a few of its parts is contemplated because there are some properties of systems that can only be treated adequately from a holistic point of view. These properties derive from the *relationships* between parts of systems: how the parts interact and fit together. In an imperfectly organized system, even if every part performs as well as possible relative to its own objectives, the total system will often not perform as well as possible relative to its objectives.

Despite the importance of systems concepts and the attention that they have received and are receiving, we do not yet have a unified or integrated set (i.e., a system) of such concepts. Different terms are used to refer to the same thing and the same term is used to refer to different things. This state is aggravated by the fact that the literature of systems research is widely dispersed and is therefore difficult to track. Researchers in a wide variety of disciplines are contributing to the conceptual development of the systems sciences, but these contributions are not as interactive and additive as they might be. Fred Emery has warned against too hasty an effort to remedy this situation:

[1] For excellent extensive and intensive discussions of 'systems thinking', see F. E. Emery [3] and C. W. Churchman [2].

SOURCE: Ackoff, Russell L., "Towards a System of Systems Concepts," *Management Science*, Vol. 17, No. 11, July, 1971, pp. 661–671.

It is almost as if the pioneers [of systems thinking], while respectfully noting each other's existence, have felt it incumbent upon themselves to work out their intuitions in their own language, for fear of what might be lost in trying to work through the language of another. Whatever the reason, the results seem to justify the stand-offishness. In a short space of time there has been a considerable accumulation of insights into system dynamics that are readily translatable into different languages and with, as yet, little sign of divisive schools of thought that for instance marred psychology during the 1920s and 1930s. Perhaps this might happen if some influential group of scholars prematurely decide that the time has come for a common conceptual framework [3, p. 12].

Although I sympathize with Emery's fear, a fear that is rooted in a research perspective, as a teacher I feel a great need to provide my students with a conceptual framework that will assist them in absorbing and synthesizing this large accumulation of insights to which Emery refers. My intent is not to preclude further conceptual exploration, but rather to encourage it and make it more interactive and additive. Despite Emery's warning I feel benefits will accrue to systems research from an evolutionary convergence of concepts into a generally accepted framework. At any rate, little harm is likely to come from my effort to provide the beginnings of such a framework since I can hardly claim to be, or to speak for, "an influential group of scholars."

The framework that follows does not include all concepts relevant to the systems sciences. I have made an effort, however, to include enough of the key concepts so that building on this framework will not be as difficult as construction of the framework itself has been.

One final word of introduction. I have not tried to identify the origin or trace the history of each conceptual idea that is presented in what follows. Hence few credits are provided. I can only compensate for this lack of bibliographic bird-dogging by claiming no credit for any of the elements in what follows, only for the resulting system into which they have been organized. I must, of course, accept responsibility for deficiencies in either the parts or the whole.

SYSTEMS

1. A *system* is a set of interrelated elements. Thus a system is an entity which is composed of at least two elements and a relation that holds between each of its elements and at least one other element in the set. Each of a system's elements is connected to every other element, directly or indirectly. Furthermore, no subset of elements is unrelated to any other subset.

2. An *abstract system* is one all of whose elements are concepts. Languages, philosophic systems, and number systems are examples. *Numbers* are concepts, but the symbols that represent them, *numerals,* are physical things. Numerals, however, are not the elements of a number system. The use of different numerals to represent the same numbers does not change the nature of the system.

In an abstract system the elements are created by defining and the relationships between them are created by assumptions (e.g., axioms and postulates). Such systems, therefore, are the subject of study of the so-called "formal sciences."

3. A *concrete system* is one at least two of whose elements are objects. It is only with such systems that we are concerned here. Unless otherwise noted, "system" will always be used to mean "concrete system."

In concrete systems, establishment of the existence and properties of elements and the nature of the relationships between them requires research with an empirical component in it. Such systems, therefore, are the subject of study of the so-called "nonformal sciences".

4. The *state of a system* at a moment of time is the set of relevant properties which that system has at that time. Any system has an unlimited number of properties. Only some of these are relevant to any particular research. Hence those which are relevant may change with changes in the purpose of the research. The values of the relevant properties constitute the state of the system. In some cases we may be interested in only two possible states (e.g., off and on, or awake and asleep). In other cases we may be interested in a large or unlimited number of possible states (e.g., a system's velocity or weight).

5. The *environment of a system* is a set of elements and their relevant properties, which elements are not part of the system but a change in any of which can produce[2] a change in the state of the system. Thus a system's environment consists of all variables that can affect its state. External elements that affect irrelevant properties of a system are not part of its environment.

6. The *state of a system's environment* at a moment of time is the set of its relevant properties at that time. The state of an element or subset of elements of a system or its environment may be similarly defined.

Although concrete systems and their environments are *objective* things, they are also *subjective* insofar as the particular configuration of elements that form both is dictated by the interests of the researcher. Different observers of the same phenomena may conceptualize them into different systems and environments. For example, an architect may consider a house together with its electrical, heating, and water systems as one large system. But a mechanical engineer may consider the heating system as a system and the house as its environment. To a social psychologist a house may be an environment of a family, the system with which he is concerned. To him the relationship between the heating and electrical systems may be irrelevant, but to the architect it may be very relevant.

The elements that form the environment of a system, and the environment itself, may be conceptualized as systems when they become the focus of attention. Every system can be conceptualized as part of another and larger system.

Even an abstract system can have an environment. For example, the metalanguage in which we describe a formal system is the environment of that formal system. Therefore, logic is the environment of mathematics.

7. A *closed system* is one that has no environment. An *open system* is one that does. Thus a closed system is one which is conceptualized so that it has no interaction with any element not contained within it; it is completely self-contained. Because systems researchers have found

---

[2] One thing (x) can be said to produce another (y) in a specified environment and time interval if x is a necessary but not a sufficient condition for y in that environment and time period. Thus a producer is a "probabilistic cause" of its product. Every producer, since it is not sufficient for its product, has a coproducer of that product (e.g., the producer's environment).

such conceptualizations of relatively restricted use, their attention has increasingly focused on more complex and "realistic" open systems. "Openness" and "closedness" are simultaneously properties of systems and our conceptualizations of them.

Systems may or may not change over time.

8. A system (or environmental) *event* is a change in one or more structural properties of the system (or its environment) over a period of time of specified duration, that is, a change in the structural state of the system (or environment). For example, an event occurs to a house's lighting system when a fuse blows, and to its environment when night falls.

9. A *static (one-state) system* is one to which no events happen. A table, for example, can be conceptualized as a static concrete system consisting of four legs, top, screws, glue, and so on. Relative to most research purposes it displays no change of structural properties, no change of state. A compass may also be conceptualized as a static system because it virtually always points to the Magnetic North Pole.

10. A *dynamic (multi-state) system* is one to which events happen, whose state changes over time. An automobile which can move forward or backward and at different speeds is such a system, or a motor which can be either off or on. Such systems can be conceptualized as either open or closed; closed if its elements react or respond only to each other.

11. A *homeostatic system* is a static system whose elements and environment are dynamic. Thus a homeostatic system is one that retains its state in a changing environment by internal adjustments. A house that maintains a constant temperature during changing external temperatures is homeostatic. The behavior of its heating subsystem makes this possible.

Note that the same object may be conceptualized as either a static or dynamic system. For most of us a building would be thought of as static, but it might be taken as dynamic by a civil engineer who is interested in structural deformation.

System Changes

12. A *reaction* of a system is a system event for which another event that happens to the same system or its environment is sufficient. Thus a reaction is a system event that is deterministically caused by another event. For example, if an operator's moving a motor's switch is sufficient to turn that motor off or on, then the change of state of the motor is a reaction to the movement of its switch. In this case, the turning of the switch may be necessary as well as sufficient for the state of the motor. But an event that is sufficient to bring about a change in a system's state may not be necessary for it. For example, sleep may be brought about by drugs administered to a person or it may be self-induced. Thus sleep may be determined by drugs but need not be.

13. A *response* of a system is a system event for which another event that happens to the same system or to its environment is necessary but not sufficient, that is, a system event produced by another system or environmental event (the *stimulus*). Thus a response is an event of which the system itself is a coproducer. A system does not have to respond to a stimulus, but it does have to react to its cause. Therefore, a person's turning on a light when it gets dark is a response to darkness, but the light's going on when the switch is turned is a reaction.

TABLE 1 Behavioral Classification of Systems

Type of system	Behavior of system	Outcome of behavior
State-maintaining	Variable but determined (reactive)	Fixed
Goal-seeking	Variable and chosen (responsive)	Fixed
Multi-goal-seeking and purposive	Variable and chosen	Variable but determined
Purposeful	Variable and chosen	Variable and chosen

14. An *act* of a system is a system event for the occurrence of which no change in the system's environment is either necessary or sufficient. Acts, therefore, are self-determined events, autonomous changes. Internal changes —in the states of the system's elements —are both necessary and sufficient to bring about action. Much of the behavior of human beings is of this type, but such behavior is not restricted to humans. A computer, for example, may have its state changed or change the state of its environment because of its own program.

Systems all of whose changes are reactive, responsive, or autonomous (active) can be called reactive, responsive, or autonomous (active), respectively. Most systems, however, display some combination of these types of change.

The classification of systems into reactive, responsive, and autonomous is based on consideration of what brings about changes in them. Now let us consider systems with respect to the kind of changes in themselves and their environments their reactions, responses, and actions bring about.

15. A system's *behavior* is a system event(s) which is either necessary or sufficient for another event in that system or its environment. Thus behavior is a system change which initiates other events. Note that reactions, responses, and actions may themselves constitute behavior. Reactions, responses, and actions are system events *whose antecedents are of interest*. Behavior consists of system events *whose consequences are of interest*. We may, of course, be interested in both the antecedents and consequences of system events.

Behavioral Classification of Systems

Understanding the nature of the classification that follows may be aided by Table 1 in which the basis for the classification is revealed.

16. A *state-maintaining system* is one that (1) can react in only one way to any one external or internal event; but (2) it reacts differently to different external or internal events; and (3) these different reactions produce the same external or internal state (outcome). Such a system reacts only to changes; it cannot respond because what it does is completely determined by the causing event. Nevertheless it can be said to have the *function* of maintaining the state it produces because it can produce this state in different ways under different conditions.

Thus a heating system whose internal controller turns it on when the room temperature is below a desired level, and turns it off when the temperature is above this level, is state-maintaining. The state it maintains is a room temperature that falls within a small range around its setting. Note that the temperature of the room which affects the system's behavior can be conceptualized as either part of the system or part of its environment. Hence a state-maintaining system may react to either internal or external changes.

In general, most systems with "stats" (e.g., thermostats and humidistats) are state-maintaining. Any system with a regulated output (e.g., the voltage of the output of a generator) is also state-maintaining.

A compass is also state-maintaining because in many different environments it points to the Magnetic North Pole.

A state-maintaining system must be able to *discriminate* between different internal or external states to changes in which it reacts. Furthermore, as we shall see below, such systems are necessarily *adaptive*; but unlike goal-seeking systems they are not capable of learning because they cannot choose their behavior. They cannot improve with experience.

17. A *goal-seeking system* is one that can respond differently to one or more different external or internal events in one or more different external or internal states and that can respond differently to a particular event in an unchanging environment until it produces a particular state (outcome). Production of this state is its goal. Thus such a system has a *choice* of behavior. A goal-seeking system's behavior is responsive, but not reactive. A state which is sufficient and thus deterministically causes a reaction cannot cause different reactions in the same environment.

Under constant conditions a goal-seeking system may be able to accomplish the same thing in different ways and it may be able to do so under different conditions. If it has *memory,* it can increase its efficiency over time in producing the outcome that is its goal.

For example, an electronic maze-solving rat is a goal-seeking system which, when it runs into a wall of a maze, turns right and if stopped again, goes in the opposite direction, and if stopped again, returns in the direction from which it came. In this way it can eventually solve any solvable maze. If, in addition, it has memory, it can take a "solution path" on subsequent trials in a familiar maze.

Systems with automatic "pilots" are goal-seeking. These and other goal-seeking systems may, of course, fail to attain their goals in some situations.

The sequence of behavior which a goal-seeking system carries out in quest of its goal is an example of a process.

18. A *process* is a sequence of behavior that constitutes a system and has a goal-producing function. In some well-definable sense each unit of behavior in the process brings the actor closer to the goal which it seeks. The sequence of behavior that is performed by the electronic rat constitutes a maze-solving process. After each move the rat is closer (i.e., has reduced the number of moves required) to solve the maze. The metabolic process in living things is a similar type of sequence the goal of which is acquisition of energy or, more generally, survival. Production processes are a similar type of sequence whose goal is a particular type of product.

Process behavior displayed by a system may be either reactive, responsive, or active.

19. A *multi-goal-seeking* system is one that is goal-seeking in each of two or more different (initial) external or internal states, and which seeks different goals in at least two different states, the goal being determined by the initial state.

20. A *purposive system* is a multi-goal-seeking system, the different goals

of which have a common property. Production of that common property is the system's purpose. These types of systems can pursue different goals, but they do not select the goal to be pursued. The goal is determined by the initiating event. But such a system does choose the means by which to pursue its goals.

A computer which is programmed to play more than one game (e.g., tic-tac-toe and checkers) is multi-goal-seeking. What game it plays is not a matter of its choice, however; it is usually determined by an instruction from an external source. Such a system is also purposive because "game winning" is a common property of the different goals which it seeks.

21. A *purposeful system* is one which can produce the same outcome in different ways in the same (internal or external) state and can produce different outcomes in the same and different states. Thus a purposeful system is one which can change its goals under constant conditions; it selects ends as well as means and thus displays *will*. Human beings are the most familiar examples of such systems.

Ideal-seeking systems form an important subclass of purposeful systems. Before making their nature explicit we must consider the differences between goals, objectives, and ideals, and some concepts related to them. The differences to be considered have relevance only to purposeful systems because only they can choose ends.

A system which can choose between different outcomes can place different values on different outcomes.

22. The *relative value of an outcome* that is a member of an exclusive and exhaustive set of outcomes, to a purposeful system, is the probability that the system will produce that outcome when each of the set of outcomes can be obtained with certainty. The relative value of an outcome can range from 0 to 1.0. That outcome with the highest relative value in a set can be said to be *preferred*.

23. The *goal* of a purposeful system in a particular situation is a preferred outcome that can be obtained within a specified time period.

24. The *objective* of a purposeful system in a particular situation is a preferred outcome that cannot be obtained within a specified period but which can be obtained over a longer time period. Consider a set of possible outcomes ordered along one or more scales (e.g., increasing speeds of travel). Then each outcome is closer to the final one than those which precede it. Each of these outcomes can be a goal in some time period after the "preceding" goal has been obtained, leading eventually to attainment of the last outcome, the objective. For example, a high-school freshman's goal in his first year is to be promoted to his second (sophomore) year. Passing his second year is a subsequent goal. And so on to graduation, which is his objective.

Pursuit of an objective requires an ability to change goals once a goal has been obtained. This is why such pursuit is possible only for a purposeful system.

25. An *ideal* is an objective which cannot be obtained in any time period but which can be approached without limit. Just as goals can be ordered with respect to objectives, objectives can be ordered with respect to ideals. But an ideal is an outcome which is unobtainable in practice, if not in principle. For example, an ideal of science is errorless observations. The amount of observer error can be reduced without limit but can never be reduced to zero. Omniscience is another such ideal.

26. An *ideal-seeking system* is a purposeful system which, on attainment of any of its goals or objectives, then seeks

another goal and objective which more closely approximates its ideal. An ideal-seeking system is thus one which has a concept of "perfection" or the "ultimately desirable" and pursues it systematically, that is, in interrelated steps.

From the point of view of their output, six types of system have been identified: state-maintaining, goal-seeking, multi-goal-seeking, purposive, purposeful, and ideal-seeking. The elements of systems can be similarly classified. The relationship between (1) the behavior and type of a system and (2) the behavior and type of its elements is not apparent. We consider it next.

RELATIONSHIPS BETWEEN SYSTEMS AND THEIR ELEMENTS

Some systems can display a greater variety and higher level of behavior than can any of their elements. These can be called *variety increasing*. For example, consider two state-maintaining elements, A and B. Say A reacts to a decrease in room temperature by closing any open windows. If a short time after A has reacted the room temperature is still below a specified level, B reacts to this by turning on the furnace. Then the system consisting of A and B is goal-seeking.

Clearly, by combining two or more goal-seeking elements we can construct a multi-goal-seeking (and hence a purposive) system. It is less apparent that such elements can also be combined to form a purposeful system. Suppose one element A can pursue goal G_1 in environment E_1 and goal G_2 in another environment E_2; and the other element B can pursue G_2 in E_1 and G_1 in E_2. Then the system would be capable of pursuing G_1 and G_2 in both E_1 and E_2 if it could select between the elements in these environments. Suppose we add a third (controlling) element which responds to E_1 by "turning on" either A or B, but not both. Suppose further that it turns on A with probability P_A where $0 < P_A < 1.0$ and turns on B with probability P_B where $0 < P_B < 1.0$. (The controller could be a computer that employs random numbers for this purpose.) The resulting system could choose both ends and means in two environments and hence would be purposeful.

A system can also show less variety of behavior and operate at a lower level than at least some of its elements. Such a system is *variety reducing*. For example, consider a simple system with two elements one of which turns lights on in a room whenever the illumination in that room drops below a certain level. The other element turns the lights off whenever the illumination exceeds a level that is lower than that provided by the lights in the room. Then the lights will go off and on continuously. The system would not be state-maintaining even though its elements are.

A more familiar example of a variety-reducing system can be found in those groups of purposeful people (e.g., committees) which are incapable of reaching agreement and hence of taking any collective action.

A system must be either variety-increasing or variety-decreasing. A set of elements which collectively neither increase nor decrease variety would have to consist of identical elements, either only one of which can act at a time or in which similar action by multiple units is equivalent to action by only one. In the latter case the behavior is nonadditive and the behavior is redundant. The relationships between the elements would therefore be irrelevant. For example, a set of similar automobiles owned by one person does not constitute a system because he can drive only one at a time and which he drives makes no difference. On the other hand, a radio with two speakers can provide stereo sound; the speakers each do a different thing and together they do something that neither can do alone.

ADAPTATION AND LEARNING

In order to deal with the concepts "adaptation" and "learning" it is necessary first to consider the concepts "function" and "efficiency."

27. The *function(s)* of a system is production of the outcomes that define its goal(s) and objective(s). Put another way, suppose a system can display at least two structurally different types of behavior in the same or different environments and that these types of behavior produce the same kind of outcome. Then the system can be said to have the function of producing that outcome. To function, therefore, is to be able to produce the same outcome in different ways.

Let C_i ($1 \leq i \leq m$) represent the different actions available to a system in a specific environment. Let P_i represent the probabilities that the system will select these courses of action in that environment. If the courses of action are exclusive and exhaustive, then $\Sigma_{i=1}^m P_i = 1.0$. Let E_{ij} represent the probability that course of action C_i will produce a particular outcome O_j in that environment. Then:

28. The *efficiency* of the system with respect to an outcome O_j which it has the function of producing is $\Sigma_{i=1}^m P_i E_{ij}$.

Now we can turn to "adaptation."

29. A system is *adaptive* if, when there is a change in its environmental and/or internal state which reduces its efficiency in pursuing one or more of the goals that define its function(s), it reacts or responds by changing its own state and/or that of its environment so as to increase its efficiency with respect to that goal or goals. Thus adaptiveness is the ability of a system to modify itself or its environment when either has changed to the system's disadvantage so as to regain at least some of its lost efficiency.

The definition of "adaptive" implies four types of adaptation:

29.1. *Other-other adaptation:* A system's reacting or responding to an external change by modifying the environment (e.g., when a person turns on an air conditioner in a room that has become too warm for him to continue to work in).

29.2. *Other-self adaptation:* A system's reacting or responding to an external change by modifying itself (e.g., when the person moves to another and cooler room).

29.3. *Self-other adaptation:* A system's reacting or responding to an internal change by modifying the environment (e.g., when a person who has chills due to a cold turns up the heat).

29.4. *Self-self adaptation:* A system's reacting or responding to an internal change by modifying itself (e.g., when that person takes medication to suppress the chills). Other-self adaptation is most commonly considered because it was this type with which Darwin was concerned in his studies of biological species as systems.

It should now be apparent why state-maintaining and higher systems are necessarily adaptive. Now let us consider why nothing lower than a goal-seeking system is capable of learning.

30. To *learn* is to increase one's efficiency in the pursuit of a goal under unchanging conditions. Thus if a person increases his ability to hit a target (his goal) with repeated shooting at it, he learns how to shoot better. Note that to do so requires an ability to modify one's behavior (i.e., to display choice) and memory.

Since learning can take place only when a system has a choice among alternative courses of action, only systems that are goal-seeking or higher can learn.

If a system is repeatedly subjected to the same environmental or internal change and increases its ability to maintain its efficiency under this type of change, then it *learns how to adapt*. Thus adaptation itself can be learned.

ORGANIZATIONS

Management Scientists are most concerned with that type of system called "organizations." Cyberneticians, on the other hand, are more concerned with that type of system called "organisms"; but they frequently treat organizations as though they were organisms. Although these two types of system have much in common, there is an important difference between them. This difference can be identified once "organization" has been defined. I will work up to its definition by considering separately each of what I consider to be its four essential characteristics.

(1) An organization is a purposeful system that contains at least two purposeful elements which have a common purpose.

We sometimes characterize a a purely mechanical system as being well organized, but we would not refer to it as an "organization." This results from the fact that we use "organize" to mean, "to make a system of," or as one dictionary puts it, "to get into proper working order", and "to arrange or dispose systematically". Wires, poles, transformers, switchboards, and telephones may constitute a communication system; but they do not constitute an organization. The employees of a telephone company make up the organization that operates the telephone system. Organization of a system is an activity that can be carried out only by purposeful entities; to be an organization a system must contain such entities.

An aggregation of purposeful entities does not constitute an organization unless they have at least one common purpose: that is, unless there is some one or more things that they all want. An organization is always organized around this common purpose. It is the relationships between what the purposeful elements do and the pursuit of their common purpose that give unity and identity to their organization.

Without a common purpose the elements would not work together unless compelled to do so. A group of unwilling prisoners or slaves can be organized and forced to do something that they do not want to do, but they do not constitute an organization even though they may form a system. An organization consists of elements that have and can exercise their own wills.

(2) An organization has a functional division of labor in pursuit of the common purpose(s) of its elements that define it.

Each of two or more subsets of elements, each containing one or more purposeful elements, is responsible for choosing from among different courses of action. A choice from each subset is necessary for obtaining the common purpose. For example, if an automobile carrying two people stalls on a highway and one gets out and pushes while the other sits in the driver's seat trying to start it when it is in motion, then there is a functional division of labor and they constitute an organization. The car cannot be started (their common purpose) unless both functions are performed.

The classes of courses of action and (hence) the subsets of elements may be differentiated by a variety of types of characteristics; for example:
(a) by *function* (e.g., production, marketing, research, finance, and personnel, in the industrial context),
(b) by *space* (e.g., geography, as territories of sales offices),
(c) by *time* (e.g., waves of an invading force).

The classes of action may, of course, also be defined by combinations of these and other characteristics.

It should be noted that individuals or groups in an organization that *make* choices need not *take* them, that is, carry them out. The actions may be carried out by other persons, groups, or even machines that are controlled by the decision makers.

(3) The functionally distinct subsets (parts of the system) can respond to each other's behavior through observation or communication.[3]

In some laboratory experiments subjects are given interrelated tasks to perform but they are not permitted to observe or communicate with each other even though they are rewarded on the basis of an outcome determined by their collective choices. In such cases the subjects are *unorganized*. If they were allowed to observe each other or to communicate with each other, they could become an organization. The choices made by elements or subsets of an organization must be capable of influencing each other, otherwise they would not even constitute a system.

(4) At least one subset of the system has a system-control function.

This subset (or subsystem) compares achieved outcomes with desired outcomes and makes adjustments in the behavior of the system which are directed toward reducing the observed deficiencies. It also determines what the desired outcomes are. The control function is normally exercised by an executive body which operates on a feed-back principle. "Control" requires elucidation.

31. An element or a system *controls* another element or system (or itself) if its behavior is either necessary or sufficient for subsequent behavior of the other element or system (or itself), and the subsequent behavior is necessary or sufficient for the attainment of one or more of its goals. Summarizing, then, an "organization" can be defined as follows:

32. An *organization* is a purposeful system that contains at least two purposeful elements which have a common purpose relative to which the system has a functional division of labor; its functionally distinct subsets can respond to each other's behavior through observation or communication; and at least one subset has a system-control function.

Now the critical difference between organisms and organizations can be made explicit. Whereas both are purposeful systems, organisms do not contain purposeful elements. The elements

---

[3]In another place (Ackoff [1]), I have given operational definitions of 'observation' and 'communication' that fit this conceptual system. Reproduction of these treatments would require more space than is available here.

of an organism may be state-maintaining, goal-seeking, multi-goal-seeking, or purposive; but not purposeful. Thus an organism must be variety increasing. An organization, on the other hand, may be either variety increasing or decreasing (e.g., the ineffective committee). In an organism only the whole can display will; none of the parts can.

Because an organism is a system that has a functional division of labor it is also said to be "organized." Its functionally distinct parts are called "organs." Their functioning is necessary but not sufficient for accomplishment of the organism's purpose(s).

CONCLUSION

Defining concepts is frequently treated by scientists as an annoying necessity to be completed as quickly and thoughtlessly as possible. A consequence of this disinclination to define often is research carried out like surgery performed with dull instruments. The surgeon has to work harder, the patient has to suffer more, and the chances for success are decreased.

Like surgical instruments, definitions become dull with use and require frequent sharpening and, eventually, replacement. Those I have offered here are not exceptions.

Research can seldom be played with a single concept; a matched set is usually required. Matching different researches requires matching the sets of concepts used in them. A scientific field can arise only on the base of a system of concepts. Systems science is not an exception. Systems thinking, if anything, should be carried out systematically.

REFERENCES

1. **Ackoff, R. L.,** *Choice, Communication, and Conflict,* a report to the National Science Foundation under Grant GN-389, Management Science Center, University of Pennsylvania, Philadelphia, 1967.
2. **Churchman, C. W.,** *The Systems Approach,* Delacorte Press, New York, 1968.
3. **Emery, F. E.,** *Systems Thinking,* Penguin Books Ltd., Harmondsworth, Middlesex, England, 1969.

PART II
Techniques for Analyzing Systems

Introduction to SECTION I

Overview of the Four Generations of System Analysis

Although general systems theory evolved from the fields of physical chemistry and biophysics and was widely applied in the "hard" sciences, in recent years it has been proven valuable in the "soft" sciences. Examples were given by von Bertalanffy, who wrote of systems theory application to the study of:

> ecosystems, the disturbance of which results in pressing problems like pollution; formal organizations like bureaucracies, educational institutions, or armies; socioeconomic systems, with their grave problems of international relations, politics and deterrence. . . . There can be no dispute that these are essentially 'system' problems, that is, problems involving inter-relations of a great number of 'variables.' The same applies to narrower objectives in industry, commerce and armament.*

The "narrow" objectives of industry and commerce are, nevertheless, complex. A wide range of techniques has evolved for handling systems in those arenas. The theoretical background presented by von Bertalanffy and Ackoff provide the system perspective necessary for understanding the evolution of specific techniques for analysis. The introductory paper of Part II, "Evolution of Business System Analysis Techniques," concentrates on the arenas of commerce and industry. Techniques for analyzing manual systems prior to 1950 are compared with those developed for analyzing systems for conversion to computer processing:

> In the 1950's only subsystems were computerized, such as the payroll system. Today, in the era of integrated

---
*From article by von Bertalanffy in Part I.

systems, the scope of the system is enlarged many times. Payroll is a module in the accounting subsystem, which is only one of several subsystems in the finance system. . . . The expansion in scope and sophistication of systems increases the complexity of system analysis and design.

Until very recently, evolution of techniques for analyzing computer-based systems lagged the evolution of computing hardware by one full generation. During the first generation of business-oriented computers, in the 1950's, system analysts continued to use unit-record oriented analysis and design techniques. Between 1960 and 1970, computer-oriented techniques for system analysis were developed. The gap has narrowed to half a generation. Third generation techniques began to emerge six years after the first installation of third generation computers. . . . However, that situation has changed. Recent developments suggest that the gap . . . will be closed by the advent of the fourth generation of computers.*

The four generations of system analysis techniques are covered by papers in Part II, categorized into four sections according to generation. Since the techniques of the first generation are no longer in use, those techniques are merely summarized in the introductory paper, in Section I. Second generation techniques are described in Section II: SOP, ADS, information matrix, decision tables, gridcharting, and flowcharting. Section III covers third generation techniques: decision table processing, TAG, information algebra, SYSTEMATICS, PSL/PSA, SODA, and the Hoskyns System. Section IV covers the fourth generation techniques: SOP/PLAN and ISDOS.

Second generation techniques were manual techniques, designed to facilitate analysis of the existing system and to develop an improved system. Third generation techniques began to employ the computer as an aid in system analysis. Fourth generation techniques integrate the process of system analysis and design, as well as programming. The computer is used as the integral component to analyze, to design, and to prepare programs.

Part II provides the papers that explain each system analysis technique and its appropriate area of application.

---

*From following article by J. Daniel Couger.

EVOLUTION OF BUSINESS SYSTEM ANALYSIS TECHNIQUES

J. DANIEL COUGER

System analysis consists of collecting, organizing, and evaluating facts about a system and the environment in which it operates. The objective of system analysis is to examine all aspects of the system — equipment, personnel, operating conditions, and its internal and external demands — to establish a basis for designing and implementing a better system.

Understanding the role of the system analyst is facilitated by reference to the steps in the system development cycle. For the purposes of this paper, seven phases in system development are distinguished:

- Phase I   Documentation of the existing system
- Phase II   Analysis of the system to establish requirements for an improved system (the logical design)
- Phase III   Design of a computerized system (the physical design)
- Phase IV   Programming and procedure development
- Phase V   Implementation
- Phase VI   Operation
- Phase VII   Maintenance and modification

System analysis, then, is concerned with Phases I and II of the system development cycle. The product of system analysis is the logical design of the new system: the specification for input and output of the system and the decision criteria and processing rules. Phase III, the physical design phase, determines the organization of files and the devices to be used.

Today's systems are complex in development. In the 1950's only subsystems were computerized, such as the payroll system. Today, in the era of integrated systems, the scope of the

SOURCE: Couger, J. D., "Evolution of Business System Analysis Techniques," *Computing Surveys*, Sept. 1973, pp. 167–198.

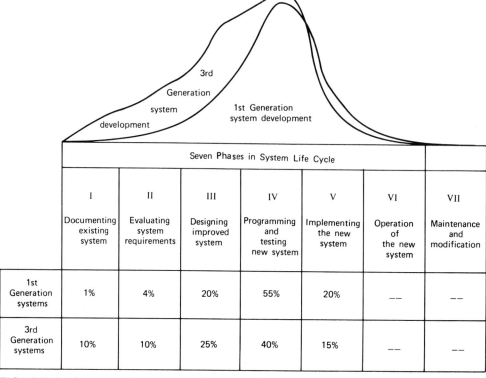

FIGURE 1 Comparative costs of 1st and 3rd generation system development.

system is enlarged many times. Payroll is a module in the accounting subsystem, which is only one of several subsystems in the finance system.

In the 1950's, *in*dependent subsystems were designed for *inter*dependent activities. The payroll application was designed as an entity when it, in reality, was a part of both the finance system and the personnel system of the firm. The payroll module of this era is redesigned to feed both of these major systems.

The systems of the 1950's were largely operational-level systems. They provided the information needed by first-level supervisors and their subordinates. Today's systems include the tactical (control) and strategic (planning) levels, as well. The thrust of system analysis/design effort in the 1970's has been to expand systems horizontally and vertically.

The expansion in scope and sophistication of systems increases the complexity of system analysis and design. There are more "front-end" costs in designing for integration.

Figure 1 shows the change in the development costs from first to third generation systems. Both the amount of cost and the distribution of resources have changed. In first generation systems, Phases I and II absorbed approximately five percent of system development cost. The expanded scope and sophistication of third generation systems have increased overall development cost, with

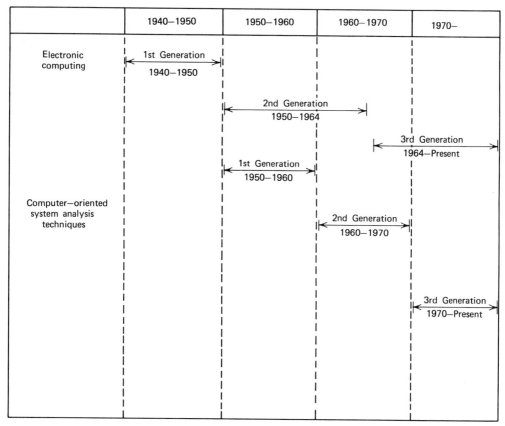

FIGURE 2 Comparison of evolution of hardware and system analysis techniques. (For consistency, the dates of each generation are taken from P. J. Denning's paper "Third Generation Computer Systems," *Computing Surveys*, Vol. 3, No. 4, Dec. 1971, p. 175.)

approximately twenty percent absorbed by Phases I and II.*

However, another reason for the increase in cost of Phases I and II is the lag in development of improved system analysis techniques.

LAG IN SYSTEM ANALYSIS TECHNIQUE DEVELOPMENT

Concomitant with the increase in complexity of systems is the need for increased capability of system analysis

---
*Figures based on the author's surveys of 50 organizations, taken first in 1965 and updated in 1972.

techniques. Yet, evolution of system analysis techniques has lagged hardware evolution by almost one full generation. System analysts continued to use techniques developed for unit-record systems during the era of first generation computers. Computer-oriented techniques were originated in the 1950's, lagging hardware evolution by one generation, as shown in Figure 2. The lag diminished only slightly from 1960 to 1970. However, new techniques have been developed in the last two years which reduce the gap significantly. The remainder of the paper will deal with the evolution of system analysis techniques from the first to the third generation. Also, predictions will be provided on fourth generation system analysis approaches.

Schematic on the Evolution of System Analysis Techniques

Figure 3 provides a schematic of the evolution of system analysis techniques through 1970. Another chart will portray the third generation of techniques. Three major categories are distinguished in Figure 3. The top portion of the schematic depicts the evolution of techniques for portraying and analyzing the *flow of information* through an organization. The central portion depicts the evolution of *mathematical and statistical* techniques for system analysis. The lower portion depicts the evolution of techniques for *recording and analyzing resources*. The number in the small circle provides an index to references on the technique. References for each technique are provided in Appendix I. Sources are listed in the bibliography (Appendix II).

Only the top chain will be discussed in this paper because the literature is rich with descriptive material on mathematical/statistical system analysis techniques and on resource analysis techniques. The references in the bibliography provide opportunity for further study of these areas.

Although this paper will concentrate on the top path, techniques for analysis of information flow and system logic, it will also emphasize the convergence of the three paths into the set of third generation system analysis techniques.

To clearly identify evolution of computer-oriented system analysis techniques, Figure 3 also depicts pre-computer techniques.

Prior to 1920

The process flowchart was used by industrial engineers in the early 1900's to show the flow of materials (Figure 4). Frederick W. Taylor and the Gilbreaths are recognized as leaders in developing techniques for process flow analysis.

The Period 1920–1950

As organizations grew and paperwork began to be a problem, the process flowchart was modified to depict forms flow. Figure 5a illustrates the shortcomings of this approach: lack of identification of data elements and volumes. With the advent of mechanical processing of information, process flowcharts were modified to portray the devices involved in data processing, shown in Figure 5b. Later in that period the tasks of processing became complicated enough to justify a system documentation package, which included tabulating procedures, process diagrams, and board wiring diagrams, shown in Figure 6.

When computers were introduced, during the latter part of this period, system analysts continued to use unit-record-oriented techniques.

The Period 1951–1960

In the 1950's, techniques especially suited for analysis of computer-based systems began to emerge. Figure 3 shows the sources through which these techniques evolved. General flowcharts and block diagrams evolved from the tabulating operation flow diagram.

Information process charts (IPC) were a combination of forms flowcharts and block diagrams (Figure 7). One line was used for each operation, with columns provided for indicating the fields of information on which the operations were performed. Certain verbs were specified and carefully defined to insure consistent understanding among all users of the charts. Although not widely used, the technique recognized the need for formal annotation which is necessary for computerized analysis.

MAP permitted a better overview of the flow of information, at the sacrifice of some detail provided in the annotation of IPC. In MAP, each horizontal level identified a type of document or file (Figure 8). A "transcription break" was used to show interrelationships among files. (The direction of the arrow shows the flow from one file to another.) Verbs were not as well defined as those used in IPC, therefore MAP contributed little in providing a foundation for computerized analysis.

The Period 1961–1970

ADS (Accurately Defined System). After several years of use internally, NCR (National Cash Register) published in 1968 the manual on ADS (Accurately Defined System). ADS was an improvement over prior techniques because it provided a well organized and correlated approach for system definition and specification.

ADS used five interrelated forms to provide the system (application) definition, shown in Figure 9. The process began with the definition of output. Next, inputs were defined—on the second form.

The third form provided the definition of computations to be performed and the rules of logic governing the computation. Interrelationship of computations were also defined on this form, as were the sources of information used in the computation. The fourth form, the history definition, specified information to be retained beyond the processing cycle for subsequent use. The fifth form provided the logic definitions, in the form of a decision table.

Within ADS, information linkage was accomplished in two ways. First, each data element was assigned a specific tag or reference. Next time the tag was used in the system, it was linked back to the previous link in the chain. All elements of data were chained from input to output, accomplished through the use of page and line numbers.

The process of chaining facilitated identification of omissions and contradictions in the system. Once the information requirements were established, the system design phase determined the appropriate hardware mix to effect the system.

Although NCR did not provide a means for computer processing of ADS, the technique paved the way for such use through its systematic approach to system definition.

Information Algebra. Information algebra was an important development because it provided a *theoretical* basis for automatic processing of system specifications.

Reference to Figure 3 shows information algebra evolving from the path of mathematical/statistical analysis techniques. The approach was developed by the Language Structure Group (LSG) of the CODASYL Development Committee.* Formed in July, 1959, the LSG's goal was to arrive at a proper structure for a machine independent problem-defining language, at the systems level of data processing.

An earlier approach involving the use of matrix representation was abandoned when rigid restrictions in form were encountered. In seeking a more general

---

*CODASYL (Conference on Data Systems Languages), ACM, 1133 Avenue of the Americas, New York, N.Y. 10036.

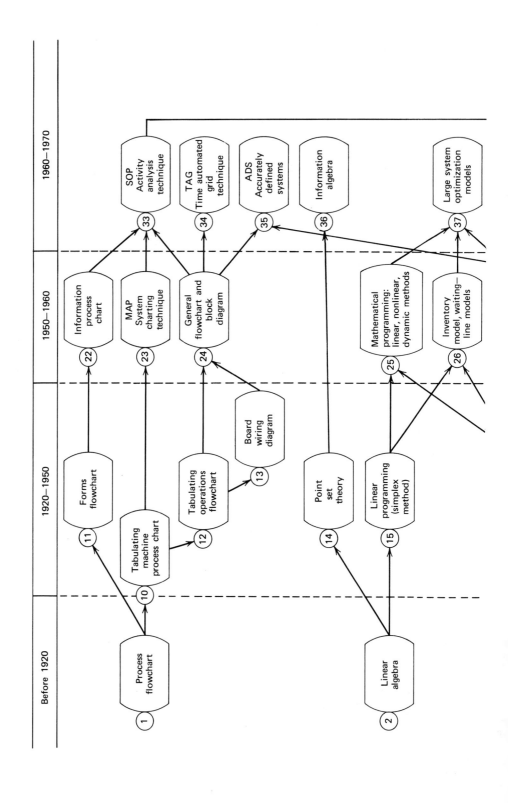

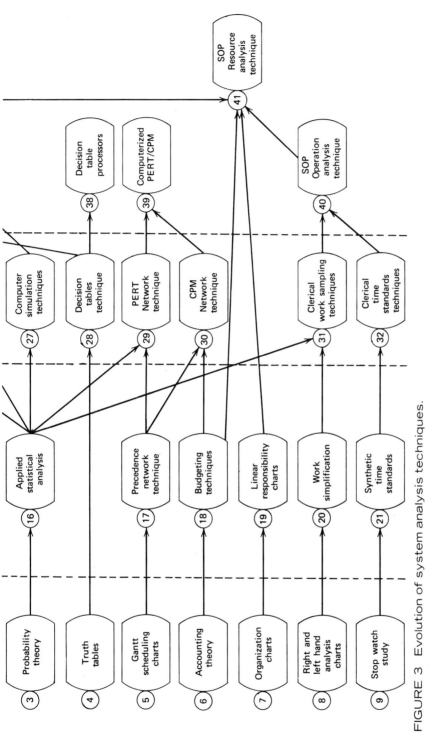

FIGURE 3 Evolution of system analysis techniques.

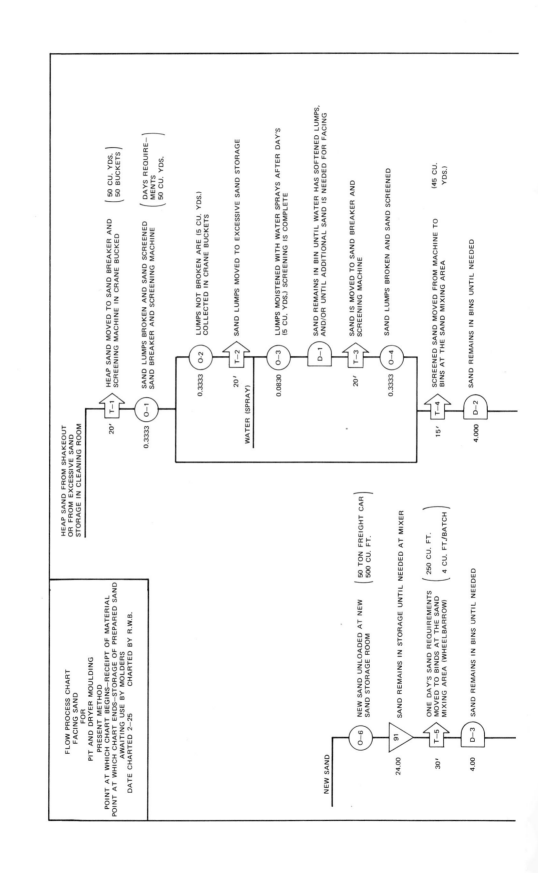

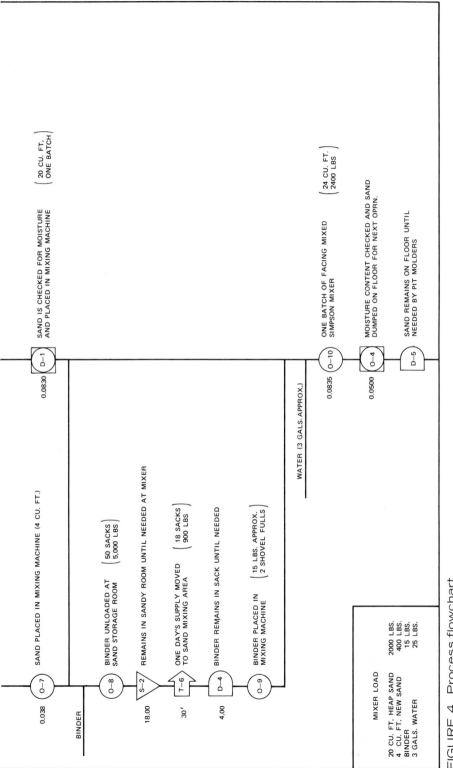

FIGURE 4 Process flowchart. (Courtesy McGraw-Hill Book Co.)

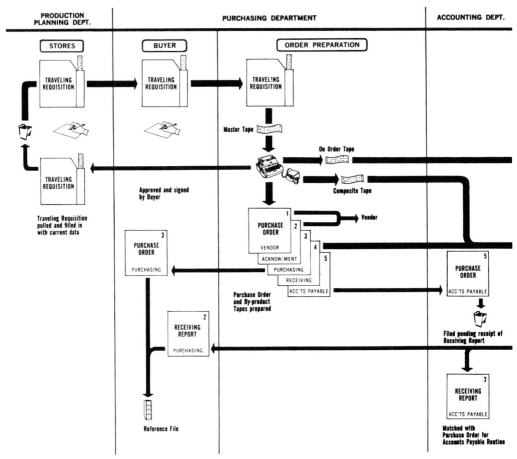

FIGURE 5a Forms flowchart.
(Courtesy Moore Business Forms, Inc.)

representation, the LSG drew upon the work of Robert Bosak of the System Development Corporation. Bosak's unpublished paper entitled "An Information Algebra" utilized the concepts of modern algebra and point set theory. After many months of work and study the language structure group expanded Bosak's work into a basic operational algebra which serves as a theoretical basis for an automatic processing of system specifications.

The underlying concepts of the Algebra have been implicitly understood for years by the business system analyst. An information system deals with those objects and events in the real world that are relevant to the task at hand. Real objects and events, called "entities," are represented in the system by data. The information system contains information from which the desired outputs can be extracted through processing. Information about a particular entity is in the form of "values" which describe quantitatively or qualitatively a set of attributes or "properties" that have significance in the system.

"Existing programming languages are inherently procedural in nature, but there are some examples in which relationships rather than procedures

Purchasing - Receiving System

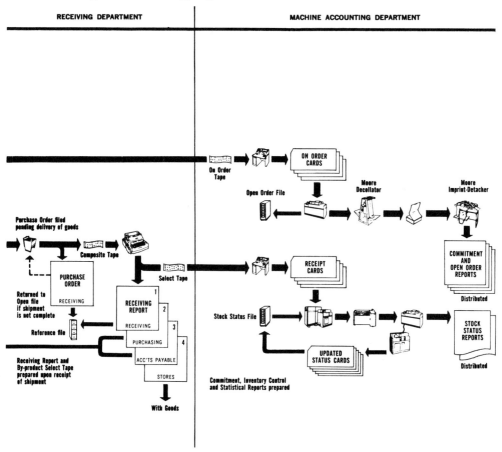

FIGURE 5a *Continued*

are specified. In particular, report writers and sort generators are of the latter type. The primary intent of the Information Algebra is to extend the concept of stating the relationships among data to all aspects of data processing. This will require the introduction of increased capability into compilers for translating this type of relational expression into procedural terms. Specification of such functions as READ, WRITE, OPEN, MOVE, and much of the procedure control definition will be left to the compiler, thereby reducing the work of the system analyst. The analyst will specify the various relevant sets of data and the relationships and rules of association by which these data are manipulated and classed into new and different sets of data, including the desired output."*

Figure 10 illustrates the Information Algebra on a sample payroll problem. The problem is to create a New Pay File from the information given in an Old Pay File, a Daily Work File, and a New Employee File. The Daily Work File

---

*This quote and the example in Figure 10 come from reference 20.

ACTION UPON RELEASE OF SCHEDULE

When a schedule is released, a schedule card is punched for each type of model scheduled. This card is punched with model number and the quantity scheduled for each month. Step by step, the following procedure is used:

1. Pull from the file the bill of materials master deck of tabulating cards for the models being scheduled.

2. Reproduce these master decks card for card and refile the master decks.

3. Sort the reproduced cards to quantity used per model and model number.

4. Control on model number and quantity. Tabulate and cut a summary card with this information only. Hold detail cards until step number 10.

5. Match, on model number, the summary cards against the schedule. Punch the quantity, by month, into each card.

6. Sort out all cards with one used — hold until step number 8.

7. Reproduce all cards two through nine used — repeat this process, each time taking off the lowest number until only the nines remain to reproduce.

8. Take all cards from steps six and seven and sort to number used by model number.

9. Cut summary card, controlling on model number and number used. Add the quantities punched in each monthly field.

10. Match the summary cards against the detail bill of materials cards held in step four.

11. Sort the detail cards to part number.

12. List the cards on the Tabulator, controlling on part number.

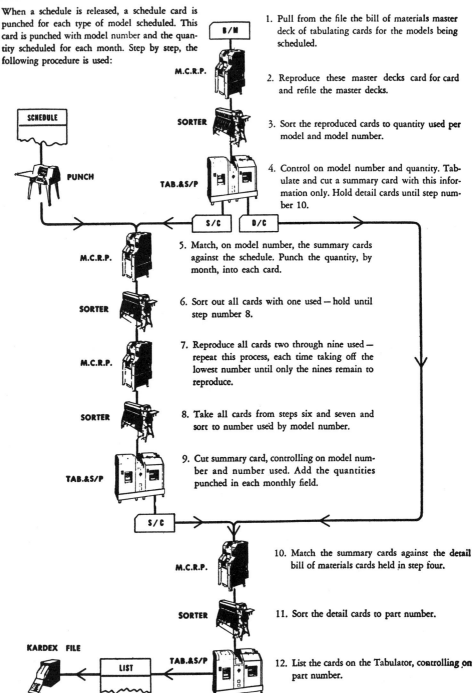

FIGURE 5b Tabulating machine process chart. (Courtesy Univac.)

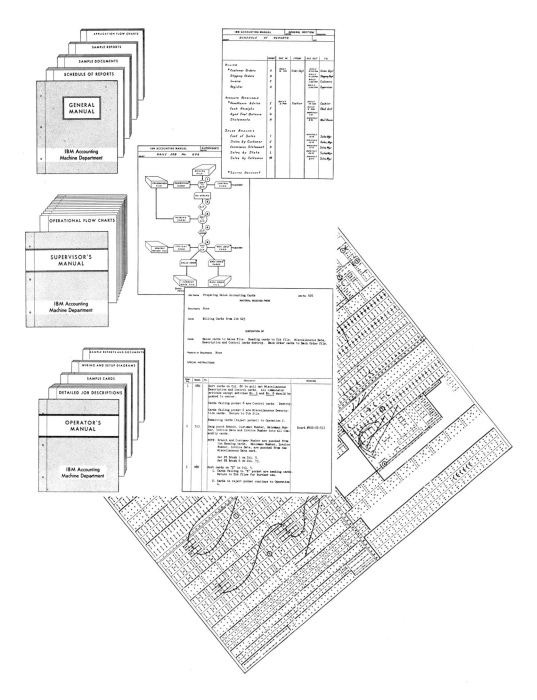

FIGURE 6 Tabulating operations documentation package. (Courtesy IBM.)

INFORMATION PROCESS CHART

ENTRIES	MAIN LINE FLOW	EXITS	SECONDARY FLOW	RECORDS	RECORD OR FIELD	FIELDS	REMARKS
2				Tool crib attendant received copy of purchase order for any tool crib rooted item being ordered.			
	AM 2			Merge	Copy of purchase order	in	Purchase order book	by	Purchase order number	
B	3			Material with one copy of receiving report arrives at tool crib.			
	SR find 4 No find		6	Search	Purchase order book	by	Purchase order number	from	Receiving report	
	+ 5			Insert	Quantity received, date received	on	Purchase order		
	CB = 6 ≠		3	Compare	Destination on purchase order	with	Tool crib number		
5	7			Compare	Material destined for a planner, engineer or foreman			
	CB = 8 ≠		4	Compare	Material	with	Special purpose tool, gauge, or fixture which has assigned tool number		
	SR No find 9 Find		7	Search	Tool number file	by	Tool number	for	Location	
	≠ 10			Create	3 copies of tool card	from	Purchase order, dimension card	Insert all fields	

FIGURE 7 Information process charting.

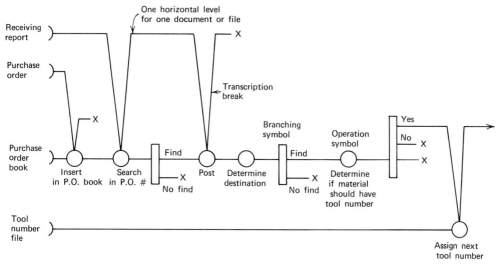

FIGURE 8 MAP diagram. (Courtesy NCR.)

contains the records of daily hours worked for each employee for the week. The Old and New Pay Files contain other information about the employees including the year-to-date totals of salary earned. The New Employee File provides rate-of-pay information about new employees for incorporation into the New Pay File. The solution is based on collecting and summarizing each employee's daily work records and on the matching of records in the Old Pay File, the summarized Work File, and the New Employee File to create the up-to-date records for the New Pay File for both old and new employees.

A property space sufficient to contain all the information is constructed on nine properties. Four areas corresponding to the four files are defined in this space. The properties, their value sets, and the relevance of each property to each area are shown in Table 1, where an "X" denotes relevant information and a "Ω" denotes nonrelevant data for that file.

The solution to this Payroll Problem is fully expressed by the following relationships. The New Pay File is expressed as the union of two areas, one derived for the old employees and the other for the new employees. Each of these areas in turn is a function of a "bundle" of two areas. The first area of each bundle is the same and is a function of a "glump" of the daily work file.*
The second area of one bundle is the Old Pay File and the second area of the other bundle is the New Employee File.

The LSG hoped that Information Algebra, being machine independent, would foster and guide the development of more universal programming languages. The group felt, also, that the Algebra could provide a step toward the goal of overall system optimization by

*One of the distinctions which may be made between a bundle and a glump is that a glumping function partitions strictly on the basis of its assigned values for points, whereas discrimination by bundling functions is contingent on the truth of some statement. (Clearly, they coincide at times.) A second distinction is that bundles can involve many areas whereas glumps are concerned with only one. A third difference is that the elements of bundles are lines (which means that the points are ordered) whereas the elements of glumps are sets of unordered points.

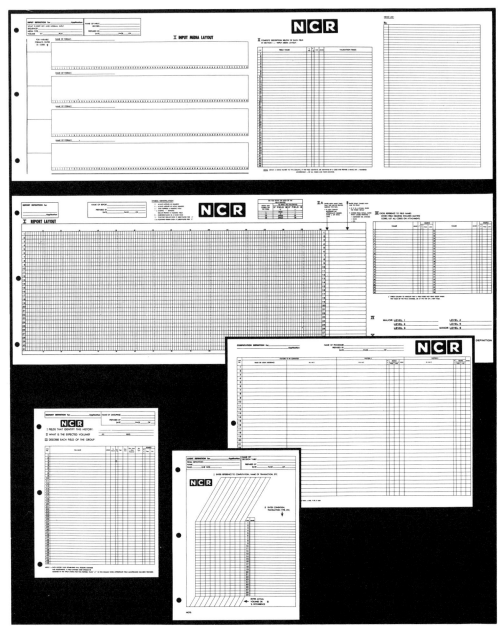

FIGURE 9 ADS interrelated forms. (Courtesy NCR.)

making it possible to manipulate the notation in which the fundamental statement of the problem is expressed. "With current programming languages, the problem definition is buried in the rigid structure of an algorithmic statement of the solution, and such a statement cannot readily be manipulated."

The LSG did not produce a user-oriented language for defining problems, nor did it specify the algorithm for translating information algebra statements into machine-language programs. The group expected that the formalism of the approach taken in developing the Information Algebra would assist in providing

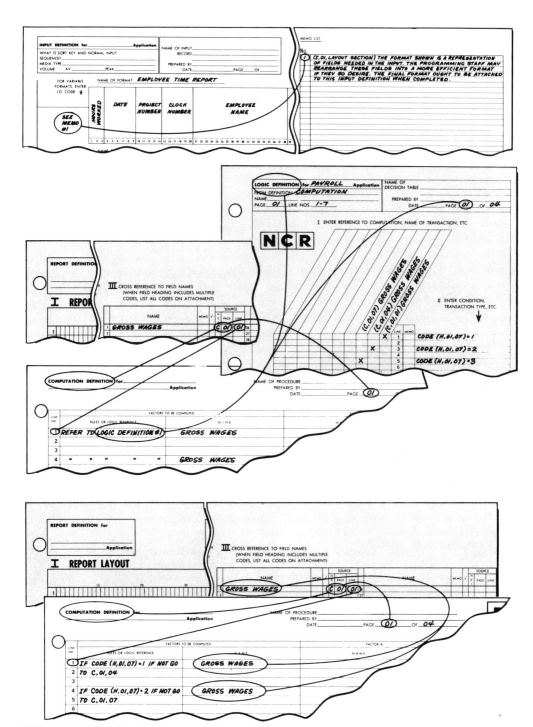

FIGURE 9 Continued

TABLE A Payroll Problem—System Information

Properties	Value Set	Areas			
		Old Pay File OP	Daily Work File DW	New Employee File NE	New Pay File NP
q_1 = File ID	PF, DW, NE	X (always PF)	X (always DW)	X (always NE)	X (always PF)
q_2 = Man ID	00000 ⋯ 99999	X	X	X	X
q_3 = Name	20 alphabetic characters	X	Ω	X	X
q_4 = Rate	00.00 ⋯ 99.99	X	Ω	X	X
q_5 = Hours	00 ⋯ 24	Ω	X	Ω	Ω
q_6 = Day number	0 ⋯ 7	Ω	X	Ω	Ω
q_7 = Total salary	00000.00 ⋯ 99999.99	X	Ω	Ω	X
q_8 = Pay period number	00 ⋯ 52	X	Ω	X	X
q_9 = Salary	000.00 ⋯ 999.00	X	Ω	Ω	X

(1) $NP = F_1[q_{12} = q_{22}, H(q_2, DW), OP]$

$\cup\ F_2[q_{12} = q_{22}, H(q_2, DW), NE]$

$$H \equiv \begin{cases} q_2' = q_2 \\ q_5' = \Sigma\ [q_5 \leftarrow (q_5 < 8) \rightarrow 1.5 * q_5 - 4] + f_1 \\ f_2 = \Sigma\ [q_5 \leftarrow (q_5 < 8) \rightarrow 8] \\ f_1 = 0 \leftarrow (f_2 < 40) \rightarrow 0.5 * f_2 - 20 \end{cases}$$

$$F_1 \equiv \begin{cases} q_7' = q_{27} + q_{15} * q_{24} \\ q_8' = q_{28} + 1 \\ q_9' = q_{15} * q_{24} \end{cases}$$

$$F_2 \equiv \begin{cases} q_1' = PF \\ q_7' = q_{15} * q_{24} \\ q_8' = q_{28} + 1 \\ q_9' = q_{15} * q_{24} \end{cases}$$

The terms of equation (1) are identified in Figure A. Let us examine each element of this equation more closely. Consider first the expression $H(q_2, WD)$. The Daily Work File (DW) consists of the daily hours-worked entries for all employees. Using q_2 (Man ID) as a glumping function, each glump element contains the entries for a single employee. $H(q_2, DW)$ is a function of a glump which calculates the values of certain properties for a set of points where each point corresponds to a single glump element. These values are defined by

$$H \equiv \begin{cases} q_2' = q_2 \\ q_5' = \Sigma\ [q_5 \leftarrow (q_5 < 8) \rightarrow 1.5 * q_5 - 4] + f_1 \\ f_2 = \Sigma\ [q_5 \leftarrow (q_5 < 8) \rightarrow 8] \\ f_1 = 0 \leftarrow (f_2 < 40) \rightarrow 0.5 * f_2 - 20. \end{cases}$$

In the definition of H, f_1 and f_2 are functions of areas. Note that $f_2 = 40$ unless $q_5 < 8$, in which case $f_2 = \Sigma\ q_5$. From this, it follows that $f_1 = 0$ if $f_2 < 0$; and $f_1 = f_2/2 - 20$ = half of excess time over 40, otherwise. The overtime rule used in q_5' states that a man is paid time and a half for each day's excess over 8, and also is paid time and a half for the excess over 40.

All other properties are undefined for this function of a glump.

The set of points obtained from $H(q_2, DW)$ constitutes an area which is the first area of the area set $[H(q_2, DW), OP]$. A bundle is defined over this area set. The bundling function for this bundle is a match on the Man ID's in each area, i.e., $q_{12} = q_{22}$. q_{12} is the property Man ID in the first area of the bundle; (q_{22} is the corresponding property in the second area of the bundle.

$F_1[q_{12} = q_{22}, H(q_2, DW), OP]$ is a function of a bundle which maps each line of the bundle into a single point for each employee. The definition of F_1 is:

$$F_1 = \begin{cases} q_7' = q_{27} + q_{15} * q_{24} \\ q_8' = q_{28} + 1 \\ q_9' = q_{15} * q_{24} \end{cases}$$

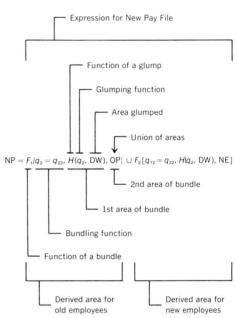

FIGURE A Identification of terms of equation (1).

The subscripts of the q-primes denote the property in the new area which is being defined by the equation. The first subscript of each unprimed q indicates the area of the bundle and the second subscript identifies the particular property. Thus, q_{27} is the property Total Salary for points in the second area of the bundle.

Each of the nonstated q-primes is understood to have the same value as the corresponding q in the last (in this case, the second) area of the bundle. Thus,

q_1' = PF $\quad q_3'$ = Name $\quad q_5' = \Omega$
q_2' = Man ID $\quad q_4'$ = Rate $\quad q_6' = \Omega$

while the remaining q-primes are defined by F_1.

Similarly, a second function of a bundle is defined for the area set $[H(q_2, DW), NE]$:

$$F_2[q_{12} = q_{22}, H(q_2, DW), NE]$$

where F_2 is defined by $F_2 \equiv \begin{cases} q_1' = PF \\ q_7' = q_{15} * q_{24} \\ q_8' = q_{28} + 1 \\ q_9' = q_{15} * q_{24} \end{cases}$

and the remaining q-primes are understood to be taken from the New Employee File:

q_2' = Man ID $\quad q_4'$ = Rate $\quad q_6' = \Omega$
q_3' = Name $\quad q_5' = \Omega$

Now two areas of up-to-date pay information have been derived. All that remains is to join them by the set operation union to form the New Pay File.

a foundation for future practical and theoretical studies in the structure of data processing languages. These expectations have been fulfilled, as will be shown in the section on third generation techniques.

SOP (Study Organization Plan). ADS and Information Algebra concentrate on specifying system requirements, that is, the final part of Phase II of the system development cycle. They presume completion of Phase I, the study of the organization and its information needs.

Several approaches have been developed for Phase I. Philips, the Netherlands based company, produced ARDI (Analysis, Requirements Determination, Design and Development, Implementation and Evaluation) through the work of Hartman, Matthes, and Proeme.* IBM produced SOP (Study Organization Plan), through the work of Burton, Grad, Holstein, Meyers, and Schmidt.

ARDI is not shown in Figure 3 because it was a handbook of techniques. SOP was a more significant contribution to the field because it pulled various techniques together into an integrated approach. The fact of its integrative quality is evidenced by the three SOP components shown in Figure 3.

SOP was designed to gather data with which to analyze the information needs of the entire organization. Information was gathered and organized into a report containing three sections, shown on the left-hand side of Figure 11.

The *General* section included a history of the enterprise, industry background, goals and objectives, major policies and practices, and government regulations.

The *Structural* section contained a schematic model of the business, describing it in terms of products and markets, materials and suppliers, finances, personnel, facilities, inventories, and information.

The *Operational* section included flow diagrams and a distribution of total resources to represent the operating activities of the business. These charts showed how the resources of a business respond to inputs, perform operations, and produce outputs.

The appendix included the detailed working documents needed to explain operations, identify documents, and define the files in which the organization's information is stored. These documents were organized into four levels of hierarchy, depicted on the right-hand side of Figure 16.

The organizational structure comprising an activity, and a cost analysis of the activity, were recorded on the Resource Usage Sheet; and the flow of the activity itself was displayed on the Activity Sheet. Operations within the activity are further described on Operation Sheets, with detailed information inputs and outputs and information resources provided on Message and File Sheets, respectively. These forms permitted the analyst to describe the flow of incoming materials or information through the internal workings of the system to output products, services, or information. The relationship of these forms is more readily understandable through the use of actual data, as shown in Figure 12.

The Resource Usage Sheet fit each system under study into its larger context. It showed the organization and structure of the business environment. It also provided a rapid analysis of costs for the organizational components being surveyed and showed the cost impact for each activity or system.

The Activity Sheet traced the flow of a single activity, breaking it down into its major operations. Each Activity

*Hartman, W., et al., *Management Information Systems Handbook* (McGraw-Hill, NY, 1968).

FIGURE 10 (*Left*) Illustrative payroll problem expressed in information algebra notation (Excerpt from *Commun. ACM*, Vol. 5, No. 4, Apr. 1962, p. 203.)

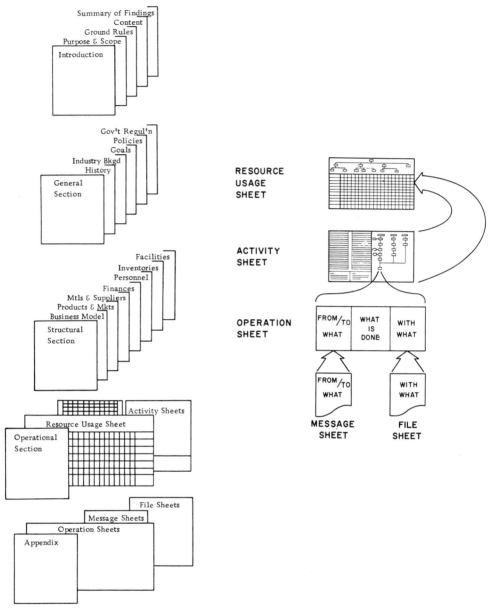

FIGURE 11 Contents of SOP documentation package (left) and documentation levels (right). (Courtesy IBM.)

Sheet presented as large a group of related operations as could be handled conveniently. It included a flow diagram of the activity with individual blocks representing various operations (each of which could be more minutely described on Operation Sheets). Key characteristics such as volumes and times were recorded in tabular form.

The Resource Usage Sheet and Activity Sheet worked together to provide a quick look at a business system.

For a closer look at the operation of a system, the analyst used forms that permit a more detailed documentation: the Operation Sheet and the Message and File Sheets.

Usually an Operation Sheet existed for each block on the Activity Sheet. It was used for recording the related processing steps that form a logical operation. It described what is done, with what resources, under what conditions, how often, to produce what specific results. Its primary purpose was to show relationships between inputs, processes, resources, and outputs.

The Message Sheet was one of two forms that supported the Operation Sheet. It described the inputs and outputs.

The File Sheet described a collection of messages, an information file. It identified the stored information the operation utilized. When describing what was done (on the Operation Sheet), the analyst could choose from several levels of detail by exercising a choice of descriptive verbs.

It is readily seen that the comprehensiveness of SOP, and the interrelated documentation techniques, enhanced the possibilities for integration of systems. However, a significant activity was missing. The relationship of the information system to the firm's master plan was not clearly distinguished.

THIRD GENERATION SYSTEM ANALYSIS TECHNIQUES

First and second generation system analysis approaches concentrated on individual systems. Although steering committees were organized to oversee feasibility studies, a computer was justified typically on one or two systems. Tangible and immediate savings could be produced in converting unit record applications to the computer. The accounting system, therefore, was one of the principal systems to justify a computer.

Emphasis in the third generation systems concept is on those activities which initiate actions, such as the market forecasting system. The forecast is automatically fed to all the other systems, insuring that all systems are working on the same forecast. Accordingly, the accounting system is designed to insure that the other systems are producing according to managerial performance standards.

Third generation systems philosophy concentrates on studying the organization as a whole, to avoid suboptimization. Third generation systems priority is on computerization of "lifestream" information systems. Computerization of lifestream operations of a firm produces considerably more benefit to the firm than applications in the administrative functions. Examples of lifestream information systems are order-entry systems, production-control systems, inventory and distribution systems. The advantages of computerizing administrative applications usually are insignificant compared with lifestream applications. For example, a 1% decrease in the cost of a $10 million inventory, resulting in a savings of $100,000, is far easier to achieve than a comparable savings in the accounting department. Lifestream projects have significant impact on the success or failure of the enterprise.

Increase in scope and sophistication of systems required a concomitant improvement in system development techniques. The third generation approach to refining system techniques employed the computer as an integral component in system development. Therefore, we refer to the third generation as the era of computer-aided system analysis.

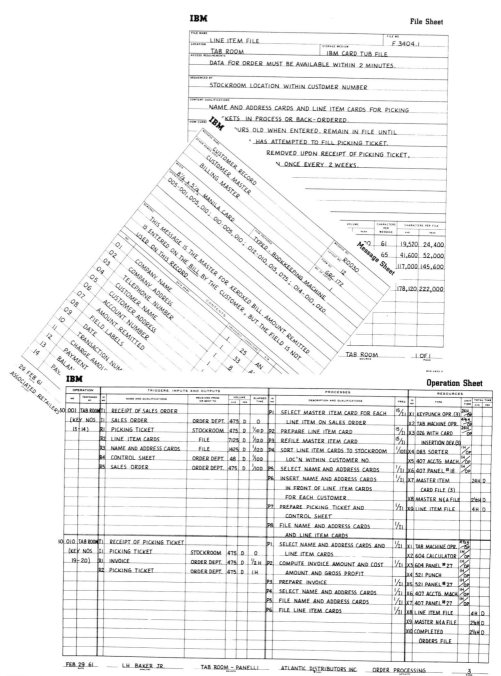

FIGURE 12 Illustration of the documentation levels of SOP. (Courtesy IBM.)

Automating Existing Techniques

The early approach was to computerize existing techniques. Decision table processors were developed in the mid-1960's. ADS processors were developed in the late 1960's. TAG (Time-Automated Grid) was developed in 1962 and automated in 1966. These three approaches will be described

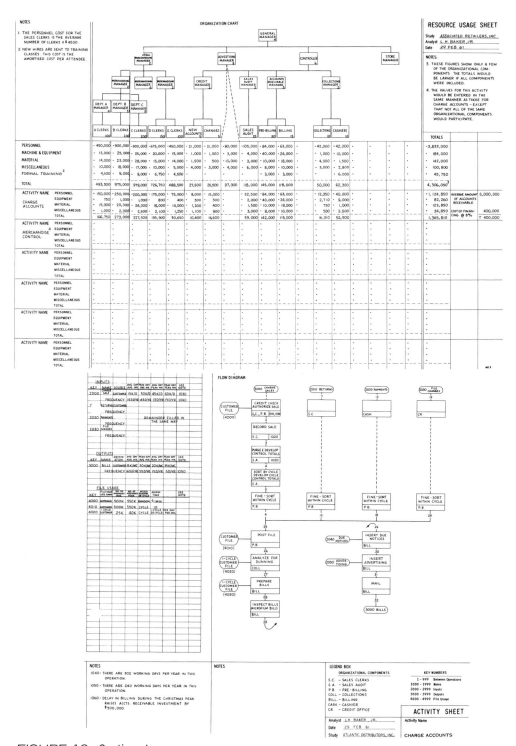

FIGURE 12 Continued

before proceeding to discussion of techniques designed specifically for computer-aided system analysis.

Decision Table Processors

The logical approach inherent in a decision table made this technique readily adaptable for computer processing. In May 1959, CODASYL began a project to enable computer processing of decision tables. In September, 1962, the product of this effort, DETAB-X, was made public. It consisted of a language supplement to COBOL-61, for use within the framework of decision tables. However, little use was made of the product. Finally, in 1965 the Special Interest Group for Programming Languages (SIGPLAN) of the Los Angeles Chapter of the Association for Computing Machinery appointed a working group to develop a decision table processor. Written in a subset of COBOL, the processor (DETAB/65) accepted decision tables coded in COBOL and converted them to COBOL source code. It was implemented on a variety of CDC and IBM computers.

However, the conversion algorithm was inefficient and the technique was not widely used until the next generation of processors evolved. The principal present-day approach consists of a preprocessor, written in COBOL, which converts decision tables containing COBOL components to code acceptable to the COBOL compiler. A few processors are written in FORTRAN.

This use of the computer did not facilitate the work of the system analyst, but it speeded up the total development process and facilitated modification of the system.

Automated ADS

Although ADS provided a well-organized approach to cross-referencing system definition documents, modifications to the system were laborious to incorporate. Also, it is difficult to assess the consequences of a modification because ADS is based on backward referencing (i.e., from output back through intermediate processing to input). The logical sequence in designing a system is forward referencing.

In automating ADS, the documentation medium is punched cards instead of paper forms. Each line of the original ADS forms is represented by one to three punched cards. The card file is analyzed by programs which 1) check for consistency and completeness, 2) check back references, 3) generate back references that have been omitted, 4) produce a dictionary of names, 5) generate incidence and precedence matrices, 6) flag errors, and 7) produce diagnostics.

The inevitable modifications in system can readily be incorporated. Cards for individual lines in the ADS definition can be revised and replaced and the effects assessed by computer analysis. The automated dictionary gives forward as well as backward referencing to aid the system analyst in redesigning the system instead of serving primarily as a documentation tool. Incidence and precedence matrices permit analysis of the structure of the system and facilitate planning of implementation.

TAG (Time Automated Grid). Developed in 1962 by D. H. Meyers, of IBM's System Research Institute, TAG was automated in 1966. To use the Time Automated Grid system, the analyst first recorded the system output requirements. Inputs were examined during later iterations of the program.

Once the output data requirements were fed into the system, TAG worked backward to determine what inputs were necessary and at what point in time. The result was the definition of the minimum data base for the system. With the aid of the reports generated by TAG, the analyst systematically resolved the question of how the required inputs were to be entered into the data flow. This approach enabled him to concentrate on pertinent input elements and to bring them into the system at the proper place. Superfluous or repetitious data were identified and eliminated from the system. Discrepancies in the use of any data element were corrected.

When both inputs and outputs were defined to TAG, the next iteration of the program produced file format and systems flow descriptions. File contents and data flow were both based upon time — the time at which data elements entered the system and the time at which they were required to produce output.

To TAG, the elapsed time between these two moments created the need for files. The files defined by TAG indicated what data must be available in each time period to enable the system to function. The job definition depicted the flow of these files, as well as of the inputs and outputs, within and between time cycles. This approach provided an overview of the system, showing the interrelationship of all data in the system.

Knowing these interrelationships made it possible for the system analyst to determine whether the outputs desired were quickly and easily obtained, and thus economically justified. With knowledge of the availability of data elements in given time periods, he determined where additional useful outputs might be obtained.

The upper portion of Figure 13 shows the principal form used for TAG. The form was divided into two horizontal sections, one for requirement titles, the other for data names. The characteristics of the input, output, or file being described were recorded in the requirements title section. Comments on the data requirements of the input, output, or file were detailed in the data name section.

The output of the TAG system was a series of ten reports that documented input, analyzed data requirements, and provided file and dataflow definition. The key report was the time-grid analysis, which traced the appearance of each data element, by time, through all the requirements in the system (shown in the lower portion of Figure 13). The grid indicated those data elements that must be carried in files, enabling the analyst to identify the minimum data base requirements.

The other nine reports were: time/key analysis, user data, glossary of data names, document analysis, sorted list of data names, summary of unresolved conditions, serial file records, direct access records, and job definition.

The development of this semiautomated technique was significant in the evolution of system analysis techniques.

Designing Computer-Aided Techniques

As could be expected, automating existing techniques proved to be a workable but suboptimal approach. As a consequence, several organizations began research on problem statement languages designed to make optimal use of the computer's capabilities. Two principal efforts were going on concurrently, at Xerox and at the University of Michigan. The Michigan research produced a problem statement language (PSL) which was, in the words of Project Director Dr. Daniel Teichroew, "a generalization of Information Algebra, TAG and ADS."

After developing an automated version of ADS, the Xerox group decided to design its own problem statement

IBM Time Automated Grid Technique (TAG) — Input/Output Analysis Form

Page 1 of 2 — Date: Feb 1

Data Type Code	Frequency	Period	Priority	Program Sequence	Requirement Title	Peak (P) Volume	Peak Frequency	Peak Survey Period	Average (A) Volume	Avg Frequency	Avg Survey Period	Minimum (M) Volume	Min Frequency	Min Survey Period	Designed For
O			2		INVOICE	10,800	1	M	9700	1	M	9200	1	M	P

Comment	Data Name	Size	A/N	Class Use	Ratio	Sequence	Format	Remarks	Signal
	CUSTOMER-ORDER-NO.	6	N	F I	1		2		
	CUSTOMER-NAME	25	A	F I	1		4		
	CUSTR-SHIP-TO-ADDRESS	75	A	F I	1		6		
	OUR-ORDER-NO.	6	N	F I	1	1	1		
	PART-CODE-NO.	6	A	F F	5	2	10		
	PART-SIZE-AND-BRAND	3	N	F F	5	3	11		
	PART-NAME	10	A	F I	5		12		
	QUANTITY-ORDERED	3	N	V F	5		7		
	QTY-SHIPPED	3	N	V F	5		8		
	QTY-OUT-OF-STOCK	3	N	V R	5		9		
C	SUBTRACT QTY-SHIPPED FROM QUANTITY-ORDERED GIVING QTY-OUT-OF-STOCK								
	INVOICE-NO.	8	N	F I	1		3		
	SOLD-TO-ADDRESS	75	A	F I	1		5		
	PRICE	5	N	V F	5		1 \$\$.\$\$\$		
	LINE-EXT	6	N	V R	5		1 \$2,\$\$\$.\$\$		
C	COMPUTE LINE-EXT ROUNDED = QTY-SHIPPED * PRICE								

DATA TYPE (Column 1) = INPUT-I, OUTPUT-O, FILE-F

IBM Time Automated Grid Technique (TAG) — Input/Output Analysis Form

Page 2 of 2 — Date: Feb 1

Comment	Data Name	Size	A/N	Class Use	Ratio	Sequence	Format	Remarks	Signal
	DISCOUNT-AMT	5	N	V R	1		1 \$2,\$\$\$.\$\$		
C	SUM OF LINE-EXT * DISC-RATE = DISCOUNT-AMT								
C	IF SUM OF LINE-EXT GREATER THAN DISC-QUALIFICATION-AMT MOVE								
C	DISC-RATE-1 TO DISC-RATE ELSE MOVE DISC-RATE-2 TO DISC-RATE								
	DISC-RATE-1	2	N	V F	1				
	DISC-RATE-2	2	N	V F	1				
	DISC-QUALIFICATION-AMT	5	N	F F	1				
	DISC-RATE	2	N	V R	1		1 \$2.\$\$		
	COD-OR-CREDIT-CODE	1	N	F F	1		18		
C	BLANK, 7, 8, OR 9 ARE INVALID								
	TOTAL-INVOICE-AMOUNT	8	N	V R	1		1 \$2,\$\$\$,\$\$\$.\$\$		
C	SUM OF LINE-EXT MINUS DISCOUNT-AMT								

DATA TYPE (Column 1) = INPUT-I, OUTPUT-O, FILE-F

```
RESULTS OF ANALYSIS BY TIME-GRID TECHNIQUE


DATA                                              DATA
NUMBER  DATA NAME           SIZE A/N USE          NUMBER  DATA NAME            SIZE A/N USE
    21  QUANTITY-ORDERED       3  N   FI              22  SHIPPING-INSTRUCTIONS  100 A   FI
    23  SOLD-TO-ADDRESS       75  A   FI              24  TOTAL-INVOICE-AMOUNT     8 N   VR


DATA NUMBER                   21   22   23   24

CYCLE


  1 CUSTOMER-ORDER             5    1    1    0
  (     1) 8300...  1 X D
  1 WAREHOUSE-ORDER            5    1    0    0
  (     2) 8300...  1 X D
  2 INVOICE                    5    0    1    1
  (     3)10300...  1 X D

SUMMARY CODES                  1    0    1    2



MEANING OF SUMMARY CODES

0 - RATIO OF INPUT = RATIO OF OUTPUT, INPUT AVAILABLE AT TIME OF OUTPUT
1 - PLURAL CYCLES - FILES
2 - SYSTEM GENERATED (VARIABLE RESULT)
3 - NO INPUT BUT OUTPUT, NOT VARIABLE RESULT
4 - NO OUTPUT BUT INPUT
5 - RATIOS NOT EQUAL
6 - OUTPUT REQUIRED BEFORE INPUT IS AVAILABLE
```

DATA NO.	DATA NAME	CODE	PAGE NO.
2	CUSTOMER-NAME	15	3
5	DATE-OF-ORDER	4	3
6	DISC-QUALIFICATION-AMT	3	3
8	DISC-RATE-1	3	3
9	DISC-RATE-2	3	3
12	INVOICE-NO	3	3
13	LINE-EXT	152	3
16	PART-NAME	13	3
18	PRICE	3	3
20	QTY-SHIPPED	153	3

```
1    1 I CUSTOMER-ORDER                 8300
              14   0   0   0   0   0   0   0
1    2 O WAREHOUSE-ORDER                8300
              14  15  17   0   0   0   0   0
2    3 O INVOICE                       10300
              14  15  17ᴬ  0   0   0   0   0
3    4 O WEEKLY-SHIPMENT-REPORT         5000
              15  17   0   0   0   0   0   0
```

FIGURE 13 (*Left and above*) TAG input form and three out of ten outputs. (Courtesy IBM.)

language, SSP. However, SSP is not operational so this paper will confine its analysis to PSL, which is operational in several organizations affiliated with the Michigan project.

PSL (Problem Statement Language) and PSA (Problem Statement Analyzer)

Using the Michigan approach the system analyst concentrates on *what* he wants without saying *how* these needs should be met. The Problem Statement Language (PSL) is designed to express desired system outputs, what data elements these outputs comprise, and formulas to compute their values. The user specifies the parameters which determine the volume of inputs and outputs and the conditions (particularly those related to time) which govern the production of outputs and acceptance of inputs.

The Problem Statement Analyzer (PSA) accepts inputs in PSL, analyzes them for correct syntax, and then:

1. Produces comprehensive data and function dictionaries,
2. Performs static network analysis to insure completeness of derived relationships,
3. Performs dynamic analysis to indicate time-dependent relationships of data,
4. Analyzes volume specifications.

The result of these analyses is an error-free problem statement in machine-readable format. The second output is a coded statement for use in the physical system design process.

Five classes of users received output from the PSA, as depicted in Figure 14. The system building process is a task undertaken under the direction of Problem Definition Management. This individual or group has the prime responsibility of defining the overall framework and structure of the problem. This task is aided by a team of Problem Definers who perform the more detailed aspects of problem definition: stating the details of individual outputs and input, and insuring that problem descriptions are accurate. The Data and Functional Administration group coordinates the activities of individual Problem Definers and monitors all data definitions. Another aspect of coordination is insuring that items common to many inputs and outputs are defined for the whole system. When this refers to data elements, it is done by DFA. When it refers to other items such as "total system requirements," it is carried out by a group called System Definers.

Once the problem has been defined and specified, the physical system design can begin. The Michigan research has produced an operational PSL/PSA system, used in several organizations which are affiliates in the Michigan research program.

Hoskyns System

Separation of the system life cycle into seven distinct phases was appropriate for three generations of system development. It is no longer appropriate. The thrust of the fourth generation is integration of the activities in system development in order to more effectively utilize the computer as an aid in system development. Phase II and Phase III of the system life cycle are being merged into one activity. This situation is aptly characterized by Professor Teichroew, "Information needs interact with the characteristics of the mechanism (speed, cost, capabilities, etc.) that will be used to satisfy them. Consequently, there must be iterative cycles between the analysis and design."* Therefore, the

---
*See reference 33 in the bibliography, p. 2.

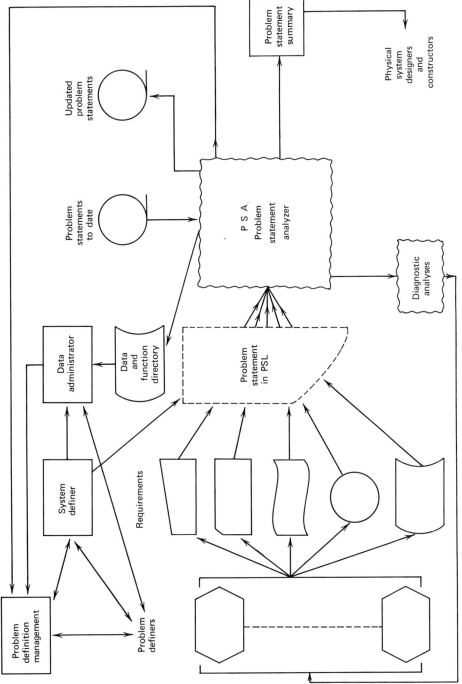

FIGURE 14 Problem statement language and analyzer.

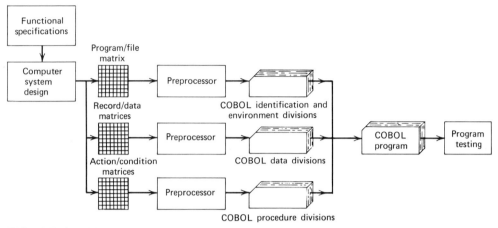

FIGURE 15 Hoskyns system.

output of the combined Phase II and III is a set of automated program specifications. An intermediate step is already operational. The Hoskyns System permits automatic translation of systems specifications into computer programs.

Figure 15 illustrates the Hoskyns System approach. A preprocessor automatically translates system specification matrices into COBOL programming statements, permitting program elements to be built and then consolidated into programs. Using the Hoskyns approach* the system is described in terms of programs and files, and the programs in terms of records and data elements. These sets of relationships are recorded in the form of matrices, as shown in the figure.

The first matrix provides the program file relationships in the system, completely defining the system flow. The second matrix states the keys by which the files may be accessed and defines which record types exist in which file. These matrices also contain such information as file organization approach (i.e., index-sequential).

Taken together, these matrices provide the information necessary to generate the identification and environment divisions for all COBOL programs in the system, as well as the File Descriptions with record layout COPY statements for the data division. These divisions represent the envelope within which the programmer must write his procedure coding. This envelope is generated automatically from the matrices by the COBOL generating processors.

The second matrix also provides the record specification. These record descriptions are held in a library, to be called into the File Descriptions of the programs by the previously described processor.

The remaining COBOL element is the procedure division. Decision tables, prepared by the system designer, list the conditions and actions of the procedure. The decision tables are input to the COBOL procedure division processor and are incorporated into the source program.

In summary, the Hoskyns system accepts system specifications and converts them to COBOL programs without manual intervention. The system was developed and implemented in three British corporations by Hoskyns Systems Research Incorporated. It was introduced in the U.S. in 1972 and is

*See reference 34 in the bibliography.

in use at Xerox, General Foods, and Allied Chemicals.

Impact of Computer-Aided System Analysis

While a majority of practitioners continue to use manual system analysis techniques, use of computer-aided techniques is growing. Some thirty organizations are affiliated with the Michigan project. ADS/PSA is operational in several organizations. TAG is used more internally by IBM than by its customers; however, several organizations are using it. A Swedish group, headed by Professor Borge Langefors, has also developed a problem statement language and analyzer, IA/1, which is close to the operational stage.*

Despite the slowness in adoption of third generation system analysis techniques, the fourth generation is beyond the drawing board stage and will be discussed next.

FOURTH GENERATION SYSTEM ANALYSIS TECHNIQUES

Fortunately, it is not necessary to prognosticate approaches to fourth generation system analysis. Enough research has transpired to recognize the direction of the next step in evolution of system analysis techniques.

Just as computer applications are being integrated, techniques for each of the phases of system development are being integrated. A natural extension of computerized problem statement is translation of those statements into programming language statements. However, to produce a complete system, not just portions of a system, a system optimizer must be included in the process. Optimizers of this type are already operational, but only as independent modules.* The fourth generation approach links these subsystems into an integrated whole.

ISDOS

The ISDOS project at the University of Michigan is designed to produce such a system. ISDOS is an acronym for Information System Design and Optimization System. While completion of the ISDOS project is some time away, a sufficient number of modules has been designed and tested to prove the validity of the approach.

The ISDOS project is formalizing the design process along the lines of the mathematical approaches pioneered by Langefors, Gross, Turnburke, and Martin.** Use of a multilevel approach, where the decision variables at one level become constraints at the next level, makes feasible evaluation of a large variety of design strategies.

Reference to Figure 16 shows the interaction of the four ISDOS modules. The Data Re-organizer accepts: (1) specifications for the desired storage structures from the physical systems design process, (2) definition of data as summarized by the Problem Statement Analyzer, (3) the specifications of the hardware to be used, and (4) the data as they currently exist and their storage structure. It then stores the data on the selected devices in the form specified. The third module, the Code Generator, accepts specifications from the physical design process and organizes the problem statements into programs recognizing the data interface as specified by the Data Re-organizer. The code produced may be either machine code, or statements in a higher level language (e.g. COBOL), or parameters to a software

---

*See reference 35 in the bibliography.

*See reference 28 in the bibliography.
**See references 29–32 in the bibliography.

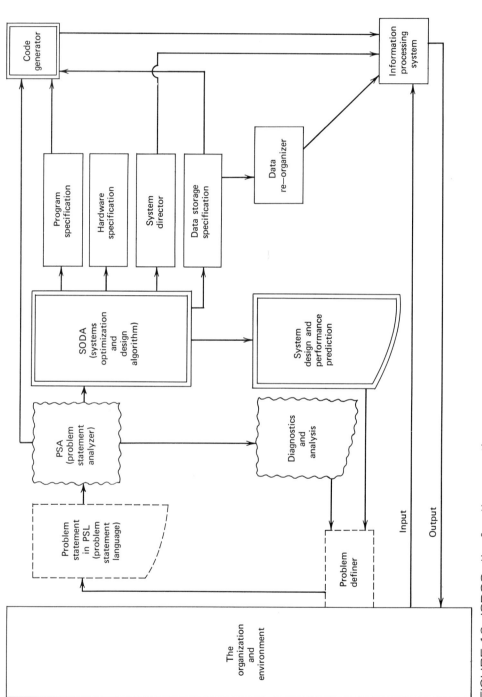

FIGURE 16 ISDOS—the fourth generation.

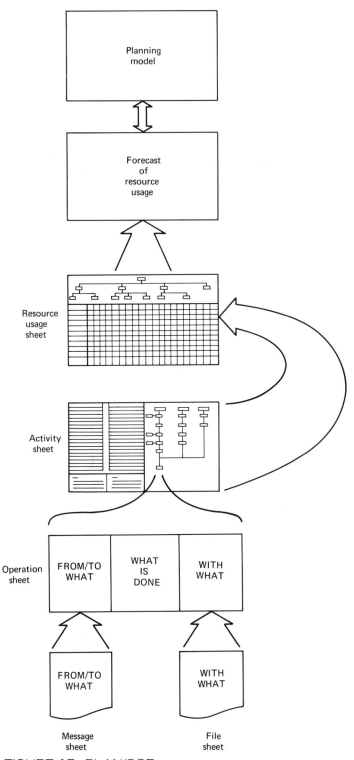

FIGURE 17 PLAN/SOP.

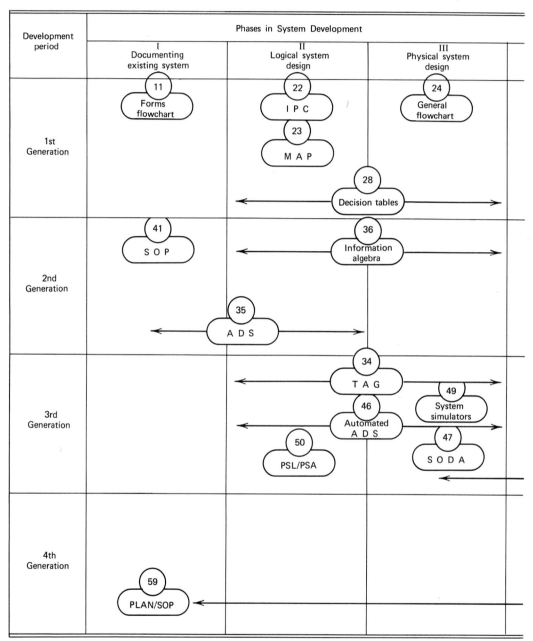

FIGURE 18 Evolution toward integration of system development techniques. (Adapted from a chart by D. Teichroew and D. Carlson.)

Phases in System Development			
IV Construction	V Test and conversion	VI Operation	VII Maintenance and modification
(24) Block diagrams (42) Programming languages (43) Compilers and assemblers	(29) PERT (30) CPM		
(44) Debugging aids (38) Decision table translators	(39) Automated PERT/CPM	(45) Operating systems	
(56) Optimizers (48) DBMS	(54) Emulators (57) Test data generators	(51) Hardward/software monitors	(52) Flowcharters (58) Librarians
	(53) Hoskyns system →		
(54) ISDOS →			

FIGURE 18 *Continued*

package. These two modules perform, automatically, the function of programming and file construction.

The final module of the ISDOS system is the Systems Director. It accepts the code generated, the timing specifications as determined by the physical design algorithm, and specifications from the Data Re-organizer, and produces the target information system. This system is now ready to accept inputs from the environment and produce the necessary outputs according to the requirements expressed in the problem statement.

PLAN/SOP

Although the objectives, policies, and practices of the organization were documented in the SOP approach, little emphasis was placed on the Master Plan for the organization—that is, the acquisition and allocation of future resources required to meet the organization's objectives.

For example, the firm may seek to become much more capital-intensive, thus drastically affecting the priority of development of subsystems. The financial subsystem would take priority over many of the labor-control subsystems for such an organization. Future plans for allocation of resources affect both priorities and contents of systems.

Also, the implication in the literature describing SOP is that the approach is used once, at the time the company decides to move from the fragmented, independent subsystem approach to the integrated systems approach. In actuality, it is an ongoing process, with updating dependent upon the dynamic characteristics of the organization.

Therefore, a need for Phase I of the fourth generation is the coupling of the planning model to the SOP, as shown in Figure 17 on page 75.

Figure 18 shows the evolution of system analysis techniques from a different perspective, with emphasis on the integration of techniques for all phases of the system development cycle. (The appendix provides references for each of the techniques, shown by the number in the small circle above each technique.) Figure 18 illustrates how the ISDOS objective is integration of all phases of the system development cycle, utilizing the computer as an integral tool in optimizing the process. The approach has great promise.

CONCLUSION

It is simply amazing that the systems profession delayed so long in using the computer as an aid in system analysis. The situation was summed up appropriately by Richard Thall, one of the University of Michigan researchers, who suggested that "it is shameful that the shoemaker's children are the last to have shoes."

The foregoing descriptions demonstrate that this deficiency is rapidly being corrected. Within the next year there should be sufficient results in each of these research efforts to evaluate their impact upon the computing community.

Progress to date suggests that the gap between development of hardware and system analysis/design techniques will be substantially narrowed by the advent of the 4th generation of computers.

APPENDIX I: REFERENCES

In anticipation that readers would want to assemble references on the techniques listed in Figures 3 and 18, the author attempted to reduce the number of sources required for reference. Therefore several handbooks are cited to enable readers to find information on the techniques. The number following each technique refers to the reference in the bibliography.

1. *Process Flow Chart*: 1, 2, 3, 4
2. *Linear Algebra*: 3, 5, 6
3. *Probability Theory*: 1, 3, 4, 5
4. *Truth Tables*: 3, 6
5. *Gantt Scheduling Technique*: 2, 3, 4
6. *Accounting Theory*: 4, 7
7. *Organization Charts*: 2, 4, 10
8. *Simultaneous Motion Charts*: 1, 2, 3
9. *Time Study (stopwatch)*: 1, 2, 3, 4, 10
10. *Tabulating Machine Process Chart*: 2, 15
11. *Forms Flow Chart*: 2, 10, 25
12. *Tabulating Operations Flow Chart*: 2, 15
13. *Board Wiring Diagram*: 2, 15
14. *Point and Set Theory*: 3, 6
15. *Linear Programming (Simplex Method)*: 2, 3, 4, 5
16. *Applied Statistical Analysis*: 1, 3, 4, 5, 10
17. *Precedence Network Technique*: 3, 5, 25
18. *Budgeting Techniques*: 1, 3, 4
19. *Linear Responsibility Charts*: 2
20. *Work Simplification Analysis*: 2, 3
21. *Synthetic Time Standards*: 2, 3, 4
22. *Information Process Chart*: 8
23. *MAP-System Charting Technique*: 9
24. *General Flowchart/Block Diagram*: 2, 10, 14, 15, 25
25. *Mathematical Programming (Linear, Non-Linear, Dynamic)*: 3, 5
26. *Inventory, Waiting Line Models*: 3, 4, 5
27. *Computer Simulation Techniques*: 3, 4, 5
28. *Decision Tables:* 11, 12, 25
29. *PERT (Program Evaluation Reporting Technique)*: 2, 3, 4, 5, 10, 25
30. *CPM (Critical Path Method)*: 2, 3, 4, 5, 25
31. *Clerical Work Sampling*: 2, 3
32. *Clerical Time Standards*: 2, 3, 4
33. *SOP (Study Organization Plan) Activity Analysis*: 16, 17
34. *TAG (Time Automated Grid)*: 18, 26
35. *ADS (Accurately Defined Systems)*: 19
36. *Information Algebra*: 20
37. *Large System Optimization Models*: 5, 21
38. *Decision Table Processor*: 11, 13, 43
39. *Computerized Production Planning*: 22, 23
40. *SOP-Operation Analysis*: 16, 17
41. *SOP-Resource Analysis*: 16, 17
42. *Programming Languages*: 37, 38, 39
43. *Compilers and Assemblers*: 38, 39, 42
44. *Automated Debugging Aids*: 37, 40, 42, 43
45. *Operating Systems*: 38, 39, 42
46. *Automated ADS*: 27, 33
47. *(SODA) Systems Optimization and Design Algorithm*: 28
48. *(DBMS) Data Base Management Systems*: 36, 42
49. *System Simulators*: 41, 42
50. *(PSL/PSA) Problem Statement Language and Analyzer*: 33, 24
51. *Hardware/Software Monitors*: 42
52. *Flowcharters*: 42, 43
53. *Hoskyns System*: 34
54. *Emulators*: 37
55. *(ISDOS) Information System Design and Optimization*: 24
56. *Optimizers*: 24, 43
57. *Test Data Generators*: 43
58. *Librarians*: 43
59. *PLAN/SOP*: 44

APPENDIX II: BIBLIOGRAPHY

1. Ireson, W., and E. Grant, eds., *Handbook of Industrial Engineering and Management,* Prentice-Hall, Inc., Englewood Cliffs, N.J., 1955.
2. Lazzaro, V., ed., *Systems and Procedures: A Handbook for Business and Industry,* 2nd Edition, Prentice-Hall, Englewood Cliffs, N.J., 1968.
3. Maynard, H. B., ed., *Industrial Engineering Handbook,* 3rd Edition, McGraw-Hill, N.Y., 1971.
4. Maynard, H. B., ed., *Handbook of Business Administration,* McGraw-Hill, N.Y., 1967.

5. Wagner, H., *Principles of Operations Research,* Prentice-Hall, Englewood Cliffs, N.J., 1969.
6. Kattsoff, L. O., and A. J. Simone, *Foundations of Contemporary Mathematics,* McGraw-Hill, 1967.
7. Garner, Paul, and K. B. Berg, *Readings in Accounting Theory,* Houghton Mifflin, Boston, 1966.
8. Grad, B., and R. Canning, "Information Process Analysis," *Journal of Industrial Engineering,* Nov–Dec., 1969, pp. 470–474.
9. *MAP-System Charting Technique,* National Cash Register Co., Dayton, Ohio, 1961.
10. *Business Systems,* Association for Systems Management, Cleveland, Ohio, 1970.
11. Pollack, S., H. Hicks, and W. J. Harrison, *Decision Tables Theory and Practice,* Wiley, 1971.
12. McDaniel, H., *An Introduction to Decision Logic Tables,* Wiley, N.Y., 1968.
13. McDaniel, H., *Decision Table Software,* Auerbach, Princeton, N.J., 1970.
14. Chapin, N., *Flowcharts,* Auerbach, Princeton, N.J., 1971.
15. Feingold, C., *Fundamentals of Punched Card Data Processing,* Wm. C. Brown Co., Dubuque, Iowa, 1969.
16. Glans, T., et al., *Management Systems,* Holt, Rinehart and Winston, Inc., N.Y. 1968.
17. *Study Organization Plan,* International Business Machines Corp., (Form No. C20-8075), White Plains, N.Y., 1961.
18. *Time Automated Grid System,* International Business Machines Corp., (Form No. GY 20-0358), 2nd Edition, White Plains, N.Y., 1971.
19. *A Study Guide for Accurately Defined Systems,* National Cash Register Co., Dayton, Ohio, 1968.
20. CODASYL Development Committee, "An Information Algebra," *Communications of the ACM,* Vol. 5, No. 4, April, 1962, pp. 190–204.
21. Lasdon, L. S., *Optimization Theory for Large Systems,* Macmillan, N.Y., 1970.
22. Miller, R. W., *Schedule, Cost and Profit Control with PERT,* McGraw-Hill, N.Y., 1963.
23. Barnetson, P., *Critical Path Planning,* Auerbach, Princeton, N.J., 1970.
24. Teichroew, D., and H. Sayani, "Automation of System Building," *Datamation,* Aug. 15, 1971, pp. 25–30.
25. Hartman, W., et al., *Management Information Systems Handbook,* McGraw-Hill, N.Y., 1968.
26. Kelley, Joseph F., *Computerized Management Information Systems,* The Macmillan Co., N.Y., 1970, pp. 364–400.
27. Thall, Richard M., *A Manual for PSA/ADS: A Machine-Aided Approach to Analysis of ADS,* ISDOS Working Paper No. 35, Dept. of Industrial Engineering, University of Michigan, Ann Arbor, 1971.
28. Nunamaker, J. F., Jr., "A Methodology for the Design and Optimization of Information Processing Systems," *Proceedings — Spring Joint Computer Conference,* AFIPS Press, Montvale, N.J., 1971, pp. 283–294.
29. Langefors, B., *Theoretical Analysis of Information Systems,* 2 Vol., Student Litteratur, Lund, Sweden, 1966.
30. Gross, M. H., "Systems Generation Output Decomposition Method," Standard Oil Co. of New Jersey, July, 1963.
31. Turnburke, V. P., Jr., "Sequential Data Processing Design," *IBM Systems Journal,* March, 1963.
32. Martin, J., *Design of Real Time Computer Systems,* Prentice-Hall, Inc. Englewood Cliffs, N.J., 1967.
33. Teichroew, Daniel, "Problem Statement Languages in MIS," *Proceedings, International Symposium of BIFOA,* Cologne, July, 1970, pp. 253–279.
34. Rhodes, John, "A Step Beyond Programming, Unpublished Paper, Copyright, J. Rhodes, Sept., 1972.
35. Lundeberg, Mats, "IA/1 — An Interactive System for Computer-Aided Information Analysis," Working Report No. 14E, ISAC, University of Stockholm, 1972.
36. CODASYL, *Feature Analysis of Generalized Database Management Systems,* Assoc. for Computing Machinery, N.Y., 1971.

37. Sammet, Jean E., *Programming Languages: History and Fundamentals,* Prentice-Hall, Inc., Englewood Cliffs, N.J., 1969.
38. Rosen, Saul, *Programming Systems and Languages,* McGraw-Hill, N.Y., 1967.
39. Donovan, John J., *Systems Programming,* McGraw-Hill, N.Y., 1972.
40. Rustin, Randall, ed., *Debugging Techniques in Large Systems,* Prentice-Hall, Englewood Cliffs, N.J., 1971.
41. Joslin, Edward, *Computer Selection,* Addison-Wesley, Reading, Mass, 1968.
42. *Software Reports,* Auerbach Computer Technology Reports, Philadelphia, Pa.
43. Canning, R. D., "COBOL Aid Packages," *EDP Analyzer,* May, 1972.
44. Couger, J. Daniel, "PLAN/SOP," unpublished paper.

Introduction to SECTION II

Second Generation System Analysis Techniques

Although third generation system analysis techniques are operational, many organizations continue to use second generation techniques. Some organizations are willing to be at the "cutting edge," while others are unwilling to take the risk associated with being one of the innovators. The situation is like that in the chemical industry where DuPont spends billions of dollars on new product development while others in the industry feel it more cost/effective to let DuPont market the new products then purchase a license from DuPont to produce the proven product. A parallel exists in the computing field where some users are willing to serve as a test center for a new version of the computer manufacturer's software in order to reap the benefits of the improved system; other users are unwilling to suffer the headaches associated with debugging the new software, although they may wait for six months to get the officially-released version.

Similarly, in 1973 less than 100 of the 50,000 U.S. computer installations were utilizing third generation system analysis techniques. Second generation techniques were adequate, just as second generation computers were adequate for the computation needs of some organizations.

With the exception of Murdick's article, the papers in this section describe second generation system analysis techniques. Murdick's paper is included because it provides a taxonomy of views on the phases in system development. It also amplifies the seven phases of system development into a pictorial schematic that is valuable for understanding the full process. While many papers treat system development solely in terms of the smallest module of the system, Murdick's paper shows the seven-phase development cycle appropriate for the overall MIS. This view is consistent with those of von Bertalanffy and Ackoff, but shows a practical application — to the Management Information System.

The second paper describes an approach for documenting the system, as it exists. The activity is frequently referred to as the System Survey. While several approaches have been developed for conducting this first phase of the system development cycle, the *Study Organization Plan (SOP)* developed by IBM is the most comprehensive. Using this approach, the business is described in broad, general terms; then it is detailed with an integrated set of documentation forms. Information is gathered on: (1) history and framework, (2) industry background, (3) goals and objectives, (4) policies and practices, and (5) government regulations. Five documentation forms are used to describe the overall system and its subsystems, in a level of detail to permit design of an improved system: (1) the organizational structure and associated budgets appear on the Resource Usage Sheet; (2) the flow of information associated with a given organizational function is depicted on the Activity Sheet; (3) operations within the function, or activity, are described on Operation Sheets; (4) details on information inputs and outputs are provided on Message Sheets; and (5) files in the processing of data are described on File Sheets.

Through use of a highly coordinated approach to documenting the existing system, such as the SOP approach, Phase II is greatly facilitated.

A variety of techniques are available for Phase II — analysis of the existing system to develop an improved system. One category of techniques is used to depict system logic, using a machine-independent orientation. The system is then analyzed for feasibility of computerization. Then the logic is further detailed into machine-oriented charts.

The most commonly used techniques for depicting system logic are flow charts and decision tables. The *flowcharting* technique has been used since the early days of computing. However, lack of standardization in the technique resulted in difficulty in translating logic by a person other than the one who prepared the flowchart originally. The paper by Chapin provides a tutorial on flowcharting, using the ISO and ANSI[1] standards to establish guidelines in flowchart development. Use of this approach greatly facilitates communication of system logic.

The *decision table* is another approach for displaying system logic. The paper by Fergus was selected from the pervasive literature on decision tables for the following reasons: (1) it

---

[1] International Standards Organization and American National Standards Institute.

provides definitions for use and development of decision tables and (2) it explains the advantages of decision tables compared with flowcharts. Much of the literature implies that flowchart and decision table techniques are used interchangeably, based on the preference of the analyst. However, each technique has its particular advantages. For example, the analyst needs to consider how his system logic will be interpreted. He validates system logic with the user of his system—the manager. Therefore, the analyst must consider which of the two techniques better communicates system logic to the manager. However, the analyst must also consider which technique most effectively communicates system logic to the programmer. By stating frequency of conditions through a decision table, the analyst aids a programmer in reducing the number of instructions. Fergus provides criteria for determining when to use flowcharts or decision tables.

In considering the interrelationship of systems, the *gridcharting* technique is of value. A gridchart is a formalized representation of the relationships between two groups of system elements. In complex cases, where the number of interrelated system elements is high, the gridchart is very useful as a starting point for approximation of design requirements. The chapter from the book by Hartman, et al., describes the gridcharting technique.

An *information matrix* aids in relating input, files, and output of a system. The matrix depicts the natural flow of information through the system, showing updating of the master files or generation of output. Use of the information matrix is illustrated on a sales order processing system in the Honeywell publication, *Business Information Systems Analysis and Design*.

The ADS (Accurately Defined System) technique utilizes a set of five interrelated forms to provide the system definition: (1) output definition, (2) input definition, (3) definition of computations to be performed, (4) history definition (information to be retained beyond the processing cycle for subsequent use), and (5) the logic definitions, in the form of decision tables. The advantage of ADS over prior techniques is the linkage of information. Each data element is assigned a tag, or reference, and is linked throughout, from input to output. ADS is described in the paper by Lynch.

The majority of the techniques described in the papers in this section are still in use, a testimony to their effectiveness. However, third generation techniques are beyond the developmental stage. The present-day practitioner must be well-versed in both second and third generation system analysis techniques.

MIS DEVELOPMENT PROCEDURES

ROBERT G. MURDICK

The recent emphasis on Management Information Systems (MIS) would seem to indicate that all a company needs to succeed is a set of managers (decision makers) and an information system. Many speak of the design of information systems as if they were independent of anything else in the firm. The fact is that the goals of the company, adding value to materials and ideas and filling a need in society, are accomplished by operational systems consisting of people, machines, and energy. The late Sherman C. Blumenthal (3) recognized this important concept by a taxonomy of systems which parallels operational systems and information systems. The

SOURCE: Murdick, R., "MIS Development Procedures," *Journal of Systems Management*, Dec., 1970, pp. 22–26.

top level of the hierarchies of the systems are:

1. Logistics
2. Physical assets
3. Financial
4. Manpower

This concept of parallelism has great significance from the standpoint of the design of MIS. MIS should not be designed in vacua; operational systems and MIS should be designed in *parallel*. Either type of system is meaningless by itself.

An interesting aspect of MIS design procedures is that there are some clearly identifiable events which may terminate a project, but there is little agreement on where a project starts. Termination, for example, may consist of the day of the cutover or some specified period of debugging after the cutover. Project initiation may be considered to start with a request for a preliminary study of an existing system, a specific set of information requirements put forth by management, an operational problem to be solved, or a grand and fundamental analysis of the business and its environment. Fig. 1 shows a survey of how

Ackoff (1)
1. Analyze the decision system
2. Analyze the information requirements
3. Aggregate decisions with the same information requirements
4. Design procedure for information processing
5. Design for control of the control system

Barnett (2)
1. Design and Installation Committee—appoint members
2. Appoint MIS Steering Committee
3. Establish policies and decision rules
4. Prepare preliminary installation plan
5. Prepare detailed systems design specifications
6. Prepare final installation plans and schedules
7. Program (software)
8. Write procedures
9. Train

Blumenthal (3)
1. Preliminary analysis
2. Feasibility assessment
3. Management consideration
4. Systems implementation and control
   a. General: project planning and control
   b. Phase 1: completion of functional requirements
   c. Phase 2: preparation of system specifications
   d. Phase 3: programming and testing
   e. Phase 4: conversion and cutover (to full operation)
5. Data processing organization activities
6. Performance evaluation

Chorafas (4)
1. Applications study phase
2. Requirements determination phase
3. System design phase
4. System evaluation phase

Ellis (5)
1. Mission objective
2. Mission design
3. Requirements analysis
4. Systems specification
5. Capabilities analysis
6. Systems design
7. Fine structure design

Feigenbaum (6)
1. System analysis—search for needs and opportunities
2. Systems programming—establish operation objectives
3. Systems design
4. System hardware and software
5. Systems installation and checkout
6. Systems service

Fisher (7)
1. Identification of the problem
2. Description of the problem
3. Design of the system to solve the problem
4. Programming the system
5. Implementation of the system
6. Support of the system

Glans (8)
1. Study and design
   a. Problem recognition
   b. Determination of objectives
   c. Study present system
   d. Determine system requirements
   e. Design new system
   f. Propose solution
2. Implement and install
   a. Detail system design
   b. File design
   c. Develop test criteria and data
   d. System test
   e. Conversion
3. Operate, evaluate, and modify
   a. Operation
   b. Efficiency analysis
   c. System modification and maintenance

Heany (9)
1. Develop or refine an information requirement
2. Develop gross system concepts
3. Obtain approval
4. Detail the design
5. Test
6. Implement
7. Document
8. Evaluate

FIGURE 1 MIS development.

Hopeman (10)

System conceptualization
1. Evaluate the problem and state objectives
2. Define the environmental set
3. Define the system encompassing the basic transformation processes associated with the system
4. Define the subsystems

Model construction and simulation
5. Apply appropriate techniques to optimize the subsystems
6. Test subsystem models
7. Link subsystems to form the system
8. Apply appropriate techniques to optimize the system
9. Test the system model
10. Modify the model for environmental influence

Documentation
11. Document logic and state parameters within which the dynamic system is reliable and valid

Hopkins (11)
1. Discovery and definition of functional objectives
2. Definition of inputs, outputs, boundaries
3. Derive conceptual requirements
4. Derive specific requirements
5. Partition into subsystems
6. Develop quantitative specs
7. Build model and test

Kast (12)
1. Conceptual (recognition of need)
2. Definition (of problem and set of requirements)
3. Acquisition (analysis, search, synthesis)
4. Operational
5. Obsolescence

Optner (13)
1. Analyze the present system
2. Develop a conceptual model
3. Test the model
4. Propose a new system
5. Pilot installation of the new system
6. Full installation of the new system

Salzer (14)
1. Define the problem
2. Visualize the bigger framework
3. Define the subsystems
4. Analyze the subsystems
5. Study the interrelationships
6. Decide the implementation sequence
7. Design the subsystems
8. Reexamine the system requirements
9. Feed back the design results
10. Continue the design cycle

Thome (15)
1. Policy definition cycle — select policy and broad objectives
2. Mission definition cycle — select missions, requirements, timing
3. System definition cycle — select system, elements, and requirements
4. System elements definition cycle — system element design approach development, requirements
5. Implementation methods cycle — select implementation methods

Wilson (16)

Preliminary design phase
1. Conceive and state problem
2. Prepare specifications
3. Prepare file (of information about available resources)

Feasibility established
4. Synthesize — i.e., determine feasibility
5. Analysis — study each subsystem and system characteristic and economics to choose among alternatives

Final design phase
6. Decide on best alternatives
7. Define
8. Check

Young (17)
1. Problem raising
2. Problem investigation
3. Search for solutions
4. Evaluation and selection of solution
5. Consensus
6. Authorization
7. Implementation
8. Direction
9. Audit

FIGURE 1 Continued

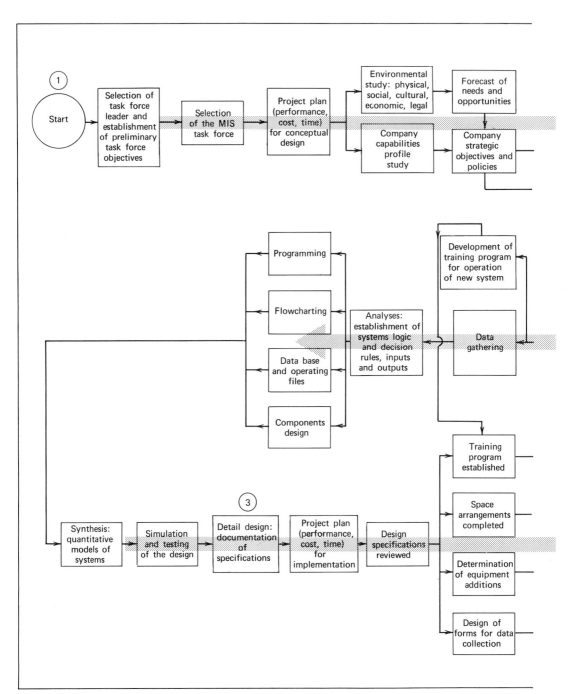

FIGURE 2 MIS development.

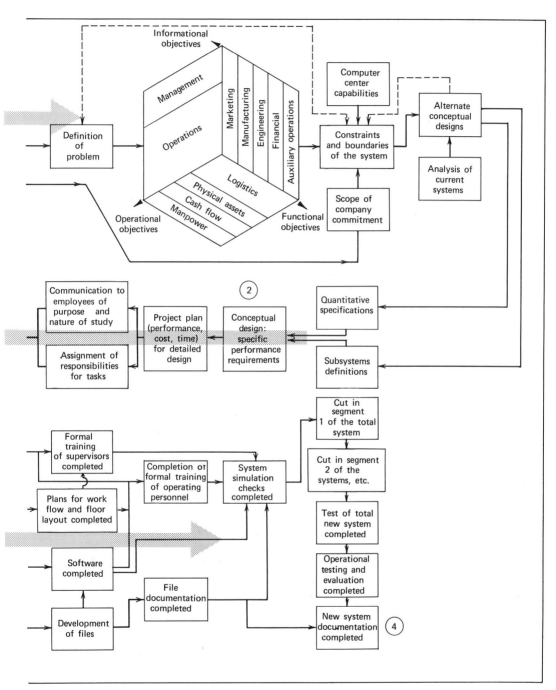

FIGURE 2 *Continued*

various authors view the design or development of systems, usually with application to MIS.

As several of these authors point out, a step-by-step description is not really appropriate. Many activities are carried out in parallel and there is much iteration or recycling to refine the design. The listings serve a purpose in highlighting the principal tasks in their approximate order.

Another problem in attempting to describe MIS design is the variability of scope of the project. One project may take the operational system to be pretty much fixed and concern itself with improving an existing MIS. Another project may be concerned with an entire marketing operation and marketing MIS with minor adjustments to interfacing systems permitted. A more complex project such as a logistics system or manpower system cuts across functional organizations of a company. An overhaul of all systems with the objectives of increased effectiveness and efficiency of operations management is another type of development. Finally, the MIS may be developed subsequent to a review of the grand strategy and the totality of operational systems so that the total systems approach is used.

In addition, even for a single project, there are options with regard to the order of work. Many analysts recommend a preliminary survey. Others emphasize the need for planning the project in detail first. Another possibility is planning at two times, prior to "Phase I," the feasibility or conceptual design, and again just prior to "Phase II," the detailed design. A third time for planning which should be considered is prior to implementation. The advantages of planning at a number of points throughout MIS is that more information is available so that more precise plans can be made. The advantage of a total plan at the beginning of a project is that costs, schedules, and performance may be scoped to keep the total project under control. Management sees what it is buying at the start before it gets the company involved in a piece-by-piece venture where each phase gets progressively costlier.

AN ECLECTIC PROCEDURE

Much thought has been applied to diagramming a network of events for the accomplishment of MIS development. It is worth an effort to bring some of these together to present a "reference" procedure for designers and researchers. Fig. 2 is an attempt to make this combination. It does not emphasize the details of computerization, but rather the team effort of management and business systems designers.

Some problems of nomenclature arise, but an examination of Fig. 1 seems to indicate the following synonyms:

1. Investigation = preliminary survey = problem definition = define need or mission objective = analyze the present system.
2. Feasibility study = conceptual design = establishment of performance specifications = gross design = Phase I design.
3. Detailed design = develop system operating specifications = systems definition = analysis and synthesis = systems acquisition = Phase II design.
4. Implementation = installation = system construction.

REFERENCES

1. Ackoff, Russel L., "Management Misinformation Systems," *Management Science,* Dec., 1967, pp. B153–B155.
2. Barnett, Joseph I., "How to Install a Management Information and Control System," *Systems and Procedures Journal,* Sept.–Oct., 1966, pp. 12–14.

3. Blumenthal, Sherman C. *Management Information Systems: A Framework for Planning and Development.* Prentice-Hall, 1968, pp. 111–112.
4. Chorafas, Dimitris N. *Control Systems Functions and Programming Approaches,* Vol. A. Academic Press, 1966, pp. 42–45.
5. Ellis, David O., and Fred J. Ludwig. *Systems Philosophy.* Prentice-Hall, 1962, pp. 39–40.
6. Feigenbaum, Donald S., "The Engineering and Management of an Effective System," *Management Science,* Aug., 1968, p. B727.
7. Fisher, D. L., "Management Controlled Information Systems," *Datamation,* June, 1969, p. 53.
8. Glans, Thomas B., et al. *Management Systems.* Holt, Rinehart and Winston, 1968, p. 4.
9. Heany, Donald F. *Development of Information Systems.* Ronald Press, 1968, p. 46.
10. Hopeman, Richard J. *Systems Analysis and Operations Management,* Merrill, 1969, p. 41.
11. Hopkins, R. C., "A Systematic Procedure for Systems Development," *IRE Trans. on Engineering Management,* June, 1961, p. 84.
12. Kast, Fremont E., "A Dynamic Planning Model," *Business Horizons,* June, 1968, p. 64.
13. Optner, Stanford L. *Systems Analysis for Business Management.* Prentice-Hall, 1960, p. 39.
14. Salzer, J. M., "Evolutionary Design of Complex Systems" in Donald P. Eckman (ed.), *Systems: Research and Design.* Wiley, 1961, p. 203.
15. Thome, P. G., and R. G. Willard. "The Systems Approach: A Unified Concept of Planning," *Aerospace Management,* Fall/Winter, 1966, p. 32.
16. Wilson, Ira G., and Marthann Wilson. *Information, Computers, and System Design.* Wiley, 1967, p. 196.
17. Young, Stanley. *Management: A Systems Analysis.* Scott, Foresman and Company, 1966, p. 48.

STUDY ORGANIZATION PLAN DOCUMENTATION TECHNIQUES

IBM

Three levels of language can be used to describe a business system: overview, systems view, and detail view. All of them are applicable at various times in studies for complete reconstruction, for improvement, or for mechanization. The language level is determined by the particular requirements of the study.

An overview provides broad understanding for the complete creative reconstruction of a business. This broad view of a large business discloses critical areas where the most dollars are being spent, or where the most profit is being made. In an adver-

tising agency, for example, a preliminary overview might disclose that of four activities—films, technical literature, space advertising, industrial promotions—space advertising alone may be responsible for 80 percent of agency income. If profit figures for the other three are not out of line with capital investment and costs, space advertising becomes the important area for further study.

In addition, a preliminary overview may disclose parallel operations which have much in common; therefore a study of one operation would be representative of the whole. In one distribution business, for example, there were 31 warehousing sites; three of these had limited assembly functions while the other 28 simply stocked products. Of the 28, 20 carried a single product line; the other eight were multiple-line warehouses supplying major population centers. Studying one or two single-line warehouses, one multiple-line warehouse, and one limited assembly function produced an accurate and complete picture of the entire business.

Successively more depth of detail is represented by systems and detail views, respectively. Most studies require

---

SOURCE: IBM, *Study Organization Plan Documentation Techniques*, (Manual C20-8075), 1961, pp. 1–26; (Manual F20-8136), 1963, pp. 2, 7–11. Reprinted by permission from International Business Machines Corporation. © 1961 and 1963 by IBM.

occasional "buttonhooking" down into these levels for operations that must be described more fully. The computer program in a mechanization study is an example of description at the detail view.

Report Structure

The general organization of the Phase I report is shown graphically in Figure 1 (layout and content of individual sections can vary considerably from business to business).

There are three major sections to the report: General, Structural, and Operational. The body of the report is preceded by an Introduction; an Appendix is added for important support documents.

The General Section contains a history of the enterprise, a statement of goals and policies, and an assessment of its position in relation to the competition. The Structural Section describes the inputs, outputs, and resources of the business. The Operational Section details the operating dynamics of an enterprise with emphasis on the flow of events, time cycles, and costs. It is in this final section that the business is first considered in terms of activities.

The General Section of the report clearly and concisely conveys the major features of a business in narrative form to management, the study team, and other interested readers alike. It is composed of five parts: history and framework, industry background, goals and objectives, policies and practices, and government regulation.

History and Framework

The present goals and practices of a business are often shaped from the important events in its history. This part of the report is an identification of major milestones of progress from the past which have influenced the present direction of a business.

Type of Data. Important historical information includes ideas, attitudes, and opinions of key management and research personnel, excerpts from the original charter, reasons for starting the company, mergers and spinoffs, expansion or curtailment of product lines and services, and reasons for changes in name or products. In addition, the growth of the physical plant and number of employees over the years is mentioned, along with a very general identification of products and services.

Example. The history and framework of Butodale Electronics* quoted below, is an example of how the information can be displayed:

> The Butodale Electronics Company was established in 1946, incorporated in the State of Massachusetts. It was founded by four engineers and scientists who had worked together for a number of years in a large corporation on advanced government project work. Their main objective was to aid research laboratories and manufacturers in design and production of the latest radar, radio, and other electronic equipment.
>
> It is significant that the corporation sales have increased in thirteen years from $170,000 in 1947 to 15.9 million in 1959. Some of the major milestones in the last five years were:
>
> 1. Established the Worcester Computation Center to develop new

---

*Examples of the various sections of a Phase I report are adapted for this text from three actual case studies. Names and identifying locations have been changed, but the substance of the material quoted here is taken from the final reports produced by these studies. The three companies are called: *Butodale Electronics,* a 15-year-old Massachusetts corporation specializing in analog computers; the *National Bank of Commerce,* a moderate-sized bank in Topeka, Kansas; and *Custodian Life Insurance Company,* once a stock company, now a mutual company in Rockford, Ill.

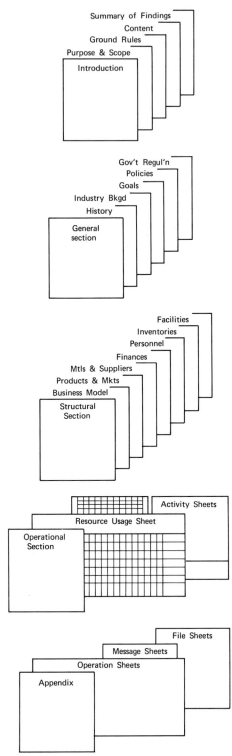

FIGURE 1 MIS development.

fields of application for the analog computer (e.g., heat transfer, nuclear engineering, management control engineering).
2. Established the Long Beach and Rio de Janeiro Computation Centers to extend what was started at Worcester and to educate prospective customers in the use of analog computer techniques.
3. Opened additional sales offices in Chicago and Fort Worth.
4. Greatly expanded and modernized the plant in Danvers.
5. Instituted a major drive to secure overseas business, particularly in South America.

Butodale's major product lines have expanded during this period to include small general-purpose analog computers, instruments, and data plotters. (None of these new lines has developed to more than 10% of the annual sales dollar.)

The above material implies certain objectives (e.g., expansion of overseas operations), and even provides reasons for starting the company. The remainder of this statement goes on to describe recently added product lines.

Butodale history runs one and one-half pages, while the history of National Bank of Commerce requires three and one-half pages. Length is not critical, as long as the narrative reveals currently useful facts.

Sources of Information. In the initial meetings with management, certain general information is requested: annual reports, current and back issues; a prospectus, if one exists; copies of speeches made by management personnel about the business; employee orientation handbooks; published texts which discuss company history. From such material, a history of key events about a company can be put together. In addition, large commercial banks such as New York's Chase Manhattan, First National City, and Manufacturers Hanover Trust, California's Bank of America, or Philadelphia's First Pennsylvania, often publish reviews of important industries which are available as source data. These summaries are oriented mostly toward the investing public, but are handy references for history and background as well as other subjects.

Standard sources of financial and operating data such as Standard & Poor's, Dun & Bradstreet, and other industrial and commercial registers, can be scanned for general information. Biographical registers like "Who's Who in Commerce and Industry" are valuable when the personality of one man strongly affects the enterprise.

Industry Background

This part of the report places the business into perspective within its industry. Comparative data which indicates why one company is successful and another is not, is included, along with facts on the entire industry. Areas of concentration, strengths and weaknesses, and market potential of the major companies are assessed here.

Type of Data. The nature of the industry is briefly summarized, showing demand for its products and services, technological developments leading to progress, growth characteristics, and growth trends. Among individual companies, comparable statistics can be prepared on sales volume, product and service likenesses and differences, territories served, profit margins, and other factors. This is sometimes difficult to do, since many multi-product line companies do not release statistics by divisions.

Example

Butodale finds itself in the electronics industry and specifically in the analog computer area. Analog computers, in the sense of their use today, are only ten to fifteen years old and fall into two categories, general-purpose and special-purpose computers. There is considerable competition in this industry; some of the biggest competitors are ABC Instrument, Jones Instrument, and National Systems, Inc. The company feels there will be continued growth for the general-purpose computer, but this growth may not be at the same rate as in the past. Naturally, arriving at these conclusions, there is considerable stress put on to find new markets and new products. In order to uncover these areas and products the company has set up a New Products Committee and Market Analysis Section. It is the specific purpose of these groups to plan future growth and to direct engineering effort towards this growth in order that the company may maintain a planned growth pattern of 20% per annum, or greater.

Some of the product areas under scrutiny are instruments, special-purpose computers, and process control. Likewise, industry statistical analyses by marketing areas are developed in order to concentrate effort in the proper industries. There has been no designed plan to integrate this company through component manufacture; however, it is not opposed to this type of growth if necessary to insure reliable source of supply, and if excess capacity can be sold profitably. Recently the company absorbed the Premium Capacitors Company and is now building high-quality capacitors.

Sources of Information. A useful source of background information is the industry's technical paper or magazine. Almost every industry is served by at least one such publication; some are quite objective and informative. Editorial and research staffs of these publications frequently have files of industry statistics; a few publications issue an annual statistical review which summarizes the state of the industry or field of operations. "Electrical Merchandising" and "Aviation Week" each publish a special issue every year which gives statistics on the retail electrical appliance trade and the aviation industry, respectively. "Electronics" publishes, in the first issue of the year, a special report on the state of the electronics business, with projections for the year to come.

The Department of Commerce publishes a wealth of material on U.S. industry and trade. Government data as a rule is more objective, and at the same time less current, than information found in business publications. The Census of Business and Manufacture, for example, maintains diversified statistics on industry which can be used to verify other data. The Department of Agriculture publishes data on the food processing industry; information on the alcoholic beverage trade can be secured from the Treasury Department; data on the drug business may be requested from the Department of Health, Education, and Welfare.

Goals and Objectives

A clear understanding of business goals is necessary before activities can be properly formulated, and the emphasis is on specific rather than general goals.

Type of Data. The goal definition is a relatively brief list of a half dozen or more specific statements.

Examples. In Butodale, the initial goal statement read:

> A major objective of this corporation is to expand sales and profits which will guarantee a proper return to stockholders and offer continued opportunity to employees.
>
> The present sales goal is to increase approximately 20%; a net profit goal of approximately 7% of sales and 15% of net worth has been established. These seem to be readily attainable goals, based on past performance; they have been charted against other companies in the industry, both large and small, seem to be definitely in line, and may very well be exceeded.

However, statements like "expand sales" and "guarantee a proper return to stockholders" were considered much too vague and general. In subsequent discussions it was revised to read:

1. Manufacture and sell standard computer equipment and accessories.
2. Design and manufacture special computer models and accessories to satisfy individual specifications and requirements.
3. Offer computation services and engineering consultation on a fee basis to industry, commerce, and schools, among others.
4. Manufacture spare parts and components for sale to the trade.
5. Repair and maintain installed equipment.
6. Conduct research on new products and services to support present lines and initiate new ones within Butodale's area of knowledge and proficiency.
7. Compensate employees and suppliers for services, and provide a satisfactory return for investors.
8. Demonstrate competence and quality in every product to clearly show advantage over competitive equipment.

This second statement was far more definitive, and reveals the goal structure of the business.

In Custodian Life, most goals were identified with a standard of attainment.

> Competition is the dominant factor in the insurance industry and reaches into many different areas of a life insurance company. Since World War II, ordinary life insurance in force in the U.S. has tripled, while group and credit insurance have shown even greater increases. Custodian Life confronts this highly competitive, rapidly expanding marketplace with these goals and objectives:
> - New business production each year to equal 16% of the insurance in force at the beginning of the year.
> - A net gain in insurance in force of 10% each year.
> - A termination rate not greater than 6% per year of the insurance in force at the beginning of the year.
> - Development of the accident and health insurance business by an increase in the annualized premium of 49% over the previous year (1959).
> - Expansion of operations into seven additional states in 1960.
> - A well balanced operation with proper consideration given to all groups within the company.

This statement could have been more specifically directed toward the individual services the company offers and the markets it serves, but is adequate as it stands.

Sources of Information. While directives, management statements, and other internal publications offer clues to business goals and objectives, the ultimate sources of information are the personal views of owners or top-level management. As the Butodale example pointed out, managers often express goals quite generally and not in the specific terms required for the study. A rather searching self-examination may therefore be necessary to produce adequate goal statements.

Policies and Practices

The goals and objectives of a business are implemented by its policies and practices. Some are common to the industry or field of concentration; others will depart from industry practice as suits the requirements of a specific business.

Type of Data. Policies or courses of corporate action are characterized by a code of ethics, a plan for expanding into new territory, the approach to advertising and publicity, attitude toward employees and promotion, and the like. Policies are ideas, attitudes, and philosophies, as distinguished from procedures or methods; and the analyst must keep these differences in mind as he compiles the policy statement.

Examples. Butodale's policies are heavily employee-oriented:

> Some of the major policies instituted by Butodale have unquestionably helped the company attain its position of eminence in the analog computer industry. One of these policies is the corporation's attitude toward its employees. Butodale has developed a labor philosophy in which it endeavors not to infringe on the private lives of its people, while offering liberal employee fringe benefits, including educational opportunities. The company makes a strenuous effort to keep layoffs to an absolute minimum. This policy has resulted in a fine labor-management atmosphere. It has made itself felt in pride of workmanship and company loyalty which are hard to equal in modern industry.

On the other hand, policies and practices for National Bank of Commerce are more detailed:

> For individual and Commercial
> (customers and prospects)
> 1. Accessible, flexible facilities for deposit and receipt of cash, checks, bonds, drafts, and other negotiable documents.
> 2. Interest paying system to encourage time deposits.
> 3. Safekeeping facilities for valuable records.
> 4. Personal, confidential, knowledgeable consultation on all financial matters.
>
> For Correspondent Bank
> (customers and prospects)
> 1. Direct sending service and fast collection of cash items.
> 2. Full draft collection service.
> 3. Fast currency and coin shipment service.
> 4. Valuable document safekeeping facilities.
> 5. Assistance on large loans and advice on trust matters.
>
> For Loan
> (customers and prospects)
> 1. Facilities and experienced personnel available for consultation and financial advice on all loan matters.
> 2. Readily accessible facilities for the closing of (and payment on) personal, commercial, or mortgage loans.
> 3. Extensive advertising program to attract loan prospects to the bank for consultation.

4. Specialist available with a broad knowledge of income-producing investments.
5. Specialists available having detailed financial status information on local individuals and businesses.
6. Analysts available who are well informed on relative valuations of all types of property.
7. Flexible interest charging structure to encourage large loans and rewards for those who pay when due.

Planned practices to meet goals are:
1. Expand advertising program to reach more potential customers.
2. Enlarge drive-up banking facilities.
3. Increase emphasis on "Installment" type loans.
4. Modernize and reorganize physical and manpower facilities as necessary for most efficient operation.
5. Establish an electronic data center using the latest data processing equipment for processing paperwork; offer such services to local industry at a minimum cost.

Sources of Information. Most companies document and publish standing operating procedures, operating and policy instructions, directives, and other internal declarations stating corporate goals, standards, and attitudes. Published information of this type should always be supplemented with statements from top management to verify current application and proper interpretation. Company advertisements and publicity releases also reflect the corporate personality, indicating areas in which the company currently operates or seeks to become established. House publications, too, frequently discuss policies, practices, and attitudes; but they cannot be considered as totally objective in their viewpoints.

Government Regulation

Government regulations at federal, state, and local levels influence the way a business is conducted.

Type of Data. The discussion on regulations should answer three basic questions:

- Which government regulations help the company do business (e.g., charters, tariffs, franchises, enabling acts, subsidies)?
- Which restrict its business activities (e.g., consent decrees, utility regulations, regulations on financial enterprises)?
- Which affect its recordkeeping practices?

For example, rulings of the Civil Aeronautics Board determine form and content of some airline reports; rules of the Federal Communications Commission require certain reports from communications facilities in specific form; rulings of the Securities & Exchange Commission affect the recordkeeping of brokerage houses. However, the general requirements of the SEC affecting stock issue for publicly held corporations would not be spelled out in the report, nor would laws regarding monopolistic practices and restraint of trade. There may be informal government regulation as well as that formalized in laws, such as attitude of the local government regarding industrial waste. This should be noted here as well.

Examples. Both permissive and restrictive regulations are mentioned in the National Bank of Commerce report:

National Bank of Commerce, organized in 1891, was chartered for business under the National Bank Act of

1864. The National Bank Act created a Bureau of Controller of Currency in the Treasury Department. The Controller, who is Director of the Bureau, has the power to charter national banks and is responsible for the examination, supervision, and rules relating to the operation and powers of such banks. Where state banking regulations are in conflict with national regulations, the national bank is normally required to comply with the state regulation.

National Bank of Commerce, like all member banks, must operate within the limits of the 22 regulations of the Federal Reserve System. Responsibility for Federal Reserve policy and decisions rests on the Board of Governors, who are appointed by the President and approved by the Senate for a term of 14 years; the twelve Federal Reserve banks; and the Federal Reserve Open Market Committee. All national banks must be members of the Federal Reserve System, hold Federal Reserve stock, and maintain legal reserve moneys on deposit in their district Federal Reserve Bank. National banks must furnish a financial report when requested by Federal Reserve, and are members of the Federal Deposit Insurance Corporation, which guarantees each depositor against loss up to a maximum of $10,000.

National Bank of Commerce must constantly adjust policies and operational procedures to meet the requirements of new Federal Reserve regulation interpretations.

The statement then lists the 22 areas covered by the regulations of the Federal Reserve System.

Regulations may not always have a direct effect on an enterprise, but still may influence business policies. A brief statement in the Butodale report illustrates the point:

> Government regulations do not play a major role in company plans. However, a very high percentage, perhaps 60% of sales, are subject to renegotiation. Since the government presently thinks 14% of sales is a fair return, this has an effect on profit objectives.

Monopoly and antitrust laws, labor laws, fair-employment laws, and income tax regulations are noted if they uniquely affect business operations.

Sources of Information. General information on government regulations can be found in industry textbooks and trade journals; annual reports and other company publications sometimes disclose their effect on a business.

Commerce and financial laws enacted by the various states, for example, are specifically restrictive on banking and insurance operations; utilities are closely regulated by state utility commissions. Consent decrees entered into by a business are another example of restrictive regulation. Franchises are examples of permissive regulations, as are federal laws which grant subsidies to industries. Summaries of this legislation, or perhaps the laws themselves, should be read to gain an accurate appraisal of their impact on a business. The company counsel's assistance may be solicited in making this appraisal.

SYSTEM LIFE CYCLE

The life of a business system moves through three stages. The first is study and design: an existing system is reviewed to gain understanding, its true requirements specified, and a new system designed to meet those requirements. The second stage is implementation: the system as designed in the first stage is detailed into a working system; for an information system, this stage would include computer programming, detailed message layout, form design, testing, debugging, site preparation, and so forth. The third stage is operation: the day-to-day routine of running and managing the system and modifying it as new requirements or equipments emerge.

The implementation stage is described in existing literature; many books detail how to flow-chart procedures, program a computer, install a system, and so forth. The long-term operation of a system has also been described; books exist which detail how to modify a system, how to manage the various types of computer installation, and so forth. But the first stage, study and design, has not previously been widely described; many isolated bits of specific information relating to particular problems exist in the literature, but a total approach is nonexistent.

Obviously, before a new system can be implemented it must be designed; any prior system which may form the basis of the new system must be completely understood. Two of the functions of the first stage are expressed in the name: study and design. Hidden between these functions is another that is frequently overlooked: the analysis of the true requirements of the system. If this phase is skipped, the new system will often be just a bigger, better, faster way of repeating existing mistakes.

These are the three phases of stage one in the life of a system (in the order in which they occur): (1) understanding the existing sytem, (2) determining the true requirements of the system, and (3) designing and describing a new system to fulfill the true requirements.

In the first phase, the aim of the methods man, systems engineer—whoever is observing and analyzing the system—is to understand what the business or system does, and to a degree how it is done, in terms of activities that thread through the business. In the second phase, the observer analyzes what is logically necessary (or, alternatively, what is specified as necessary by the controlling management) to accomplish the goals of the specific activity under study. In the third phase, a system which can meet those requirements is designed and then described.

The specific aim of this paper is to demonstrate how a related group of recording forms can be used for documenting the three phases: study, analysis, and new system description. The same forms are used in phases one and three, and as support information in phase two. Additional forms are used to display the requirements resulting from the analysis in phase two. All can be used in different ways, for different purposes, at different levels of detail, with different inputs and different supporting data.

The documents described, together with the rules for filling them out, are in effect a language for describing systems. They are useful for full-scale systems studies in depth, and it is for this purpose that they were developed. But they are equally useful for quick, short-term surveys of existing systems, for system analyses, for describing new systems, or for case studies of advanced applications.

These recording forms and the techniques for using them work together. They provide a clear, clean, formal description of what is involved in a system or an application. The fields in the various descriptive forms guide the observer in making a logical review of existing practices and procedures. Other forms define and display true system requirements, and this in turn assists in the design and description of the new system.

These standard documents also substantially ease the problem of communicating the characteristics of a system or installation. If the forms are filled out as prescribed, anyone familiar with this text can read them and quickly determine what the system does, as a necessary preliminary to a trouble-shooting, improvement, or modification. Further, the use of these documentation techniques prepares the observer for their further use in making system studies in depth; the more comprehensive Study Organization Plan, evolved by International Business Machines Corporation, uses the same recording forms for more sophisticated systems studies.

DOCUMENTING A SURVEY

The desire to examine an existing business, activity, or system usually arises from some other requirement. Generally stated, the basic purpose of a survey is to determine quickly what is done, from what source, with what resources, to achieve what results.

More specific purposes include the need to understand how a system or activity fits into a larger organization, that is, into its environment. A coherent description of present operation is needed as a basis for further analysis of how an existing system meets present or projected demands. Quick-look surveys can also determine exact costs of personnel, machines and equipment, material and facilities; they can provide a clear picture of overall efficiency in terms of elapsed times in a sequence of operations, unit operations times, and operating volumes.

A system survey seeks only the information needed to provide understanding of the system or activity for the particular purpose. Its aim is to show how men, materials, and information, using such resources as buildings, machines, and inventories (or files), respond to inputs to produce outputs and results.

Documentation forms for describing a system must be easy to use if they are to be valuable for quick-look surveys. They must also be logically organized, and should clearly segregate and identify the key information needed for understanding of an existing system in operation.

Five documentation forms have been developed for describing an existing system. The relationship among these forms is illustrated in Figure 2. The organizational structure comprising an activity, and a cost analysis of the activity, appear on the Resource Usage Sheet, and the flow of the activity itself is displayed on the Activity Sheet. Operations within the activity are further described on Operation Sheets, with details on information inputs and outputs (from or to what the operation works) and information resources (with what the operation works) provided on Message and File Sheets. These forms permit the observer to describe the activity in motion; they reveal the flow of incoming materials or information through the internal workings of the system to output products, services, or information.

The Resource Usage Sheet fits each system under study into its larger context. It shows the organization and structure of the business environment. It also provides a rapid analysis of costs for the organizational components being surveyed and shows the cost impact for each activity or system.

The Activity Sheet traces the flow of a single activity, breaking it down into its major operations. Each Activity Sheet presents as large a group of related operations as can be handled conve-

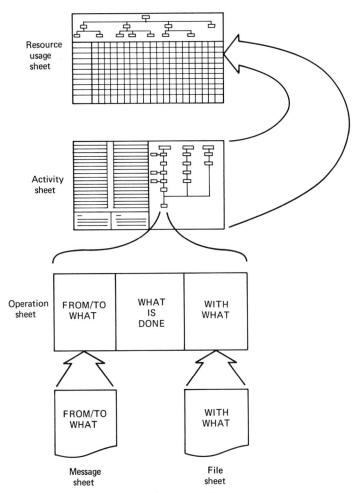

FIGURE 2 Documentation structure.

niently. It includes a flow diagram of the activity with individual blocks representing various operations (each of which can be more minutely described on Operation Sheets). Key characteristics such as volumes and times are recorded in tabular form.

The Resource Usage Sheet and Activity Sheet work together to provide a quick look at a business system. In order to dig down for a closer look at the operation of a system, the observer uses forms that permit a more detailed documentation. These forms are the Operation Sheet and the Message and File Sheets.

There is usually an Operation Sheet for each block on the Activity Sheet. It is used for recording the related processing steps that form a logical operation. It describes what is done, with what resources, under what conditions, how often, to produce what specific results. Its primary purpose is to show the relationships between inputs, processes, resources, and outputs.

The Message Sheet is one of two forms that support the Operation Sheet. It describes the inputs and outputs — in effect, it displays, to the level of detail desired, from what source or toward what result an operation is working.

The File Sheet describes a collection of messages, an information file; it is the second of the two forms supporting the Operation Sheet, and shows with what stored information the operation works.

These five forms work together to provide a coherent description of a business system. Moreover, they compose an inspecting instrument of variable magnifying power, since they can be used at several levels of detail. The Resource Usage and Activity Sheets together display the dynamic mechanism of an activity; together with the Operation Sheet they provide a fairly detailed operational view. The Operation Sheet by itself permits a look at critical operations, and when supported by the Message and File Sheets, provides a closeup of information used in or produced by such operations. Message and File Sheets by themselves permit observation and analysis of information inputs, outputs, and resources, with again a variable level of detail as desired by the observer.

When describing what is done (on the Operation Sheet, in the field reserved for process description and qualifications), the observer can choose from several levels of detail by exercising a choice of descriptive verbs. Such verbs as *prepare, reproduce, update* are descriptive of processes at one level of detail. The language of a computer compiler such as COBOL or Commercial Translator is an example of a far more detailed language.

The amount of detail to be entered on these forms is somewhat discretionary, and depends on what level of understanding is needed as well as on the time allotted for the survey. Information can be omitted; the observer can single out critical operations from the Activity Sheet and avoid making up Operation Sheets for every operation. Short-cut methods can be used for entering information. Many of the fields on the Message Sheet can be left blank, since a sample of the completed message form is customarily attached. Qualification of content may not be needed on the File Sheet, nor even description of content.

A NONMECHANIZED SYSTEM

The charge account system of a department store is typical of activities that are almost independent of other parts of the related business. The sale of goods is the department store's chief activity, with stock buying, advertising, inventory management, and personnel as other typical activities. Within the sales activity, charge accounts may be considered a major subdivision, with cash sales and time purchases as other subdivisions.

The charge account activity in many department stores remains a nonmechanized clerical system; our first example of documenting a survey will deal with such an activity.

The section of the business which includes the charge account system is displayed on the Resource Usage Sheet. This basic form shows the economics both of the system and of its immediate environment.

In the Resource Usage Sheet in Figure 3, the organizational structure of a major section of the department store is graphically illustrated in the organization chart. In each box on the chart is shown the total number of people directly employed in the organizational component represented by the box. Employees in any lower-level components not shown on the chart are totaled into the proper lowest-level box on the chart. Upper-level boxes show only their own immediate employee totals, not including components subordinate to them.

The lower section of the Resource Usage Sheet is a cost tabulation. Costs of personnel, machines and equipment, materials, and so forth are summarized for each organizational component in the bottom tier of boxes in the chart. Higher-level components may also be summarized in this tabulation, as is the advertising manager's component in Figure 3. Costs for the higher-level organization components are usually prorated among the departments under the jurisdiction of the higher-level manager. Thus the personnel costs tabulated on the Resource Usage Sheet are comprehensive for the entire organization table as drawn. Unusual items of significant cost, such as an insurance policy carried on the president's life, would not be prorated but would be separated out in the unlabeled columns at the right side of the sheet.

Costs used in this tabulation may be budget figures for the current year, actual figures from the previous year, or client estimates. Note that in Figure 3 the formal training of personnel was considered important enough to be separated as a cost factor in the summary tabulation. Total costs for each organizational component are tabulated immediately below the related box on the organization chart, and grand totals for the section of the business covered by the Resource Usage Sheet are developed in the TOTALS column at the right.

The costs are then broken down by activities which usually cut across organizational lines. In Figure 3, the specific activity that this survey is considering, the charge account system, is broken out in a separate tabulation. The costs shown for each activity are part of the summary costs listed in the top section of the cost tabulation. Thus the total cost budgeted for Department A and its 100 clerks is $493,500; of this, $166,750 is the cost included in the charge account activity.

Merchandise control and cash sales are other activities that might be examined in the same manner as charge accounts; the same organizational components would not necessarily participate. Conversely, note that the advertising department does not directly participate in the charge account system, and so no costs are entered in this block.

Two unmarked columns at the right of the tabular section can display various data such as the average amount of money tied up in inventories or accounts receivable. Where possible, the cost of having the money tied up would also be shown. In this case, the first of the two unheaded columns describes the item and the second shows the amount. The average investment in accounts receivable and the cost of maintaining this investment are tabulated in Figure 3.

By summing horizontally, the observer can develop the total costs for each activity. In Figure 3, the totals indicate that charge accounts involve $1.36 million in annual costs, about 31 percent of the total cost of $4.3 million.

The Resource Usage Sheet permits a rapid analysis of the structure into which the system fits, and of the costs of the activities that make up the system. Its principal function is to document the organizational and economic information from a survey. In the example of the charge account system of a department store, it provides a quick look at the way in which the charge account activity fits into the department store organization.

The Activity Sheet, shown in Figure 4, includes a flow diagram which shows the sequence of operations performed by the various departments to provide the charge account service, and a tabular

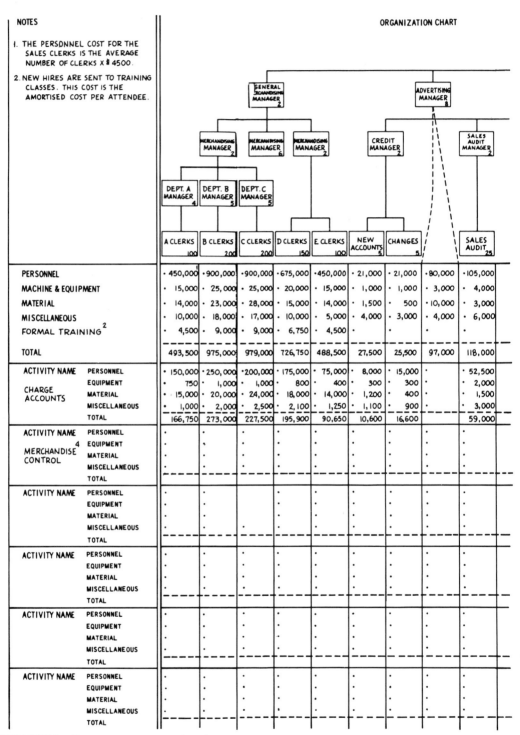

FIGURE 3 Resource usage sheet.

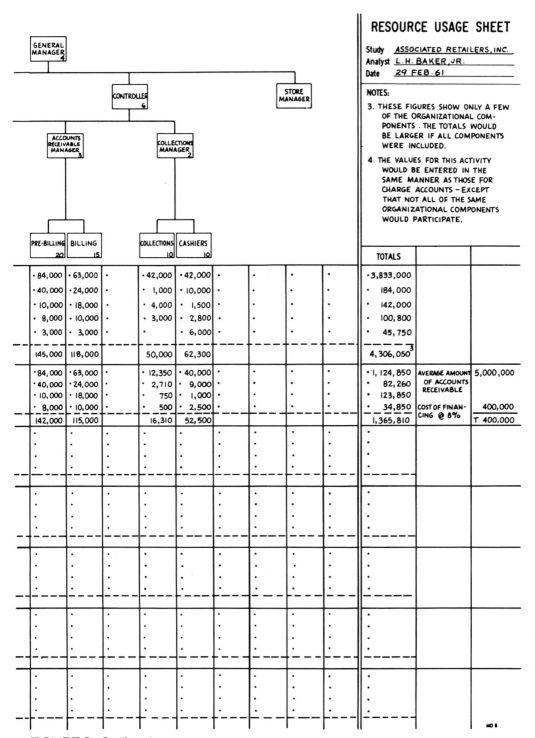

FIGURE 3 Continued

FUNCTION: POSTING A/R								INPUTS							
FREQUENCY: DAILY								KEY	NAME	SOURCE	AVG. DAY AVG. WK.	PEAK DAY AVG. WK.	AVG. DAY PEAK WK.	PEAK DAY PEAK WK.	SEE NOTE
INPUTS: 2000								2000	CHARGE SALE	CUSTOMER	15K/D	30K/D	45K/D	60K/D	1030
CONDITIONS									FREQUENCY		143D/YR	143D/YR	13D/YR	13D/YR	1040
DAY	AVG.	AVG.	PEAK	PEAK	AVG.	AVG.	PEAK	PEAK	2010	RETURNS	CUSTOMER				
WEEK	AVG.	AVG.	AVG.	AVG.	PEAK	PEAK	PEAK	PEAK	2010	FREQUENCY					
CYCLE BEING BILLED?	NO	YES	NO	YES	NO	YES	NO	YES	2020	PAYMENTS		REMAINDER FILLED IN THE SAME WAY			
KEYS 1-3	1D	1D	1D	1D	1D	1D	1D	1D		FREQUENCY					
3-6	1/2 D	1/4 D	1D	1/2 D	1D	3/4 D	2D	1 1/2 D	2030	FILE CHANGES					
7-12	1 1/2 D	1 1/4 D	2D	1 1/2 D						FREQUENCY					
13-18															
19-24			REMAINDER FILLED		IN THE										
25-26				SAME WAY					OUTPUTS						
									KEY	NAME	DESTIN-ATION	AVG. DAY AVG. WK.	PEAK DAY AVG. WK.	AVG. DAY PEAK WK.	PEAK DAY PEAK WK.	SEE NOTE
									3000	BILLS	CUSTOMER	15K/MC	30K/MC	20K/MC	35K/MC	
FUNCTION: BILLING									FREQUENCY		165D/YR	55D/YR	15D/YR	5D/YR	1050
FREQUENCY: 20 DAYS PER MONTH															
OUTPUTS: 3000															
CONDITIONS									FILE USAGE						
DAY	AVG.	PEAK	AVG.	PEAK					KEY	FILE NAME USG. NAME	NO. OF MESSAGES AVG.	NO. OF MESSAGES PEAK	ACCESS RANDOM OR CYCLE	ACCESS TIME		SEE NOTE
WEEK	AVG.	AVG.	PEAK	PEAK					4000	CUSTOMER	500K	550K	RANDOM	≤ 1 MIN.		
KEYS 26-27	1D	1 1/2 D	2D	2D					4010	CUSTOMER	500K	550K	CYCLE	1D		
27-28	1D	2D	2D	3D					4020	1 CYCLE CUSTOMER	25K	40K	CYCLE	1 CYCLE PER DAY 20 CYCLES PER MO.		
28-32	1D	1D	1 1/2 D	2D												
TOTAL 26-32	3D	4 1/2 D	5 1/2 D	7D												
26-13 (AVERAGE PAYMENT RECEIVED)	18D	19 1/2 D	22 1/2 D	25D												
SEE NOTE		1060	1060												

NOTES

1010 – UNIT AUTHORIZER IS SUPERVISOR OR CREDIT MANAGER IF NECESSARY, DEPENDING ON AMOUNT OF SALE.

1020 – UNDER PRESENT SYSTEM, MEDIA FROM A REGISTER INCLUDES CASH SALES, CHARGES, COD, LAYAWAYS, VOIDS, ETC. ALL MEDIA FROM REGISTER ARE BALANCED TO THE REGISTER TOTAL; TOTALS ARE DEVELOPED FOR MERCHANDISE AND NON-MERCHANDISE CLASSES, AND MEDIA SEPARATED BY TYPE. CAUTION SHOULD BE USED IN ANY ATTEMPT TO SEPARATE THESE OPERATIONS FROM THE TOTAL JOB.

1030 – AT PRESENT, ONLY 60% OF CHARGE SALE TRANSACTIONS ARE AUTHORIZED WITHOUT APPROVAL OF UNIT AUTHORIZER.

NOTES

1040 – THERE ARE 302 WORKING DAYS PER YEAR IN THIS OPERATION.

1050 – THERE ARE 240 WORKING DAYS PER YEAR IN THIS OPERATION.

1060 – DELAY IN BILLING DURING THE CHRISTMAS PEAK RAISES ACCTS. RECEIVABLE INVESTMENT BY $500,000.

FIGURE 4 Activity sheet.

FLOW DIAGRAM

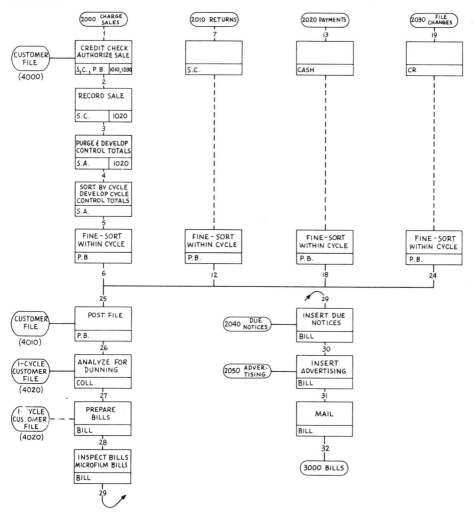

FIGURE 4 Continued

section in which amplifying information is listed. The flow diagram starts with an action from the external environment—the customer's purchase of merchandise—and carries through the mailing of bills.

The Activity Sheet flow diagram is a simple balloon-and-box type. The individual operations are shown separately in boxes; the balloons represent resources, and the small ovals are used for inputs and outputs. The flow path is kept narrow, and operations not directly related to the activity goal are omitted. Organizational lines of the business are ignored; the purpose of the Activity Sheet is to trace a series of operations irrespective of departmental boundaries. The flow diagram demonstrates overall sequence and flow, and points out the significant inputs, resources, and outputs used in or produced by the operations. The inputs and outputs shown on the Activity Sheet are only those which enter from or go into the external environment.

To keep the flow path narrow, Figure 4 shows only operations on charge sales. Returns, payments, and file changes are parallel paths in the charge account activity, feeding into the posting and billing sequence. Details are omitted from this example; in an actual survey, these three sequences would be entered in the same detail as charge sales.

The tabular section at the left of the Activity Sheet displays amplifying information about the operations that make up the activity. This grid is unlabeled, without reserved spaces for specific information. Use of the grid is therefore at the discretion of the observer. In Figure 4, the far left section displays elapsed times, volumes, and frequencies for critical operations. Elapsed time is shown for each of eight sets of conditions. The other section is used to show detail information on volumes and frequencies of the inputs and outputs.

Special attention is given to information resources, the files that are referenced or changed during the various operations. The flow diagram shows which operations use which files; specifics of file usage are then detailed in the grid section. In Figure 4, the details on file usage include pertinent statistical information on the average and peak activity of the file and its access time.

The file name (usage name) describes the file in its specific usage. Thus, in Figure 4, the same physical file is referenced in keys 4000, 4010, and 4020. In the first case it is subject to random reference in less than a minute for checking credit limit and authorizing the sale. It is next referenced sequentially every day for posting the day's business. In key 4020, one-twentieth of the file is pulled out each day for the monthly billing cycle.

In the legend box at the lower right-hand corner of the Activity Sheet, the observer enters the names (and locations, if important) of the departments involved in the activity; the full-name entries explain the abbreviations used in the flow diagram. A numbering system is specified in this box which is useful for cross-referencing the flow diagram to the tabular grid section and to footnotes. Boxes marked NOTES (below the grid) provide space for footnotes to amplify and explain peculiarities of the activity being described; in Figure 4, for instance, the analyst has noted the various media emerging from the RECORD SALE operation (note 1020) and the number of working days for sales (note 1040) and preparing bills (note 1050).

The Activity Sheet, with its flow diagram and tabulations, displays the functional mechanism of a single goal-directed activity and traces its operational path. Together with the Resource Usage Sheet (Figure 3) it provides a quick look into the dynamics of a busi-

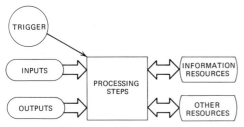

FIGURE 5 Operation schematic.

ness system. For closer inspection of the system, the operation blocks on the Activity Sheet may be analyzed on Operation Sheets, and the inputs, outputs, and files detailed on Message and File Sheets.

The Operation Sheet is the principal means for collecting and displaying operational data. It explodes the operation boxes on the Activity Sheet to describe in detail what is done, from or to what, with what. Figure 5 is a graphic illustration of the information displayed on the Operation Sheet.

Included in the information entered on the Operation Sheet are triggers, inputs, and outputs; the description and qualifications of the processing steps; and the resources used in the various steps. When a single sheet is used for more than one operation, as is frequently the case, all the material for each operation is segregated from the one following by a horizontal line drawn across the page.

The Operation Sheet in Figure 6 is thus subdivided. Note too in this illustration that there need not be any connection between a trigger, input, or output on the one hand and a process or resource entered on the same line; the three major sections of the sheet are tabulated without horizontal reference to each other excepting within heavy lines separating operations.

Inputs and outputs are identified by sequences of numbers starting respectively with I1 and R1 (for *result*, to avoid the confusion with zero that would arise if O were used). Inputs to an operation are items entering from the external environment or transferred from a previous operation. Outputs are items that are produced by the operation to go into the external environment or to a subsequent operation. The name of an information input or output must correspond to an identifying name on a corresponding Message Sheet which further describes the item.

Triggers are identified by the letter T. Each operation is started by one and only one trigger. A trigger may be the arrival of an input item or items; it may be a time of the day or day of the month; it may be a combination of a time plus the availability of input items. In the Operation Sheet in Figure 6, it is the accumulation of ten customer sales records that triggers the file-posting operations; although ten documents are required, this is still considered one trigger, in this case a multiple-input trigger.

The organizational component from which an input is received, or to which an output is sent, is also listed. The volume figures on inputs and outputs are listed, together with the time elapsing from the arrival of the first input to the arrival of each output at its destination. In the example in Figure 6, the outputs go to the next department and so arrive all at the same time. In some cases, one output may be required locally while another may be used in a remote location; elapsed times for the various outputs include transfer periods embodying the average time required for delivery by messenger, mail, and so forth.

THE DESCRIPTION AND QUALIFICATIONS field lists the processing steps that make up each operation. The process description consists of a verb and its objects plus necessary modifiers. Verbs should be selected which are

IBM Operation Sheet

OPERATION		TRIGGERS, INPUTS AND OUTPUTS					PROCESSES			RESOURCES			
NO.	PERFORMED BY	ID NO	NAME AND QUALIFICATIONS	RECEIVED FROM OR SENT TO	VOLUME AVG	PER	ELAPSED TIME	ID NO	DESCRIPTION AND QUALIFICATIONS	FREQ	ID NO	TYPE	UNIT TIME	TOTAL TIME AVG	PER
04.010 BILLING (KEY NOS. 27-28)	BILLING	T1	RECEIPT OF 10 I1					P1	COUNT TRANSACTION SLIPS	1/I1	X1	CLERKS (5)	20 MIN		
		I1	CUSTOMER RECORD	COLLECTIONS	20K	D	0	P2	COMPARE COUNT TO NUMBER OF TRANS-	1/I1	X2	XEROX COPIER (2)		13H	D
		I2	TRANSACTION SLIP		160K	D	0		ACTIONS ON CUSTOMER RECORD						
		R1	CUSTOMER RECORD	ANALYSIS SECTION OF BILLING	20K	D	7 MIN	P3	RETURN CUSTOMER RECORD AND TRANS-	1/50 I1					
		R2	BILL		20K	D	7 MIN		ACTION SLIPS TO COLLECTIONS						
		R3	TRANSACTION SLIP		160K	D	7 MIN		DEPT FOR ERROR TRACING AND						
		R4	ERROR SHEET	COLLECTIONS SUPVR	5	D	8H		CORRECTION.						
								P4	ENTER ACCOUNT NUMBER AND NATURE	1/50 I1					
									OF ERROR ON ERROR SHEET.						
								P5	XEROX BILL FROM CUSTOMER RECORD	1/I1					
								P6	ATTACH TRANSACTION SLIPS TO BILL	1/I1					
								P7	ENTER INITIALS ON CUSTOMER RECORD	1/I1					
								P8	SEND CUSTOMER RECORD, BILL, AND	1/10 I1					
									TRANSACTIONS SLIPS TO ANALYSIS SECTION						
								P9	SEND ERROR SHEET TO ANALYSIS SUPVR.	1/D					
04.020 BILLING (KEY NOS. 28-29)	BILLING	T1	RECEIPT OF 10 I1					P1	INSPECT BILLS FOR LEGIBILITY AND	1/I1	X1	CLERKS (2)	2 MIN		
		I1	CUSTOMER RECORD	FILE SECTION OF BILLING	20K	D	0		ATTACHED TRANSACTION SLIPS		X2	MICROFILM CAMERA		4H	D
		I2	BILL		20K	D	0	P2	RETURN CUSTOMER RECORD, BILL, AND	1/700 I1					
		I3	TRANSACTION SLIP		160K	D	0		TRANSACTION SLIPS TO FILE SECTION.						
		R1	CUSTOMER RECORD	COLLECTIONS	NOTE 1		1H	P3	ENTER FILE CLERK INITIALS AND NATURE	1/700 I1					
		R2	BILL	MAILING SECTION OF BILLING			5 MIN.		OF ERROR ON ERROR SHEET.						
		R3	TRANSACTION SLIP				5 MIN.	P4	MICROFILM BILL	1/I1					
		R4	ERROR SHEET	FILE SECTION SUPVR	2	D	8H	P5	SEND BILL AND TRANSACTION SLIPS	1/10 I1					
			① VOLUME HAS CYCLIC FLUCTUATION:						TO MAILING SECTION.						
		FREQ	165D/YR	55D/YR	150/YR	5D/YR		P6	SEND CUSTOMER RECORD TO COLLECTIONS.	1/250 I1					
		VOLUME	15K/D	30K/D	20K/D	35K/D		P7	SEND ERROR SHEET TO FILE SECTION SUPVR	1/D					

DATE: 29 FEB 61 ANALYST: L H BAKER JR SOURCE: BILLING DEPT. – SCHULTZ STUDY: ASSOCIATED RETAILERS INC. ACTIVITY: CHARGE ACCOUNTS PAGE: 6

FIGURE 6 Operation sheet.

broad enough to make details unnecessary but clear enough to avoid ambiguity. The object of the verb should answer the question "What?" in reference to inputs, outputs, and resources. Each process is assigned a number prefixed by P.

The frequency with which the process is performed is also displayed; where possible, it should be expressed in terms of the number of executions per input, output, or operation. Where this means of expression is unfeasible, the frequency per unit of time is valid for processes occurring cyclically; but in these cases the cycle period should be shown (monthly cycle, weekly cycle, etc.).

The resources section of the Operation Sheet is concerned with people, equipment, information, and facilities that are used in the various process steps that make up the operation. Resources are identified by a sequence of numbers starting with X1. Classifications of operating personnel and the number of each class are entered; names or classifications, or both, of raw materials, blank forms, and so forth, are listed with the approximate volume of each.

Entries under UNIT TIME show how much of a particular resource is used per operation, or per input or output. Unit time is not specified for materials, nor generally for files; it is commonly shown for equipment and personnel. When it is specified, it should cover the total resource named even if a multiple-unit resource; thus, five clerks who each contribute four minutes make a total contribution of 20 minutes of unit time. This figure is clarified by an entry CLERKS (5) which would appear in the TYPE column of the resources section.

Entries under TOTAL TIME show the average usage of the resource per process or per operation. Unit of time selected should be consistent with cost information. This column should be left blank for material resources, and generally for files as well.

The fine details of information inputs and outputs are recorded on Message Sheets. A message is considered to be any notice or communication entering or leaving an activity, regardless of the medium of transmission. It may be recorded or unrecorded. Signals, for instance, are unrecorded messages of a transitory nature such as a telephone call. The contents of a formal message, on the other hand, are fixed in nature, order, and relative length; and the message is recorded on some more or less permanent medium.

The Message Sheet displays the name or names by which the message is identified. In Figure 7A, the Message Sheet for the customer record indicates that this form is also called a customer master or a billing master. The message medium is shown and the method by which it is prepared. The Message Sheet displays both the operations which originate data in the message and the operations that merely use the message. The form in Figure 7A supports the two operations — prepare bills, and inspect and Microfilm bills — displayed in Figure 6; in both, the customer record is involved, but neither originates data on the record. Consequently the numbers of these operations appear in the field OPERATIONS INVOLVED IN but not in the field ORIGIN.

Data elements and arrangement of fields are tabulated in the CONTENTS section of the Message Sheet.

To further illustrate this display, a copy of the filled-in customer record, Figure 7B, is associated with the Message Sheet. In many cases, the attachment of the source forms eliminates the need for filling in the CONTENTS

IBM		Message Sheet

MESSAGE NAME		MESSAGE NO
CUSTOMER RECORD		R0030
OTHER NAMES USED		LAYOUT NO
CUSTOMER MASTER		12
		FORM NO
BILLING MASTER		GRC 172
		NO OF COPIES
		1
MEDIA	HOW PREPARED	
8½ x 5½ MANILA CARD	TYPED - BOOKKEEPING MACHINE	
OPERATIONS INVOLVED IN		
005-001, 005, 010 ; 010-005, 010 ; 012-010, 015, 075 ; 014-010, 020.		

REMARKS
THIS MESSAGE IS THE MASTER FOR XEROXED BILL. AMOUNT REMITTED IS ENTERED ON THE BILL BY THE CUSTOMER, BUT THE FIELD IS NOT USED ON THIS RECORD.

CONTENTS

NO	DATA NAME	FREQUENCY	CHARACTERS	A/N	ORIGIN
01	COMPANY NAME	1	25	AN	ON FORM
02	COMPANY ADDRESS	1	33	AN	ON FORM
03	TELEPHONE NUMBER	1	8	AN	ON FORM
04	CUSTOMER NAME	1	50	AN	012-075
05	CUSTOMER ADDRESS	1	150	AN	012-075
06	ACCOUNT NUMBER	1	7	N	012-075
07	AMOUNT REMITTED	1	6	N	SEE REMARKS
08	FIELD LABELS	1	35	AN	ON FORM
09	DATE	1-22 AVG 8	6	N	010-005
10	TRANSACTION NUMBER	1-22 AVG 8	2	N	010-005
11	CHARGE AMOUNT	1-22 AVG 7	6	N	010-005
12	PAYMENT / RETURN AMOUNT	1-22 AVG 1	6	N	010-005
13	BALANCE AMOUNT	1-22 AVG 8	6	N	010-005
14	PAYABLE INFORMATION	1	79	AN	ON FORM

DATE: 29 FEB 61 ANALYST: L H BAKER JR. SOURCE: COLLECTIONS PAGE: 1 OF 1
STUDY: ASSOCIATED RETAILERS INC.

FIGURE 7a Message sheet.

section of the Message Sheet; the observer can satisfy the survey requirements by identifying the message, describing the medium and preparation method, listing the operations which supply information for the message, and attaching a sample copy.

Signals are displayed on Message Sheets when they are critical inputs to an operation. For a customer-service or telephone-order department, incoming telephone calls are typical input signals that would require description on a Message Sheet.

A File Sheet is used to identify, locate, and describe each ordered collection of messages that is used as an information resource in an operation.

```
             ASSOCIATED RETAILERS INC.
               1107 FOURTH STREET
                 HARTFORD CONN.
                    WG 5 1045


               JOHN W. CUSTOMER III
               7212 E. PLAZA COURT
               HARTFORD, CONN.
                                              $
                     173-4201           _____
                                         AMOUNT REMITTED
   DATE     T    CHARGES   PAYMENTS/RETURNS   BALANCE
  02-29-61  B                                  27.00
  03-01-61  1     12.77                        39.77
  03-05-61  2      3.09                        42.86
  03-06-61  3                   27.00          15.86
  03-09-61  4      3.79                        19.65
  03-14-61  5      8.45                        28.10
  03-23-61  6      2.04                        30.14
  03-25-61  7      5.27                        35.41
  03-26-61  8      2.80                        38.21




              PAYMENT DUE WITHIN 30 DAYS OF INVOICE RECEIPT
   GRC172        MAKE CHECKS PAYABLE TO
                    ''ASSOCIATED RETAILERS''
```

FIGURE 7b Sample customer record.

The upper sections of the File Sheet identify the file by name, number, location, and storage medium; display access requirements; describe the type of material in the file; and outline the retention characteristics. If pertinent, they show who is and is not allowed in the file, how long the file is open, how often or how quickly it must be referenced, and similar operating characteristics.

The File Sheet in Figure 8 describes the customer file of the charge account activity. This file serves three purposes in the charge account operation: (1) it must be quickly referenced for credit authorization when the customer is in the store to buy, (2) it must be posted daily with new charges (and also with credits from payments and returns), and (3) one-twentieth of the total file must be billed daily for a monthly billing cycle. These details are entered in the ACCESSS REQUIREMENTS field. In other fields are entered content qualifications (further explained by the entry in RETENTION CHARACTERISTICS), and information on the immediacy of the file data and the method of sequencing the records in the file. Entries regarding the file growth or peculiarities of usage appear under REMARKS.

The lower section — under CONTENTS — provides space for identifying and characterizing the documents stored

IBM	**File Sheet**

FILE NAME		FILE NO.
CUSTOMER FILE		F 0400
LOCATION	**STORAGE MEDIUM**	
COLLECTIONS DEPT.	TUB FILE	

ACCESS REQUIREMENTS
① ≤ 1 MIN (CREDIT CHECK); ② ≤ 1 DAY (POSTING); ③ CYCLE : 1 CYCLE / DAY , 20 CYCLES / MO. (BILLING)

SEQUENCED BY
ACCOUNT NUMBER

CONTENT QUALIFICATIONS
CUSTOMER RECORDS HAVING ACTIVITY WITHIN LAST 6 MONTHS.

HOW CURRENT
DATA UP TO 1 DAY OLD WHEN ENTERED.

RETENTION CHARACTERISTICS
CUSTOMER RECORDS NOT HAVING ACTIVITY WITHIN LAST 6 MONTHS ARE MOVED TO INACTIVE FILE.

LABELS
——

REMARKS
FILE USED FOR 3 PURPOSES : CREDIT CHECK , POSTING , BILLING .

CONTENTS

SEQUENCE NO.	MESSAGE NAME	VOLUME		CHARACTERS PER MESSAGE	CHARACTERS PER FILE	
		AVG	PEAK		AVG	PEAK
01	CUSTOMER RECORD	500K	550K	553	276,500K	304,150K

DATE	ANALYST	SOURCE	PAGE
29 FEB 61	L H BAKER JR	COLLECTIONS - ALDRICH	1 OF 1

STUDY: ASSOCIATED RETAILERS INC.

FIGURE 8 File sheet.

in the file. Data are entered both for average and peak volumes of documents and for average and peak volumes of characters; this last is of only incidental importance in a manually maintained file (as in the charge account activity), but becomes critically important in mechanized filing systems.

The five descriptive forms work together to enable an observer to document the critical characteristics of an existing system. In the charge account activity which has been discussed, the Resource Usage Sheet provides a graphic illustration of the way in which the activity fits into a section of the department store, and of the cost of that activity. The Activity Sheet then displays the flow of operations that make up the servicing of charge accounts. The Operation Sheet permits closer analysis of two operations selected for their critical effect on the activity. Message and File Sheets permit a closeup of inputs to and outputs from the activity and the information resources used by it.

A MECHANIZED SYSTEM

The order-processing activity of a wholesale distributor provides a good example of a mechanized business system. The fact that machinery is used to perform many chores of the activity does not alter the requirements of the survey in any substantial way, although it does place greater restrictions on some types of information: machinable unit records, for instance, may need to be more rigorously described. But the main purpose of the survey remains: to gain a coherent understanding of the system. The same forms are used.

The use of these forms to describe a mechanized system makes it easy to arrive at a clear understanding of the dynamics of system operation. The mass of detail data which conventionally emerges from a survey of a mechanized system is considered as supporting documentation, if at all. The structure and dynamic flow of the activity are clearly exposed in the five recording forms.

The information entered in the Resource Usage Sheet is the same in both the mechanized and nonmechanized cases: an organization chart of the business structure (in this case, the wholesale distributor) which contains all components affecting the activity of interest; and a summary tabulation of costs for each organizational component, with costs broken out and itemized for each activity, and summed for an activity total. The Resource Usage Sheet is not shown for this mechanized example because the material is so similar to that already given in the nonmechanized system example.

The Activity Sheet in Figure 9 traces the mechanism of the order-processing activity. Note that there are three main sequences: the preparation of orders for shipment; the preparation of will-call orders; and the preparation of invoices, into which sequence both of the others feed. The orders-for-shipment sequence can be started by a salesman's order received by mail or telephone, or by customer purchase orders however received; in the case of a telephoned salesman's order or customer purchase order, a preliminary step is needed to prepare the sales order before the main sequence is begun with a credit check. Note that the straight line flow of the main sequence is not interrupted by what, in some documenting or programming systems, would be considered an exception loop: the preparation of a back order. The operation is included in the main path; if a back order is necessary, it is prepared as a normal tab room operation. If it is not needed, the step is skipped. Volume for the "short list" indicates frequency of execution.

In such an activity as order processing, input condition and frequency, and elapsed times for various processes under average and peak loads, are critical parameters; these are listed in detail in the grid section of the Activity Sheet. Peculiarities of the system and amplifying data are displayed under NOTES.

Two outputs from the activity are the invoice to the customer and the "short list" which guides the buyers. The merchandise can reasonably be considered an output; but for the purposes of this information-system example, it has not been covered.

Files are of paramount importance in the order-processing activity. File usage

ORDER PROCESSING					
FREQUENCY: DAILY					
INPUTS: 2000, 2010, 2020, 2030					
CONDITIONS					
DAY	AVG	PEAK			
FREQ.	17/MO.	5/MO.			
KEYS			SEE NOTE		
1-2	1H	1H			
3-4	1H	1H			
5-6	2H	4H			
6-7	1H	1H			
7-9	.25H	.25H			
9-10	2H	1.5H			
10-11	1H	1H			
12-13	.5H	.5H			
13-14	.75H	.75H			
14-16	.75H	1.2H			
16-18	3H	3H	1050		
12-16	2.0H	2.75H			
BILLING					
FREQUENCY: DAILY					
OUTPUTS: 3010					
KEYS			SEE NOTE		
19-20	1.2H		1060		
20-21	1.4H				
21-22	.6H				

INPUTS			MONDAY			TUE - FRI			
			7:30 AM	10:00 AM		7:30 AM	10:00 AM		SEE
KEY	NAME	SOURCE	MAIL	MAIL	PHONE	MAIL	MAIL	PHONE	NOTE
2000	SALES ORDER	MAIL	215/D	110/D	—	200/D	105/D	—	
2010	PHONE ORDER	TELEPHONE	—	—	85/D	—	—	75/D	1100
2020	CUSTOMER PURCH ORD	TELEPHONE OR MAIL	25/D	10/D	15/D	25 D	10/D	5/D	
2030	WILL CALL ORDER	TELEPHONE	—	—	50/D	—	—	35/D	1110
TOTALS			240/D	120/D	150/D	225/D	115/D	115/D	
TOTAL ORDERS				510/D			455/D		

OUTPUTS			
KEY	NAME	DESTINATION	VOLUME
3000	SHORT LIST	BUYERS	120/D
3010	INVOICE	CUSTOMER	475/D

FILE USAGE						
KEY	FILE NAME	AVG # OF MSG	PEAK # OF MSG	ACCESS	ACCESS TIME	SEE NOTE
4000	GENERAL CATALOG	31000	37000	RANDOM	≤1 MIN	
4010	CREDIT	2500	2500	RANDOM	≤1 MIN	
4020	MASTER ITEM	31000	37000	RANDOM	≤1 MIN	
4030	MASTER N E A	2400	2700	RANDOM	≤1 MIN	
4040	LINE ITEM	3200	4000	RANDOM	≤2 MIN	1140
4050	SALES ORDER	28000	40000	CYCLE	1D	1120
4060	BACK ORDER	550	750	RANDOM	30 MIN	
4070	COMPLETED ORDER	2500	3000	CYCLE	1W	1130
4080	CASH RECORD	30	50	CYCLE	1D	

NOTES

1010 — SPECIAL APPROVAL NEEDED IF CUSTOMER IS NOT IN CREDIT FILE OR IF CREDIT EXPERIENCE IS UNSATISFACTORY.

1020 — 90% OF SALES ORDERS ARE AUTHORIZED FOR PROCESSING WITHOUT SPECIAL APPROVAL.

1050 — 4 BACK ORDER LISTINGS PER DAY. 30 PICKING TICKETS PER RUN HAVE 1 OR MORE BACK-ORDERED ITEMS.

1060 — 4 INVOICE LISTINGS PER DAY. 120-150 INVOICES PREPARED PER RUN.

NOTES

1100 — "PHONE ORDERS" DO NOT INCLUDE TELEPHONED CUSTOMER P.O.'s OR WILL CALL ORDERS. PHONE ORDERS ARE RECEIVED THROUGHOUT THE DAY.

1110 — WILL CALL ORDERS ARE RECEIVED THROUGHOUT THE DAY.

1120 — SALES ORDER RECORDS ARE RETAINED FOR 4 MONTHS, THEN MICROFILMED (ON A CYCLIC CUSTOMER - REVIEW BASIS) AND DESTROYED.

1130 — COMPLETED ORDER FILE IS HELD FOR WEEKLY STOCK ANALYSIS AND COST DISTRIBUTION.

1140 — 3200 TO 4000 MESSAGES REPRESENT 275 TO 350 UNFILLED ORDERS.

FIGURE 9 Activity sheet.

FLOW DIAGRAM

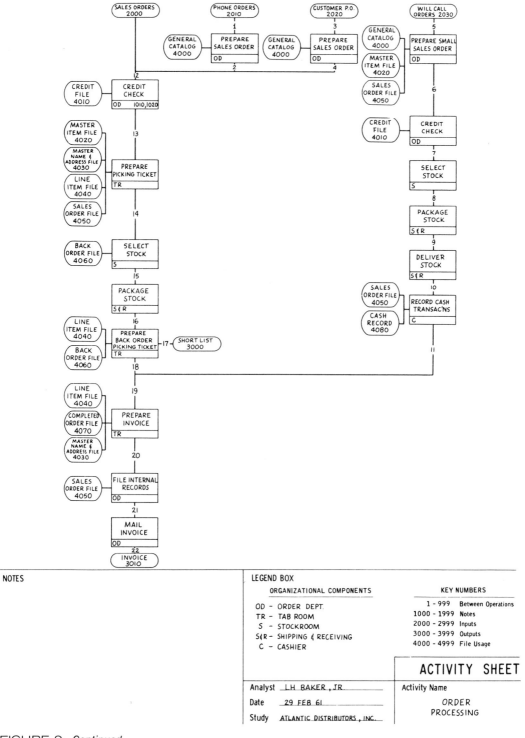

FIGURE 9 Continued

OPERATION			TRIGGERS, INPUTS AND OUTPUTS		VOLUME		ELAPSED TIME		PROCESSES		RESOURCES		UNIT TIME	TOTAL TIME	
NO.	PERFORMED BY	NO.	NAME AND QUALIFICATIONS	RECEIVED FROM OR SENT TO	AVG	PER		NO.	DESCRIPTION AND QUALIFICATIONS	FREQ	NO.	TYPE		AVG	PER
030	001 TAB ROOM	T1	RECEIPT OF SALES ORDER					P1	SELECT MASTER ITEM CARD FOR EACH	15/11	X1	KEYPUNCH OPR. (3)	/OP		
	(KEY NOS.	I1	SALES ORDER	ORDER DEPT.	475	D	0		LINE ITEM ON SALES ORDER		X2	TAB MACHINE OPR.	4½H /OP		
	13-14)	R1	PICKING TICKET	STOCKROOM	475	D	1/10 D	P2	PREPARE LINE ITEM CARD	15/11	X3	026 WITH CARD	24H /OP		
		R2	LINE ITEM CARDS	FILE	7125	D	1/12 D	P3	REFILE MASTER ITEM CARD	15/11		INSERTION DEV.(3)			
		R3	NAME AND ADDRESS CARDS	FILE	1425	D	1/12 D	P4	SORT LINE ITEM CARDS TO STOCKROOM	1/101	X4	083 SORTER	1H /OP		
		R4	CONTROL SHEET	ORDER DEPT.	48	D	1/10 D		LOC'N WITHIN CUSTOMER NO.		X5	407 ACCTG. MACH.	1H /OP		
		R5	SALES ORDER	ORDER DEPT.	475	D	1/10 D	P5	SELECT NAME AND ADDRESS CARDS	1/11	X6	407 PANEL #18	/OP		
								P6	INSERT NAME AND ADDRESS CARDS	1/11	X7	MASTER ITEM		24H	D
									IN FRONT OF LINE ITEM CARDS			CARD FILE (3)			
									FOR EACH CUSTOMER		X8	MASTER N&A FILE		2½H	D
								P7	PREPARE PICKING TICKET AND	1/11	X9	LINE ITEM FILE		4H	D
									CONTROL SHEET						
								P8	FILE NAME AND ADDRESS CARDS	1/11					
									AND LINE ITEM CARDS						
030	010 TAB ROOM	T1	RECEIPT OF PICKING TICKET					P1	SELECT NAME AND ADDRESS CARDS AND	1/11	X1	TAB MACHINE OPR.	4½H /OP		
	(KEY NOS.	I1	PICKING TICKET	STOCKROOM	475	D	0		LINE ITEM CARDS		X2	604 CALCULATOR	/OP		
	19-20)	R1	INVOICE	ORDER DEPT.	475	D	½ H	P2	COMPUTE INVOICE AMOUNT AND COST	1/11	X3	604 PANEL #27	/OP		
		R2	PICKING TICKET	ORDER DEPT.	475	D	1 H		AMOUNT AND GROSS PROFIT		X4	521 PUNCH			
								P3	PREPARE INVOICE	1/11	X5	521 PANEL #27	1H /OP		
								P4	SELECT NAME AND ADDRESS CARDS	1/11	X6	407 ACCTG. MACH.	1H /OP		
								P5	FILE NAME AND ADDRESS CARDS	1/11	X7	407 PANEL #27	/OP		
								P6	FILE LINE ITEM CARDS	1/11	X8	LINE ITEM FILE		4H	D
											X9	MASTER N&A FILE		2½H	D
											X10	COMPLETED		2½H	D
												ORDERS FILE			

DATE: FEB 29 61   ANALYST: LH BAKER JR   SOURCE: TAB ROOM – PANELLI   STUDY: ATLANTIC DISTRIBUTORS INC   ACTIVITY: ORDER PROCESSING   PAGE: 3

FIGURE 10 Operation sheet.

data are displayed in great detail. Five of the files must be randomly accessible in less than two minutes; comparison of input volumes with these requirements quickly discloses whether or not the file arrangement is bottlenecking the operation, and permits an estimate of when danger may arise.

Two of the key operations in this flow of the activity are preparing the picking ticket and preparing the invoice; these are detailed in the Operation Sheet in Figure 10. The picking ticket is used in the stockroom to select the merchandise for packing and shipment. The invoice is one of the two outputs to the activity's external environment. Both operations are mechanized sequences using tab room facilities which are listed in the RESOURCES field of the Operation Sheet.

Note that although the whole order-processing activity is considered as having two outputs (the customer invoice and the "short list"), each processing step has its own inputs and outputs. There are five outputs, for example, from the operation PREPARE PICKING TICKET: (1) the picking ticket itself, sent to the order department, (2) line item cards and (3) name and address cards, to be filed for use in the invoice-preparation sequence, (4) a control sheet used by the order department for picking ticket control, and (5) the sales order (not prepared in this operation, but rather the input to it) which is sent back to the order department.

The sales order is the only input to the operation, and its arrival is the trigger. This is an example of a single-input trigger; it indicates that the operation is performed for each input. The observer should be alert for such a trigger, since possible increases in efficiency may be obtained by batching input items and using a multiple-input trigger of six or ten orders or more. Noticing these conditions during the preliminary survey sometimes eases subsequent design problems.

In the invoice-preparation operation, again a single input—the picking ticket received after use in the stockroom—triggers the sequence. The sole output besides the picking ticket is the invoice itself; both are sent to the order department.

The process descriptions include the individual steps which the machines perform, and also the multiple references to files. These process steps, incidentally, are described by verbs which are very like the verbal language of a commercial compiler: *sort, prepare, compute*. The steps of a tab room procedure are similar to computer subroutines of the type that are triggered by compiler pseudocode.

Frequency data listed as a process qualification should be fairly precise estimates, as should the times required of the various mechanical, material, and informational resources.

The supporting documentation illustrated for this mechanized example consists of the line item card described in the Message Sheet in Figure 11, and the file in which these messages are stored, described in the File Sheet in Figure 12.

In Figure 11, the fields of the line item card (R2 from the first operation in the Operation Sheet) are precisely delineated in the CONTENTS section of the Message Sheet; a second page is required to list the information. The descriptive data in the upper half of the sheet display the operations with which the unit record is involved; typically, a punched card will be used in multiple operations.

In Figure 12, retention characteristics of the file are listed on the File Sheet. In this case, the line item card and its associated customer name and customer address cards constitute a transient record, needed only until the picking ticket is returned to the tab room after the stockroom has filled the order. This information is shown on the File Sheet, plus the added intelligence that the file is purged in a special run every two weeks to make sure that no order or back order has been overlooked.

Message characteristics are taken off the Message Sheet, multiplied by average and peak character volumes per file. In the case of a multiple-record file (as in the present example) the total volumes and character counts should be recorded in order to display the overall characteristics of a mechanized file.

As in the preceding example of a nonmechanized system, the five reporting forms permit a clear picture to be developed of the workings of a business system. They answer the question: What is being done? To a great extent, they also permit the observer to find out how it is done, in terms of the activities and operations that thread through the business system.

DEVELOPING AN UNDERSTANDING OF A BUSINESS ACTIVITY

Having demonstrated the use of the five reporting forms in displaying characteristics of both a nonmechanized and a mechanized business activity, it now becomes advisable to discuss briefly the procedure by which an observer can arrive at an understanding of a business with the help of the forms. This procedural discussion may illuminate the way in which these forms guide the observer to an understanding of an existing system.

In using these forms, it is necessary that an observer think in terms of activities rather than in terms of sets of

IBM Message Sheet

MESSAGE NAME		MESSAGE NO.
LINE ITEM CARD		R 3008
OTHER NAMES USED		LAYOUT NO.
PARTS CARD		02
		FORM NO.
		D 17130
		NO. OF COPIES
		1

MEDIA	HOW PREPARED
IBM CARD	KEYPUNCHED – FROM MASTER ITEM CARD ①

OPERATIONS INVOLVED IN
030-001, 030-007, 030-010, 030-017, 030-020, 030-021,
AND ACCOUNTING DISTRIBUTION
AND STOCK ANALYSIS

REMARKS
① ORIGINALLY PREPARED FROM MASTER ITEM CARD (OPERATION 030-001).
LATER ENTRIES ARE KEYPUNCHED OR CALCULATED.

CONTENTS

NO.	DATA NAME	FREQUENCY	CHARACTERS	A/N	ORIGIN
01	ITEM NUMBER	1	6	N	030-001
02	ITEM NAME	1	10	AN	030-001
03	DEPARTMENT CODE	1	1	AN	030-001
04	COMMODITY CODE	1	1	AN	030-001
05	UNIT OF MEASURE	1	2	A	030-001
06	QUANTITY ORDERED	1	5	N	030-001
07	QUANTITY FILLED	.20	5	N	030-007
08	QUANTITY BACK ORDERED	.20	5	N	030-007
09	UNIT COST	1	5	N	030-001
10	TOTAL COST	1	6	N	030-010
11	UNIT GROSS MARGIN	1	4	N	030-001
12	TOTAL GROSS MARGIN	1	5	N	030-010
13	UNIT SELLING PRICE	1	5	N	030-001
14	TOTAL SELLING PRICE	1	6	N	030-010
15	CUSTOMER NO.	1	4	N	030-001

DATE	ANALYST	SOURCE	PAGE
29 FEB. 61	L H BAKER JR.	TAB ROOM	1 OF 2

STUDY: ATLANTIC DISTRIBUTORS INC.

FIGURE 11 Message sheet.

machinery or blocks on a table of organization. The characteristics of an activity that are implicit in the use of the term throughout this text are that it is self-contained and goal-directed; that is, an activity is a set of operations aimed at a single goal or small number of related goals, with only a few connections to any other activity. Three or four paths all aimed at a single goal may be grouped into a single activity, but branching paths with multiple goals are seldom single activities.

Activities may vary widely in size and complexity depending on the approach of the survey and the level at

IBM — Message Sheet

MESSAGE NAME	MESSAGE NO.
– CONTINUED –	R 3008
OTHER NAMES USED	LAYOUT NO. 02
	FORM NO. D 17130
	NO. OF COPIES 1

MEDIA	HOW PREPARED

OPERATIONS INVOLVED IN

REMARKS

CONTENTS

NO.	DATA NAME	FREQUENCY	CHARACTERS	A/N	ORIGIN
16	SALESMAN NO.	1	2	N	030-001
17	STOCKROOM LOCATION	1	4	AN	030-001
18	TRANSACTION CODE	1	1	AN	030-010
19	DATE PICKED	1	3	N	030-010

DATE	ANALYST	SOURCE	PAGE
29 FEB 61	L H BAKER JR.	TAB ROOM	2 OF 2

FIGURE 11 *Continued*

which understanding is sought. Thus the sales system in a department store, or the materials system in a manufacturing enterprise, may each be considered as an activity; in the latter case, the system includes everything from the procurement of raw materials to the disposal of the finished article. On the other hand, invoicing—which is a small subsystem logically a part of both department store sales and the manufacturing enterprise—may also be considered an activity.

The procedure for developing a survey of an activity starts and ends with the preparation of a Resource Usage

IBM   **File Sheet**

FILE NAME: LINE ITEM FILE	FILE NO.: F 3404.1
LOCATION: TAB ROOM	STORAGE MEDIUM: IBM CARD TUB FILE

ACCESS REQUIREMENTS: DATA FOR ORDER MUST BE AVAILABLE WITHIN 2 MINUTES.

SEQUENCED BY: STOCKROOM LOCATION WITHIN CUSTOMER NUMBER

CONTENT QUALIFICATIONS: NAME AND ADDRESS CARDS AND LINE ITEM CARDS FOR PICKING TICKETS IN PROCESS OR BACK-ORDERED.

HOW CURRENT: 1 TO 5 HOURS OLD WHEN ENTERED. REMAIN IN FILE UNTIL STOCKROOM HAS ATTEMPTED TO FILL PICKING TICKET.

RETENTION CHARACTERISTICS: DATA NORMALLY REMOVED UPON RECEIPT OF PICKING TICKET, SPECIAL PURGE RUN ONCE EVERY 2 WEEKS.

LABELS: ———

REMARKS: ———

CONTENTS

SEQUENCE NO.	MESSAGE NAME	VOLUME AVG.	VOLUME PEAK	CHARACTERS PER MESSAGE	CHARACTERS PER FILE AVG.	CHARACTERS PER FILE PEAK
01	CUSTOMER NAME CARD	320	400	61	19,520	24,400
02	CUSTOMER ADDRESS CARDS	640	800	65	41,600	52,000
03	LINE ITEM CARDS	2250	2800	52	117,000	145,600
	TOTALS	3210	4000	178	178,120	222,000

DATE: 29 FEB 61   ANALYST: LH BAKER JR.   SOURCE: TAB ROOM   PAGE: 1 OF 1

STUDY: ATLANTIC DISTRIBUTORS INC.

FIGURE 12 File sheet.

Sheet, representing the structure of the business into which the activity fits. The first step is to lay out, on the Resource Usage Sheet, the organization chart for the whole business or for that part of it which includes the system or activity under study.

The second step is to acquire the summary cost figures for each organizational component on the bottom tier of the chart. Cost figures for the preceding fiscal period are often available in the accounting department of the company. In many companies, fairly precise budget figures for the current fiscal period may be available; these may be more valid than historical cost figures if one section of the business is growing at a different rate from others.

If neither accurate budget figures nor historical cost figures are readily available, reaonable estimates by department managers may be used. Interviews with department managers may be necessary in any case to apportion departmental budgets among personnel, machines and equipment, materials, training, and other costs called for in the tabulation on the Resource Usage Sheet.

The next step, actually a series of steps, is critical to a successful survey. The observer must trace the activity he is studying through the organization to find all the departments which affect the activity and to determine exactly what each contributes to the end result. Here again interviews with operating personnel and managers in each department will be necessary. The observer must be particularly alert to unravel from the rest of the business those operations and processes connected with the activity of interest to him. For example, the department store sales audit department purges charge-sale records and develops control totals, but also may handle time sales and other transactions; it is necessary to determine fairly precisely how much time and effort this department expends on charge sales, and exactly what the department does for the charge account activity.

The processing steps in each operation are recorded on Operation Sheets. The associated inputs, outputs, and triggers are recorded; and the effort expended by the department is apportioned as accurately as available information will allow. Messages and files are given special attention on the Message and File Sheets that support each Operation Sheet. Fields on these three working documents guide the observer in his search for and collection of data.

The operations are next assembled in logical sequence. A flow diagram prepared from these operations in sequence is then laid out on the Activity Sheet; the process of preparing the flow diagram will disclose any skips in the operation sequence. Information on volumes, times, elapsed times, operation cycles, characteristics of inputs and outputs, and file usage is then entered in the grid section of the Activity Sheet from the data on the various Operation Sheets. The information on the Operation Sheets regarding the apportionment of departmental resources, equipment, materials, and manpower is next used to fill in the blocks on the Resource Usage Sheet where the costs of the activity are summarized. The refinement and summing of these figures, and the addition of notes to explain peculiarities of the system, are the final steps in the procedure.

The end product should be a coherent description of the system as it exists. Its basic dynamics are displayed on the Resource Usage and Activity Sheets, and the details of the processing steps are described on the Operation Sheets; Message and File Sheets provide close-ups of information inputs, outputs, and resources.

With this description completed and refined, it becomes possible to proceed with analysis of the true requirements of the system. This is the second phase of stage one in the life of a business system.

FLOWCHARTING WITH THE ANSI STANDARD: A TUTORIAL

NED CHAPIN

HISTORICAL DEVELOPMENT

Flowcharting is a means of graphically stating ways of solving information handling problems. Flowcharting, as people use the term in working with computers, must be distinguished from other graphic aids. For stating clerical procedures, such as those used in systems and procedures work, people use a graphic means which has also been the subject of a standard [4]. But it is quite different from the standard under discussion here. Logic designers also use graphic aids for stating the character of the machines they design for handling information. These too have been the

SOURCE: Chapin, Ned, "Flowcharting with the ANSI Standard: A Tutorial," *Computing Surveys*, Vol. 2, No. 2, June, 1970, pp. 119–143. Copyright © 1970 by Ned Chapin and drawn from his book, *Flowcharts*, Auerbach Publishers, 1971. Reprinted with author's permission.

subject of a standard [3, 19]. The emphasis in this tutorial paper is on stating information handling problems where the information handling is done at least in major part with the aid of the automatic computer [7].

The intellectual father of flowcharting is John von Neumann. He and his associates at Princeton University's Institute for Advanced Study were the first to use graphic aids systematically for this purpose and publish their use [11]. Even though the details of the flowcharting as the standard specifies it today differ considerably from what they advocated, the spirit, the philosophy, and the rationale remain much as they presented them.

For their own internal purposes and for dealing with customers, each of the major computer manufacturers has over the course of the years developed, adopted, published, modified, and advocated flowcharting conventions (see, for example, [13, 15]). These have differed from vendor to vendor, in part deliberately as an attempt to distinguish one vendor from the competing vendors, and in part out of a sincere attempt to reflect what each has felt to be unique differences in their philosophy and approach to information processing problems.

Users of computers have individually and collectively made decisions on flowcharting conventions. Most small and medium and many large computer users have adopted the conventions presented to them by the vendor of the computer they have elected. But a few larger users have chosen to go their own way, and have developed their own internal standards. The United States Air Force, for example, has developed its own standards for this purpose and in practice has had a multiplicity of them (see, for example, [21]).

Users of computers acting collectively through the user groups have sometimes addressed themselves to the problem of standards for flowcharting. Since these user groups normally have been composed of users of only one vendor's computer, the effect usually has been to recommend modifications or suggestions to the computer vendor for changes in the vendor's standard. But a few user groups, for example SHARE, have independently presented their own standards and have advocated them for general adoption [10].

Individuals with competence and standing in the computer field have for their own use sometimes deviated from the practice of the various computer vendors and from the other sources. They have presented their own recommendations for flowcharting. These recommendations have been effective through their author's publications (see, for example, [7, 12, 16, 18]). Some of these recommendations, however, have ceased to be significant with the advent of formal standardization.

Using the computer itself to produce flowcharts has had relatively little influence on the development of standards, although programs for doing it have been in use since 1957. To assist the process, a Systems Flowchart Language (SFL) was even developed [17]. Of more practical significance in popularizing the *de facto* vendor standards has been the vendors' practice of providing free plastic templates to aid in drawing flowcharts (see, for example, [14]).

During the 1960's a committee attempted to develop a standard for flowcharting. Working through the Business Equipment Manufacturers Association and the American Standards Association,[1] with the committee members drawn from computer vendors and a few major computer users, the committee after the usual compromises drew up a proposed standard and circulated it for reaction. With revisions, it was approved in 1963 and published as an American Standard. The Association for Computing Machinery and other groups published this standard in their periodicals, giving it considerable publicity (see, for example, [1]). This standardization effort in the United States paralleled a similar effort conducted for the International Standards Organization (ISO).

Subsequently, in 1965 and again in 1966, 1968, and 1970, the American Standard was revised. The 1965 revision was major, but the 1966 and 1968 revisions were only minor. The 1970 revision extended the standard to match more closely the ISO standard. The standard as it is presented in this tutorial paper is the ANSI 1970 revision [2]. Since this is a tutorial paper, it does not cover every part of the standard, nor every variation in the use of the standard.

ANSI STANDARD

Outlines

The ANSI standard consists, in the first place, of a series of graphic outlines or boxes, which the standard terms "symbols." The standard advances these

---

[1] As of 1969, the American National Standards Institute (ANSI), 1430 Broadway, New York, N.Y. 10018.

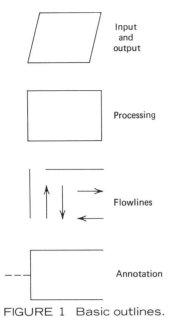

FIGURE 1 Basic outlines.

flowchart outlines in three groups: the basic, the additional, and the specialized. Complete flowcharts can be drawn using only the basic and the additional outlines. The use of the specialized outlines is optional. If they are used, however, they should be used in a manner consistent with the standard.

For the outlines in each group, the standard specifies the shape, but not the size. The shape is specified in two ways: by the ratio of the width to the height and by the general geometric configuration. This means that the user of the standard is free to draw outlines of any size to fit his own convenience. He may vary the size from time to time during the course of his flowchart, but he is to observe the ratio and general configuration specified. For those who do not have ready access to the standard, the Appendix to this paper summarizes the ratios and configurations.

The standard implicitly advances the use of a single width or weight of line for drawing the outlines. It also implicitly advances a single orientation or positioning of the outlines with respect to each other. In particular, portions of the outlines shown horizontally oriented in the standard are to be drawn that way.

Basic Outlines

The basic outlines specified in the standard are the input-output, the process, the flowline, and the annotation outlines. These are illustrated in Figure 1 and described in the Appendix.

The input-output outline indicates an input or output operation, or input or output data. It is defined for use irrespective of media, format, equipment, and timing. Some specialized outlines may be substituted for this outline.

The process outline is the general purpose outline. It is the *de facto* default outline for use when no other outline is specified by the standard. The process outline indicates data transformation, data movement, and logic operations. Some specialized outlines may be substituted for this outline.

The flowline outline is an arrow of any length which connects successive other outlines to indicate the sequence of operations or data (the "direction of flow"). It is defined for use in an alternating fashion with the other outlines. As such, it also indicates the sequence in which the other outlines are to be read. To specify the direction of flow or reading, open arrowheads may be used on any flowline as shown in Figure 1.

The normal direction of flow is the normal direction of reading for people trained in the English language: from top to bottom and from left to right. Where the flow follows this normal pattern, no open arrowheads are needed to remind the reader. In the event of any significant deviation from this pattern, arrowheads are required to signal the deviation to the reader's attention. Whenever the direction of flow might be ambiguous to a reader, arrowheads should be used to provide clarification. Bidirectional flow may be indicated by dual arrows each with open arrowheads, or less preferably by open arrowheads in both directions on single flowlines.

The annotation outline provides a way to supply descriptive information, comments, and explanatory notes. Its dashed line indicates the outline to which this explanation or clarification applies.

Additional Outlines

The additional outlines are for the convenience of the reader, and not for the purpose of describing data-processing action. These symbols provide for handling the limitations of pages of various sizes, and make it more convenient to show connections in the sequences of flow. These outlines are shown in Figure 2.

The connector outline, a circle, must in practice be used at least in pairs. To that end, the standard advances two varieties, the inconnector or entry connector, and the outconnector or exit connector. An inconnector or entrance has a flowline leaving it but none entering it; an outconnector or exit has a flowline entering it but none leaving it. Each inconnector may have from zero through any number of outconnectors associated with it. However, each outconnector must have exactly one inconnector associated with it. One function of the connector outline is to enable a long sequence of outlines (a "flow") to be broken into pieces to fit conveniently on a page. The connector outline also provides ways of joining together convergent lines of flow that fan in to some particular point. And, it provides a way of identifying divergent lines of flow.

The terminal connector outline serves to indicate a beginning, an end, or a

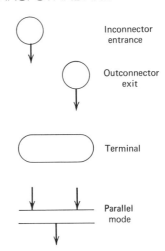

FIGURE 2 Additional outlines.

break in the usual line of flow. In the first two uses, it substitutes for an ordinary connector at the beginning and the end of major portions of a sequence of outlines (a "flow"), particularly when these portions are identified by a name, as, for example, for a closed subroutine. In its third use, it may represent a start, stop, halt, delay, pause, interrupt, or the like. For this use, it has both an entry and an exit flowline.

The parallel mode outline is a pair of horizontal lines with one or more vertical entry flowlines and one or more vertical exit flowlines. It is used to indicate the start or end of simultaneous operations.

Specialized Outlines

Groups. The specialized outlines fall into three groups. One group permits specification of the data-carrying media (see Figure 3). Another permits specification of the peripheral equipment type (see Figure 4). A third permits specification of selected types of processing action (see Figure 5). In each case, where no specialized outline has been provided, the standard specifies that the basic outline covering the situation should be used. Thus, for any media or equipment, it should be the input-output outline. For

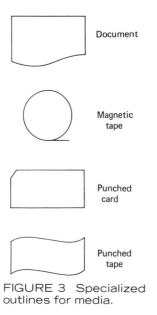

FIGURE 3 Specialized outlines for media.

any processing, it should be the process outline. The one exception is the communication link, for which the basic outline is the flowline.

Media Outlines. The document outline is the most commonly used of all the specialized media outlines. This outline, a stylization of a torn piece of paper, represents data in the form of hard copy input or output of any type. For example, it may represent data taking the form of printing on paper produced by a high speed printer, or of marks on cards read by an optical reader, or of a graph produced by a data plotter, or of a page of typing produced on a terminal.

The magnetic tape outline is a circle with a horizontal rightward pointing line tangent to the bottom. This outline represents data in the medium of magnetic tape.

The punched-card outline represents data in the medium of a punched card of any style, size, or punching, such as Hollerith punched cards, binary punched cards, Binary Coded Decimal (BCD) punched cards, fifty-one column cards, and the like. Thus, a time card which only has on it printed numbers and never is punched with equivalent or even unrelated information does not qualify for representation with the punched-card outline (it takes a document outline in-

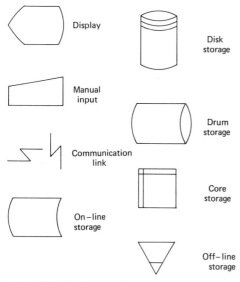

FIGURE 4 Specialized outlines for equipment.

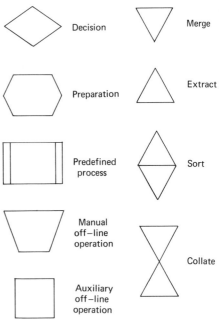

FIGURE 5 Specialized outlines for processing.

stead). Two further specialized forms of the punched-card outline are described below.

The punched-tape outline represents data in the medium of any punched continuous material, such as paper tape, punched plastic tape, punched metal tape, etc. The requirements to be met are that the medium be indefinite in length and that the data be represented by punch patterns. The outline is a stylization of a partial fold of paper tape.

Equipment Outlines. The display output outline is a stylization of a cathode-ray tube (CRT) with the face of the tube to the right and the neck of the tube to the left. This outline represents any kind of transitory data not in hard copy form, as for example CRT displays, console displays, and the like. But the standard also advances the display outline for intermediate output data used during the course of processing to control the processing. Common examples are the data produced on console printers and time-sharing terminals if the human user is expected to utilize immediately the data presented.

The manual input outline represents data acquired by human control of manually operated online equipment. Examples are data from the operation of keyboards, light pens, console switch settings, pushbuttons, transaction recorders, tag readers, and the like, where the human operator provides the timing.

The communication link outline is represented by a zigzag flowline. This is appropriate because typically data communication done with the aid of equipment provides a flow of data from one place, or from one medium or equipment, to another. As such, even though the outline is equipment oriented, it is used as a specialized flowline. Where necessary, open arrowheads may be used to indicate the direction of flow, in the same manner described previously.

The on-line storage outline represents data held in any on-line intermediate and external storage device of any type, as for example magnetic disks, magnetic tapes, magnetic drums, magnetic cards, additional banks of magnetic core storage, microfilm, etc. For data on some of these devices, the standard provides more specialized outlines when more precise specification is desired, as described below. A more specialized outline for data on magnetic tape was noted above in the media group.

The disk storage outline represents data stored on a disk device of any type, especially a magnetic disk. The outline is a stylization of a cylinder standing on end. The outline is a further specialization of the online storage outline.

The drum storage outline represents data stored on a drum device, especially a magnetic drum. The outline is a stylization of a drum lying on its side. The outline is a further specialization of the online storage outline.

The core storage outline represents data stored in a magnetic core or similar high speed device that is *not* the primary internal storage for the computer. It might be, for example, an auxiliary on-line bulk core device, or a remote computer connected on-line with the computer doing the main processing. The outline is a stylization of two drive lines in a magnetic core array.

The off-line storage outline is an equilateral triangle with a small bar. This outline represents any data stored off-line regardless of the medium and regardless of the equipment used. In common practice, it is used for manually maintained data files.

Process Outlines. The most common of the specialized process outlines is the decision outline. It indicates comparison, decision, testing, or switching operations which determine or select among a variety of alternative flows (sequences of

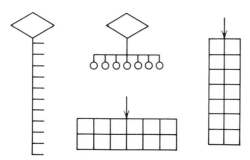

FIGURE 6 Outlines for large numbers of decisions.

operations). As such, the number of flowlines leaving a decision outline must always be greater than one.

Sometimes the number of flowlines leaving a decision outline (the number of exits) exceeds three. In this case, the standard advances, as shown in Figure 6, several alternatives which are equally acceptable. One is the organizational chart tree pattern of flowlines from a single flowline leaving a decision outline. Another alternative used in the same way is a vertical flowline which has a number of horizontal flowlines from it. To save space, formal outconnectors may be omitted from these two forms. Quite a different alternative is a "branching" table in the form of a series of pairs of rectangles, packed together in a double row or a double column. The upper or left portion replaces the decision outline; the lower or right portion replaces the usual outconnectors. The tables use their first pairs of boxes for explanation of the branching.

The preparation outline indicates operations on the program itself. They are usually control, initialization, cleanup, or overhead operations not concerned directly with producing the output data, but usually necessary to have done. Three examples are setting the limiting value as an iteration control, decrementing an index register, and setting the value of a program switch. By convention, when either a decision or a preparation outline could be used (such as for testing a switch), the common practice is to use the decision outline.

The predefined process outline indicates or identifies one or more operations which are specified in more detail elsewhere, as in a booklet or in a different flowchart (but not in another part of this same flowchart). Examples of a predefined process are a named closed subroutine or a routine from the operating system for the computer.

The manual operation outline indicates any off-line input or output producing operation which has its speed determined by the speed of the human operator, as for example entering data off-line by means of a keyboard as in a keyboard-to-magnetic-tape operation, or finding a folder in a file cabinet drawer.

The auxiliary operation outline indicates any off-line operation performed on equipment which operates at its own speed or a speed determined by something other than the speed of its human operator. Examples of auxiliary operations are cardsorting operations, punched-card interpreting operations, and the like. Auxiliary operations are performed by equipment, not by human beings.

The merge outline indicates the creation of one set of items from two or more sets having the same sort sequences. The outline may be used for both on-line and off-line operations.

The extract outline indicates the reverse of the merge. That is, it indicates the creation of two or more sets of items from and in the same sort sequence as the original set. The outline may be used for both on-line and off-line operations.

The sort outline indicates the sorting of a set of items into some sequence on the basis of some (usually specified) key. The outline may be used for both on-line and off-line operations.

The collate outline indicates a combination of merge and extract. Thus, this outline requires more than one entrance flowline and more than one exit flowline. This definition of collate used in the standard is not fully consistent with the usual definition of collate. The outline may be used for both on-line and off-line operations.

Standard Conventions

Striping. The standard specifies limited use for either horizontal or vertical striping within an outline. The vertical striping has already been covered in the special outline for predefined process (see Figure 7). Other uses of vertical striping are not specified by the standard.

Horizontal striping is advanced by the standard as one alternative way of indicating a reference to another part of the flowchart which provides a more detailed representation, as for example, of a sub-routine. A horizontal line may be drawn from the left edge to right edge in the upper portion of an outline, except for the flowline, communication link, and additional outlines. The upper area thus enclosed is used to refer to some other part of the flowchart. The lower enclosed area is used in the usual manner, as shown in Figure 7. Wherever a horizontal striping is used within a symbol, the portion of the flowchart referred to must in turn be represented on the flowchart as beginning with and ending with terminal outlines as described previously. Both the detailed representation and the striped outline must have location cross-references, as described below.

Cross-References. In order to make cross-referencing easy between parts of the flowchart, the standard advances two conventions. One is to use or to assign names to portions of the flow represented by the flowchart. These names often are the same names used in the program or system. These may be the same as, or different from, the identifying names used for connectors.

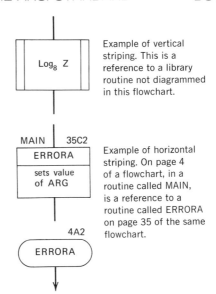

FIGURE 7 Conventions for striping and references.

An alternative convention (not mutually exclusive with the other) is to identify a location on each physical piece of the flowchart, as for example, in terms of page, row, and column, as in the manner of map coordinates. An example is the reference to page 4, row A, column 2 cited for the terminal outline in Figure 7. The standard leaves open the exact manner of composing such location references.

The ANSI standard is still in conflict with the ISO standard and with previous usage in the United States on the handling of references. The ISO and general American usage has been to place the identifying name immediately above and to the left of the outline (such as MAIN in Figure 7), and to place the coordinate reference above and immediately to the right of an outline (such as 35C2 in Figure 7). The ANSI standard advances exactly the opposite convention, but recognizes and cites the deviation from the ISO standard. In this paper, the ISO convention is used since it is also a common usage in the United States and since the ANSI standard explicitly recognizes the ISO position.

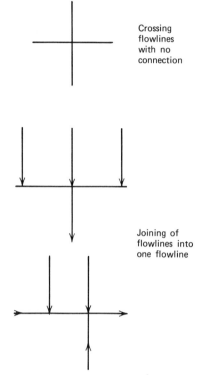

FIGURE 8 Conventions for flowlines.

Crossing Flowlines. The standard makes specific provision for connectors and cross-references. These can be used to avoid the necessity of having crossing flowlines.

If it is desired to use crossing flowlines, then the standard specifies that the flowlines shall have no arrowheads in the vicinity of the crossover. The standard specifies that the presence of arrowheads on the flowlines is to indicate a conjunction or coming together of the flow. The direction of flow at such points of joining flowlines is to be designated by the position and direction of the arrowheads, as shown in Figure 8.

Multiple Outlines. The standard makes specific provision for the possibility of multiple instances for the specialized media outlines. These take two forms, one specifically for punched-card media and the other for media generally. The one applicable to punched cards only advances a convention for representing a deck of cards and a card file, as shown in Figure 9. These find their most common use in flowcharts for systems implemented with punched-card handling equipment.

The standard also advances a more general convention for use when multiple forms of specified media have different identifications and uses. For example, in punched-card installations, it is common to have a master or header card followed by several detail or trailer cards. To represent this situation, the standard advances that the main or first medium outline should be drawn in full. Following it and partially obscured by it in any clockwise position from it may be drawn in sequence a partial but closed outline of other instances of the same medium, as shown in Figure 9.

The standard is not clear with regard to the handling of flowlines for multiple outlines. The standard advances the convention that flowlines may enter or leave from any part of the multiple outline group. This in effect treats the group as though it were a single outline simply having inner lines marking off interior portions. But for those common cases where the group is to be broken into component parts and the component parts processed differently, the standard is silent. Perhaps the designers of the standard assumed all group formation and separation to be operations requiring a process outline, although this is contrary to long-standing custom.

To avoid this problem, this tutorial paper is based on the position that if a single flowline enters and a single flowline exits, then the multiple media shown are treated as a group. If multiple flowlines are used on either the entrance or exit sides, then the multiple flowlines apply only to the specific media outlines to which they individually connect.

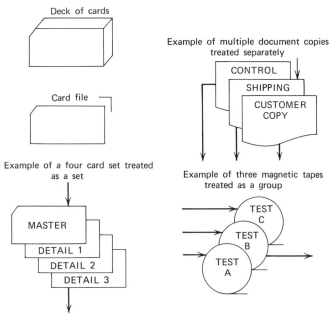

FIGURE 9 Multiple symbol conventions.

USE OF THE STANDARD
Situations

The ANSI standard flowchart symbols for information processing cover two major situations. One situation is for representing algorithms, especially those for execution by a computer. The other is for representing systems without indicating the character of the component algorithms. Some other situations are noted briefly later in this paper. The term "flowchart," as used in the standard, may therefore refer to either of these situations.[2] Hereafter in this paper a clear distinction is necessary between flowcharts of systems and flowcharts of algorithms. Hereafter, "flow diagram" designates a flowchart of an algorithm, and "system chart" designates a flowchart of a system.

Other terms are also current in the field for these two situations. Thus other terms sometimes used for flow diagram are block diagram, logic chart, and process chart, as well as flowchart. For system chart, other terms are run diagram, procedure chart, and flowchart.

The distinction between the flow diagram and the system chart is vital because the use of the standard differs considerably for these two. In the case of the system chart, the focus is upon the inputs and the outputs produced by the sequences of runs, programs, or procedures. In contrast, the focus in the flow diagram is upon the sequences of data transformations needed to produce an output data structure from an input data structure. The flow diagram tells "how." Whereas a system chart identifies programs, runs, or procedures by name and data structures by name, the flow diagram identifies individual operations on portions of data structures. The flow

---

[2]The standard's definition of the term "flowchart" is a subversion, well supported by popular usage, of a far older definition. The term "flowchart" has a history predating the use of computers. In the field of systems analysis, it historically has designated a graphic aid to analysis quite different from that contemplated in the standard. This older use of the term is illustrated in [7, 1963 ed., pp. 237–239], and three forms of flowcharts following this older definition are illustrated in [6, Ch. 5].

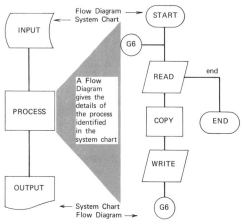

FIGURE 10 Relationship between flow diagram and system chart.

diagram is usually an elaboration of what is indicated by a single process outline in a system chart (see Figure 10).

In the remainder of this paper, system chart conventions and system chart guidelines are considered first. These use a greater variety of outlines, but the logical complexity is relatively low. Then flow diagram conventions and flow diagram guidelines are discussed. Flow diagrams can become logically complex even though the number of different outlines utilized is typically fewer.

System Chart Conventions

Basic Format. The basic format of the system chart follows a sandwich rule; that is, it is composed of alternating layers of data identifications and process identifications. The data identifications are equivalent to the bread of the sandwich, and the process identifications are equivalent to the filling in the sandwich. Just as sandwiches may be of the Dagwood type, so the output produced from one process operation may serve as the input for a following process operation (a compound system chart). But a system chart must always begin with inputs (data identifications) and must always end with outputs (data identifications).

To see this sandwich rule in use, consider the creation of a system chart using only the basic outlines. Assume that the input available is a set of data about the ages of employees. Assume that the output desired is a single number, the average age of employees. Using the basic outlines, no attention is needed to the media or to the equipment. Hence, as shown in Figure 11, the systems chart begins with an input-output outline for the input. Connected to that by a flowline is a process outline. Connected to that by a flowline and ending the system chart is an input-output outline for the output.

To summarize, the basic format of the system chart is a sandwich. It always begins with the bread of input and ends with the bread of output. The sandwich filling is the processing which converts the input into the output. But the system chart does not tell how the processing is accomplished; it only identifies what processing is done.

Identifications. The bare outlines shown in Figure 11 are meaningful to someone who has the identification of

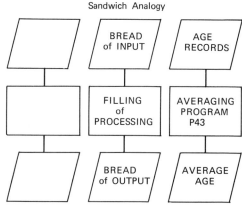

FIGURE 11 A simple system chart using only the basic outlines.

the input, the output, and the processing clearly in mind. But it has less communication value to others because, even though it tells that input is to be converted into output, it does not identify which particular input, what particular processing, or what particular output. To improve the communication value of the system chart, therefore, a common convention is to indicate within the outlines the identification of each input, each output, and each process.

For this purpose, the usual convention is to use the names normally assigned at that installation to the input and output. If the system chart is likely to be read by persons not conversant with those names, then the English-language equivalent in full may be written out within the outlines to provide the identification. Thus, Figure 11 also provides a restatement that incorporates the identifications absent from the system chart on the left in Figure 11.

This same system chart is also present in the top part of Figure 12. But Figure 12 also illustrates more, since it shows a compound system chart rather than just a simple one. The average which was the output of the first processing operation in turn serves as input for another operation. Here the average age of employees is to be combined with previously calculated data on the average age of employees to produce a chart showing the trend of the average age of employees over the course of time. A preparation of this chart is a separate processing operation from a computation of the current average age. The other output of the trend program is the updated record of the prior averages.

Here, as before, clear identification of each of the inputs, output, and processings is provided, but no indication of the nature of the medium or equipment is provided in the choice of outlines. Note the Dagwood sandwich structure.

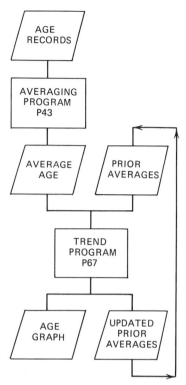

FIGURE 12 Compound system chart using only the basic outlines.

Specialized Outlines. In order to improve the communication value of the system chart still more, specialized outlines may be used in place of the basic outlines already presented. Thus, Figure 13 presents a redrawing of Figure 12 using the specialized outlines. Figure 13 shows that the data on the age of employees is on cards, and that the output of the average age is put onto a magnetic disk or other external storage device where it serves as input to the trend program.

The other input to this trend program is from a magnetic tape, which has recorded on it the previously computed average ages. A magnetic tape also receives the output from this program so that it can be recycled to serve again as input if necessary.

The hard copy output from the second program is a time series graphic plot.

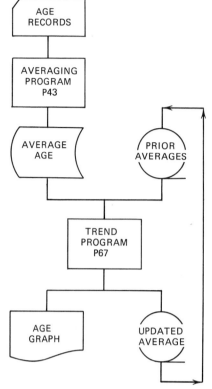

FIGURE 13 Compound system chart using the specialized outlines.

Assuming that the stress is upon the hard copy aspects of this output, then the document outline is the appropriate specialized outline to use. If this plot, however, were displayed on a CRT, then a display outline would be the appropriate choice.

Use of Connectors. If a compound system chart requires more space on the page to represent it than is available, then connectors may be used to break the chart into parts and to indicate the connection between the parts on the separate pages. This procedure is consistent with the standard but does not improve the communication value of the chart. An alternative procedure also consistent with the standard gives superior results.

The matter can be likened again to the Dagwood sandwich. If one makes a Dagwood sandwich that is too large to bite, one procedure is to break the sandwich into two or more parts. Whenever doing this, however, one must add an additional slice of bread for each break because each complete sandwich must begin with a slice of bread and end with a slice of bread. Since, by analogy, input or output data serve as slices of bread, one may break a long system chart into shorter charts and still show a connection between them by the choice of representing input or output data. To see this, imagine breaking the system chart in Figure 13 into two separate system charts.

To make the communication value of the system chart high, it is important to show all of the inputs and outputs for each part of the system. If one of these appears at one place as an output, and at another as an input, then it is only necessary to repeat the outline with the identical identification and appropriate cross-reference. This is illustrated in Figure 14. Notice that the disk output from the first run which serves as an input to the second run is shown twice. This is considerably more illuminating to someone who studies any part of the chart than would be the alternative of using a connector outline, illustrated in Figure 15, where two different ways of doing it are shown.

In summary, the most effective way to break a system chart into parts in order to fit it on limited size pieces of paper is to repeat the representation for selected inputs or outputs, identifying and cross-referencing them appropriately. In this way the material shown on each page is complete in itself.

Annotation. When breaking a system chart, a problem arises on identifying the source and use of data. The clearest convention is as shown in Figure 14, by the use of cross-references with an exact repetition of the data identification. This also serves well for multiple uses of an output as inputs to several process outlines.

One alternative way of handling the situation is by the use of the wording within the symbols themselves. But this clutters the space conventionally used for identification, making it a dual-use space rather than a single-use one. This decreases the communication value of the chart, and is not illustrated here.

Another alternative way of maintaining the identification of the source and use of data is by the use of an annotation outline. For example, one could be inserted at the bottom of the first page indicating the page to look on to find the additional use of this output as an input, as shown in Figure 16. On the page on which it appears as an input, the annotation outline could again be shown with an indication in it where this input came from as an output. It is clear from even a casual examination of Figure 16 that the excessive use of annotation outlines clutters the system chart and can decrease its communication value.

To avoid the clutter of the annotation outlines yet provide the annotation, an alternative is to add a column of annotation to the right (or the left) of the system chart. This can provide information that is frequently helpful in interpreting system charts; see Figure 17 for an example. The information most commonly shown in such annotation columns is the volume of the input and output, the timing of the availability of input or output, the control procedures, the equipment configurations required, the personnel complements, and the geographic locations. These items are normally not conveniently shown in the system chart itself because they clutter it up too much, and are not of concern to

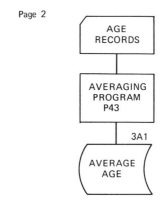

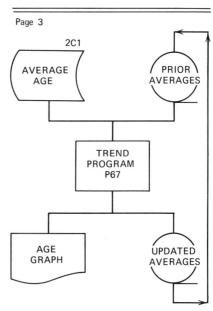

FIGURE 14 Example of good practice in breaking a system chart.

all readers. The use of this column alternative is an elaboration of the standard.

Guidelines for System Charts. In preparing system charts, the following guidelines have been found helpful from experience. A first guideline is to choose the wording within the outlines (the data and process identification names) to fit the needs of the readers of the system chart. A chart using something approaching the English language, as illustrated in Figure 17, can be widely understood.

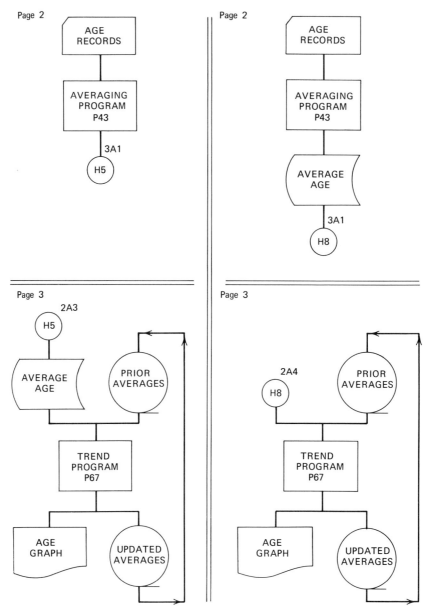

FIGURE 15 Example of poor practice in breaking a system chart.

A chart which uses only specialized names can be fully understood only by those who know the specialized names. Thus, the system chart shown in Figure 18 is less easily understood than the same system chart shown in Figure 17.

A second guideline is to use the data and process identification names consistently and to keep them brief. If the same name appears more than once anywhere in the system chart, it should always identify the same thing.

A third guideline is to use a relatively small size for the outline. It improves the communication value of the chart because it enables a more compact layout, which allows the reader's eyes to take in more at one glance.

A fourth guideline is to leave blank space around major unconvergent flows. This visually sets them off, and makes their role in the system more easily comprehended, as in Figure 19. By contrast, in a uniformly packed or tightly spaced system chart, even simple straight lines of flow are difficult to see clearly.

A fifth guideline is to collect incoming flowlines and outgoing flowlines so that the flowlines actually entering and leaving a processing outline are kept to the minimum. This is illustrated in Figure 17.

A sixth guideline is to minimize crossing flowlines. Crossing flowlines can be eliminated by repeating input or output outlines with appropriate cross-references as to source or destination.

A seventh guideline is to use the specialized outlines wherever possible (compare Figures 12 and 13). Their use improves the communication value of the system chart. They are only slightly more difficult to draw.

An eighth guideline is to use cross-referencing and annotation generously, but not to excess. The more the system chart can tell the reader quickly and easily, the more valuable it is. Ways of providing cross-referencing and annotation are mentioned above.

A ninth guideline is to give particular attention to the processing that affects data prior to the time that the data become input to a computer program or run. These are most often manual operations and auxiliary operations. Failure to specify them in full is one of the most common shortcomings of system charts. (Example: How did the data get punched into the cards in the system shown in Figure 18?)

A tenth guideline is to begin with what is known and well understood. Prepare the system chart describing that. Then extend the system chart in each direction. That is, for each of the inputs

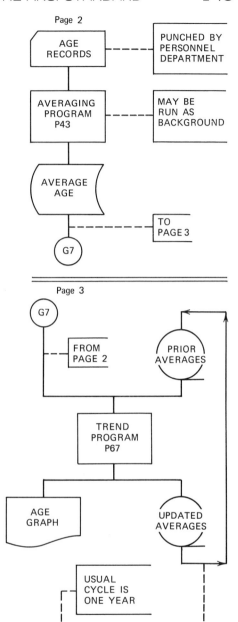

FIGURE 16 Annotation symbols in system charts.

shown, find the processing that produces it and what its input is. Then, regarding those inputs as outputs, continue the procedure for each of those inputs. The same can be done in the other flow direction for the outputs. In this way the system chart can be extended to cover the

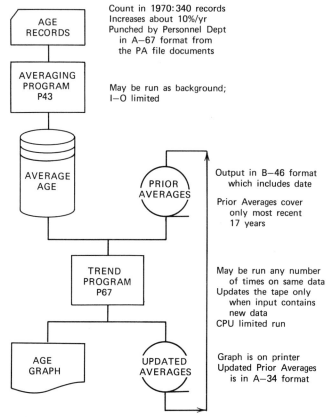

FIGURE 17 Annotation column for system charts.

entire system, as well as covering this system's tie-in with other systems. Typically, particular attention is needed to the manual and auxiliary processing of data in the system.

An eleventh guideline is to make no violations of the standard, and to shun deviations. This requires a clear distinction between a violation, a deviation, and an elaboration. An example may clarify each. Using closed arrowheads on flowlines is an example of a violation. Another example is using a circle with a short right-of-center horizontal lime touching the bottom center, as an outline for data on punched cards. This is a violation because the standard assigns a specific significance to such a circle, and provides a specific outline for data on punched cards.

Three examples of deviations from the standard are provided by IBM [14]. IBM advances an outline for "keying." This is a serious deviation since the standard already provides two outlines for data from key-driven equipment, or its operation. These are the manual input (on-line) and the manual operation (off-line). IBM advances an outline for a "transmittal tape." This is a kind of document. The standard provides an outline for documents generally. IBM advances an outline for an "off-page connector." The standard provides a connector outline and specifies the use of cross-references to indicate the location.

An example of an elaboration of the standard is the use of a five-pointed star outline to represent data acquired by the

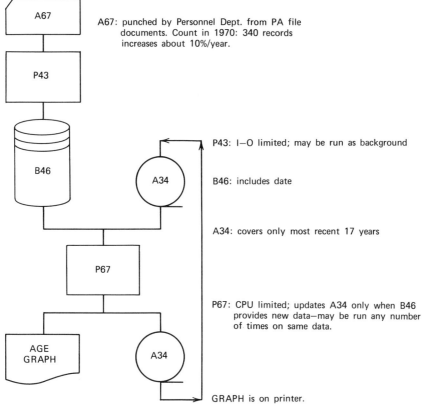

FIGURE 18 Example of specialized wording within the outlines of a system chart.

on-line operation of an optical detector of particle tracks in a spark chamber. Such equipment is not generally available, and hence the standard provides no specialized outline. The basic input-output outline is applicable, and hence no specialized outline is really necessary.

Flow Diagram Conventions

Function. The flow diagram describes the algorithm for transforming input data structures into output data structures. As such, the primary emphasis is upon depicting the sequence of operations that tell how data are transformed. The secondary emphasis is upon identifying the portions of the data structures affected and the operations performed. Questions of media or equipment typically become trivial.

Since commonly the operations to transform data structures consist of long sequences of actions, the character of the flow diagram differs considerably from that of the system chart. In the system chart, a sandwich rule describes its basic structure. No such convenient rule serves in the case of the flow diagram. Since it is in effect an elaboration and extension of what is usually shown as a single outline on a system chart, the flow diagram requires many more outlines and a much more extensive presentation of details than does the system chart.

The general character, therefore, of a flow diagram is of a sequence of alternating flowline and process outlines. Somewhere in the early portion of this sequence will usually appear one more

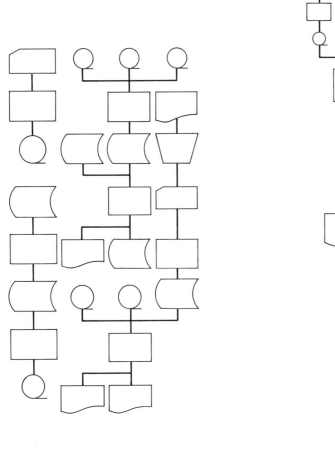
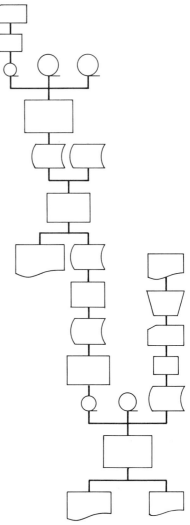

FIGURE 19 Example of a tight and an open layout of a system chart for the same system.

input-output outlines to indicate the input of a data structure. Near the end of the sequence will usually appear one or more input-output outlines to indicate the output of a data structure.

Because of its greater length, the flow diagram must be broken into parts, as a practical matter. For this reason, connector usage and cross-referencing become important considerations in the creating and reading of flow diagrams.

Basic Format. Flow diagrams may be drawn with the basic and additional outlines described previously. For example, consider the program to find the average age of the employees shown in the system chart in Figure 17. Using the basic and the additional outlines, the flow diagram can be stated as shown in Figure 20.

Some comments are in order on this example. First, the beginning and the end of the flow are marked with the

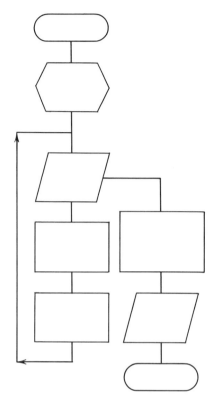

FIGURE 20 A flow diagram using basic and additional outlines, with one specialized outline.

Just as in the case of the system chart, a flow diagram using only the basic outlines provides some information but not as much as it can when augmented with written identifications within the outlines. These identifications are important to indicate the portions of the data structures affected and the operations. A long sequence of process outlines which do not designate the portions of the data structures affected soon becomes ambiguous. For this reason it becomes, as a practical matter, important to specify in detail in a flow diagram exactly what portions of the data structures are affected and in what way.

To this end, the flow diagram in Figure 20 can be redrawn as shown in Figure 21. Here the outline choice and sequence are identical to those in Figure 20, but

termination outline. This is unlike the case of the system chart, where the start or the end of anything was an input-output outline.

Second, the sequence shown follows the common pattern of read-transform-write. Since that transformation cannot usually take place until the input data have been read, the input operation precedes the process operation. Both precede the output operation.

Third, a very common feature of algorithms prepared for implementation on a computer is the use of iteration. This commonly appears in a flow diagram as a loop of flow, as shown in Figure 20. Note that, in order to indicate a section of flow contrary to the normal rule, a long flowline together with open arrowheads has been used.

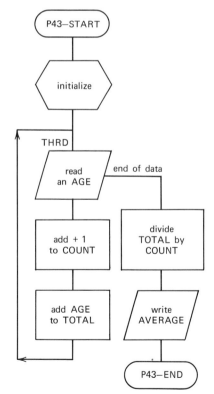

FIGURE 21 A flow diagram with identifications.

now the outlines each contain identifying information. The identifications consist of four things: the names of parts of the algorithm as implemented (entry points, usually, such as THRD), the names of operations (such as "add"), the names of conditions (such as "end of data"), and the names of operands (parts of the data structures affected, such as TOTAL).

Specialized Outline. To improve the communication value of a flow diagram, the usual practice is to use the specialized process outlines wherever possible. The specialized process outlines available for this purpose are the decision, preparation, predefined process, merge, extract, sort, and collate outlines. The manual and auxiliary operation outlines are not usually applicable because the flow diagram commonly represents only processing operations carried on within a computer.

Using these outlines, Figure 21 can be redrawn as shown in Figure 22. Here the decision outline provides an explicit end-of-input-data test. The striped preparation outline refers to the short flow sequence called INITIAL.

Also available in the standard, as described previously, are specialized input-output outlines such as those to indicate media or equipment. As mentioned previously, the main attention in a flow diagram is not upon the media or the equipment but upon the logical character of the data transformations. Use of the specialized input-output outlines clutters the flow diagram with additional shapes which tend to distract from the main focus of attention. Further, with present-day operating systems, the equipment and media for input and output can be altered at any time for operator convenience, and hence are neither statically nor dynamically determined by the character of the data. For these reasons, the use of specialized input-output outlines in a flow diagram is not recommended. Their use is not a violation of the standard, but contributes little to the communication value of a flow diagram and can even detract. For this reason, their use is not illustrated in this paper.

One outline that is illustrated is the annotation outline. This outline can be very helpful in the flow diagram to describe values and to provide explanation. Thus, in Figure 22, an annotation outline has been used to indicate, for the iterative loop, the expected number of times the loop will be executed. This information, it should be noted from Figure 21, is not available from the outlines or from the normal identification information supplied within the outline. Another use of annotation outlines is the warning about the need for the accurate date of the run, as shown in Figure 22.

Connectors and Cross-References. Connectors and cross-references are important in flow diagrams because of the common length of flow diagrams and because of convergent and divergent flows. Flow diagrams are almost always too large to represent on one sheet of paper. Usually they include alternative flow paths. Convergence points (fan in) and divergence points (fan out) must be presented and the flows clearly identified.

To make the communication value of a flow diagram high, it is desirable to have the flow pattern shown in as linear or straight-line a form as possible. The more cut up, chunky, bunched, or branched the flow pattern is, typically the more difficult it is for a person to comprehend. For this reason, the use of connector outlines and of cross-referencing normally helps give a smoother, more linear appearance to the flow.

If the flow diagram shown in Figure 22 could not be represented all on one page but had to be broken into parts, it could

FLOWCHARTING WITH THE ANSI STANDARD 149

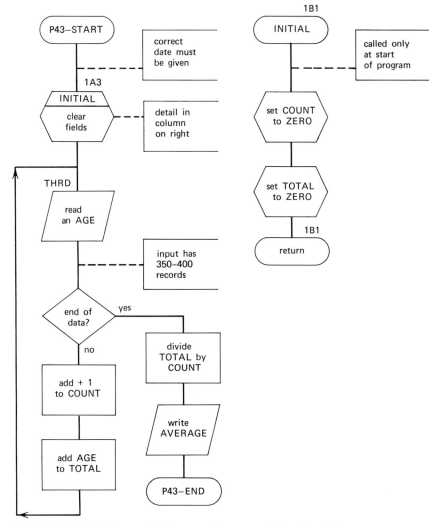

FIGURE 22 A flow diagram using specialized outlines.

be broken at any point. One way of doing it is shown in Figure 23. The connector outline does not substitute for any other outline but instead serves as an additional outline, in effect specializing the flowline symbol.

As one reads the line of flow down the page 8 part of Figure 23, one encounters an exit or outconnector A3. The reader can then search Figure 23 for an entry or inconnector having the same identifying set of characters A3 within it. This entry or inconnector is the continuation of the flow. The identification within the entry and exit connectors must permit unequivocal and unique identification of the one appropriate entry connector to be associated with each exit connector.

To facilitate finding the entry connector for each exit connector, the common practice is to use cross-references in the manner described previously. If any name has been assigned to a portion of the algorithm or program (that is, the program data structure), such as THRD, then this name may be entered above

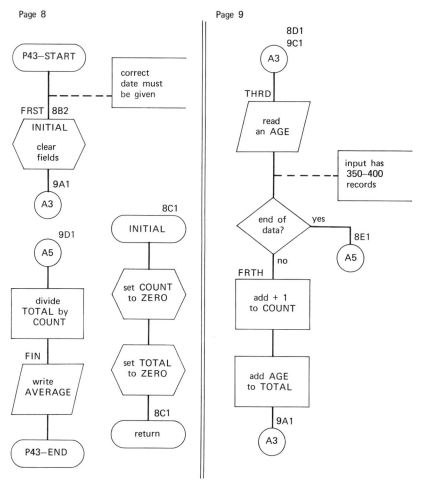

FIGURE 23 A flow diagram with connectors and cross-references.

and immediately to the left of the connector, or within the connector itself. If a coordinate plan has been established for identifying portions of the flow diagram, then the location indication can be entered above and immediately to the right of the connector, as shown in Figure 23.

The standard does not specify any particular coordinate convention for use specifically in flow diagrams. As a practical matter the most common convention used is a page number followed by a row and then a column designation (see Figure 24 for an example). Some omit the row designation and some omit the column designation. Some use a sequential count to provide an index of the position within a row or within a column of a connector position. Anything that begins with a digit causes no confusion since names in most programming languages may not start with a numeral.

The symbolic name and cross-reference notations may be used with any outline, not just with connectors. They are needed for all horizontally striped outlines, as a minimum, as shown in Figure 23. The names also provide a way of cross-referencing any part of the flow

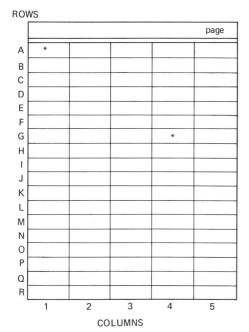

FIGURE 24 One cross-referencing scheme. *Example: thus, 9A1 is on page 9, row A, column 1; and 9G4 is on page 9, row G, column 4.

sequence shown in the flow diagram to the actual program coding. This is illustrated in Figure 23 by the names FRTH and FIN, for example, which refer to portions of the program respectively, even though they are not entry points.

Notation. The standard does not specify any particular language, symbols, or notational scheme for use within the outlines to identify the data or to name operations. Ordinary English prose tends to be too verbose to be readily accommodated within small-size outlines. Yet using small-size outlines yields a more easily comprehended flow diagram. A notational scheme or set of symbols that is compact and that permits easy representation of the common situations is highly desirable.

To this end, some years ago the American Standards Association (now ANSI) circulated a working paper advocating a notation or set of characters (graphics) for such a use (references). Many users of flow diagrams prefer to use the same notation they use in the programming language. Although this works well with some computationally oriented languages on computationally oriented jobs (such as FORTRAN or PL/I in engineering or scientific work), it falls short of the need for string operations, complex operations on arrays, and manipulations of all types of structures in logical terms. To meet these problems, several motational schemes have been advanced in the literature, of which the Iverson notation is probably the most widely known [16].

An eclectic list drawn from these three major sources is offered in Figure 25. This list is composed from graphics included in the ASCII, EBCDIC, and IBM BCD codes.[3] Hence, computers can print these in computer-drawn flow diagrams. The one exception is the arrow, since none of these codes includes arrows. The arrow, however, is still widely found where communications and display equipments are used, and has a history of use in programming work. It provides a neater alternative, especially for publication, than does a double graphic.

Of particular importance are the "is replaced by" symbols. Two alternative symbols are common, and serve to indicate that the symbol on the left has its value determined by what is on the right, as illustrated in Figure 26. For this purpose, the equal sign has sometimes been used. Such usage is inconsistent with mathematical practice, and gives a dual role to the equal sign, the other role being to indicate equality, as for example, from a comparison.

[3]American Standard Code for Information Interchange, published by ANSI; Extended Binary Coded Decimal Interchange Code; and Binary Coded Decimal.

Symbol	Meaning	Symbol	Meaning
:= , ←	is replaced by	≠ , ¬=	is not equal to
+	plus, or addition	≮ , ¬<	is not less than
−	minus, or subtraction		
*, ×	multiplication	─ , ⏉⏊	underline, blank; absolute value
**, ↑	exponentiation	¬ , ~, −	negation
/	division		
EOF	end of file	|	logical OR
:	comparison	&	logical AND
>	is greater than	' '	literal (zero level address)
=	is equal to	( )	grouping, level of address
<	is less than	A( )	address constant
≯ , ¬>	is not greater than		

FIGURE 25 Symbols for use in flow diagrams.

For comparison, a colon is common. The variable of comparison is shown on the left, and the standard or constant of comparison is shown on the right side of the colon. The exit flowlines from a decision outline must be provided with an indication of the basis for their choice, expressed in terms consistent with the notation used within the decision symbol outline, as shown, for example, in Figure 27.

Parentheses can indicate grouping, a usage borrowed from mathematics. Another use is to indicate levels of addressing. Most literals are enclosed within prime marks (single quote marks) to indicate the zero-level addressing status. Numeric literals, when they are to be used in arithmetic operations, are sometimes shown without enclosing prime marks. Nonnumeric character combinations appearing without the prime marks are assumed to be first-level addresses—that is, the names of items of data, such as a field or variables.

The notation for second and higher levels of addressing is to enclose in successive pairs of parentheses, one for each additional level of addressing desired. A special variant of this is the address constant—that is, something whose value will be determined by its machine language address at the time of execution. An A in front of the parenthesis can serve this purpose.

The use of a terse notation such as summarized here permits considerably greater amounts of material to be shown within each outline, or smaller size outlines to be used in a flow diagram. In both cases, an improvement in the communication value of the flow diagram typically results.

Guidelines for Flow Diagrams. A first guideline is to chose the wording or symbols within the outlines to fit the readers of the flow diagram. This depends in a major part upon the level of detail to be shown. The more summary (less detailed. this is, the more difficult is it to find a satisfactory wording or symbols to use within the outlines. As a general rule, whatever is chosen should be terse in order to permit the use of small size outlines.

A second guideline is to be consistent in the level of detail shown in the flow diagram. If some parts of the flow diagram are in great detail and others are only sketchy, the statement of the algorithm is distorted. A consistent level of

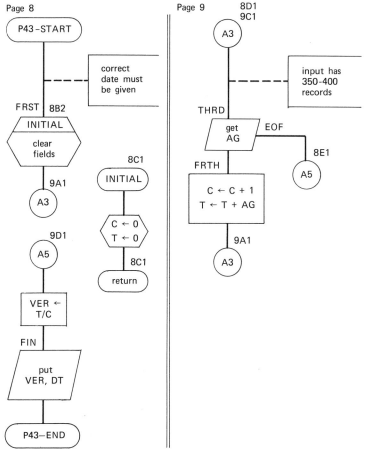

FIGURE 26 The flow diagram of Figure 23 redrawn using symbols within the outlines.

detail provides a sounder basis for making judgments about the algorithm and presents a better basis for making estimates of computer time, programming time, and conversion difficulties and for debugging than does a fluctuating level. Maintaining a consistent level of detail is simple only when the level of detail matches the implementing programming language. The difficulty comes with flow diagrams at summary and intermediate levels of detail.

A third guideline is to use identifying names consistently. Given the type and level of symbols for use within the outline, names for data and operations should be used uniformly and accurately.

Figures 21, 26, and 27 illustrate good practice, for example.

A fourth guideline is to use cross-references liberally in the flow diagram. Cross-references both to the program and to locations in the flow diagram improve the communication value of the flow diagram. To keep connectors small and to use space efficiently, it is helpful to use a location cross-reference inside the connector outline, and the program name cross-reference outside the connector, as illustrated in Figures 26 and 27.

A fifth guideline is to keep the flow diagram simple and clean. Clutter and lack of "white space" decreases the

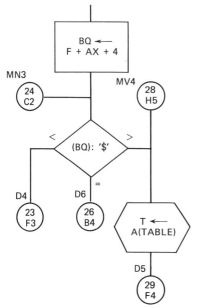

FIGURE 27 Entry and exit flowlines in a flow diagram.

communication value of the flow diagram. For convenience and clarity, spacing the diagram so it can be typed (if it is prepared by manual means) is a real assistance.

A sixth guideline is to keep clearly separate the operations to be performed on program data structures from those to be performed on operand data structures. That is, operations on the program itself, such as switch settings, indexing, initialization of program control variables, and the like, should be shown in preparation outlines, separate from operations that transform input into output data. This guideline helps make a flow diagram more easily understood, and it improves debugging work. To facilitate a clear separation, some people add a mark to the basic outlines that specify operations on the program data structure. If used, the mark should be one that is easily detected in a quick scan of the flow diagram. Examples are a large round spot within the outline, or a bold left edge for the outline, or a shaded upper left or center corner of the outline. None of these techniques are sanctioned by the standard; only the second is a violation of the standard, and hence should not be used.

A seventh guideline is to avoid using successive connector outlines. If more than one connector should properly appear in a series (as when multiple names are assigned to one entry point, or when the program calls for consecutive unconditional transfers of control), good practice is to collect the connectors to the left or above the line of flow. A tree arrangement (like an upside-down version of that used for multiple exits from a decision outline, as shown in the top center of Figure 6) can also be used.

An eighth guideline is to observe consistently the general flow pattern from top to bottom and from left to right. When this guideline is observed, arrowheads may be omitted from the flowlines that conform to the guideline. Clarity in a flow diagram is improved by arranging the main flow to conform as much as possible to the guideline. Minor and alternative flows may then deviate from normal, and by this deviation can be identified as not part of the main flow.

A ninth guideline is to draw entrances at upper left and exits at lower right. Entry and exit connectors are most noticeable if they are in a consistent position. Thus, the usual practice is to place an entry either above or to the left of the line of flow it is to join, and to place an exit either below or to the right of the line of flow it comes from, as shown in Figure 27. For both cases, when set to the left or right, then the flowline goes from left to right. If set above or below, the flowline drops down to or from the line of flow. Partial exceptions may be made to maintain symmetry, as when a decision outline has three exits, as exampled by exit D4 in Figure 27.

A tenth guideline is to draw flowlines so that they enter and exit at the visual centers of the outlines. The outlines usually possess either vertical or horizontal symmetry, typically about their center points. If therefore entrance and exit flowlines be drawn vertically and horizontally so that they appear to point toward or emerge from the center point of the symbol, the visual appearance of the flow diagram is improved.

An eleventh guideline is to use connectors and cross-references to avoid excessive crossing flowlines. The reasons for this have already been presented above.

A twelfth guideline is to draw only one entrance flowline per outline symbol. This is especially important in summary and intermediate levels of flow diagram. When more than one operation is specified or implied within an outline symbol, multiple entrance flowlines could mean that the sequence of events within the outline symbol may be different for the respective entrances, or even that not all the operations are to be performed for one or more of the entrances. To avoid this ambiguity, the usual practice is to make the flowlines join prior to entry into the outline, as shown in Figure 27. An alternative approach lacking conformity to the standard and not free of ambiguity is to subdivide an outline (see, for example, [9]). When the notation within the outline symbols is at the same level of detail as the program itself, no practical difficulty arises. For consistency, however, this guideline should be observed at all levels of detail.

A thirteenth guideline is to draw, with four exceptions, only one exit flowline for each outline. One exception is the decision outline which by definition must have multiple exits. An example is shown in Figure 27. The second is the input-output outline when an end-of-file or end-of-data condition may result in a failure to read input data. The notation within the outline should specify clearly the basis for the choice of the exit. The other exceptions are the extract and collate outlines in system charts.

A fourteenth guideline is to identify clearly all multiple exits. This is an extension of the previous guideline, and is illustrated in Figure 27.

A fifteenth guideline is to make no violations of the standard, and to shun deviations. This guideline is explained above.

Preparation of Flow Diagrams. The preparation of flow diagrams is primarily a heuristic matter. It is one for which no hard and fast rules can be presented.

A basic heuristic approach to the preparation of a flow diagram uses the "in-do-out" pattern. In order for the computer to produce output data, it must have some input data to operate on. It cannot perform the processing until the input data have been made available. Hence, the general character of a flow diagram is a sequence involving the following steps: preparing to read input data, reading input data, processing the input data read, producing the output data, and ending the processing actions.

This is sometimes referred to as initialize, process, and cleanup. But in turn, each stage itself has these elements usually also present. Thus, in order to do the output operation, it is necessary to do some initialization, to do the actual output, and then to do a cleanup.

A second common heuristic approach to assist in preparing flow diagrams is to assume that a job is really much simpler than it actually is. Then a flow diagram is prepared for this simpler job. After that, more realism is added by dropping some of the simplifying assumptions, and the flow diagram is redrawn to reflect this closer approximation to the real job. This two-step process is repeated as many times as needed. Each successive flow diagram comes closer to the actual job.

A third common heuristic approach is to draw a series of flow diagrams, each one at a greater level of detail. The first flow diagram is very general and provides only an overview (a "first cut" at a job). Then each successive flow diagram amplifies and provides additional detail. Each expands each of the outline symbols in the earlier flow diagram into a sequence of outline symbols. The final diagram, then, is near or at the same level of detail as the programming language that is to be used.

PROBLEMS IN THE USE OF THE STANDARD

The evidence is clear that thus far, people in the computer field do not use the standard. For example, the author of this paper made a survey of the 1968 and 1969 issues of five periodicals, sixteen books published in 1968 and 1969, a sample of a dozen computer installations, and a sample of five flowcharting software packages. The evidence from each supported the others.

The periodicals surveyed had surprisingly few articles using flowcharts. The periodicals were the *Communications of the ACM, Computers and Automation, Datamation, Business Automation,* and the *Journal of Data Management.* The books included texts on programming, and conference proceedings, as well as specialized and general survey books. The installations covered a range of sizes from small to very large. The software packages included the best-known flowcharting programs.

The conclusion was that none of the installations even attempted to use the standard, and none of the periodicals had any article free of violations of or deviations from the standard. Only one of the books used the standard free of significant deviations, but even it used the 1963 version of the standard. Only one of the flowcharting software packages offered the standard, and that only as a non-default option. Most flowcharts observed in the survey followed the computer vendor recommendations, especially the 1964 recommendations from IBM.

This lack of use of the standard suggests the existence of a number of problems. These fall in four main areas: ignorance, construction, conception, and applicability. A brief look at each may help clarify the reasons behind some of the material presented earlier in this paper.

Ignorance of the standard is widespread. This ignorance has generally been abetted by the professional societies, the computer vendors, and ANSI. For example, ACM, which even has a policy of supporting standardization work, has published for the information of its membership and for the guidance of the professional community, only the 1963 version of the standard. It has published to date none of the revisions. Nor does it require that all flowcharts in its publications conform to the standard. The action of the computer vendors is noted above. It is noteworthy that IBM in 1969 began the issue of a new template [14] that can be easily used to prepare flowcharts conforming to the standard. ANSI, by its refusal to grant free reprint rights and by its high charges per page for copies of the standard, severely inhibits the spread of knowledge about the standard. This is a case where ANSI's need for income to support its work results in hindering its work.

Some of the construction problems are closely related to the ignorance problem just noted. The two most common violations of the standard are in the ratio dimensions of the outlines and in the use of deviant outline shapes. Some would-be users of the standard are just ignorant; others in an attempt to make better use of space are willful violators. In the latter respect, the two most common complaints are that the standard wastes space and that it fails to provide enough space. People cite the "seven-page program that takes a twenty-page flow diagram" and claim it is faster and easier to read the program listing. Although cross-referencing in the diagram helps, it is not the entire solution. Summarization is a solution, but this gets into conception problems. On the charge of wasting space, many would-be users of the standard fail to draw outlines of differing size (while preserving the ratio dimensions) to fit the wording or symbols that go within the outline. This is especially true for decision outlines and connectors, as a practical matter.

The conception problems are more serious. Thus, it is widely agreed that a flow diagram is most useful if it is not as detailed as (is more summary than) the program it describes. To create these requires compressing, condensing, and eliminating details. But which ones? And how many? A poor choice can render the resulting flow diagram nearly useless. Being more summary also increases the difficulty of providing useful cross-referencing between parts of the flow diagram and between the diagram and the program.

This level of detail problem affects the rigor and completeness of a flowchart. Prepared in full detail, everything must be present in its proper place. All ties together. When full detail is absent from the flowchart, it becomes difficult or even impossible from the flowchart itself to determine whether a particular process or operation is correctly shown or is essential. This difficulty arises particularly for decision outlines in flow diagrams and for process outlines in system charts.

The applicability problems are major. A minority of people in the field hold that a flowchart is a waste of time. Since neither programmers nor analysts, they claim, think in terms of flowcharts, why prepare them? Preparing and trying to read them just divert people from a more productive use of their time. It is this author's observation that this indeed is true for some people, but the documentation value of flowcharts for other people's use still justifies their preparation and maintenance. Along this same line, others claim that alternative techniques offer superior documentation, for example, decision tables [8], logic flow tables [20], Ivenson notation [16], or networks or directed graphs. Further, conceptually the standard could be applied to prepare other types of graphic aids than the system chart and the flow diagram. So far, these have yet to be defined in the literature in clear form.

These objections and alternatives have indeed a measure of truth. This can be seen in several aspects of the flowchart. First, the flowchart is weak in showing the timing of processing and of data availability or need. This lack of showing "when" is very serious for communications oriented systems and programs, for example. Second, the flowchart does not directly tell "why"; it tells instead "what" and "how." It leaves the "why" to human inference. Third, the flowchart does not tell "how much." This cannot usually even be inferred. Fourth, the flowchart does not fully tell what or who does or is to do something. It has elaborate outlines for identifying peripheral equipment and media, but no way of indicating people, other computers, or parallel or alternative parts of a given computer as performers of actions on data. It is doubtful if any attempt should be made, beyond the use of annotation, to extend the flowchart beyond what it is already. It cannot be all things to all analysts and programmers; but it does have a place.

CONCLUSION

Both the system chart and the flow diagram provide important tools for the analyst and the programmer in preparing and in documenting work for implementation on a computer. Carefully prepared, flowcharts can enhance the rigor with which the analyst and programmer think through the systems, programs, and associated procedures. This greater rigor in turn typically reduces the cost of debugging. But the major contribution of the system chart and the flow diagram are to communicate some essential aspects of information processing work from one human being to another. The quality of this communication is enhanced by a consistent use of the standard.

APPENDIX
DRAWING OUTLINES

Terms

Ratios. The dimensional ratio of the outlines defined in the standard is determined by the following procedure. Construct a rectangle circumscribing the outline. The rectangle must be formed from vertical and horizontal lines, and each line should just touch the inscribed outline. The ratio is then determined by measuring the horizontal and the vertical sides of the circumscribing rectangle, and is usually defined as the height as a proportion of the width taken as unity.

Sizes. The standard permits outlines of any size as long as the dimensional ratio and general configuration are maintained.

Lines. The standard indicates the use of any uniform width or weight of line for all of the outlines, irrespective of their sizes. Lines need not be continuous, but may be created by the close spacing of discrete symbols.

Configuration. The standard specifies that the outlines in shape should conform sufficiently close to the configurations shown in the standard to permit rapid and unambiguous identification. The curvature of lines and the magnitude of angles may vary from those shown in the standard, provided that the shapes still be clearly recognizable. To that end, since the angles and curvatures shown in the standard are sometimes difficult to use in drawing, the configurations and description given in this paper sometimes simplify or round off to make the drawing easier, within the restrictions imposed by the standard.

Orientation. The figures in this paper illustrate the general orientation specified by the standard for each outline. Outlines may not be turned. Each outline except the connector and the decision outlines has at least one straight line which must be either vertical or horizontal, as illustrated in the figures. The connector outline has no specified orientation; the decision outline is horizontally oriented along its widest dimension. Flowlines should normally be vertical or horizontal, and may make right-angle bends. Flowlines deviating from horizontal or vertical are neither recommended nor proscribed by the standard.

Outlines

Input-Output. This is a parallelogram with its base edge 15 degrees further to the left than its top edge. It has a width-to-height ratio of 1 to $\frac{2}{3}$, as illustrated in Figure 1.

Process. This is a single rectangle with a width-to-height ratio of 1 to $\frac{2}{3}$, as shown in Figure 1.

Flowline. This may be of any length. The spread between the barbs of the arrowhead and the length of the arrowhead are each approximately ten times the width of the flowline, as shown in Figure 1. The angle of the barbs is $26\frac{1}{2}$ degrees from the flowline—i.e. the entire arrowhead just fits within a square.

Annotation. This is a rectangle with the right side missing, but with a width-to-height ratio of 1 to $\frac{2}{3}$, as shown in Figure 1.

Connector. This is a circle, as shown in Figure 2. In practice, it should be kept as small as possible. A size just large enough to accommodate four characters of wording is about typical.

Terminal. This has a width-to-height ratio of 1 to $\frac{3}{8}$ and looks like a slim rectangle but with half-circle ends, as shown in Figure 2.

Parallel Mode. This is two horizontal parallel lines of any equal length spaced about ten or so line widths apart, as shown in Figure 2. Entering the upper one may be one or more vertical flowlines positioned anywhere along the line. Leaving the lower one may be one or more vertical flowlines positioned anywhere along the line. The number of entrance and exit flowlines may not both be equal to one and neither may be equal to zero—i.e. one must have two or more flowlines, and one must have one or more flowlines.

Document. The top three edges are a portion of a rectangle, but the bottom edge is a curved line to represent a break in the paper. The width-to-maximum-height ratio is 1 to $\frac{2}{3}$, as shown in Figure 3. The width-to-height ratio of the rectangle portion is 1 to $\frac{1}{2}$. The radius of the left curve is one-half the width, and the center is one-fourth of the width in from the left edge. The radius of the right curve is one and one-quarter of the width, and the center is straight below the right edge of the outline.

Magnetic Tape. This is a circle with a horizontal rightward pointing line tangent to the bottom. The tail line, as shown in Figure 3, extends over to the point of intersection with an imaginary vertical line tangent to the rightmost point on the edge of the circle.

Punched-Card. The punched-card symbol has a width-to-height ratio of 1 to $\frac{1}{2}$ and appears generally like an upper left corner cut punched card, as shown in Figure 3. The corner cut has an angle of about 30 degrees and cuts off about one-sixth of both the width and height.

Punched-Tape. This has a maximum-width-to-maximum-height ratio of 1 to $\frac{1}{2}$, as shown in Figure 3. The centers of the curves are at the one-fourth points of the width; the radius of the curves are three-quarters of the width. The end lines are vertical. The top left corner and the bottom right corner are not more than 10 percent of the height in from imaginary horizontal lines tangent to the maximum points of the arcs.

Display. As shown in Figure 4, the width-to-height ratio is 1 to $\frac{2}{3}$. The radius of the curved lines is one-half of the width. The neck curves join the horizontal lines one-third of the width in from the left ends.

Manual Input. This outline, as shown in Figure 4, is a stylization of a cross section of a keyboard with the sloping surface (at an angle of about 10 degrees) of the keyboard having its lowest point to the left. This cuts about one-third off the height of the left vertical edge. The width-to-maximum-height ratio is 1 to $\frac{1}{2}$.

Communication Link. As shown in Figure 4, the zigzag has an angle of 45 degrees. The distance between the parallel line segments is from ten to fifteen times the width of the flowline. The lines may be of any length, and one or more zigzags may be located anywhere along the lines.

On-line Storage. This is a stylization of a portion of a cylinder with a convex end at the left and a concave end at the right, and having a width-to-height ratio of 1 to $\frac{2}{3}$ as shown in Figure 4. The ends are arcs with a radius equal to one-half of the width.

Disk Storage. This outline shown in Figure 4 is a reorientation of the on-line storage outline, but with three lines added, one convex on the top, and two to mark off bands. These are arcs with a diameter equal to the width of the equivalent on-line storage outline. The spacing of the bands is about one-tenth to one-twelfth of the width of the equivalent on-line storage outline. The overall width-to-height ratio is 1 to $\frac{5}{3}$.

Drum Storage. This outline, as shown in Figure 4, omits the two band arcs from the disk storage outline, and reorients it to match the on-line storage outline. The overall width-to-height ratio is $\frac{5}{3}$ to 1.

Core Storage. As shown in Figure 4, this is a square and hence has a width-to-height ratio of 1 to 1. The two lines, each parallel to a side, are in one-eighth of the width from the edge of the outline.

Off-line Storage. This outline is an equilateral triangle standing on a point. Since it is equilateral, it has a width-to-height ratio of 1 to 0.866 (about 15 to 13), as shown in Figure 4. The small line drawn about six-tenths or eight-thirteenths of the distance from the top to the bottom tip of the triangle is a required part of the outline.

Decision. As shown in Figure 5, this is a diamond outline with a width-to-height ratio of 1 to $\frac{2}{3}$.

Preparation. The horizontal lines, as shown in Figure 5, are two-thirds of the total width. This makes the angle of the four sloping sides fall at $26\frac{1}{2}$ degrees from the vertical, since one-sixth of the total width is missing from each end of the horizontal lines. This gives a width-to-height ratio of 1 to $\frac{2}{3}$.

Predefined Process. This is a rectangle with vertical bars in about one-eighth to one-sixth of the width from the left and right ends. Figure 5 shows the lines at the one-eight position. The overall shape is the same as the basic process outline with a width-to-height ratio of 1 to $\frac{2}{3}$.

Manual. This is a keystone (trapezoidal) shaped outline, as shown in Figure 5, with a maximum width-to-height ratio of 1 to $\frac{2}{3}$. The slope of the sides is about 15 degrees. This makes the length of the bottom horizontal line equal to one-half of the width.

Auxiliary. This outline as shown in Figure 5 is a square.

Merge, Extract, Sort, and Collate. These are shown in Figure 5. All are constructed from equilateral triangles. Those that use two abutting triangles must use triangles of the same size for the two parts.

Card deck. As shown in Figure 9, this is an extension of the punched-card outline. The left corner outline is extended to give a height about one-fifth greater than that of the punched-card outline. The topmost horizontal line is extended about one-eighth of the width beyond the right end of the punched-card outline embedded in this outline, and the new right vertical is made slightly shorter, as shown in Figure 9. This gives an overall width-to-height ratio of about $\frac{5}{4}$ to $\frac{2}{3}$.

Card file. This is like the card deck outline but with parts of the top, left slant (corner cut), right slant, and right vertical erased. The position of the top and right lines is determined in the same manner as for the card deck outline. The remaining upper right right angle has lines equal in length to about one-half of the height of the punched-card outline embedded in this outline.

REFERENCES

1. American Standards Institute.[4] Proposed American Standard flowchart symbols for information processing. *Comm. ACM* 6, 10 (Oct. 1963), 601–604.

[4] As of 1969, American National Standards Institute (ANSI).

2. ANSI. Standard flowchart symbols and their use in information processing (X3.5). American National Standards Institute, New York, 1970 (20 pp.).
3. ———. Standard graphic symbols for logic diagrams (Y32.14). American National Standards Institute, New York, 1962 (12 pp.).
4. ———. Standard method of charting paperwork procedures. American National Standards Institute, 1959 (12 pp.).
5. USA Standards Institute.[4] Graphic symbols for problem definition and analysis—a standards working paper. *Comm. ACM 8,* 6 (June 1965), 363–365.
6. Association for Systems Management. Charting. In *Business Systems,* Assoc. for Systems Management, Cleveland, Ohio, 1963, Ch. 5.
7. Chapin, Ned. *An Introduction to Automatic Computers.* Van Nostrand, Princeton, N.J., 1957 (525 pp.) and 1963 (503 pp.).
8. ———. An introduction to decision tables. *Data Proc. Manag. Ass. Quart.* 3, 3 (April 1967), 2–33.
9. Forsythe, Alexandra I., et al. *Computer Science.* Wiley, New York, 1969 (533 pp.).
10. Fritz, W. Barkley, et al. (SHARE's ad hoc Committee on Flow Chart Symbols). Proposed standard flow chart symbols. *Comm. ACM 2,* 10 (Oct. 1959), 17–18.
11. Goldstine, H. H., and von Neumann, John. *Planning and Coding Problems for an Electronic Computing Instrument, Vols. I, II, III.* Van Nostrand, Princeton, N.J., 1947, and 1948.
12. Gruenberger, Fred J., et al. *Introduction to Electronic Computers.* Wiley, New York, 1963 (167 pp.).
13. IBM Corporation. Flowcharting techniques, C20–8152. IBM Corp., New York, 1964 (34 pp.).
14. ———. Flowcharting template, X20–8020. IBM Corp., New York, 1969 (one plastic cutout drawing guide in a printed envelope).
15. ———. Problem planning aids, IBM Type 650. IBM Corp., New York, 1956 (18 pp.).
16. Iverson, Kenneth E. *A Programming Language.* Wiley, New York, 1962 (286 pp.).
17. Lewis, F. David, et al. Program 1401–2.0.019. IBM Corp., New York, 1963 (18 pp.).
18. McCracken, Daniel D., et al. *Programming Business Computers.* Wiley, New York, 1959, pp. 26–43 and *passim.*
19. RCA Service Company. The Language and Symbology of Digital Computer Systems. RCA, Camden, N.J., 1959 (114 pp.).
20. Self, Sidney B. Logic flow table. *J. Data Manag.* 5, 12 (Dec. 1967), 30–36.
21. United States Air Force. AFICCS Documentation Standard Manual TEM-AF-2. US Air Force, Washington, D.C., 1968 (126 pp.).

[4] As of 1969, American National Standards Institute (ANSI).

DECISION TABLES —WHAT, WHY AND HOW

RAYMOND M. FERGUS

PURPOSE OF THE PAPER

The purpose of this paper is to present a pragmatic discussion of the nature, value, and construction of decision tables. This will be a discussion for those people who want to consider in everyday terms such items as:

- What are tables and how are they read?
- Where can tables be used and how will they affect application systems and programming?
- How can decision-table use be undertaken in an organization? Steps to take, problems to expect, and benefits to gain.

- Some solid illustrations of what we are talking about, and perhaps just a little blue sky as to how tables might be of value in the future.

In short, this paper is intended for you if you want to consider in a very practical sense, from an application point of view, what decision tables are, why they are of value, and how they can be used.

INFORMATION SOURCES

Many of the ideas given here reflect the findings of a decision-table research project I conducted while at Collins Radio Company. The research project involved several different study techniques:

1. Literature search. Over thirty different published and unpublished documents were reviewed to gain a basic familiarity with decision tables and to advance our knowledge concerning them.
2. User contact. Experienced users of decision tables were contacted for their suggestions and comments. In general, the people contacted had extensive experience with de-

SOURCE: Fergus, R. M., "Decision Tables— What, Why and How," *Proceedings, College and University Machine Records Conference,* University of Michigan, 1969, pp. 1–20.

cision tables since the early 1960's. The individuals were extremely helpful and their contributions proved invaluable. (While I owe gratitude to many, I would especially like to recognize and thank Roger Bornemen, Orren Evans, Donald Fisher, Catherine McDermott, Peter Piechocki, and Sol Pollock.)
3. Testing. About a dozen people with varying degrees of data processing experience were involved in using decision tables to either document or understand simulated and real problems. Emphasis was placed on identifying and diagnosing difficulties in the use of tables.
4. Thinking. The results of the literature search, user contact, and testing were organized, reviewed, and evaluated in order to produce meaningful results.

This presentation, then, represents a synthesis of the thoughts and experiences of many people, including myself. I hope that our observations can be of real practical value to you.

WHAT IS A DECISION TABLE?

A decision table is a tabular representation of

1. Conditions: factors to consider in making a decision.
2. Actions: steps to be taken when a certain combination of conditions exist.
3. Rules: specific combinations of conditions and the actions to be taken under those conditions.

These items are contained within the body of a decision table, which is divided into four major sections by double or heavy vertical and horizontal lines. The various parts normally making up a table are

1. Condition statements: statements which introduce one or more conditions.
2. Condition entries: entries that complete the condition statements.
3. Action statements: statements which introduce one or more actions.
4. Action entries: entries that complete the action statements.
5. Rules: unique combinations of conditions and the actions to be taken under those conditions.
6. Header: a title and/or code identifying the table.
7. Rule identifiers: codes uniquely identifying each rule within a table.
8. Condition identifiers: codes uniquely identifying each condition statement/entry.
9. Action identifiers: codes uniquely identifying each action statement/entry.
10. Notes: comments concerning the contents of the tables. Notes are not required, but might be used to clarify some items recorded in the table.

The parts of a decision table are diagrammed in Figure 1.

DECISION TABLE EXAMPLE

An example of decision table format is presented here by showing a procedure documented in narrative form and then in decision table form.

Narrative Form:
Credit Order Approval Procedure

If a customer's credit limit is okay, then the credit request may be approved. If the credit limit is not okay but the customer's pay experience is favorable,

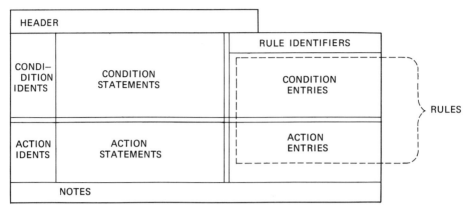

FIGURE 1 Decision table parts.

then the credit request may be approved. However, if the customer's credit limit is not okay and the pay experience is unfavorable, credit should not be approved and the order should be returned to the sales department. (Be certain that the latest credit information is used.)

Decision table form

As seen in Figure 2, rules (vertical columns R1, R2 and R3) are the guides for a decision table user. Conditions are matched against those stated in the table, allowing a specific rule to be located which then identifies the actions to be taken.

TYPES OF DECISION TABLES

There are three types of decision tables:

Limited Entry Table

In a limited entry table (Figure 2) the condition and action statements are complete. The condition and action entries merely define whether or not a condition exists or an action should be taken.

The symbols used in the condition entries are:

Y	Yes, the condition exists.
N	No, the condition does not exist.
– (or blank)	Irrelevant, the condition does not apply, or it makes no difference whether the condition exists or not.

CREDIT ORDER APPROVAL PROCEDURE				
		R1	R2	R3
C1	CREDIT LIMIT OKAY	Y	N	N
C2	PAY EXPERIENCE FAVORABLE	–	Y	N
A1	APPROVE CREDIT	X	X	–
A2	RETURN ORDER TO SALES	–	–	X
NOTE: BE CERTAIN THAT THE LATEST CREDIT INFORMATION IS USED.				

FIGURE 2 Decision table example (limited entry).

FIGURE 3 Extended entry table.

CREDIT ORDER APPROVAL PROCEDURE		R1	R2	R3
C1	CREDIT LIMIT	OK	NOT OK	NOT OK
C2	PAY EXPERIENCE	–	FAVORABLE	UNFAVORABLE
A1	CREDIT ACTION	APPROVE	APPROVE	DON'T APPROVE
A2	ORDER ACTION	–	–	RETURN TO SALES

FIGURE 4 Mixed entry table.

CREDIT ORDER APPROVAL PROCEDURE		R1	R2	R3
C1	CREDIT LIMIT OKAY	Y	N	N
C2	PAY EXPERIENCE	–	FAVORABLE	UNFAVORABLE
A1	CREDIT ACTION	APPROVE	APPROVE	DON'T APPROVE
A2	RETURN TO SALES	–	–	X

The symbols used in the action entries are:

- X   Execute. Take the action specified by the action statement.
- – (or blank)   Don't execute the action specified by the action statement.

Extended Entry Table

The condition and action statements in an extended entry table (Figure 3) are not complete, but are completed by the condition and action entries.

Mixed Entry Table

The third type of decision table is the mixed entry form (Figure 4) which combines both the limited and extended entry forms. While the limited and extended entry forms can be mixed within a table, only one form may be used within a condition statement/entry or an action statement/entry.

DECISION TABLE FORMATS

Two basic formats for decision tables are the vertical format and the horizontal format. The vertical format has the rules displayed vertically on the right side of the vertical double or heavy line. The tables illustrated in Figures 1–4 have vertical formats. A horizontal format displays the rules horizontally below the horizontal double or heavy line. Figure 5 shows a limited entry table in a horizontal format.

The logic expressed in the two formats is exactly the same. The vertical format is well suited for the display of many conditions and actions. The horizontal format allows the reader to read the tables from left to right. Vertical is probably the more popular format.

GUIDELINES AND EXAMPLES FOR TABLE CONSTRUCTION

Some general guidelines and examples for constructing decision tables are given in this section.

CREDIT ORDER APPROVAL PROCEDURE				
	C1	C2	A1	A2
	CREDIT LIMIT OKAY	PAY EXPERIENCE FAVORABLE	APPROVE CREDIT	RETURN ORDER TO SALES
R1	Y	—	X	—
R2	N	Y	X	—
R3	N	N	—	X

FIGURE 5   Decision table—horizontal format.

Table Size

When people start to use decision tables, they tend to make the tables too large. Often several small tables are better than one large table. Experience will help an individual to decide what the "best" size is for the job he is doing. As a tool of communication the table should be of a size that facilitates understanding. In one case that might mean a table of only one or two conditions and actions, while in another case, ten or more conditions and actions might be desirable.

Rules

Several points concerning decision rules in tables are:

1. Every rule must have an action entry associated with it. That is, a rule must state the action to be taken under a given set of conditions.
2. Tables can be just action tables consisting of a single rule; i.e., the only condition satisfied is that the table is entered. Such a table is illustrated in Figure 6. (The condition "enter" implies that a decision maker/reader has entered the table from somewhere outside the table. That being the case, the indicated actions are then taken.)
3. Rules are unique and independent; they do not duplicate or contradict one another. Therefore, only one rule will apply in any given situation.
4. The sequence of rules is immaterial, since only one rule can satisfy the conditions in a given situation. As will be discussed later, conditions and rules might be sequenced to aid coding efforts. However, rules would remain unique and independent and could actually be considered in any order.
5. Within a rule an "and" relationship exists among applicable conditions and actions. That is, for a rule to be satisfied the first applicable condition *and* the second applicable condition *and* the third . . . must exist. Likewise all of the applicable actions must be taken.
6. Within a rule an "if-then" relationship exists between conditions and actions. That is, a rule implies that *if* certain conditions exist *then* certain actions are to be taken.

END OF JOB ROUTINE		
		R1
C1	ENTER	Y
A1	CLOSE FILES	X
A2	PRINT EOJ	X
A3	HALT	X

FIGURE 6   Action table.

SWITCH/CODE TEST				
		R1	R2	R3
C1	SWITCH	ON	ON	OFF
C2	CODE	1	2	1 or 2
A1	SET INDICATOR	A	B	C

FIGURE 7 Example of an "or" condition.

INDICATOR SET			
		R1	R2
C1	CODE A	1	3
C2	CODE B	1	3
A1	SET INDICATOR	1	2

FIGURE 8 Example of an "or" condition.

"Or" Relations

There are methods for expressing "or" relationships among conditions within a rule. One method is to use an extended entry format (R3/C2 of Figure 7).

Another method is to place rules with "or" conditions next to a second set of double lines. Within those rules an "or" relationship is assumed between the conditions. Therefore, whenever one of the conditions is met the rule applies. For example, rule 1 in Figure 8 states that if Code A and Code B are both 1, set indicator 1. Rule 2 states that if Code A *or* Code B is 3, set indicator 2.

"Not" Conditions

Sometimes it is advantageous to express a negative condition. One method of expression is to place a bar (—) over the negative entry. Rule 5 in Figure 9, for example, states that if the quantity is greater than 100 and the item value is *not* $0.01–$9.99, then the special order procedure should be utilized. (Note how multiple values for condition C2 are shown in this mixed entry table.)

"Else" Decision Rule

The "Else" decision rule specifies the actions to be taken when none of the other rules in the table apply (Figure 10). This rule might be used to handle exceptions or some rarely occurring situations that require the same actions.

Sequence of Actions

The sequence in which actions are performed within any rule can be specified in several different manners. Normally in a vertical format table actions are executed in the top-to-bottom order in which they appear. Occasionally certain rules require that a different sequence be followed. This requirement can be met by writing the affected action statements two or more times (Figure 11). The requirement can also be met by using numeric action entries rather than X's to indicate actions to be taken and their sequence (Figure 12).

Multi-Applicable Entries

Condition or action entries applying to two or more adjacent rules can be shown by having the entry cross through the appropriate rules (Figure 13).

Multiple Tables

Often multiple tables will be needed to document an application. Some action statements that might be used are:

1. GO TO. An action command that tells what table to go to next to continue the process. (This command might send you back to the same table.)
2. DO (or PERFORM). A command that directs you to another table from which you will return to the next sequential action in the originating table.

ORDER PROCEDURE		R1	R2	R3	R4	R5
C1	QUANTITY > 100	N	Y	N	N	Y
C2	ITEM VALUE (1) $0.01–$9.99 (2) $10.00–49.99 (3) $50.00–99.99	1	1	2	3	1
A1	AUTOMATIC ORDER	X		X		
A2	SUPERVISED ORDER		X		X	
A3	SPECIAL ORDER					X

FIGURE 9 "Not" condition example.

3. RETURN. A command that directs you back to the next sequential action following a DO command in an originating table.

Figure 14 presents four tables which illustrate these commands while giving a procedure for crossing the street and going in a house. A narrative of the procedure follows:

Crossing Street and Going in House Procedure. If the walk light is not on wait at the curb until it is on. Once the walk light is on check for moving traffic. If traffic is moving wave your arms, blow your whistle, jump up-and-down, and again check for moving traffic. Continue this activity until traffic stops, at which point step down from the curb, walk across the street, step up on the curb, and check to see if the house door is open. If the door is not open wave your arms, blow your whistle, jump up-and-down, and again check the door. Continue this activity until the door is open, then step in the house and relax.

SOME SYSTEMS/ PROGRAMMING CONSIDERATIONS

Decision Tables Compared with Flowcharts

Since the purposes of decision tables and flowcharts are somewhat similar, some pertinent features of the two tools should be compared. Flowcharts are especially useful for straight-forward programs where very little branching occurs; for showing the sequence of events where sequence is important; and for showing flow control from one element to another. Some of the criticisms directed at flowcharts include:

1. They are difficult, clumsy, and laborious to draw and are heavily influenced by personal preference and jargon.

FILE ACCEPTANCE		
	R1	ELSE
FILE IDENTIFICATION	CORRECT	–
PROCESS DATE	CORRECT	–
PROCESS DATE	X	–
REJECT FILE	–	X

FIGURE 10 "Else" rule of example.

CHANNEL REQUEST				
		R1	R2	R3
C1	CHANNEL 1 REQUEST	NORMAL OR PRIORITY	PRIORITY	NORMAL
C2	CHANNEL 2 REQUEST	NORMAL	PRIORITY	PRIORITY
A1	SERVICE CHANNEL 1	X	X	
A2	SERVICE CHANNEL 2	X	X	X
A3	SERVICE CHANNEL 1			X

FIGURE 11 Action sequence difference—action statement repeated.

CHANNEL REQUEST				
		R1	R2	R3
C1	CHANNEL 1 REQUEST	NORMAL OR PRIORITY	PRIORITY	NORMAL
C2	CHANNEL 2 REQUEST	NORMAL	PRIORITY	PRIORITY
A1	SERVICE CHANNEL 1	1	1	2
A2	SERVICE CHANNEL 2	2	2	1

FIGURE 12 Action sequence difference—numeric action entries.

CODED REQUEST RESPONSE							
		R1	R2	R3	R4	R5	R6
C1	CODE	1	2	3	1	2	3
C2	REQUEST	NORMAL			PRIORITY		
A1	SERVICE	NORMAL			PRIORITY		
A2	SET SWITCH A	ON	OFF		OFF		

FIGURE 13 Multi-applicable entries.

2. A diagram of a complex procedure is itself complex.
3. Clarity may not exist. Individual decision/action steps can be simple, but a series of such steps can be difficult to follow. The path from beginning to end through a flowchart can be confusing, since the action resulting from a particular condition depends on all prior condition/action segments.
4. It is difficult to determine if the total problem is covered, if all logical elements are defined and analyzed, and if superior alternative methods exist.

5. Flowcharts are difficult to revise.
6. Because of flowchart deficiencies, analysts/programmers accept poor problem statements, and stop trying to find "better ways." Programming of the computer is performed within that framework.

Decision table features that overcome some of the flowchart deficiencies are:

1. A standard format that is easy to draw.
2. Conditions and actions are presented in a clear and orderly manner.
3. Applications involving complex interactions of variables are documented more clearly than with flowcharts. In fact, as the complexity of an application increases, the advantage of decision tables increases. (At the same time tables can become clumsy if action segments are based only on a few simple conditions.)
4. Because condition testing is separated from action taking, it is easier to follow an exact path to completion and to check the table for completeness.
5. Tables are easy to revise and maintain.
6. Tables facilitate clear communication among the various people involved in the design and programming of a system. This promotes better problem definition and the seeking of "a better way" before computer programming actually occurs.

Decision tables are not the answer to all documentation and communication problems. However, tables do offer certain advantages that overcome some of the drawbacks of the flowchart technique.

TABLE CONSTRUCTION

Decision tables specify a set of conditions followed by a series of actions.

TABLE 1

		R1	R2
C1	WALK LIGHT ON	N	Y
A1	WAIT AT CURB	X	–
A2	GO TO TABLE 1	X	–
A3	GO TO TABLE 2	–	X

TABLE 2

		R1	R2
C1	TRAFFIC MOVING	Y	N
A1	STEP DOWN FROM CURB	–	X
A2	WALK ACROSS STREET	–	X
A3	STEP UP ON CURB	–	X
A4	DO TABLE 3	X	–
A5	GO TO TABLE 2	X	–
A6	GO TO TABLE 4	–	X

TABLE 3

		R1
C1	ENTER	Y
A1	WAVE ARMS	X
A2	BLOW WHISTLE	X
A3	JUMP UP-AND-DOWN	X
A4	RETURN	X

TABLE 4

		R1	R2
C1	DOOR OPEN	N	Y
A1	DO TABLE 3	X	–
A2	GO TO TABLE 4	X	–
A3	STEP IN HOUSE	–	X
A4	RELAX	–	X

FIGURE 14 Multiple table example.

TABLE 1		R1	R2	R3	R4	R5	R6
C1	SWITCH ON	Y	Y	Y	N	N	N
C2	CODE VALUE	1	2	3	1	2	3
A1	ACTION 1	X		X	X		X
A2	ACTION 2		X			X	

FIGURE 15 Unique condition combination example.

Flowcharts differ in that they generally consist of a decision followed by action(s) followed by a decision followed by more action(s), and so on. Therefore, decision tables are developed in a different manner than flowcharts.

To construct a table, start by writing or outlining the problem in detail, or by drawing a rough flowchart, or by making notes on the conditions and actions to be considered. Next, identify and list the conditions and actions as statements in table format. Finally, complete the table by making condition and action entries to form the table rules.

TABLE ANALYSIS

Decision tables can be analyzed to promote completeness and accuracy, and to detect excessive rules, contradictions, and redundancies.

Completeness

Tables can be analyzed for completeness by determining if the table accounts for the correct number of independent rules to cover all combinations of conditions that are possible. Since a rule covers a unique combination of conditions, the number of rules to be accounted for is equal to the number of unique condition combinations possible. Within a table, the number of possible condition combinations is equal to the product of the number of significant values each condition might assume ($v_1 \times v_2 \times v_3 \ldots =$ number of unique condition combinations). As an example, assume a table (Figure 15) which contains two condition statements. Assume that the first condition has two possible values ("yes" or "no"), and that the second condition has three possible values (Codes "1, 2, or 3"). This example then has the potential for six unique condition combinations or rules ($2 \times 3 = 6$).

If all the conditions given in the table have the same number of values, the number of rules needed to satisfy all condition combinations possible is the number of values raised to the power of the number of conditions ($v^c =$ number of unique condition combinations). For example, a table containing four conditions, each of which has only two possible values, has sixteen possible condition combinations ($2^4 = 16$).

When calculating the number of condition combinations/rules that a table should account for, several factors should be kept in mind:

1. Each rule that has all condition statements completed with explicit entries, counts as one independent rule.
2. Each rule that contains irrelevant condition entries (blanks or dashes) is equivalent to several rules. The number of equivalent rules for a stated rule is equal to the product of the number of values possible for the irrelevant condition entries given in the stated rule ($v_1 \times v_2 \ldots$

TEMPERATURE TEST		R1	R2	R3	R4
C1	TEMPERATURE < 40°	Y	N	N	Y
C2	TEMPERATURE > 80°	N	Y	N	Y

FIGURE 16 Impossible situation.

= equivalent number of rules). For example, assume a table with a stated rule contains four condition entries with two of the entries being dashes (irrelevants). If one of the irrelevants could have had a "yes or no" value (2 values) and the other irrelevant could have had a code 1, 2, 3, or 4 value (4 values), the stated rule is equivalent to eight rules ($2 \times 4 = 8$).

3. The "else" decision rule represents the difference between the number of possible condition combinations and the number that actually is expressed as stated or equivalent rules in the table.
4. It is not always feasible to construct a table containing all possible condition entry combinations. The table might become too unwieldy, and/or contain entries that would be illogical or unreal relative to the "real world" situation being documented. For example, the situation displayed in Figure 16 has four possible condition combinations. Obviously the condition combination stated in R4 is impossible. Only the first three rules are required to satisfy the "real world" system, but the four rules would be required to satisfy table logic. Even though there can be drawbacks, table completeness might best be promoted in some situations by initially displaying all condition combinations and then removing those combinations which are inappropriate.

Accuracy

Accuracy is checked by testing the rules of the table with real and simulated situations. This checking might take the form of desk checking, for instance. Analysts will be pleased to find that unlike flowcharts, decision tables can often be employed easily and directly with data processing users in order to check the accuracy of the documented logic.

Value of Analysis

It is important to realize that while table analysis provides an extensive examination of a decision table, it is only a partial analysis of the system represented by the table. For example, while table completeness checking promotes logical table completeness it does not assure logical real world completeness/accuracy. As an illustration, assume that a decision table describing the boiling of water gives the boiling temperature as 50 degrees Fahrenheit and makes no mention of atmospheric pressure. Table completeness checks would not uncover a temperature statement error or the omission of the atmospheric pressure condition. This does not mean that completeness checks are valueless, but it is important to recognize their limitations. The completeness check is a valuable tool and does provide a definite means for promoting and improving the logical completeness of decision tables. However, the check is performed within the framework of conditions as given in the table.

Another example of a possibly undetected flaw is table entries that can result in endless looping. In our earlier multiple table example we saw that the person wanting to cross the street or get in the house should jump up-and-down, blow his whistle, and wave his arms until traffic stops or the door is opened. If for some reason either traffic will not stop or the door will not open, the individual will be jumping, blowing, and waving endlessly. The answer is, of course, to see that the system and its decision table description contains actions that alter conditions such that endless loops do not develop. The real point here is that decision table checking might not reveal the real world problem of endless looping.

Decision table analysis can uncover table errors and will help reveal invalid or incomplete system logic or definition. However, analysis does not assure one hundred percent correctness. Analysis techniques promote completeness and accuracy; they do not guarantee it.

Program Conversion

Three methods used for converting decision tables to computer programs are: manual coding from tables, pre-processors that convert tables into source languages for input into existing compilers, and compilers that translate tables directly into machine language. An infrequently employed fourth method, the use of programs for interpreting decision tables, might prove of interest in future on-line situations.

Methods other than manual coding require special programs, and have generally been viewed as producing running programs that are inefficient in execution time and/or memory requirements. Various individuals and groups have been working on techniques for overcoming these deficiencies, and are apparently meeting with some degree of success. If the methods can be refined, they offer the potential for shortening and improving the line of communication which translates management thinking into operation computer programs.

Examples of decision table compilers are TABSOL and LOGTAB developed by General Electric; and FORTAB, a FORTRAN-oriented compiler developed by the Rand Corporation. A number of pre-processors have been written including:

- *Decision Logic Translator.* An IBM program that translates decision tables into FORTRAN source statements on the IBM 1401.
- *Autocoder Decision Table Assembly.* A program developed at United Gas of Shreveport, Louisiana, which translates to Autocoder Statements on the IBM 7070.
- *DETAB.* A joint venture of Systems Development Corporation and a special interest committee of the Association for Computing Machinery produced DETAB, which is a pre-processor for COBOL compilers.
- *PET* (*P*reprocessor for *E*ncoded *T*ables). This program, used by the Bell Telephone Company of Canada, was discussed by Bell Canada at the GUIDE 24 Conference in New York in May 1967. Using a PL/1 type decision table language, the program produces PL/1 source statements. Bell's studies indicate PET produces about one statement for every three of DETAB, and the efficiency of resultant programs is approximately comparable to those written by an above-average coder.

Manual coding is probably the most frequently used method of converting decision tables to computer programs, at least for new decision table users. For that reason, we will discuss some details of the manual coding technique.

Manual Coding

While some programmers code exclusively from decision tables, others view tables as a tool that can seldom if ever be used for coding. When programmers do not feel that coding from a table is feasible, they will often draw a flowchart based on the decision table and then perform the coding step.

When coding from a decision table, one of two approaches is usually followed. The first approach is scanning, in which each rule is tested separately with the most frequently employed rules tested first.

The second approach is condition testing. Here the conditions and rules in the tables are sequenced to facilitate testing. (Generally the rules are sequenced from left to right and the conditions are sequenced top to bottom.) The program is then coded to test the first condition variable; based on the test results, it branches to another test. Condition testing and branching continue until the appropriate rule is identified and the specified actions are taken. Each rule in the table takes advantage of previous condition tests; so a particular rule is usually not reached unless the conditions in previous rules in a sequence have not been satisfied. In other words, the table is constructed with an order to the rule and condition sequences that approximates the condition testing sequence that would appear in a flowchart. As experience is gained with tables, programmers find it easier to construct them in a manner that considers a condition testing sequence, thereby facilitating their later coding efforts.

Figure 17 shows a table constructed randomly. Figure 18 shows the same table sequenced for ease of programming, so that coding of the program might test the first condition, then the second, and so on until the appropriate rule is reached.

ADVANTAGES/ DISADVANTAGES AND USES OF DECISION TABLES

Up until this point we have mainly been discussing the "what" of decision tables by looking at what tables are and of what they consist. Now let us turn our attention to the "why and how" of decision tables by consideration of table advantages, disadvantages, and areas of use.

Advantages/Disadvantages

Some of the advantages of decision tables are that tables:

1. Display situations concisely.
2. Promote completeness and accuracy.
3. Can have completeness and accuracy checking techniques applied to them.
4. Permit a complex situation to be more easily grasped.
5. Have allowed the study and design of systems previously too complex to be handled by other methods.
6. Facilitate modularity.
7. Are relatively easy to construct, modify, and read.
8. Are not limited to computer applications.
9. Improve customer/analyst communication.
10. Are easily understood by both business and scientific personnel.
11. Are more easily understood than flowcharts.

TABLE 1		R1	R2	R3	R4	R5
C1	END OF FILES A OR B	–	–	Y	–	–
C2	PART NO. A B	Y	–	N	–	N
C3	START OF PROGRAM	N	Y	N	N	N
C4	PART NO. A B	–	–	N	–	Y
C5	PART NO. A = B	N	–	N	Y	N
A1	CLOSE FILES A AND B			X		
A2	READ FILE A		X		X	X
A3	PROCESS RECORDS				X	
A4	END OF JOB			X		
A5	GO TO TABLE 1	X	X		X	X
A6	READ FILE B	X	X		X	

FIGURE 17 Unsequenced table.

TABLE 1		"R2"	"R4"	"R1"	"R5"	"R3"
		R1	R2	R3	R4	R5
C1	START OF PROGRAM	Y	N	N	N	N
C2	PART NO. A = B	–	Y	N	N	N
C3	PART NO. A B	–	–	Y	B	N
C4	PART NO. A B	–	–	–	Y	N
C5	END OF FILES A OR B	–	–	–	–	Y
A1	PROCESS RECORDS		X			
A2	READ FILE A	X	X		X	
A3	READ FILE B	X	X	X		
A4	CLOSE FILES A AND B					X
A5	END JOB					X
A6	GO TO TABLE 1	X	X	X	X	

FIGURE 18 Sequenced table.

12. Document applications involving complex interactions of variables more clearly than flowcharts.
13. Can be used as a computer source language.

Some of the disadvantages of decision tables are:

1. People experienced in drawing flowcharts find it initially difficult to create decision tables.
2. It would sometimes be helpful to be able to refer to a specific rule in one table while working within another rule in the same or a different table (e.g., branch to rule 2 of Table 7). However, this violates one of the current conventions of table usage.
3. Tables can become unwieldy if action segments are predicated on just one or two simple conditions rather than many complex conditions. However, an alert table creator can usually prevent this situation.

Where and How Tables are Used

Decision tables can be used by many different levels and functions of an organization. The policies and directives of top management can often be expressed in tabular form, as can the ideas of other organizational levels. Engineering, accounting, mathematics, manufacturing, personnel, etc., are all areas to which tables can be applied. Tables might be employed in such computer related activities as documenting existing systems, developing and documenting program logic, program writing, and program operating instructions. Just where and how tables can actually be used somewhat depend upon available table handling techniques and the user's ingenuity.

Our initial uses of decision tables have mainly been to improve communications in three categories:

1. Communications between system analysts and data processing users.
2. Communications between application analysts and programmers.
3. Communications between software programmers and the application programmers who utilize the software.

While our use so far has been limited, our results have been excellent. Our intention is to make much more use of tables in the near future, as we implement a number of changes in our software, hardware, documentation, and systems. From what we have learned through our study and our own experiences to date, we feel that decision tables, while not a panacea for all systems definition problems, are definitely a tool that can contribute much to improving the communications link between man and the computer.

Potential Uses

At the beginning of this talk I said that we would deal with a little blue sky as to how tables might be of value in the future. Perhaps now is the time to look through the tinted glasses and see where decision tables might fit in on cloud nine.

First of all, we see that decision tables can be rather easily maintained under computer program control. As such they can be quite easily displayed and changed. The problems associated with having a computer maintain tables are certainly fewer than having a computer maintain flowcharts. The advantages/disadvantages of having the computer do this job, as opposed to humans and non-computer tools, is a separate subject with questions peculiar to each organization. I am only saying that for those interested in maintaining tables under computer control the assignment is not extremely complex. (There are a number of organizations that are doing this today.)

Our next hazy view really comes from within the fluffy sections of cloud nine. Now we see decision tables combined with the fairly general desire for an integrated decision documentation plan. The plan we see covers the spectrum and relates the elements from top management decision-making to computer program modules. Some of the features of this plan are:

1. Decision tables are used throughout the organization to document all decision-making deemed worthy of documentation. This includes top level policy, middle management directives, and computer program routines or modules.
2. All decision tables and their rules are cross-referenced, as appropriate. For example, assume a top management policy is documented in a table consisting of ten rules. Each of those rules might be expanded and documented at the middle management level with separate decision tables, each of which has its own ten rules. And, of course, each of the rules at all levels is cross-referenced. Thus the effect of a top management policy at the computer program level can be traced; or vice versa, a computer program module can be traced back to its parent top management policy.
3. All data elements used in the organization are cataloged and coded. These codes are used in the decision tables whenever data elements are stated.
4. All decision tables, data element codes, and cross-referencing information are maintained under computer program control.

As a result of the aforementioned features, this plan allows:

1. Instant tracing of the effect of decisions or decision changes throughout the documented structure.
2. Improvement in the study and design of large, integrated systems while providing a total view of the entire organization's data/information flows and requirements.
3. Easier application of advanced techniques for systems simulation and information flow studies.
4. The rapid reflection and implementation of decision rule changes at all related lower levels. For example, a change in top management policy quickly results in a change of the related and affected program modules.
5. On-line display of documented decision making.

While this picture from cloud nine might seem pretty far out, especially from a practical point of view, I suggest that it is not really too unrealistic or impractical. Many of the elements required by the plan are already in use or under development. The plan does not depend on some major new discoveries or breakthroughs, but rather the serious and intelligent efforts of knowledgeable people. Certainly the plan could not be completely developed and implemented overnight. However, I do feel that this "blue sky" plan for shortening and improving the communication's channel that joins management and the computer is practical and could be accomplished in a reasonable manner.

Implementing Decision Table Use

Another "how" aspect of decision tables that we will consider is how a data processing organization can implement decision table use. As with most anything, the best way to implement decision table use is through a plan tailored to your particular situation. However, I would like to outline some steps that you might follow, steps that might serve as suggestions as you evolve your own plan. The steps are seven in number:

1. *Get preliminary knowledge.* By this, I mean that you or some individual in your organization should acquire a brief broad picture of what the decision table area is all about. The exposure, such as we are covering today, should include the basic what, why, and how of tables so that a framework is established for your future developmental efforts.
2. *Develop an "expert."* Have at least one individual in your organization become very conversant with the decision table area. This person can be your main reference point so that problems can be better anticipated, solutions can be more quickly reached, and potential table values to your organization can be more fully realized. The brief reference listing at the end of this paper (Appendix A) is intended to be of initial value to the person attempting to become an "expert". The listing is strictly an initial guide; it is definitely not exhaustive.
3. *Anticipate problems.* It is important that you, your management, and your staff are aware that problems will be encountered in undertaking table use. Some people will object to the tables, instances will arise (especially at first) where people will have difficulty knowing how to express their particular thoughts in table format, etc. I believe quite a few people and organizations have failed with tables just because they jumped right into decision table use without having a plan, or at least not recognizing with their plan that there would be problems in starting up with tables.
4. *Have a reference document.* As your people start to use tables they should have a reference document, preferably supported by an orientation session. Ideally, the document would be tailored to your own needs; but it might be a more general document.
5. *Encourage users.* Management must provide the backing to get tables into use. Whether that backing takes the form of coercion, suggestion, or some point in-between is a matter you can best decide. Whatever method is used, the decision table undertaking will probably die without management's encouragement of the users.
6. *Explain outside the data processing area.* To really gain a better start-up of a decision table program, explain tables to the data processing users. Your goal is to help the users understand how to read tables and to see that tables help them clarify their communications with the data processing department. As a result, the users will react positively to your analysts when tables are employed, and will even seek to have their problems defined in table format.
7. *Follow-up.* Finally, you will want to follow-up on your decision ta-

ble program. You will want to identify and correct problems; learn of people and/or areas not using tables and determine why; and keep abreast of developments within your organization and elsewhere that might suggest changes in your way of employing tables. Your follow-up might suggest that you drop your decision table program. Conversely, and hopefully, your follow-up might suggest that you should expand your program and move further into the cloud nine situation we were describing earlier.

CONCLUSION

It was stated at the beginning that this paper was intended for the application oriented who wanted to consider in a very practical sense what decision tables are, why they are of value, and how they can be used. If that was your interest, I hope this presentation has been of value to you.

APPENDIX A

Related Documents

Canning, Richard G., "How to Use Decision Tables," *EDP Analyzer,* 4, No. 5 (May, 1966), 1–14.

Chapin, Ned, "An Introduction to Decision Tables," *DPMA Quarterly,* 3, No. 3, (April, 1967), 2–23.

"Decision Tables": A Systems Analysis and Documentation Technique," Form Number F 20-8102, White Plains, N.Y.: IBM Corporation, 21 pp.

Devine, Donald, "Decision Tables as the Basis of a Programming Language," *Data Processing,* 7, Park Ridge, Ill.: Data Processing Management Association (1965), 461–466.

Fisher, D. L. "Data, Documentation and Decision Tables," *Communications of the ACM,* 9, No. 1 (January, 1966), 26–31.

Grad, Burton, "Engineering Data Processing Using Decision Tables," *Data Processing,* 7 Park Ridge, Ill.: Data Processing Management Association, (1965), 467–476.

Kirk, H. W., "Use of Decision Tables in Computer Programming," *Communications of the ACM,* 8, No. 1 (January, 1965), 41–43.

Meyer, H. I., "Decision Tables as an Extension to Programming Languages," *Data Processing,* 7, Park Ridge, Ill.: Data Processing Management Association (1965), 477–484.

Pollack, Sol, "Analysis of Decision Rules in Decision Tables," *Federal Clearinghouse* report AD-407719 (May, 1963), 69 pp.

Pollack, Sol, "Decision Tables for Systems Design," *Data Processing,* 7, Park Ridge, Ill.: Data Processing Management Association (1965), 485–492.

Pollack, Sol, "How to Build and Analyze Decision Tables," *Federal Clearinghouse* report AD-425027 (November, 1963), 17 pp.

"The Decision Logic Table Technique," AFP 5-1-1, Department of the Air Force (September, 1965), 42 pp.

GRIDCHARTING

W. HARTMAN
H. MATTHES
A. PROEME

PURPOSE

A gridchart is a formalized representation of the relationships between two groups of system elements. In this context, the term "system element" means any defined operation or physical obobject that is part of the system.

Even when the number of system elements involved is high, representation by means of a gridchart remains clear. In addition it is amenable to analysis and manipulation. It is thus a valuable tool for use in circumstances where it is necessary to consider rearrangement, or reduction in the number of system elements. In many instances, gridcharting can serve as a useful design check.

SOURCE: Hartman, W., H. Matthes, and A. Proeme, "Gridcharting," *Management Information Systems Handbook,* N. V. Philips-Electrologica, 1968, Section 6–23, pp. 1–6.

AREAS OF APPLICATION

Gridcharting finds application at all phases of a project.

GENERAL

The following list, giving data-element pairs whose gridcharts might be useful, is intended to be suggestive rather than exhaustive:

- departments — functions
- departments (personnel) — reports (forms)
- departments — data elements (fields)
- reports — data elements
- input forms — output forms
- input forms — data elements
- reports (forms) — files
- functions — data elements
- functions — files
- programs — files
- data elements — files
- programs (procedures) — data elements
- programs (procedures) — reports

The first example of the use of gridcharts is not drawn directly from the field of systems engineering; in order to

sa.\comp.	A	B	C
M	4	2	–
N	–	10	5
P	3	2	4

CHART 1

sa.\ass.	A	B	C
1	1	1	3
2	–	2	3

CHART 2

the composition of assemblies 1 and 2 in terms of these subassemblies.

It is obviously possible from these two charts to discover how many of each component are present in each assembly. Assembly 1 contains one "A", contributing four "M's", one "B", contributing two "M's", and three "C's", with no "M's"—a total of six "M's". Similarly, the entire component content of each assembly can be derived, leading to the following gridchart (Chart 3).

ass.\comp.	1	2
M	6	4
N	25	35
P	17	16

CHART 3

achieve simplicity, we shall deal with an imaginary manufacturing situation in which assemblies 1 and 2 are constructed from subassemblies A, B, and C, which in turn contain components M, N, and P.

Chart 1 shows the number of components M, N, and P required for the manufacture of each of the subassemblies A, B, and C, while Chart 2 gives

To permit a simple and systematic procedure for this sort of manipulation, the following type of layout is suggested (Chart 4):

		sub-assemblies			assemblies	
		A	B	C	1	2
components	M	4	2	–	6	4
	N	–	10	5	25	35
	P	3	2	4	17	16
assemblies	1	1	1	3		
	2	–	2	3		

CHART 4

The procedure is now as follows: For the number of "M's" in assembly 1, take corresponding boxes in the "M" and "1" rows, multiply and add as shown

$$1 \times 4 + 1 \times 2 + 3 \times 0 = 6,$$

and enter the result in the appropriate box on the right, in the 1 column. Similarly, for the number of "P's" in assembly 2,

$$0 \times 3 + 2 \times 2 + 3 \times 4 = 16.$$

Whilst no special mathematical knowledge is necessary in order to be able to combine gridcharts in this manner, it will be noted by those familiar with matrix methods that the operation is equivalent to the multiplication of two matrices.

MANIPULATION AND INTERPRETATION

Hierarchical Relationship

The manipulation of gridcharts as shown above is useful for obtaining an explicit set of relationships which was implicit but not obvious from the original two gridcharts. It is applicable only when the two sets of system elements related by the first chart, say X and Y, and those related by the second chart, say Y and Z, are in a continuing relationship of the "hierarchical" or "continuous inclusion" type. This is true of the relationship between components—subassemblies—assemblies, as in the above example, and also for such groupings as data elements—files—functions, because functions access data elements via files. In groupings with this type of relationship, a gridchart relating the first two sets of system elements can be combined with one relating the last two, to eliminate the middle set of system elements, thus providing a gridchart relating the first and last sets.

Example

The next example shows a chart of data elements against files combined with a chart of files against functions to produce a chart of date elements against functions (Chart 5).

It would be physically possible to combine by multiplication the chart of data elements against functions with that of functions against files; but since in this case the ordering of the sets of system elements would not be correct, the resulting chart would not represent the relationship between files and data elements. Indeed, the result would, in the context of systems analysis, be meaningless.

The hypothetical system dealt with in this set of charts comprises the material requirements calculation, the production planning, the external order planning, and the goods inward inspection with respect to the manufacture of assembly "A", which is constructed from components "P" and "Q". Component "P" is bought out, and component "Q" is manufactured in plant from material "R".

It will be noted, that the set of files dealt with is oriented more to a "paper" system than to an automated one; this has been done in order to keep the example clear and simple.

Interpretation

Where a figure 1 appears in the chart of functions against data elements, this implies that the data element in the row concerned is accessible to the function in the associated column via the proposed file design. If a 0 appears where access to a data element is required, the file design must be modified accordingly: inspection, particularly in a case where more than one element is missing,

		files					functions			
		stock	article	capacity	customer Order	external Order	mat. reg. calculation	production planning	extonal order planning	goods inward inspection
data elements	code number of assembly "A"	1	1	1	1	0	3	4	1	0
	number of "A's" in stock	1	0	0	0	0	1	1	1	0
	number of "A's" on order	0	0	0	1	0	1	1	0	0
	code number of component "P"	1	1	0	0	0	2	2	1	0
	number of "P's" in stock	1	0	0	0	0	1	1	1	0
	number of "P's" per "A"	0	1	0	0	0	1	1	0	0
	code number of component "Q"	1	1	0	0	1	3	3	2	1
	number of "Q's" in stock	1	0	0	0	0	1	1	1	0
	number of "Q's" on order	0	0	0	0	1	1	1	1	1
	number of "Q's" per "A"	0	1	0	0	0	1	1	0	0
	code number of material "R"	1	0	0	0	1	2	2	2	1
	quantity of "R" in stock	1	0	0	0	0	1	1	1	0
	quantity of "R" on order	0	0	0	0	1	1	1	1	1
	quantity of "R" per "P"	0	1	0	0	0	1	1	0	0
	reorder level for "Q"	0	0	0	0	1	1	1	1	1
	reorder level for "R"	0	0	0	0	1	1	1	1	1
	machine group (MG) code	0	0	1	0	0	0	1	0	0
	total capacity per MG	0	0	1	0	0	0	1	0	0
	committed capacity per MG	0	0	1	0	0	0	1	0	0
	capacity required per "A"	0	0	1	0	0	0	1	0	0
functions	material requirements calculation	1	1	0	1	1				
	production planning	1	1	1	1	1				
	external order planning	1	0	0	0	1				
	goods inward inspection	0	0	0	0	1				

CHART 5

will often reveal the most economical modification.

If a figure greater than 1 appears, redundancy may be indicated. In the example given, the figures greater than 1 do not indicate redundancy so long as the number of files and their named purposes remain as shown; they appear only in the rows of those data elements which are identifiers, and thus necessary at least once in each of the files which contain data referring to the system elements which they identify.

Value in Design

The restriction of gridchart "multiplication" to those cases which involve the elimination of the middle set of a group of three ordered sets of system elements means that the technique is in general more applicable to design checking than to design itself. To "divide" one gridchart by another is not normally possible. However, first approximations to design requirements are often possible by means of a critical examination of the appropriate gridcharts. An examination of the functions against data elements and functions against files charts of the above example, entry by entry, will not permit reconstruction of the files against data elements chart; but it will give a valid, although not necessarily the most economical, version of it. In complex cases, however, the version thus found may be of considerable value in that it offers a workable starting point.

BUSINESS INFORMATION SYSTEMS ANALYSIS AND DESIGN

HONEYWELL

During recent years the use of electronic data processing equipment and methods has grown to the extent that modern businesses would be hard-pressed to function without them. Computers were first introduced into the business world for applications that required large amounts of gathering, maintaining, and reporting data, mainly financial. The use of computers has spread beyond the accounting departments to where, today, many segments of a business are linked together in Total Information Systems.

Whereas these systems would be impossible without the computer, they owe their very existence to the humans who design, implement, and operate them. Systems analysts perform the task of translating management's goals and objectives into a business system that meets those needs. Trained analysts are a valuable asset to any company.

BISAD represents a methodology of performing the task of *B*usiness *I*nformation *S*ystems *A*nalysis and *D*esign. It was developed because of the general lack of training available within this critical area. Reference books and courses were shown to be little more than a collection of techniques, such as how to flow chart, design a file, and present the results of a study. Modern analysts need much more. Today's analyst must be a businessman in addition to being technically proficient.

An analyst performs definite tasks in his efforts to analyze a business and to design an information system that will respond to the needs of management. In BISAD these tasks have been organized into six logical steps which may be applied to obtaining the solution to any information problem.

The first step, background analysis, is the foundation of accurate problem

SOURCE: Honeywell, *Business Information Systems Analysis and Design,* (144.0000.0000.0-954), 1968, pp. ii, iii, 3-11 through 3-15. Reprinted with permission of Honeywell, Inc.

definition. The analyst gathers information, classifies it into logical groups, and develops his knowledge of the study area.

Functional analysis divides the total business into logical groups of task centers, allowing the analyst to study each segment of the business as a unit, describing the information requirements of each function and developing the relationship of each segment to the other areas of the total business.

System Design Prototype develops a model of the business system by progressing from *what* must be done to *how* it must be done. The model is developed for each function and then these individual models merged to produce the total system. The development of the prototype system is accomplished in minimum time. The result is a picture of the system hierarchy described in business terms.

After the business system model has been approved, the selection of a priority area can be made as the particular situation dictates. The analyst is then concerned with the design detail necessary to convert the prototype to a working system, involving both media and sequence of activities.

Once the working system has been developed, BISAD stresses the importance of planning to make the system operational. Utilizing such aids as PERT (*Program Evaluation and Review Technique*), the analyst is involved in the development of a plan and the controls necessary for the successful implementation of the system. Representing the system specifications, documentation gathered during the study must then be presented for the final approval of management.

The analyst examines each of the activities; and when all the activities are completed, the functional model is formed as shown in Figure 1. It will be noted that the analyst has introduced two master files which are being updated in this function, the Customer Master File and the Open Order Master File. As the analyst examines the input and the output for each activity, he determines the need for the master files. For example, in the Order Entry activity the analyst is interested in determining the credit status of each customer. From the interface diagram it has been determined that information about the credit status of each customer is available from Financial Control. In the Order Entry and Order Billing activities there is information about each customer's generated order, i.e., dollar amounts and billing dates, which is to be sent to Financial Control. It has also been determined that all this type of information and such other information as customer name and address may be combined into a customer master file, which may be used by both the Order Processing and the Financial Control functions. This step eliminates the need for maintaining two identical files for each function.

The other master file shown in Figure 1 (Open Order File) is created for a somewhat different set of reasons. The Open Order Master is an internal master file used by sales order processing to maintain the current status of orders that have been entered into the system. In the Order Entry activity, acknowledgment is sent to the customer telling him of the receipt of his order and the scheduled production date. Before these acknowledgments can be generated, Production Planning and Control must schedule the order. This means that orders received in one cycle of the order process must be processed by another function (Production Planning & Control) and the results re-entered into the next cycle of order processing in the form of production schedule dates. For this reason a master file is maintained containing all of the details of orders for which make orders have been

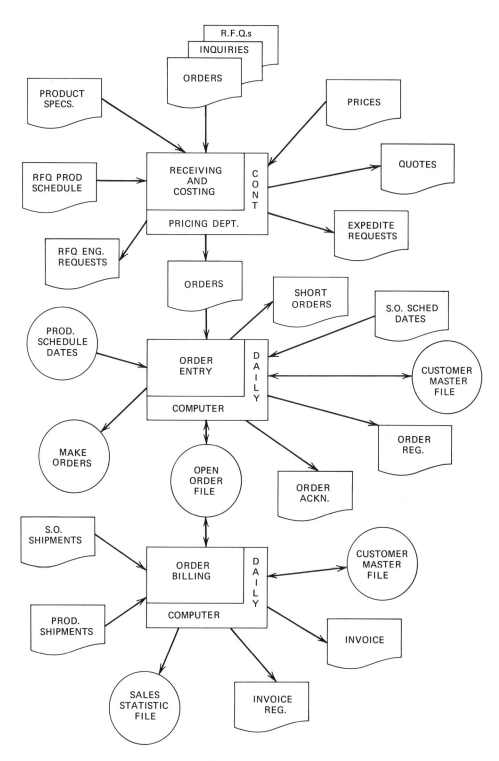

FIGURE 1 Silgol Corporation prototype model sales order processing.

issued. As each order is scheduled, the Open Order Master is updated to show its scheduled date, and acknowledgments are generated. The Open Order Master File is also used in the Order Billing activity. The Silgol Corporation operates on a post-billing principle. That is, invoices are not sent to the customer until the products ordered are manufactured and shipped. In the Order Billing activity, as notification is received from Production Planning and Control that the ordered products have been shipped, the corresponding orders are deleted from the Open Order Master and invoices issued to the customer.

INFORMATION FLOW

Having developed the functional model as described above, the analyst must now check the model to see that it is complete. The best method for doing this is to follow the natural flow of information through the model. The following narrative describes the flow of information through the functional model just developed for Silgol's Sales Order Processing function.

Inquiries, requests for quotations (RFQ), and orders are received from Silgol customers. Inquiries are requests for information about possible additions, deletions, or changes in required dates for orders which have been received at an earlier date. Such inquiries are verified and sent to the Production Planning and Control function as expediting requests.

Requests for quotations are actually requests to produce a product not previously manufactured. Before such a product is made, determinations about production specifications, prices, and possible schedule dates must be made and sent to the interested customers as a quotation. After customer approval, quotations are returned and handled as regular orders. Upon receipt of an RFQ, engineering requests are generated and sent to Design and Engineering for development of product specifications. Design and Engineering forwards the product specifications to Production Planning and Control for determination of possible production schedule dates. Both the product specifications and the scheduled production dates are returned to Sales Order Processing and are used to produce customer quotations. Orders are verified as far as existing customer, product description, and price. This information is coded for computer processing.

In the Order Entry activity the coded orders are entered into the system. They are checked against the Customer Master File for identification and credit limits, and the order summary information is placed on the Master File. The details of the order, such as the individual line items, are placed on the Open Order File. An Order Register is printed, listing each order, order number, date received, date requested, and dollar amount. Orders are separated as to regular production orders and "short orders." Short orders are sent to Design and Engineering for manufacture and for request of a schedule date. The regular production orders are sent to Production Planning and Control to be included in the master schedule. Scheduled dates are from Production Planning and Control, and Design and Engineering for the last cycle's orders. At this time an Order Acknowledgment may be printed for those orders to include the scheduled date. The Open Order File is also updated with the scheduled dates.

The Order Billing activity is triggered by receiving the shipped notices. Those orders shipped are deleted from the Open Order file, and the Customer Master is updated to indicate shipped orders. Both files are needed to generate the invoice: the Open Order File for products, prices, and quantities ordered,

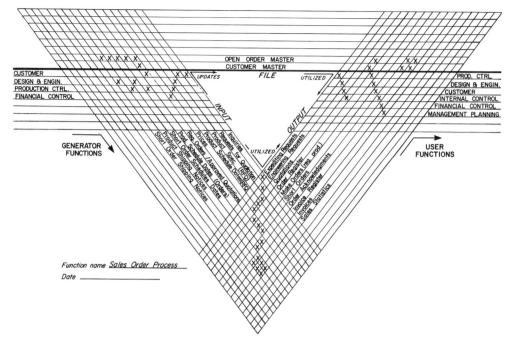

FIGURE 2 Information matrix and flow.

and the Customer Master for customer name and address and credit terms. A register of all invoices issued is also printed. Sales statistics are generated for each order shipped, and the file will be used by Management Planning for sales analysis and forecasting. The Customer Master will be used by Financial Control in the Accounts Receivable activity. Both order and invoice registers are part of the internal control and audit trail system.

INFORMATION MATRIX

The functional model just discussed represents graphically the activities of a data processing system. From the model, one can see the input to an activity, the files used, and the output generated. What is not shown is the way the input and the files are used within the activity to produce the output. This void is one reason for stressing that a complete descriptive narrative must accompany each flow diagram. Such is still true of the functional model. However, there is another document which will help the analyst graphically represent the prototype. This is the information matrix. Figure 2 represents an information matrix for Silgol's Sales Order Processing function.

The information matrix shows five connections between the input, the files, and the output of the functional model. The matrix shows the internal or external source of the input to the model (Generator functions). By following the natural flow of information, the input within the model may be shown as updating the master files used in the model or generating output from the model, or it may be used in conjunction with a master file to generate output. The output may then be shown as being used by a particular function within the company or by a function outside the company. These relations are shown by means of a check mark in the appropriate grid location on the matrix. When complete, the matrix may be used to trace the flow of information through the activities.

ADS: A TECHNIQUE IN SYSTEMS DOCUMENTATION

HUGH J. LYNCH

ADS (Accurately Defined Systems) is a technique for communication and establishment of a working discipline in the definition of objectives, criteria, and specifications for systems being designed for EDP processing. The ADS method is not oriented to a particular brand of hardware. It approaches system definition by starting with specification of the report form or document which is to be output. With the report as a starting point, separate, interrelated forms are used to specify input records, history records, computation, and logic operations. All system elements are tied together through continuing use of cross reference procedures. The result is a disciplined, universally-understandable set of documents suitable for use in programming systems on any computer hardware. Communication between programmers and users of computers is standardized and disciplined for universal validity and understanding.

THE NEED FOR ACCURATELY DEFINED SYSTEMS

Everyone who earns his livelihood in or about data processing installations has, as an occupational hazard, experienced the frustration of inadequate or misunderstood system definitions at one time or another. In each case, the problem reflects the professional viewpoints, the personalities, or the inadequacies of the individuals involved. But there are enough common denominators so that the problem can be typified:

- Frequently, the person defining a system and the individual responsible for implementing it have different backgrounds and work within entirely different disciplines. For example, there are semantic differences in interpretation of terms and objectives, and differences in basic assumptions.

SOURCE: Lynch, H. J., "ADS: A Technique in Systems Documentation," *Database,* Vol. 1, No. 1, Spring, 1969, pp. 6–18. Copyright © 1969, Association for Computing Machinery, Inc., and reprinted with their permission.

- As a general rule, the person outlining the objectives of a system uses one set of criteria while the data processing man subscribes to another.
- Definitions tend to be loosely written and freely interpreted. Gaps are filled in on the basis of seemingly logical assumptions which are inevitably interpreted differently by the parties involved.
- Entirely too much time is wasted; too many false starts occur; expenses are too high.

Clearly, there is room for a common-denominator discipline—a technique which channels the efforts of both parties toward a common goal.

At NCR (National Cash Register), this system definition requirement took on critical proportions in 1966, as we moved into the final stages of application programming in support of the recently introduced Century Series of data processing systems. At that particular point in time, our situation was typical for major programming efforts which have been carried out throughout the industry, among users of in-house systems, software firms and manufacturers: We had literally dozens of industry-oriented program packages to bring to the marketplace. We had industry experts working with customers and test sites all over the country defining the systems which were needed. The programming staff assigned to implement these systems was thinly spread in terms of the massiveness of the effort.

Inevitably, our biggest obstacle lay in defining systems so that our programmers could begin implementation with a minimum of wasted effort and false starts. Further, our industry specialists could not be involved with programming details. We needed a bridge between the different worlds of programmers and industry specialists so that they could exchange concepts with clarity and understanding. ADS, a technique for Accurately Defined Systems, was designed to fill this need.

Briefly, ADS is a method under which uniform discipline is applied to the communication links between system specification and implementation. As a uniform system, it leads the specifier into a more thorough specification, inherently. At NCR, ADS has become a standardized technique basic to all data processing programming efforts. Experience has shown that, although it was designed largely for business applications initially, it can work with any type of data processing system for which output documentation, data files, and input records can be pre-defined.

As we gained experience with ADS, it became apparent that we had come up with a general purpose tool. That is, although the technique was developed to solve a problem in connection with a specific line of computers, this phase of the effort is completely hardware-independent. Subsequent experience has shown that ADS techniques function with equal effectiveness for any type of computer system.

THE ROLE OF ADS

The hardware-independence of ADS stems from its highly specific area of responsibility within the system development cycle.

As illustrated in Figure 1, data processing program development can be classified as a five-phase activity. Briefly, the jobs break down this way:

1. Study the feasibility. The job under consideration can be handled profitably on a computer system.
2. Survey the application.

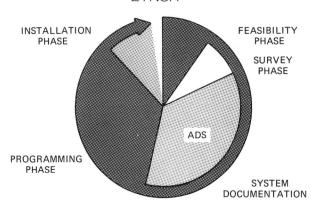

FIGURE 1 Program development cycle.

3. Next comes one of the two major time and cost factors in the development cycle—system definition. As can be seen from Figure 1, documentation is the area in which ADS has been designed to function. This is the foundation for success of system.
4. Following system definition, the computer programs are developed. This activity includes both the documentation writing and the testing of programs and operating system.
5. Finally, installation and implementation can take place.

Note that the representation of these system development jobs is done in a circular format resembling the face of the clock. The space occupied by each of the tasks has been apportioned according to relative time relationships. Thus, system documentation and programming are each represented as requiring about one-third of the time to be invested in the development of a typical system.

The information accumulated during the system research phase, including both the survey and the feasibility study, provides a basis for implementation of the system.

ADS ORIENTATION

It should be stressed that, even if the computer on which the new system will run has been designated right from the beginning, activities are pretty much hardware-independent through the first three stages of the system development cycle. Referring back to the time-oriented chart in Figure 1, note that the first major cycle is referred to as "System Documentation."

The documentary orientation is intended literally. Before any computer programming is done, the ADS discipline documents the detail system requirements. All activities within the ADS phase of a development program deal with reports and records necessary to enlarge upon the generalized conclusion derived from the research.

ADS is a discipline designed to facilitate the definition and communication of the objectives, criteria, and specifications of an EDP system. It approaches system definition by starting with specification of the report form or document which is to be output. From this starting point, separate, interrelated forms are completed to specify input records, history records, computation, and logic operations. All system elements are tied together by cross-reference tables. The result is a well-organized and precise set of system-defining documents readily understood by everyone involved in the development or use of the system.

ADS PROCEDURES

ADS procedures begin with the premise that system development should be results-oriented. That is, the logical starting point for system definition is an indication of the results to be delivered. Accordingly, ADS procedures begin with specification of the reports or documents to be generated by the EDP system.

For illustration, this presentation will use a common, typical example: The building of a report utilizing a payroll master form. The examples cited are not brought to any specific conclusion. Rather, they are carried just far enough to illustrate the principle generally.

It should be stressed, however, that ADS is designed to function in any data processing environment. The examples used were chosen in the hope that they will be universally understandable.

THE ADS REPORT FORM

ADS procedures begin with the execution of a general-purpose Report Layout form (Figure 2) which calls for entry of specifications on both layout and content for one specific document. Each ADS report form is complete and self-contained, incorporating all of the instructions, tables, and references needed for full specification of a single report.

To illustrate the self-instructional nature of ADS procedures, each report form has a Symbol Identification section (Figure 3) listing the symbols to be used for field definition.

The body of the Report Layout (Figure 4) has standard horizontal and linear spacing adaptable to the design of reports for any on-line printer. In this example, note that the employee name field has been blocked out with "X" symbols, indicating an alphanumeric field. The wages field is blocked out with "9" entries, indicating that it is a numeric field. Note also that the wages field has three leading "$" entries. This specifies use of a floating dollar sign.

As shown in Figure 4, each field blocked out on an ADS report form is designated by an arabic numeral. On all ADS working forms, all fields are numerically designated in this manner. These same numbers are used, within charts built into the forms themselves, as functional cross-references relating output requirements back to processing, history files, and input data records. The need for and the value of positive field and record identification will be covered as this discussion progresses.

Field numbers designated in Figure 4 have been applied to detail entries only. The ADS system is applicable to situations where headings are to be computer-printed or where variables are to be entered on a pre-printed form by the computer. Variable entries on title lines or column headings would be designated on the report form and numbered as fields.

The cross-reference which will interrelate each report with its data sources within the ADS system is initiated through entries into a table incorporated within the report form (Figure 5). As indicated in this example, each time a field is specified on a report layout, its name is entered on a correspondingly numbered line in the cross-reference table. This consistency of numeric designation is important—the number by the field must correspond to the line number within this table. Within any ADS-defined system, references to page and line number on the working forms will

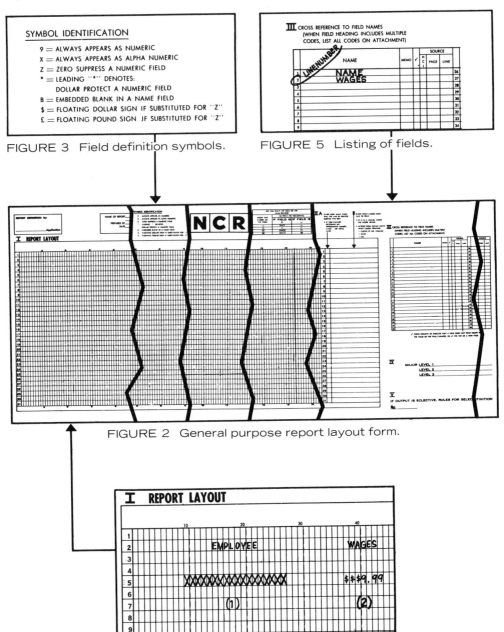

FIGURE 3 Field definition symbols.

FIGURE 5 Listing of fields.

FIGURE 2 General purpose report layout form.

FIGURE 4 Body of report layout form.

always have the same meaning to everyone associated with a given project.

The relationship between the cross-reference table and the numeric designations of fields on ADS report forms is illustrated in Figure 6.

In practice, system designers, working closely with responsible departments and individuals in organizations developing computer programs would work out the entire content of a projected report following the routines shown in

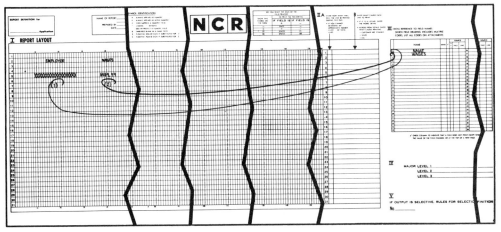

FIGURE 6 Linkage of field designation and field listings.

these brief illustrations. In addition, separate spaces are provided on the ADS report form for entry of:

- Notes qualifying and describing the operating sequences for the running of the report.
- All control breaks.
- Spacing and page numbering patterns, etc.

Conceptually, it is important to understand that:

1. The output report is the starting point for all system documentation efforts under the ADS technique.
2. The report form is generally fully executed and approved by all parties before further documentation can be implemented.
3. The purpose, at this and every other phase of an ADS project, is to apply a communication discipline which makes for universal understanding on the part of all persons affected.

ADS INPUT RECORD DESIGN

A separate ADS form (there are a total of five) is used to define input records. As with the report form, the Input Media Layout sheet (Figure 7) has corresponding areas for diagramming records and for listing and describing them in accompanying cross-reference tables. Figure 8 shows the format of a single record in the diagramming section. As shown in this example, field lengths are indicated through the ruling of vertical lines in positions which indicate the number of characters within each record segment. Field definition according to characters is employed so that, in keeping with the hardware independence of ADS, the same input documentation may be applied to any source medium: magnetic tape, punched cards, punched tape, magnetic cards, disc, or drum.

Figure 8 shows the designation of four separate input fields. These cover date, name, hours worked, and rate. After this record was laid out, the first field was assigned an identifying number, "1." At this point, none of the other fields in Figure 8 were numbered.

As indicated in Figure 9, the cross-reference table on the Input Media Layout form frequently requires more than one line of description per field. Therefore, it is best to wait until one field has been entered into this table before assigning the next field number, which must correspond with the line of initial entry on the cross-reference table.

Note that a full description is called for in the columnar format of the cross-reference table of the Input Media Layout form. In completing the table entry for field number 1, the system designer has entered the name of the field, "Date," in the wide column at the left. The next column calls for a figure indicating the percentage of the records in the input stream in which this field will appear. In this case, the date will appear in 100 percent of the records. Should the figure be less than 100 percent, the appropriate figure or the closest feasible approximation, would be entered.

The next column asks whether the data will be in alphanumeric or numeric mode.

The following column requests the size of the field, and number of characters.

The next column, for "Memo" entries, has not been used in Figure 9. This column is set aside for designation of transaction codes which might be required to identify fields. For example, if a report were dealing with retail sales, one field in the input record might be identified as containing transactions. In such a case, the memo column would be used for codes to differentiate between charge sales, cash sales, customer returns, and so on. Similarly, in a payroll application, transaction codes could identify sick leave, vacation leave, time off for jury duty, or any other status change from standard hours worked.

The final column of the field illustrated in Figure 9 is used to describe the processing rules to be applied in validating each field of information. Description of the field continues until all factors have been covered. Then, a single line is skipped and the description for the next field is initiated.

In this case, the description for the second field, on employee name, is started on Line 6. This line number is then used as the designated number for the employee name field. Figure 10 shows the entry of control numbers in the record layout to correspond with entries on the table in Figure 9.

In the completed Input Media Layout form (Figure 11), the same relationship exists between the designated record field and the descriptive table as was established in the Report Layout form.

ADS CROSS-REFERENCE TECHNIQUES

After the Input Media Layout form has been completed, the designated fields must be interrelated with the Report Layout form, which serves, continuously, as the working hub of system definition activity. Figure 12 shows the original cross-reference table of the Report Layout form with additional entries to indicate the source of data to be incorporated in the file report.

Note that one column is headed to indicate entries of codes "H," "C," or "I." These codes indicate, respectively, whether data for the report comes from History files, Computation, or Input records. In this example, entries indicate that the name field comes from input records while the wages result from computation. In each case, subsequent entries identify the page and line number of the ADS forms on which the data fields are originally defined.

As ADS procedures are carried forward, each data definition that has a bearing on the final report must be cross-referenced in this table. Cross-referencing not only ties the elements together, but is an auditing technique to insure that all data elements are properly related to each other. In effect, this single table represents a central focal point for the ADS system definition effort.

ADS: SYSTEMS DOCUMENTATION 197

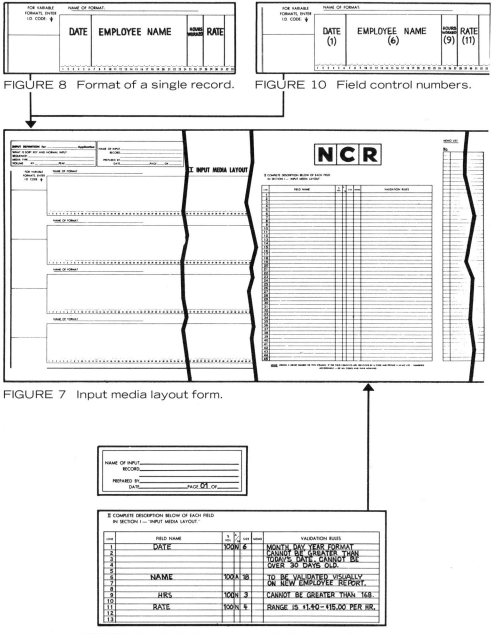

FIGURE 8 Format of a single record.

FIGURE 10 Field control numbers.

FIGURE 7 Input media layout form.

FIGURE 9 Field descriptions for input media layout.

DEFINING COMPUTATION OPERATIONS

Computation operations needed to develop data for system output are specified, within ADS, through use of the form shown in Figure 13. In connection with computation particularly, ADS techniques remain completely hardware-independent.

The computation definition form is columnar in nature, with spaces provided

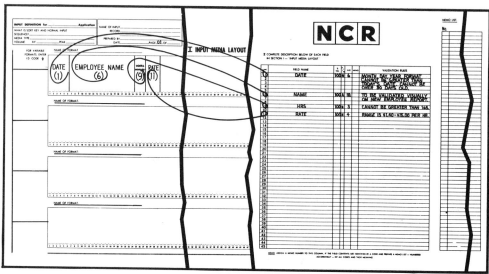

FIGURE 11 Completed input media layout form.

on each line to list two operands and a result. In addition, the form provides spaces for identifying the source of data and specifying the operations to be performed. As with other ADS forms, all data entries are identified within the system by page and line number.

Figure 14 contains an example of a computation entry. In this case, input data on hours and rate are to be multiplied to create a new field, wages. Alongside the designation for the hours field, the program planner has indicated that the information comes from Line 9 on Page 1 of the input form for this system. In the function column between the two operands, an "X" has been entered to indicate that the figures are to be multiplied. Note that the rate information is cross-referenced to Line 11, Page 1 of the input form.

Under ADS, the computed "wages" field assumes an identify of its own. It can be referenced through Line 1, Page 1, on the computation form for any further operations. Should computation be needed upon wages, this designation can simply be carried down to a new line on the same form. If the wages field is to be incorporated into an output document, the designation is entered on the cross-reference table of the output form.

DEFINING HISTORY RECORDS

Within the ADS technique, history has a special meaning. In normal EDP system development, data are summarized or retained in detail for future reference. These are, of course, files. The same can be observed about any manual sys-

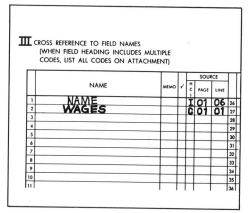

FIGURE 12 Cross-reference table on original report layout form.

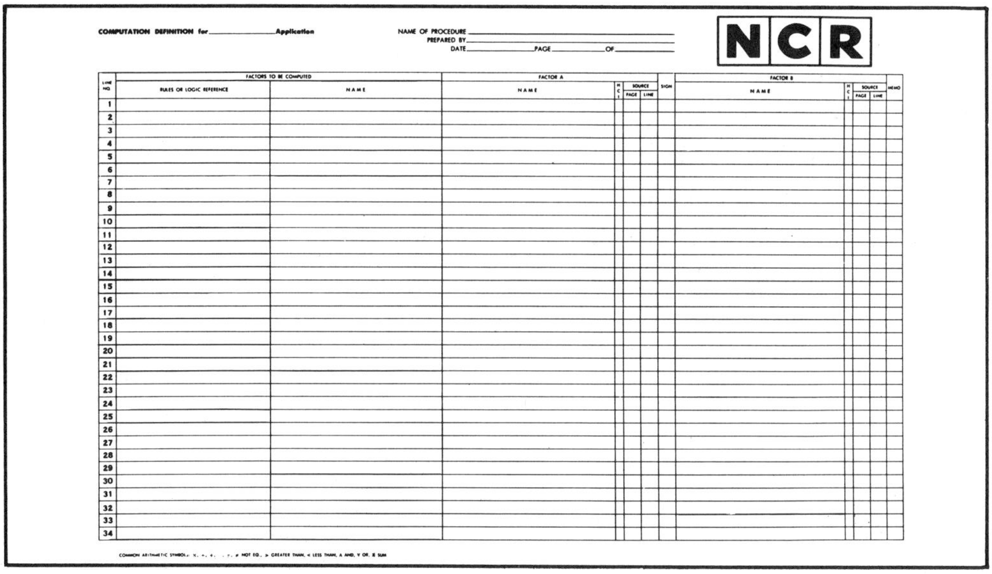

FIGURE 13 Computation definition form.

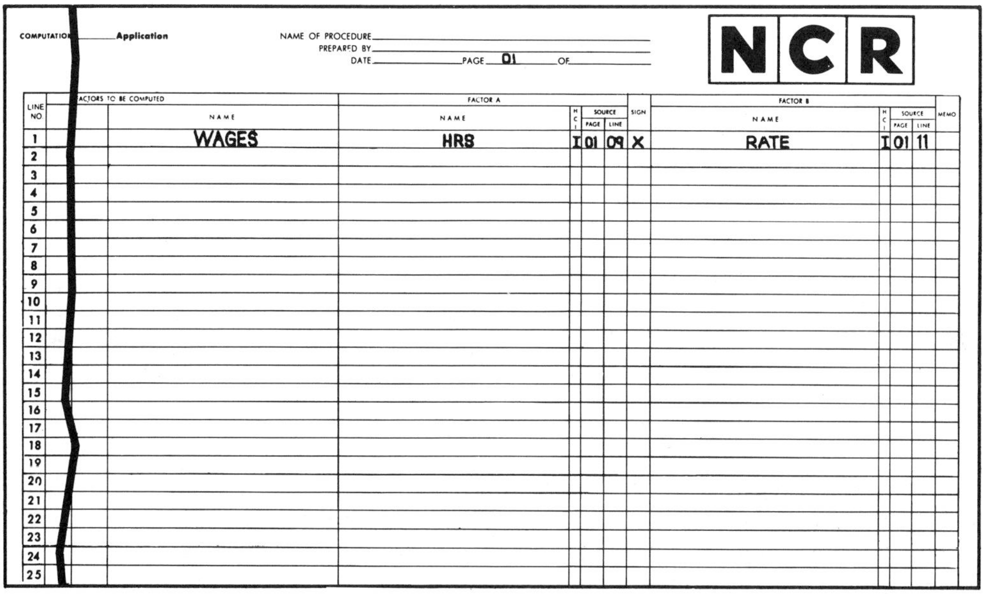

FIGURE 14 Computation form entry.

tem. The intent of history retention is to provide a means of describing data to be retained without consideration of ultimate file design.

With ADS, however, we are talking about report development programs. The History Definition form (Figure 15) is used to specify indicative or accumulated fields from other data which will be required as the basis for subsequent reporting or calculations. For example, in the case of a payroll, each worker's Social Security number could be considered as an indicative field triggering

FIGURE 15 History definition form.

a line of output on a report listing earnings on a year-to-date basis. Year-to-date earnings themselves would be an accumulated field.

Figure 16 demonstrates entries onto a History Definition form for Social Security number and year-to-date wages. Note that the entries here are similar in format to those on the table of the Input Media Layout form. Entries include a percentage of inclusion, sign for numeric fields, size of record, life cycle of the record, and source identification.

Note also that a column is provided for a check mark. Spaces of this type are commonly incorporated into ADS cross-reference tables. The check is inserted after the program planner has satisfied himself that, in addition to accounting for the source of the data, he has made provision for adjustment of "maintenance" of this item.

DEFINING LOGIC OPERATIONS

The final form used for ADS system definition deals with logic operations.

Before reviewing the form itself, it may be valuable to understand the relationship of logic operations to other segments of an ADS-defined system. As shown in Figure 17, logic is not an integral part of the ADS report generation cycle. Rather, it is a supplemental definition which supports the computation function or any of the other forms where logical rules govern selection. Logic

ADS: SYSTEMS DOCUMENTATION

III DESCRIBE EACH FIELD OF THE GROUP

Line No.	NAME	MEMO	% Occurs	A/N	Sign	Max. Size	How Long is Data Retained	See Note 1	SOURCE		
									H/C	Page	Line
1	SOCIAL SECURITY NO.		100	N		9	UNTIL DELETE	✓	1	01	07
2											
3	YTD WAGES		100	N		8	UNTIL DELETE	✓	C	01	12
4											
5											

FIGURE 16 Sample history form entries.

Definition forms are executed for only those fields where selection or decision-making are required.

An ADS Logic Definition form is shown in Figure 18. As indicated, logic is specified according to the coincidence of data and conditions. The top line of the logic form is used for entering descriptions of transaction or computation fields requiring the application of logic. The column at the right is used for describing specific conditions. The placement of "Y" or "N" marks in the appropriate grid positions expresses inclusive or exclusive combination of conditions.

An example of the use of logic within a payroll procedure is illustrated in Figure 19. In this case, the Logic Definition form is being used to designate the selection and application of wage scales for hourly and salaried workers and for regular and overtime pay. The coding of wage fields is according to ADS documentation within the same system. All three of the wages fields come from Page 1 of the ADS computation form. The wages indicated are indexed, respectively, to line 01, 03, and 06. Applicability of earnings conditions to wage fields is indicated with appropriate "Y" marks at columnar intersection points.

This, briefly, completes the description of ADS documents. Execution of these five interrelated documents has been designed to lead directly into the

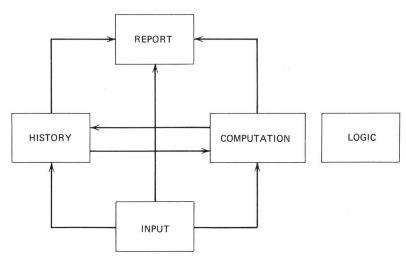

FIGURE 17 Report generation cycle.

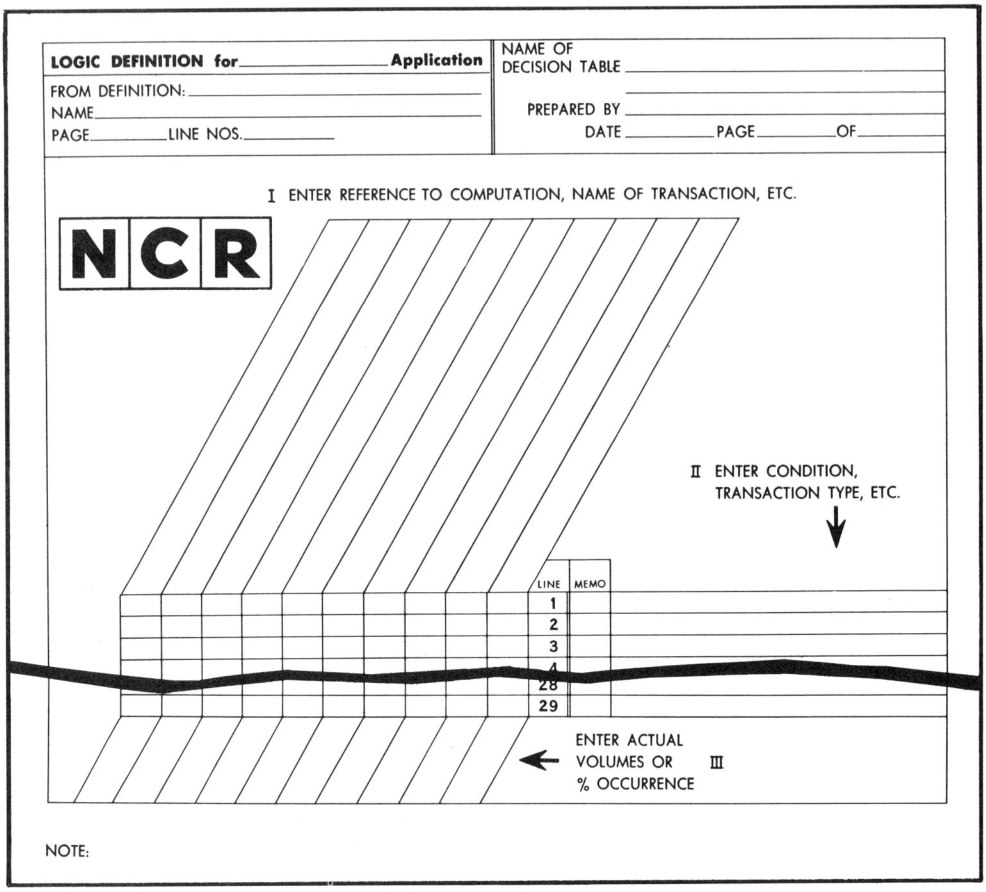

FIGURE 18 Logic definition form.

next phase of the system development cycle – programming. Since this presentation deals specifically with ADS techniques, it is not appropriate to go into the final two phases of system development in depth. However, it is pertinent to describe the remaining steps to be completed after ADS documentation is submitted for programming implementation.

USING ADS DOCUMENTATION IN PROGRAM WRITING

ADS documentation provides a sound, factual starting point. Because ADS provides a detail system definition, programming management is much better equipped to plan appropriate manpower assignments. For example, programmers can be given segments of an overall system based on individual reports and contributing procedures. Or, if a job is more urgent, the work can be further sub-divided, with one person handling form design while another designs run flowcharts.

Within NCR, in addition to the ADS procedures, a separate, similarly disciplined documentation system for controlling the programming phase of system development has been in effective use for two years. In the ADS documentation phase, it is important to have forms and procedures which will be readily understood and used by non-EDP per-

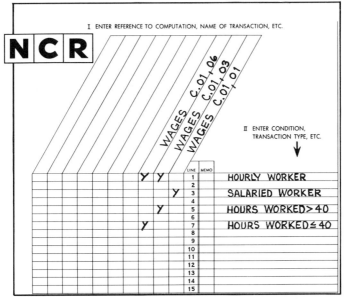

FIGURE 19 Sample logic definition form entries.

sonnel. This communication requirement was the chief reason for the elimination of any hardware dependence within ADS procedures. On the other hand, the programming documentation techniques developed by NCR are closely related to system design. Of necessity, they are intended primarily for the use of EDP-trained personnel.

Under these procedures, program development is broken down into six stages:

A-SPEC. System description relating to run flow, hardware approach, timing estimates, uniform descriptions, and specifications of planned systems tests. Each program required by the system has documented all transactions and narrative of responsibilities and an overall function flowchart. *A-SPEC* also includes a detail estimate of implementation in man-months and programming assignments.

B-SPEC. Macro flowcharts of each program, file labels and layouts, and individual test cases to be used in testing. *B-SPEC* also incorporates an updated *A-SPEC*.

C-SPEC. Coding of each program.

D-SPEC. Testing and debugging at the program level. Test data are documented and preserved for subsequent maintenance.

E-SPEC. Testing of complete system, incorporating run-to-run communications, controls, label checks, and all appropriate requirements of the operating system.

F-SPEC. This is a final edit and consolidation of documentation produced during the prior development steps.

A disciplined system of implementation, as represented by the *A-* through *F-SPEC*, provides numerous advantages to the program manager. He can measure progress and quality. He achieves flexibility of job assignments by having quality documentation available. And most important, new employees are more consistently productive because they have a guideline of the discharge of responsibilities.

ADS RESULTS

The ADS concept, coupled with the A-through F-SPEC for system definition and programming, addresses an important need by adding a number of definite values to EDP system development. These include:

- Management understanding.
- Guidance, including the ability to check system definition progress at any given time.
- Better communication.
- Task definition at each documentation level.
- Systematic approach definition.
- Effective use of personnel.
- Realistic scheduling commitments.
- Quality control over system definition.
- Status review of any job, individually and independently at any time, by system development and management personnel.
- Consistently accurate and complete documentation.
- Savings in time and money.

Introduction to SECTION III

Third Generation System Analysis Techniques

Use of the computer as an aid in system analysis characterizes the third generation of system analysis techniques. ADS processors were developed in the late 1960's, but were not used extensively because modifications to the system were laborious to incorporate. Decision table processors proved much more effective, as described in the paper by Pollack, et al. IBM's TAG system (Time Automated Grid) was developed in 1962 and automated in 1966. When the proper data regarding output of a system are fed into TAG, the system derives information specifying what inputs are necessary at what point in time. The next iteration produces file format and system flow descriptions. The second paper in this section explains the TAG system.

Although beneficial, the automating of manual system analysis techniques produced suboptimal results. While this process was occurring, theoretical work was under way to develop techniques designed specifically for computer use. Young and Kent, Grindley, Langefors, and the CODASYL committee produced papers which provided theory essential to the development of both third and fourth generation system analysis techniques. Five of these papers are included in this section.

The paper on "Problem Statement Languages in MIS," by Teichroew correlates the theory in these papers with other theoretical work. A research project at the University of Michigan, headed by Teichroew, produced a problem statement language that enables system requirements to be input directly to the computer. A second paper by Teichroew describes both PSL and PSA, the Problem Statement Analyzer which produces an automated set of program specifications.

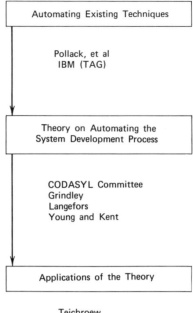

FIGURE 1 Structure of section III.

The Hoskyns System permits translation of system specifications into computer programs without manual intervention. The paper by Rhodes describes the Hoskyns System.

Next in the automation of the system development process is the optimization of the system design. SODA (System Optimization and Design Algorithm) generates an optimal system design, starting from the statement of processing requirements. Specifically, SODA is a system for selection of CPU and core size and the specification of alternative designs of program and file structure. SODA also selects auxiliary memory devices and makes a performance evaluation of alternative designs. The paper by Nunamaker explains SODA.

PSL/PSA is operational. SODA is also operational, although portions of the process require man-machine interaction. The Hoskyns System is also operational. Surprisingly, only a small set of organizations is utilizing these techniques.

In summary, the objective of the third generation of system analysis techniques was computer-aided system analysis. In the early part of this era, effort was devoted to automating existing system analysis techniques. Although adequate, this approach was suboptimal. The latter part of the era of

INTRODUCTION

the third generation produced techniques designed specifically for computer processing.

It is enigmatic that practitioners in the computer field have been slow to use the computer as a means to improve their own activities. The papers in this section show that the potential of the computer has finally been used to facilitate the tasks for system analysis and design. A summary of third generation system analysis theory and techniques is diagrammed in Figure 1.

THE LANGUAGES USED IN DECISION TABLES

S. POLLACK
H. HICKS
W. HARRISON

HISTORICAL DISCUSSION

Tables are a familiar part of everyone's life. From mathematical tables to the box score of yesterday's baseball game, they provide us with an orderly presentation of data. While such tables often assist us in making a decision, they are not decision tables. A decision table is a special form of table that codifies a set of decision rules based on a specific, clearly identified set of conditions and resulting actions. While the history of tables in general can be told in terms of centuries, decision tables are a fairly recent phenomenon.

SOURCE: Pollack, S., H. Hicks, W. Harrison, "The Languages Used in Decision Tables," *Decision Tables: Theory and Practice,* Wiley-Interscience, 1971, pp. 3-6, 63-69. Copyright © 1971 by John Wiley and Sons, Inc.

In November 1957 General Electric initiated a research effort called the "Integrated Systems Project." Its objective was the study of the manufacturing processes that occur from the receipt of a customer order through the production of the finished product and the part that computers might play in them. It soon became apparent that the available methods of describing decisions—flowcharts, formuli, narratives, and the like—were inadequate for expressing the complex logic encountered in the processes being studied. For this reason the project team began a search for a new method of expression which culminated in the development of "decision structure tables" and a computerized method for solving them. These tables had all the characteristics of what we know today as decision tables but had a format similar to the truth tables from which they originated. Examples of a truth table and a TABSOL table are shown in Figures 1-1 and 1-2. The processor for solving these tables operated initially on an IBM 702 and was successively implemented on an IBM 305, 650, and 704. An improved processor and language called TABSOL were implemented on the GE 225 in early 1961.

A	B	A ∨ B	A ∧ B
T	T	T	T
T	F	T	F
F	T	T	F
F	F	F	F

FIGURE 1-1 Truth Table indicates for each of the truth values that A and B can assume the truth value of the logical statements "A OR B," "A AND B." The table is read horizontally.

ITEM-1	ITEM-2 EQ	ITEM-3 EQ	GO TO
EQ 4	3	05	TABLE-2
EQ 6	4	10	TABLE-2
GR 7	5	15	TABLE-3

FIGURE 1-2 A decision structure table in TABSOL format. Decision rules are read horizontally; for example, the first rule reads: "If ITEM-1 EQ 4 and ITEM-2 EQ 3 and ITEM-3 EQ 05, then GO TO TABLE-2."

At approximately the same time, and independently of General Electric's efforts, the Sutherland Company developed a decision table different in form but identical in concept. Whereas General Electric developed the concept of decision tables and the computer-based solution method almost simultaneously, Sutherland developed their tables strictly as an aid to system analysis and documentation, leaving the solution of the table to the programmer. As a result, the subject matter expressed within the two forms of tables differed. General Electric utilized their tables to describe manufacturing decisions in great detail, whereas Sutherland used tables to express more general "management rules," expressions of policy independent of the eventual processing media. As was the case with General Electric, Sutherland was forced to develop decision tables out of desperation. They had expended almost one man-year of effort attempting to specify the logic of a complex file update procedure without useful results. When the effort was restarted using decision tables, it was completed successfully in about twelve man-weeks. Following this initial experience, Sutherland continued using decision tables for documenting a wide variety of systems. An example of a communication-oriented decision table is shown in Figure 1-3.

Another early user, Hunt Foods and Industries, began using decision tables as an aid in man-to-man communication in 1959. This work was described in one of the earliest published works on decision tables.[1]

In May 1959 the Conference on Data Systems Languages (CODASYL, the organization that developed COBOL) was convened. It designated one of its committees, the Systems Group, to pursue the objective of developing a machine-independent, systems-oriented language. After reviewing several approaches to this objective, they began to study decision tables—an effort that was to occupy two years and result in a decision table language known as DETAB-X (Decision Tables, Experimental).

In 1960, General Electric presented their work on decision tables at the Eastern Joint Computer Conference; and during the next two years a great deal of effort, most of it unreported in the literature, was spent in developing decision table processors. IBM guided the implementation of at least three decision table processors, one on the IBM 1401, one on the 7080 in cooperation with the Boeing Company, and one on the 7090 in conjunction with RAND

[1] Orren Y. Evans, "Advanced Analysis Method for Integrated Electronic Data Processing." *IBM General Information Manual*, F20-8047.

		L-CODE = 6		
		DESC = 8	DESC = 9	DESC = 10
CLASS = 2	TYPE = 4	ACTION-1	ACTION-1	ACTION-2
	TYPE = 3	ACTION-3	ACTION-2	ACTION-1

FIGURE 1-3 An example of a decision table used for man-to-man communication. The conditions are listed both horizontally on the left and vertically across the top, with the actions shown in the center. Just as the TABSOL table is related to truth tables, this form appears related to the Karnaugh map.

Corporation (FORTAB). Insurance Company of North America produced a decision table processor of their own called LOBOC, also on the IBM 7080, and General Electric, as noted earlier, implemented TABOSOL on the GE 225.

In September 1962 the CODASYL Systems Group held a seminar in New York to present the results of their study of decision tables to the public. The product of their effort, called DETAB-X, consisted of a language supplement to COBOL-61 to be used within the framework of decision tables. The seminar featured talks by the early developers: Sutherland, General Electric, Insurance Company of North America, RAND, and IBM. Its objective was to stimulate interest and experimentation in the use of decision tables and their translators. In spite of the enthusiasm of the Systems Group and the information content of the seminar, the experimentation and resulting exchange of information that was hoped for never took place. The Group shortly moved on to other projects, leaving as their testament the format of decision table now accepted as "standard." An example of this format is shown in Figure 1-4.

The period from the DETAB-X seminar until 1965 was marked by inactivity. Few articles were written and little expansion in the use of decision tables by new users was observable. Then, in June 1965, the Special Interest Group for Programming Languages (SIGPLAN) of the Los Angeles Chapter of the Association for Computing Machinery appointed a working group to develop a decision table preprocessor. In order to guarantee a wide distribution, the preprocessor was written in a restricted subset of COBOL and accepted decision tables coded in COBOL to convert them to COBOL source code. The preprocessor, called DETAB/65, was released free of charge and distributed through the Joint Users Group. Although it was implemented on a number of computers, including the CDC 1604, 3400, and 3600 and the IBM 7040, 7044, and 7094, its admittedly inefficient conversion algorithm and lack of maintenance led to its disuse and

Table 3	1	2	3	4	ELSE
FIELD-1 = 3 FIELD-2 = FIELD-3 =	Y 3 ZERO	Y 4 ZERO	N 10 POSITIVE	N 15 NEGATIVE	– – –
MOVE A-6 TO A-7 GO TO	X TABLE-4	– TABLE-4	X TABLE-5	X TABLE-6	– TABLE-9

FIGURE 1-4 A decision table in the "standard" format. Conditions and actions are listed on the left-hand side with decision rules read vertically from top to bottom; for example, Rule 2 reads "If FIELD-1 = 3 and FIELD-2 = 4 and FIELD-3 = ZERO, then GO TO TABLE-4."

eventual disappearance. It seems evident, however, that DETAB/65 was the ancestor of the current group of proprietary decision table preprocessors that have been developed since 1966. These preprocessors generally follow the DETAB/65 design—a preprocessor written in COBOL that converts decision tables containing COBOL components to a stream of COBOL code suitable for presentation to a compiler. The exception is IBM's System/360 Decision Logic Translator that processes decision tables coded in FORTRAN. Some of the preprocessors originally developed for COBOL offer the option of using FORTRAN.

In summary, the history of decision tables can be viewed as consisting of four eras:

1. The era of initial development, 1957–1960.
2. The first era of preprocessors, 1961–1962.
3. The era of silence, 1963–1965.
4. The second era of preprocessors, 1966–present.

A review of this history leads one to wonder why decision tables are not the universally used technique for system analysis and program development. The literature of decision tables is replete with stories of how this technique succeeded where the more widely used methods of flowcharting and narrative had failed. Indeed, the early "pioneers" were driven to invent decision tables because they could not solve their problems by any other means, and many subsequent users first turned to decision tables for the same reason. This being true, why has the use of decision tables not been wider?

Three possible causes can be identified. First, there has been limited information on decision tables available to the main body of system analysts and programmers. Many of the earlier articles represent highly technical discussions of conversion methods (which assume prior knowledge), or were published in proceedings, technical journals, or other media not generally read by the commercial practitioner. Computer manufacturers have not, as a rule, taught decision tables in their introductory programming courses, whereas they invariably teach flowcharting.

Second, the use of decision tables requires a different way of looking at problems than does flowcharting, the method of analysis learned first by analysts and programmers. Flowcharting leads one to adopt a sequential model of decision-making—a test followed by one or more actions, then another test or two, and so on. Decision tables, on the other hand, require an overall analysis of the conditions that comprise a given problem and the effect of their various combinations on the solution. It is only natural that analysts and programmers, trained in a sequential type of analysis, resist shifting their fundamental outlook to accommodate decision tables.

Third, there has been a general lack of decision table translators available to the data processing community. Most of those developed before DETAB/65 were distributed in a limited way if at all. This meant that most decision tables that were developed had to be hand translated to sequential code for input to the computer. Experience has shown that the absence of a mechanized means of translation will result in a rapid decrease of interest in decision tables on the part of the programming staff.

These three conditions are now in the process of amelioration. Six books have been published in the last two years on the subject of decision tables. In addition, seminars conducted by both ACM and several private consultants have been and are currently increasing the number of analysts and programmers who know of decision tables. This increase of knowledge, coupled with the availability of a number of preprocessors, will, one hopes, overcome the historical resistance to decision tables.

THE LANGUAGES USED IN DECISION TABLES

It has been mentioned previously that a variety of languages are used in decision tables: COBOL, FORTRAN, ALGOL, or any number of other languages including the natural languages such as English.

For decision tables used in computer programs, each particular decision table preprocessor will specify the language or languages that may be used. The manual or user's guide for a specific preprocessor will explain restrictions or extensions to the particular language which are of interest to their users.

The three most commonly used languages—English, COBOL and FORTRAN—are briefly discussed here. Each of these languages has a large vocabulary and rules of usage that are beyond the scope of this section.

Syntax and Semantics

Each of our languages (English, COBOL, and FORTRAN) may be considered as a combination of two parts: *syntax* and *semantics*.

Syntax. Syntax is the structure of a language. To illustrate, in COBOL the structure (i.e., syntax) of the MOVE statement (simplified) is,

$$MOVE \begin{Bmatrix} \text{data-name-1} \\ \text{literal} \end{Bmatrix}$$

TO data-name-2 [data-name-3] . . .

This structure means that after the word MOVE, only a data-name or a literal may appear (the two are in braces, indicating a choice). After this, only the word TO may appear. Following TO, a data-name *must* appear (data-name-2 is not in brackets indicating that it is required), and any number of data-names *may* follow that required one (being in brackets, data-name-3 is optional and the ellipsis (. . .) following indicates it may be repeated.

As allowed by the syntax, the following are legal COBOL MOVE statements:

MOVE JOE TO MARY

MOVE MARY TO JOE
MOVE 30 TO TOM
    DICK HARRY

But the following are not legal, because they violate the syntax (or structure) for the MOVE statement as it was stated above:

> MOVE JOE MARY TO
> TOWN (JOE MARY
> is neither a data-name
> nor a literal.)
> JOE MOVE TO MARY
> JOE TO MARY MOVE

To sum up, syntax has to do with the allowed order of words in a meaningful statement.

Semantics. Words, of course, are not simply strings of letters. They have meanings and the exact meaning of a word in a statement is the concern of semantics.

In the MOVE example given in the discussion of syntax, the names JOE, MARY, and others were used, in a syntactically correct manner, as data-names. If, in the COBOL program that contains "MOVE JOE TO MARY," JOE is defined as procedure-name rather than as a data-name, then the statement is incorrect because of semantics.

Languages

Both the syntax and semantics of any natural language are very involved and often not fully definable in any formal way. The *grammar* of a natural language includes the areas of both syntax and semantics.

Programming languages are artificial, rather than natural, languages. They are modeled both after the natural language and after the symbolic notation which might be called the "natural language" of mathematics. COBOL tends much more to English-like syntax while FORTRAN is closer to being mathematics-like. All programming languages were designed to aid and simplify the statement of a problem in terms that are easily translated to machine-executable instructions.

English. English is widely used in decision tables. System analysts often use English freely or define a special subset related to their problem. An example from a manufacturing system analysis is shown in Figure 1-5.

One major semantic question here is, what does *fit* mean? It could be interpreted as "inside diameter of the bearing equal to outside diameter of the shaft," or perhaps "does the threaded mounting parts of each have the same thread characteristics?" or a thousand other interpretations.

The English-language statement often relies on the reader "knowing what is meant," within a particular context. Within the particular manufacturing operation, whose analysis included the table in Figure 1-5, the meaning will be clear. Everyone involved will "know what fit means" so there would be no point in explaining matters further.

In attempting to analyze a system, or organize some information or procedure that is not intended to be directly converted to a computer program, or simply

STUBS . . .		RULES . . .	
C Will this bearing fit that shaft?		Y	N
A Use this bearing		X	—
A Try another bearing		—	X

FIGURE 1-5 Sample decision table containing English.

to help organize one's own thoughts, the English language table is exceptionally well suited for use.

COBOL. COBOL is a procedural programming language that is used to define, step by step, operations to be followed in performing systems tasks. Commercial work is its primary application; indeed the name COBOL is derived from the phrase *COmmon Business Oriented Language.* Strong in the areas of logical decision-making itself, COBOL is perhaps the favorite language utilized in decision tables that are part of computer programs.

COBOL'S basic logical structure is the IF statement whose syntax (simplified) is, IF condition THEN true-action [ELSE false-action].

The condition might be the equality of the value of two data items (IF X = Y . . .) or any number of other forms. Since COBOL language programs must be translated to machine executable instructions by a computer program known as a compiler, the full set of syntactical and semantical requirements can be obtained from the manual for the particular COBOL compiler to be used. COBOL has been standardized by ANSI, the American National Standards Institute (formerly USASI). COBOL compilers conforming to ANSI standards are called ANS COBOL compilers.

To return to the IF statement, if a condition is evaluated and found true, then the COBOL statements in the *true-action* are executed. If the condition is found false, those statements in the optional *false-action* are executed. This is similar to the mutual exclusion between rules within a single limited entry condition row of a decision table.

COBOL-oriented decision tables are often decomposed into a series of IF statements linked to one another, and to the table actions by GO TO statements. GO TO's direct the sequence of execution of the program, to permit the full checking of all conditions required to select the one rule of the decision table whose actions will be executed. Different preprocessors often generate logically similar, but physically different, coding. These differences result from the differences in the algorithms used to achieve efficiencies. One procedure for setting up this structure of IF statements is discussed under the subject of Decomposition.

Another COBOL statement of interest in connection with decision tables is the PERFORM statement which permits execution of *subroutine* that are located outside of the table or execution of other tables. Numerous other COBOL statements, including debugging aids, may be studied by those interested in COBOL-oriented decision tables.

FORTRAN. This language is known as a scientific and engineering applications language but it is also used for commercial programming.

Two types of IF statements occur in FORTRAN. The older is a simple comparison of two operands with three actions: greater, equal, or less. The newer form is a two-way compare and permits a greater variety of types of comparisons than does the older form. Its syntax (simplified) is, IF (operand-1 .xx. operand-2) THEN true-action.

.xx. represents a number of logical operators such as .EQ., .LT., etc. These FORTRAN operators should not be confused with the bound-operators of decision table conditions despite the use of the same set of operator names. The above FORTRAN comparison form, like the COBOL form, is a binary (two-way) comparison whose results are mutually exclusive.

FORTRAN also contains other operators of interest to decision table users, such as GO TO and DO. DO causes repetitive execution of a series of statements following it with control over a variable. The DO is somewhat similar to the COBOL PERFORM, but the "assigned GO TO" feature is popularly used for remote subroutine execution.

FORTRAN compilers have been provided for almost every major general-purpose computer system on the market. It has also been standardized by ANSI.

Other Languages. Almost any programming language has been or could be used in conjunction with decision tables. These include ALGOL, PL-I, BASIC, and assembly languages. New programming systems are constantly emanating from computer manufacturers, software companies, universities, and professional and industry organizations; standardization of these languages is an active area. All of these newer languages have provision for making binary decisions (e.g., IF) and for altering the flow of control (e.g., GO TO). Both of these are basic elements in the decomposition of decision tables to computer code.

COBOL Macros. COBOL is known as a *higher level language* because of its ability to represent commonly used operations and procedures in a more concise, and consequently much shorter, way than is possible using assembly language. Thus a single COBOL (or FORTRAN, or any other higher level language for that matter) statement is more encompassing than a single assembly language statement. COBOL can therefore be considered a *macro* form of assembly language.

However, certain COBOL statements are quite commonly used together; thus they may be considered as eligible for being grouped and represented by new, even higher level, macros. One good example of this is IBM's COBOL-F compiler (for the S/370 line, OS operating system) in its treatment of the Report Writer features. The compiler internally preprocesses such Report Writer statements as "GENERATE" into other COBOL code such as WRITE statements which are then compiled as if the programmer had originally written such statements. Inasmuch as these Report Writer statements are part of ANS COBOL, the term COBOL *macro* is not really appropriate to them; but they do illustrate the nature of a macro representation of a high-level language.

Macros can be defined as statements that represent a larger body of statements of a lower level language.

Macros can be either parametric or nonparametric. That is, variables may be provided which will be substituted into the macros at specified places or change the macros' structure if they are parametric, but no substitutions are permitted in nonparametric macros. One special case of nonparametric macros, known as *shorthand,* is widely used by commercial programmers to cut down on the amount of writing required in COBOL programming. Shorthand usually does just what its name implies—it serves primarily to reduce writing. The COBOL standard itself provides a limited number of shorthands (also called "abbreviations") in its Report Writer feature.

Parametric macros permit the programmer, when using them, to indicate data-names to be substituted into the structure. More advanced macros not only allow the simple substitution of names but can also restructure themselves, sometimes in drastic ways, depending on the parameters supplied. The details of this type of macro are well beyond the scope of this book; but the general approach used is somewhat similar to the macro definition facility in many assembly languages.

STUBS...		RULES...					
C	SUPPLY-OK (SHAFT, BEARING)	Y	N	N	N	N	N
C	GET-FROM (IN-HOUSE)	–	Y	Y	N	N	N
C	GET-FROM (VENDOR)	–	–	–	Y	Y	N
C	AVAILABLE (REQUIRE)	–	Y	N	Y	N	–
A	ITEM-ORDER (BEARING)	X	X	X	X	X	X
A	ORDER (FILL)	X	–	–	–	–	–
A	ORDER (MODEL-SHOP)	–	–	–	–	X	X
A	ORDER (BACK, WORK)	–	X	X	–	–	–
A	ORDER (BACK, PURCH)	–	–	–	X	–	–
A	QUALIFY (PRIORITY)	–	–	X	–	X	X

FIGURE 1-6 Decision table with macros.

Notice that the preprocessed decision table fits the stated definition for a macro; the decision table *is* a macro in a sense. Moreover, macros can often be used within the stub and rule entries of decision tables. In fact, it is quite common to use shorthand types of macros in programming decision tables; and several vendors of decision table preprocessors offer combined decision table-shorthand packages. Figure 1-6 shows a decision table containing macros in both the conditions and actions sections. These COBOL macros are compact ways of writing larger, more detailed, groups of COBOL statements which occur often in an applications area. COBOL macros are not currently widely used and their format is far from standardized. In this example, the common *assembly language macro* approach is used by putting variable parameters in parenthetical clauses following the macro *verb*. Figure 1-7 shows the same table after the macros have been expanded. The general definition of these macros and the expansion process is not given here. [However, portions of the expansion in Figure 10-3 are due to a first and second parameter.] One macro may effect the expansion of another. Such effects are illustrated in the expansions of ORDER where BEARING was set by ITEM-ORDER. Macros may be simple shorthand forms (this form is not illustrated here). However, the parameter insertions in GET-FROM are typical of multi-unit shorthand substitutions), direct insertion of parameters (illustrated by SUPPLY-OK), or change of form (illustrated by QUALIFY). These macros are meant to be close to the terminology that might be used by the system analyst.

Interactive Programming and Decision Tables

The term *interactive programming* means that the programmer communicates directly with the computer through a console, usually a typewriter or cathode ray tube (CRT) type of device. Commercial COBOL programmers are now beginning to use interactive programming. FORTRAN and other programmers have been using it for some time.

This environment is similar to the "good old days" of programming when debugging consisted of sitting at the console and displaying core locations in mid-run, altering the stored program, and the dozens of other tricks known to console debuggers. The breed of console debuggers faded because of the increasing cost of machine time; no one could afford the high percentage of idle time associated with console debugging.

But the advent in recent years of multi-programming has changed that picture. Interactive programming as

	STUBS . . .	\multicolumn{6}{c}{RULES . . .}					
C	QUANT OF SHAFT OF TRANSACTION NOT > ON-HAND OF BEARING OF MASTER-FILE-RECORD	Y	N	N	N	N	N
C	SUPPLIER OF MASTER-FILE-RECORD = 'F' OR = 'W' OR = 'G' OR PRODUCTION-CURRENT = 'Y'	–	Y	Y	N	N	N
C	SUPPLIER OF MASTER-FILE-RECORD = 'V'	–	–	–	Y	Y	N
C	LEAD-TIME OF MASTER-FILE-RECORD NOT> REQUIRED OF TRANSACTION-RECORD	–	Y	N	Y	N	–
A	MOVE "BEARING" TO ORDER-ITEM MOVE QUANT OF BEARING OF TRANSACTION TO ORDER-QUANT.	X	X	X	X	X	X
A	PERFORM FILL-ORDER.	X	–	–	–	–	–
A	PERFORM ENGINEERING-ADVICE. PERFORM RFQ. PERFORM MODEL-SHOP-ORDER.	–	–	–	–	X	X
A	PERFORM BACK-ORDER ADD QUANT OF BEARING OF TRANSACTION TO BACK-INVENTORY OF MASTER-FILE-RECORD. PERFORM WORK-ORDER.	–	X	X	–	–	–
A	PERFORM BACK-ORDER ADD QUANT OF BEARING OF TRANSACTION TO BACK-INVENTORY OF MASTER-FILE-RECORD. PERFORM PURCHASE ORDER.	–	–	–	X	–	–
A	MOVE "RUSH, TOP PRIORITY" TO ADVISE OF ORDER-RECORD.	–	–	X	–	X	X

FIGURE 1-7 Expansion of the macros in the Figure 1-6 decision table.

it is practiced today is conducted at the COBOL symbolic, rather than at the machine, level and does not permit many of the old-timer's techniques. But it does permit, typically, changing the symbolic (source) program, and inspecting results via the console, saving considerable computer turnaround time. Many systems also permit inspection of the program state in mid-execution along with other useful facilities.

Decision tables can be extremely useful in the interactive environment. They are powerful macros representing a great deal of logic in a very compact form. Their automatic logical integrity (completeness, lack of redundancy and contradiction) helps both programmers and system analysts control their logical structures. The ease with which decision tables can be changed is welcomed by the interactive programmer.

The use of macros, such as shorthand, within decision tables has the obvious added advantage of cutting down considerably the key-strokes required on the keyboard. Thus properly designed shorthand can help directly reduce errors while increasing readability.

The documentation value of the decision table is of particular value to the on-line interactive programmer.

With a relatively slow (typewriter) or limited space (CRT) display device, the compact table format delivers a lot of information at a minimum of cost.

The format of input of the decision table on the console is obviously not exactly the same as the card-oriented formats used by many preprocessors. Alignment of rules vertically has been altered by several different implementations. The authors feel that it is still too early to suggest a conventional decision-table format for use in the interactive environment.

THE TIME AUTOMATED GRID SYSTEM (TAG)

IBM

Progress has been the keynote in data processing since the introduction of the computer: improvements in hardware design have resulted in equipment capable of performing hundreds or even thousands of instructions in a single second; software packages have been created that allow the computer to monitor its own operation; complex activities in all industries have been brought under the computer's control. The trend has been, and still is, toward increased speed and sophistication of equipment and broader scope of use. Real-time systems, management-information systems, systems utilizing integrated data bases figure more and more among the processing objectives of users.

But as the computer and its functions grow in complexity, the task of systems design becomes more difficult. Effectively automating an order entry application, for example, is by no means as formidable a project as implementing a tem that involves twenty application areas, or one that is geared to respond in minutes to unanticipated inquiries about any element in the data base. The problem of defining the *best* systems solution increases with the scope of the system being developed. Recognition of this fact has led to the establishment of formalized techniques intended to aid the systems planner in his approach to the design of large-scale systems.

Valuable though these techniques are in directing the systems study, they suffer significant limitations: they neither lessen the effort nor significantly reduce the amount of time required to carry out a study. They provide a plan of action, but rely totally upon manual effort for execution of the plan. Where advanced systems concepts are to be realized, dependence on manual techniques alone

SOURCE: IBM, "The Time Automated Grid System (TAG): Sales and Systems Guide," Publication No. GY20-0358-1, 2nd Edition, May, 1971, pp. 1–12. Reprinted by permission of International Business Machines Corporation. © 1971 by IBM.

is an obstacle to design. An examination of why this is so will lead to an understanding of the benefits to be gained from use of the Time Automated Grid system (TAG).

SYSTEMS DESIGN PROBLEMS

Systems design is always subject to certain obstacles. Large-scale systems design is subject to a magnified version of these same obstacles. The difficulty of properly surveying system requirements, current and future, of drawing up design objectives, and of defining the optimum solution to a systems question is intensified when the area under study is vast. The time and staff required to perform a thorough study increase in proportion to the scope of that study. In most instances of large-scale systems design, however, both time and personnel are limited.

Systems specialists represent a very small percentage of the total staff within any company. Yet their job requires that they accurately research all, or a great portion, of the activities in which the corporation engages. The quality and the amount of work that the systems group can accomplish in a given timespan depends upon the number in the group and their experience in the systems field. This is true whether the system under consideration is restricted in scope or is all-encompassing. But designing a large, integrated system makes greater demands upon the talents of the systems group than does development of a more limited system. Therefore, where advanced concepts are to be implemented, some means of increasing productivity—of obtaining more from the study team—must be found. One way to do this is to expand the systems group to an optimum size. However, attempting to increase team size causes additional problems.

First, skilled systems analysts are not always readily available. Second, when hired, they must be trained in company procedures before they can participate efficiently in design work. Once that work is completed and a new system is implemented, personnel requirements may undergo a change. Where fewer analysts are needed to maintain a system than were required to assist in its design, the question arises of how to continue utilizing the additional staff members. Perhaps most significant of all is the economic question—how to justify and absorb the costs entailed.

If supplementing the study team is not the answer to maximizing productivity, perhaps the next best solution might be an increase in the amount of time allotted for the systems study. Yet experience has shown that extending the period of time between the start of a study and implementation of a system is usually not feasible. Need for a speedy solution to the problems that made redesign of an existing system necessary in the first place is one reason. The dynamic quality of any business organization is another. A company's goods, the services it provides, its internal structure and procedures vary with the passage of time. In a rapidly developing corporation, changes may occur in a matter of months. Changes will almost always occur over a period of years. To design a unified system with an integrated data base in an evolving environment is like trying to hit a moving target. Design objectives that are valid at the beginning of the study project may no longer be so, or may be only partially so, six months or a year later. Requirements that did not exist at the start of the design phase must be incorporated into work already completed. Evaluation and communication of changes and of their effects on the final system design must take place. All this requires additional time and can push implementation of the system still further into the future.

Obviously, if increased productivity cannot be achieved by augmenting either systems personnel or the time to be given over to the systems study, the technique employed to direct the study should provide a means of attaining this goal. Conventional study-directing techniques cannot do so because, being manual, they require time-consuming, laborious effort on the part of the systems planner.

Documentation and reduction of data, and the organization of it into a systems flow, are all done by hand. Currently, the clerical portion of systems design work can require as much time as its creative or decision-making aspects, if not more. Given this situation, plus limitations in staff and study time, it is highly possible that important design questions may be neglected or glossed over. Failure to probe deeply enough adversely affects the quality of the systems solution. Ideally, the study team should be able to weigh the value of all the data it collects, search out discrepancies and redundancies in the data base, and examine all the alternative solutions to the systems question and the effects of each solution on the total design. In reality, systems planners are often forced to accept a less than optimum design. Sacrifices in quality are made to ensure completion of the systems task within a reasonable time-span. For example, we know that functional areas within a business organization may overlap one another. Data pass between them. What is output in one area may be input to another. In order to design a unified system, the analyst must be aware of when and how data elements cross the barriers between functional and operational areas. Developing this awareness can demand an expenditure of time that the systems planner cannot afford. Consequently he may choose to study each activity within his company as though it were a self-contained entity isolated from all other activities. Maintaining an overview of the system then becomes virtually impossible. The concept of system unity, or integration, is lost.

The key to effective design lies in giving over more of the available study time to the decision-making processes and making it easier for the systems planner to reach conclusions about data requirements. If the systems analyst must rely on his own data-recording and organizing abilities (as he has had to do in the past), his productivity cannot be as great in the purely inventive area of design. What is needed is a technique to relieve him of his clerical duties while directing him in his creative efforts. That technique is TAG.

TAG-ASSISTED SYSTEMS DESIGN

Originally developed as a manual systems design tool, TAG was automated late in 1966. TAG is a general-purpose technique applicable to the design of any data-processing system in the commercial environment. Its use is not limited to a particular industry, application, or hardware configuration. With TAG, the user systematizes his study effort while reducing study time and maximizing the creative utilization of personnel. Although TAG is not an "instant" systems design tool—that is, it does not eliminate the need for systems analysts—it is of significant assistance in all areas of systems work: data collection, analysis of data requirements, and definition of data flow.

Use of TAG begins with transcription of the system's output data requirements on an Input-Output Analysis Form. Because the form is simple, those who fill it out need not be experienced systems personnel. The user can draw upon his operational, programming, and control staff in the survey effort, leaving his systems staff free to review the completed work and to direct their attention to questions that require further investigation. Employment of personnel over and above the systems team affords greater speed in the data collection effort, and means that the analysis and design phases of the systems study can begin earlier than they otherwise would. Standardized documentation of systems requirements and upgrading of personnel in systems operation are by-products of the TAG procedure.

The tasks of analysis and design occur earlier in a TAG-assisted study for a second reason: initially, only a portion of the systems information available need be studied before analysis may begin. TAG looks first at the user's output requirements only. Inputs are examined during later iterations of the program.

Once the output data requirements have been fed into the TAG system, TAG works backward from the output to determine what inputs are necessary and at what point in time. As a result of its output-oriented approach, TAG is able to define a *minimum* data base for any system. With the aid of the reports generated by TAG, the user can systematically resolve the question of how the required inputs are to be entered into the data flow. He is assured of defining only pertinent input elements and of bringing them into the system at their proper place, all with minimum effort on his part. Superfluous or repetitious data can be identified and eliminated from the system. Discrepancies in the use of any data element can be corrected.

When both inputs and outputs have been defined to TAG, the next iteration of the program produces file format and systems flow descriptions. File contents and data flow are both based upon time, the time at which data elements enter the system and the time at which they are required to produce output. To TAG, it is the elapsed time between these two moments that creates the need for files. The files that TAG defines indicate, in detail, what data must be available in each time period to enable the system to function. The job definition depicts the flow of these files, as well as of the inputs and outputs, within and between time cycles. The user obtains an overview of his system, showing the interrelationship of all data in the system. Knowing these interrelationships makes it possible for the system planner to determine whether the outputs desired are quickly and easily obtained, and thus economically justified. With knowledge of the availability of data elements in given time periods, he can readily see where additional useful outputs might be obtained. His creative ability is enhanced. The information provided by TAG allows the analyst to do a more thorough, knowledgeable job of systems design.

With the job definition before him, the user then decides how he wishes to proceed. He may define further data requirements or he may begin optimizing the system flow. The user studies as much or as little information as he chooses at any one time, a few application areas or the operations of the entire business organization. The size of the study is restricted in no way by the use of TAG. TAG is an iterative tool; its function is to develop an integrated systems flow and to maintain that integration no matter how many changes or how much additional data the user introduces. And TAG will do so accurately and at computer speed. Limited time need not be the obstacle to effective design that it is when manual tech-

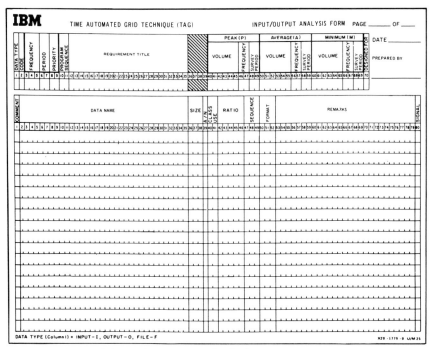

FIGURE 1 TAG input/output analysis form.

niques alone are used. An analyst can experiment with any number of alternative solutions, for a single application or a total system, and arrive at the best one in much less time than he could using manual methods. Users of TAG can convert systems in parts, according to the timetable most convenient for them, without encountering difficulty in tying in other application areas later. If a company's data requirements change at any point during the systems study, new requirements can rapidly be analyzed by TAG and incorporated into the existing job definition. In short, the TAG user obtains, in systems analysis and design, the same flexibility he seeks when he automates application areas.

To summarize, use of TAG permits a reduction in the time and effort required to go from problem definition to systems implementation, maximizes the creative use of systems personnel, and produces as an automatic by-product standardized, up-to-date documentation.

USE OF TAG

TAG Input/Output Analysis Form

All user data to be analyzed by the Time Automated Grid system is recorded on the Input/Output Analysis Form (X20-1779). An illustration of the form appears in Figure 1 (see pp. 68–69 for examples of completed forms and output).

The form is divided into two horizontal section, one dealing with requirement titles, the other with data-names. The characteristics of the input, output, or file being described are recorded in the requirement title section. Comments and the data requirements of the input, output, or file in question are detailed in the data-name section.

Requirement Title Section. Title information is recorded in column 1–70. The name field is left-justified; all other fields are right-justified. Leading zeros are not required.

Column	Field name	Contents
1	Data type	I (input), O (output), F (file).
2	File code	Present only if column 1 is F. R indicates a reference file (input); A indicates an audit file (output).
3–5	Frequency	Number of times this title name is processed.
6–7	Period	Shortest clock or calendar time period within which this data is processed. Possible periods and their codes are second (S), minute (MI), hour (H), day (D), week (W), month (MO), quarter (Q), year (Y). (Frequency plus period establishes a time interval.)
8–9	Priority	Sequence of inputs and outputs within a time interval. Ranking is from 1 to 99, with 1 being highest priority. All requirements not having a priority indicated will be assigned the lowest priority, 99.

TAG uses the frequency, period, and priority information to establish what is called a cycle number. TAG begins with the smallest interval of time designated by the user (e.g., 999 times per second) and works up to the largest, incrementing the cycle number by one each time there is a change in frequency, period, or priority. All requirements with the same frequency, period, and priority will be assigned the same cycle number. (Note that initial assignment of priority cannot be made by the user until all documents for a given frequency and period have been reviewed together. Therefore, in most cases, it is suggested that priorities not be assigned until after the first iteration of TAG.)

Column	Field name	Contents
10–11	Program sequence	Not used with TAG.
12–35	Requirement title	A unique name for the input, output, or file (column 12 may not be blank).
36–39		Blank.
40–69		Peak, average, and minimum volumes, with their associated frequency of occurrence and survey periods.
40–45	Volume	Quantity of inputs or outputs noted during the survey for the time period indicated in columns 6–7.
46–47	Frequency	Number of times the volume indicated occurred during the survey period.
48–49	Survey period	Study period or time during which the volume information was gathered.
50–55	Volume	See 40–45
56–57	Frequency	See 46–47
58–59	Survey period	See 48–49
60–65	Volume	See 40–45
66–67	Frequency	See 46–47
68–69	Survey period	See 48–49
70	Designed for	The volume figure for which the design is intended (P-peak, A-average, M-minimum)
71–80		May contain user ID

Data Name Section. The data-name information is recorded in columns 1–80 of the I/O analysis form. The name field is left-justified; all other fields are right-justified. Leading zeros are not required.

Column	Field name	Contents
1	Comments	When the letter C is entered in column 1, TAB treats columns 2–80 as a comment. Comments may be in any form: an English statement, a FORTRAN formula, a COBOL sentence. If there is no C, it is assumed to be a data description line.
2–35	Data name	Data field name.
36–38	Size	Total number of characters in the data element.
39	Alphabetic/ numeric	N for an all-numeric item; A for all others.
40–41	Class-use	Describes the characteristics and the use of the data element. There are four categories into which data may fall: fixed informational, fixed functional, result of calculation, factor in a calculation. When coding Class-Use, refer to Figure 2.
42–46	Ratio	The average number of times the data element is required in a unit of the title requirement. *Ratio will vary depending upon the unit of measure used to express volume.**
47–49	Sequence	Used for data elements that control the sequence of the input or output. Sequence is indicated as 1 (major control field), 2 (intermediate control field), etc. A maximum of 15 control fields will be accepted. *Sequence should be indicated for at least one data element within a title name. Sequence elements must be listed whether or not they appear on the output.*
50–52	Format	Indicates the order (left to right and top to bottom) in which the data elements appear on the requirement. If format is unimportant, enter a P in column 52. On an output, if the data element is not to appear, leave columns 50–52 blank.
53–79	Comments	May be used for explanatory comment.
80		0, 1, 2 used to modify summary codes. Use with caution!!!!

*For example, assume that a payroll register contains one printed line per employee and that there are 1000 employees. The volume figure for the payroll register may be expressed in terms of the number of lines in the register (1000), or in terms of the number of pages in the register (assume 20), or in terms of the number of registers to be produced (1). If each employee's name is to appear on the register, the ratio figure for "employee name" would be 1 if volume = 1000 lines, 50 if volume = 20 pages, or 1000 if volume = 1 register.

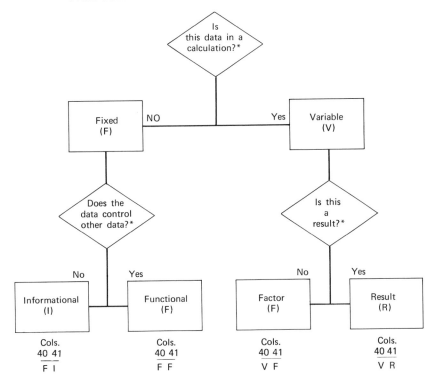

*With respect to this requirement only

Note: For data names coded VR, all factors needed to calculate that result should be listed as data elements even when they do not print on the document. If possible, show on a separate line, as a comment following each VR, the formula used or computations required to arrive at the result.

FIGURE 2 Chart for coding of class-use.

TAG Reports

The output of the Time Automated Grid technique is a series of ten reports that document the user's input, analyze his data requirements, and provide file and data-flow definitions. All reports and their suggested method of utilization are explained below.

Report 1: User's Data. The first report generated by TAG is a listing of the cards keypunched from the data on the I/O analysis form. All information, including comments, is shown. Accompanying each requirement title is a sequential number called an entry number, which is assigned to the requirement as it is read into the TAG system. In addition, each card processed is given a unique card number. This listing provides the user with a record of data analyzed to date.

Report 2: Glossary of Data-Names. Each of the unique data-names submitted to TAG by the user is listed, in alphabetic order within the glossary. The size, characteristics, and reference number associated with each field are printed beside the field name. The glossary provides documentation and a guide for standardization of terms.

Report 3: Document Analysis. Once all requirement titles and their data elements have been fed into the TAG system for analysis, TAG sequences the user's inputs and outputs by priority (01–99) within time interval (999 times per second to once per year). A cycle number is then assigned to each unique time-priority designation. Next, the requirement titles are sorted to type (input, output, reference file, audit file) within cycle, and are given a number by the TAG system called a "header" number. Report 3, the document analysis, is in header-number sequence with a cross-reference to I/O entry number. It summarizes all the requirement titles that have been analyzed, and describes the systems study in terms of the number of headers being examined, their volume, the cycle in which they occur, and the number of alphabetic and numeric characters they contain.

Report 4: Sorted List of Data Names. This listing records, in alphabetic sequence by data element, each usage of an element the way it originally appeared on the I/O analysis form. The cycle, the header number, and the header type for which the data-name is required appear with it. For ease of reference, the card number associated with each element is printed beside it. The sorted list may be used to resolve discrepancies and redundancies in data-names, field sizes, and contents. Duplicate data-names can be deleted; number and type of characters contained in data fields can be standardized with the aid of this report.

Report 5: Time-Grid Analysis. To produce Report 5, TAG reduces the Sorted List of data-names to unique elements. The data numbers previously assigned to the elements are then printed across the page, 20 per page, and the headers in which the data elements appear are printed, in sequence by time, down the page. The figures shown within the grid describe the number of appearances of the element within a particular header. Below each grid, following the last header analyzed, is a line of summary codes, one or more for each data element within the body of the grid. A legend explaining the codes is printed at the bottom of each page of the report.

Use of this grid makes it possible to trace the appearance of each data element, by time, through all the requirements in the system. The grid indicates those data elements that must be carried in files (summary code 1). Thus the time-grid analysis enables the reviewer to begin identifying the minimum data base his system requires.

Elements that may be generated by the system itself are also pointed out by this report (summary code 2). Where these variables are the result of calculations, a check should be made by the reviewer to ensure that all the factors required to produce those results have been included in the system.

Report 6: Summary of Unresolved Conditions. All data elements needing further investigation are listed here. These include items

(a) for which there is no input (summary code 3).
(b) which are not required to produce output (summary code 4).
(c) whose ratio varies from requirement to requirement (summary code 5).
(d) which must be produced before the necessary input is available (summary code 6).
(e) which are duplicates of other data elements required by the same header (summary code 7).

Using this information, the analyst must now determine which of the conditions listed are truly error conditions and decide how he wishes to resolve them. Only he can determine, for example, whether variation in ratio is an error or simply the way in which a data element is used, whether an unused input item is superfluous or will be required by an output not yet submitted to TAG, or whether a needed element should be introduced in a file or on a document.

Report 7: Time/Key Analysis. The time/key analysis lists each header and its characteristics: volume, type, the cycle during which the header is inputted or created, and the key elements associated with it. By examining the time and sequence requirements for each header in relation to all other headers, the analyst can easily see where sequencing problems exist or sorts are required. If he uses the time/key analysis in conjunction with the time-grid analysis, he can determine the best possible source from which to draw missing input elements, since the two reports taken together provide a simple means of matching outputs and inputs having like key elements and time requirements.

On the basis of the information contained in the preceding reports, the analyst makes corrections and changes to the data originally submitted to TAG. Once the modifier user information has been fed into the TAG system, a second series of reports is produced. The first seven will be the same as those described above, updated to reflect all changes. If all input requirements have been defined, TAG constructs working files to communicate between the time periods in the system.

TAG utilizes two approaches to file design, the serial method and the direct access method.

Report 8: Serial File Records. In the serial approach, TAG limits the amount of data in any one record to a required minimum. Although TAG has no way of applying volume and activity figures to the files it constructs, the assumption is made that passing superfluous data through a cycle is costly in terms of time. The underlying philosophy is that the fewer the characters in a single data record, the less read/write time is required to process it, and the shorter a record, the greater the number of records that can be contained in a block of given size. Therefore, TAG attempts to minimize the data base within each individual file. It does this by building a new serial file whenever a sequence change in the required data occurs or a time change between the availability and use of data takes place in the system. This approach results in a multiplicity of files, and can be considered the worst solution to the problem of file format definition.

Report 9: Direct-Access Records. The unrestricted use of direct-access files is the second design alternative outlined by TAG. In this solution, no attempt is made to minimize record size, the concept being that inactive records need not be passed when files are retained on direct-access devices. Therefore, more data than are required in any given time period may be maintained on a file without significantly degrading system performance within a cycle. Using this approach, TAG constructs files on the basis of sequence changes only. Time requirements are ignored. The result is fewer files, but larger data records than exist in the serial-record solution.

Report 10: Job Definition. All working files, inputs, and outputs for each time interval (cycle) in the system are brought together in Report 10. A job definition, or data flow, is produced for each of the two file types, serial and direct-access.

The result of TAG's analysis is a basic working system which has been divided, by time intervals, into several subsystems. Although it is by no means an optimum solution and is not based on actual volume figures and hardware considerations, nevertheless it does provide a definition of the minimum data base required to satisfy output objectives in each cycle. Using his knowledge of record volume and activity, and of the equipment available to him, the analyst must now optimize the data flow. Content of files and time requirements for files, in detail, have been established for him by TAG. He may now work out the most feasible solution to the requirements of the total system or to any one of the subsystems. Each of the subsystem solutions may then be reassembled into a total systems design, using as interfaces between them the working files described by TAG. At any point in the optimization process, the analyst can again submit his data to TAG and receive an updated set of reports showing the effects of the changes he has made upon the entire system.

SYSTEMS STUDY IMPLEMENTATION

The Time Automated Grid technique can be used to best advantage in the development of the management information system. TAG is particularly well suited to the design of large-scale systems, where diversified activities, requiring numerous outputs, are to be brought together and supported by an integrated data base. Effective utilization of TAG, however, is by no means restricted to this specific design situation alone. The technique offers valuable assistance to the analyst wherever the need to define a system's minimum data requirements exists. The "system" in question may represent a single activity area with a high volume of output; it may comprise multiple applications; it may be manual or mechanized. No matter what the nature of the system, TAG will aid the analyst in defining, or re-fining, a data base and information flow. For example, when TAG is applied to the development of a paper-flow system, the output generated by the tool provides a reference guide to the inventory of information to be maintained and the documents to be processed within the system. If the design problem under consideration is one of cutover from unit-record equipment to a computer, TAG alerts the user to the existence within his present system of the redundant data elements that frequently characterize a card-oriented installation.

Use of TAG need not be limited to the area of design. By altering the type of information designated as input or output, the systems analyst can employ TAG to study subjects such as file conversion or the feasibility of implementing a particular application package, given the contents of existing files. TAG is a flexible tool; there are no rigid rules about how it may be employed.

Evaluation of TAG

It is recommended that, prior to undertaking a full-scale study, the potential TAG-user evaluate the technique using data of his own choosing. He should plan to conduct a small pilot study involving approximately a dozen outputs with varying time requirements. The pilot will serve to familiarize the user with TAG's concepts and will afford him the opportunity to compare the technique to other design methods. A careful examination of the results of the study, in terms of effort expended and work accomplished, should show the systems planner whether or not TAG can be of benefit in his particular design environment.

After TAG has been evaluated and accepted as the study-directing technique, the systems project proceeds as it would were conventional design methods used. It is important to remember that TAG does not eliminate any of the essential "ingredients" of the design process. Data collection, analysis, and system definition are components of any design project, whether TAG-assisted or not. However, TAG makes it possible to accomplish these tasks more quickly and more systematically than do "paper-and-pencil" techniques.

Utilization of TAG

Systems Team. The services of at least one experienced systems analyst are required to implement a TAG-directed study. During the survey phase of the systems project, his task is to delegate responsibility for the collection of data, to resolve questions about those data, and to coordinate and monitor the work of the staff at his disposal.

The personnel who make up the survey team may be drawn from any functional area the systems planner desires. The major criterion for effective participation in the study at this point in the project is a familiarity with some phase of the activity being reviewed. If, for example, a department employs clerks who regularly handle certain documents, the clerks could be asked to transcribe the requirements of the documents on I/O analysis forms, for later review by an analyst.

The size of the survey team that an analyst can effectively supervise will, of course, depend upon the competence and systems experience of the individuals in the team. Non-systems personnel, for example, might require more direction from the analyst than systems trainees; and, therefore, a group made up of nonsystems people might, of necessity, be smaller than one composed of systems personnel. Nevertheless, no matter what the characteristics of the study team, TAG can significantly enhance the ability of the analyst in charge to delegate the "clerical" or nondesign portion of systems work to the other members of the team.

Upon completion of the data collection effort, full responsibility for the remainder of the design project reverts to the experienced analyst, whose job it then becomes to study the reports generated by TAG, investigate unresolved conditions, and redesign a system that is in keeping with specific hardware and file considerations.

Design Steps. The first step in the design process, after selection of the activity or activities around which development of a new system is to begin, is the gathering of known facts about systems output requirements. The details of what is needed to create these outputs are coded on I/O analysis forms, reviewed by an analyst, keypunched, and submitted to TAG. The initial analysis of the user's data will pinpoint errors and discrepancies in data definition, such as inadvertent duplication of data-names or incorrect variations in the descriptions of data fields and their contents. In all likelihood, the usage of certain terms will require standardization; and the study team may wish to use the glossary provided by TAG for this purpose.

Following the correction of error conditions and the resubmission of output data requirements to the TAG sytem, the analyst begins assigning priorities to system outputs. Priorities can be assigned only after all of the outputs for a given time period have been reviewed together. With the aid of the reports created by TAG, the analyst redefines the time intervals at which output must be produced. Since he is aware of the over-all relationship among data items and outputs, he is in a position to settle any conflicts that may arise over relative priorities of output. For the same reason, he can also recognize possibilities for new and improved outputs.

When outputs, their characteristics, and their time requirements have been firmly established, TAG directs the user to the items needed to produce those outputs (code 3). Using the report of unresolved conditions as a checklist, the analyst now considers the question of availability of input data. It is up to him to select the most feasible method of entering required data into the system. He must analyze the nature of each input element to be introduced. Is it the kind of item that the system itself has available—a constant or the result of a calculation? If it is not, then the element must be brought into the system via a document or a reference file. The document solution will be the one chosen when an item must be reintroduced each time it is used.

Since this is obviously the most unwieldy and costly manner of entering data into the system flow, the analyst should be absolutely certain that no other answer to the input question exists. If, on the other hand, the analyst determines that a data element *can* be used over again once it has entered the system, then that element must be incorporated into a reference file; and the analyst must consider the problem of file maintenance. Once method of entry has been established, the analyst approaches the question of when in time to introduce the document or file and the problem of what other elements—key fields and additional data fields—are to be brought in with the required input items. Here again, an examination of the TAG-generated glossary and analysis will help him find valid response to these questions in a shorter time and with less effort than might otherwise be possible.

The final stage of the design project is, as always, system definition. After TAG has processed the required information on user inputs, outputs, and files, data and job description reports are created that provide a word picture of the user's system, or that portion of it that has thus far been analyzed. The analyst now refines the generated file format and data-flow descriptions, applying to them record volumes and activity ratios, and, if known, hardware considerations. He looks for possible new outputs and improvements to existing ones. He may change the contents of an output, its priority, and the frequency with which it is produced. He may consolidate duplicate reports. Since he can trace the steps needed to produce desired outputs, the analyst knows the relative ease or difficulty with which those outputs are obtained. Thus, he can justify eliminating from the system costly reports, ones that require excessive data manipulation to produce.

In designing files, the systems planner first compares the contents of existing reference files to the data requirements outlined by TAG. Where discrepancies exist, the analyst determines whether or not it is feasible to attempt to include the needed data in the system. If it is feasible, then he must locate the source from which the input is to come and make some decision about when and how to incorporate those data into new files. A study of this question may lead to the establishment of plans for future file conversion. The final criteria of design must be based on volume and activity figures for the data records in the system. Working from the format definitions supplied by TAG, the analyst must develop a data base compatible with these figures and with the hardware and configuration of the proposed system. Often, the final design choice will be one based on economic considerations. The analyst redefines file formats and data flow, working within the boundaries of each established time interval, until he has achieved an optimum system. Changes made in the original job and data definitions may, of course, be submitted to TAG for analysis. Doing so will ensure that such changes do not result in any new unresolved conditions and do not destroy the integration of the previously established data base.

As stated earlier, the steps outlined above are ones that would have been necessary whatever technique was utilized to aid the systems planner. The Time Automated Grid technique assures more rapid and more effortless accomplishment of these steps by providing direction and computerized analysis of requirements.

```
          RESULTS OF ANALYSIS BY TIME-GRID TECHNIQUE


DATA                                      DATA
NUMBER  DATA NAME        SIZE  A/N  USE   NUMBER  DATA NAME              SIZE  A/N  USE
   21   QUANTITY-ORDERED    3   N   FI       22   SHIPPING-INSTRUCTIONS   100   A   FI
   23   SOLD-TO-ADDRESS    75   A   FI       24   TOTAL-INVOICE-AMOUNT      8   N   VR


DATA NUMBER                       21   22   23   24
CYCLE

   1 CUSTOMER-ORDER          5    1    1    0
     (   1) 8300...  1 X D
   1 WAREHOUSE-ORDER         5    1    0    0
     (   2) 8300...  1 X D
   2 INVOICE                 5    0    1    1
     (   3)10300...  1 X D
SUMMARY CODES                     1    0    1    2


MEANING OF SUMMARY CODES

0 - RATIO OF INPUT = RATIO OF OUTPUT, INPUT AVAILABLE AT TIME OF OUTPUT
1 - PLURAL CYCLES - FILES
2 - SYSTEM GENERATED (VARIABLE RESULT)
3 - NO INPUT BUT OUTPUT, NOT VARIABLE RESULT
4 - NO OUTPUT BUT INPUT
5 - RATIOS NOT EQUAL
6 - OUTPUT REQUIRED BEFORE INPUT IS AVAILABLE
```

```
DATA NO.   DATA NAME              CODE   PAGE NO.

   2       CUSTOMER-NAME            15      3
   5       DATE-OF-ORDER             4      3
   6       DISC-QUALIFICATION-AMT    3      3
   8       DISC-RATE-1               3      3
   9       DISC-RATE-2               3      3
  12       INVOICE-NO                3      3
  13       LINE-EXT                152      3
  16       PART-NAME                13      3
  18       PRICE                     3      3
  20       QTY-SHIPPED             153      3
```

```
1     1 I CUSTOMER-ORDER             8300
              14    0   0    0   0   0   0
1     2 O WAREHOUSE-ORDER            8300
              14   15  17    0   0   0   0
2     3 O INVOICE                   10300
              14   15  17ᴬ   0   0   0   0
3     4 O WEEKLY-SHIPMENT-REPORT     5000
              15   17   0    0   0   0   0
```

FIGURE 3 Three of the key TAG outputs. (See p. 68 for sample TAG input forms)

AN INFORMATION ALGEBRA

CODASYL DEVELOPMENT COMMITTEE

This report represents the results of the first phase of the work of the Language Structure Group. The goal of this work is to arrive at a proper structure for a machine-independent problem-defining language, at the systems level of data processing. The report is based, for the most part, on a mathematical model called "An Information Algebra" developed primarily by R. Bosak.

It is hoped that this report will be read (a) with avid interest by programming language designers and implementors, and all those interested in developing a theoretical approach to data processing; (b) with interest and understanding by professional programmers and systems analysts; and (c) with appreciation by the businessman-analyst-manager.

The authors have not attempted an exhaustive discourse in this report. Rather, they have tried to present a philosophy to the professional people who are vitally concerned with providing a working language for the systems analyst's use. They trust that the ideas in this report will stimulate others to think along similar lines.

Questions and comments will be welcomed, and can be addressed to any of the members of the Language Structure Group:

Robert Bosak, System Development Corporation
Richard F. Clippinger, Honeywell EDP Division
Carey Dobbs, Remington Rand Univac Division
Roy Goldfinger (Chairman), IBM Corporation
Renee B. Jasper, Navy Management Office

SOURCE: CODASYL Development Committee, "An Information Algebra Phase I Report," *Communications of the ACM,* Vol. 5, No. 4, April 1962, pp. 190–204. Copyright © 1962, Association for Computing Machinery, Inc., and reprinted with their permission.

William Keating, National Cash
  Register
George Kendrick, General Electric
  Company
Jean E. Sammet, IBM Corporation

INTRODUCTION TO
THE REPORT

The Language Structure Group (LSG) of the Development Committee of CODASYL was formed in July, 1959 to study the structure of programming languages for data processing problems and to make recommendations which would guide the development of future programming languages. The LSG quickly agreed that a single unifying conceptual framework—a well formulated theory of data processing—was needed so that programming languages could be designed in a less intuitive fashion. The LSG has *not* produced such a theory, but many of the elements in this report could contribute to such development.

Current programming languages tend to have three primary limitations: (1) they are machine dependent to varying degrees; (2) they require ordered procedure statements; (3) segmenting a program into discrete runs is required. The LSG decided to try to devise a basic operational algebra to serve as the theoretical basis for truly automatic programming systems. An earlier approach involved the use of a matrix representation, which was abandoned when rigid restrictions of form were encountered. In seeking a more general representation, the LSG drew upon the concepts of Modern Algebra and Point Set Theory. The concepts developed at this stage were formalized by Robert Bosak in a working paper entitled "An Information Algebra." The present form of the Algebra, which is the subject of this report, has evolved after many months of work and study.

In general, the underlying concepts of the Algebra have been implicitly understood for years by the business systems analyst. An information system deals with objects and events in the real world that are of interest. These real objects and events, called "entities," are represented in the system by data. The data processing system contains information from which the desired outputs can be extracted through processing. Information about a particular entity is in the form of "values" which describe quantitatively or qualitatively a set of attributes or "properties" that have significance in the system.

Existing programming languages are inherently procedural in nature, but there are some examples in which relationships rather than procedures are specified. In particular, report writers and sort generators are of this latter type. The primary intent of the Information Algebra is to extend the concept of stating the relationships among data to all aspects of data processing. This will require the introduction of increased capability into compilers for translating this type of relational expression into procedural terms. Specification of such functions as READ, WRITE, OPEN, MOVE, and much of the procedure control definition will be left to the compiler, thereby reducing the work of the systems analyst. The analyst will specify the various relevant sets of data and the relationships and rules of association by which these data are manipulated and classed into new and different sets of data, including the desired output.

It is hoped that the Information Algebra, being machine independent, will foster and guide the development of more universal programming languages. It should help the information processing community (as it has helped the members of the LSG) to clarify, understand, and appreciate the fundamental and essential features of data processing considerations. Further, the Algebra could provide a step toward the goal of overall system optimization by making it possible to manipulate the notation in which the fundamental statement of the problem is expressed. With current programming languages, the problem definition is buried in the rigid structure of an algorithmic statement of the solution, and such a statement cannot readily be manipulated.

The LSG has not produced a user-oriented language for defining problems, nor has it specified the algorithm for translating information algebra statements into machine-language programs. In fact, these goals are probably unattainable in full generality. However, their lack should not discourage systems analysts from consideration of the basic concepts of the Algebra. The isolation of these problems is expected to stimulate progress toward their solution in restricted cases which, though restricted, may be of considerable utility.

In addition to working in these two primary areas, application studies will be undertaken. These efforts should contribute to the refinement and extension of the Information Algebra leading to the addition of more functions and operators where they are needed.

It is hoped that the formalism of the approach taken in developing the Information Algebra will assist in providing a foundation for future practical and theoretical studies in the structure of data processing languages.

I. Introduction to the Algebra

The Algebra provides a framework for describing data processing problems. By virtue of the general approach taken, it is intended to be capable of describing problems which range over a broad spectrum of data processing. The rules of the algebra reflect the common characteristics of these problems and are couched in the language of mathematics, particularly Modern Algebra and Set Theory. The Algebra is built on three undefined concepts: entity, property, and value. The three concepts are related by the following two rules, which are used as the basis for two specific postulates in the algebra:

(a) Each property has a set of values associated with it.
(b) There is one and only one value associated with each property of each entity.

Although these concepts are formally undefined, certain intuitive statements can be made:

An *entity* (Webster's Collegiate: "... A thing which has reality and distinctness of being either in fact or for thought...") as used in this Algebra has the following synonyms:

(1) *Thing* (Webster's: "... Whatever exists, or is conceived to exist, as a separate entity; any separable or distinguishable object of thought; as, there is a name for every*thing*...")
(2) *Instance,* *event,* and *occurrence* (to some extent)

Data processing is the activity of maintaining and processing data to accomplish certain objectives. The data are collections of *values* of certain selected *properties* of certain selected *entities*. For example, a payroll manager has many objectives of which the primary one is the payment of his employees (the *entities*). He selects certain *properties* of these entities, such as employee number, name, sex, and hourly payrate and maintains a file in which the *values* of these properties are recorded.

Property and *value* are best illustrated by examples:

Property	Values associated with the property
Employee no.	576, 75807, 25
Sex	Male, female
Name	Bill, Bob, Carey, Dick, George, Jean, Renee, Roy
Hourly payrate	$1.15 $1.75 $2.50

II. Property Spaces

Following the practices of Modern Algebra, this Algebra deals with sets of points in a space. For each particular application, the space that is defined reflects the requirements of the problem at hand. Once the problem is plotted into the space, these requirements can be completely specified in terms of the Algebra. The following definitions, postulates, and theorems are basic to the definition of the space.

(1) Three Undefined Concepts

Property (q)
Value (v)
Entity (e)

(2) Postulate. Each property has one and only one set of values assigned to it. The set of values so assigned will be called the *Property Value Set (V)*.

The property value set (V) defines the kind and range of variations of the values of the property. As an example, for each selected property the payroll manager has chosen, he selects a property value set adequate to contain any anticipated value. Thus, the value set he selects for employee number may be the integers from 1 to 1,000,000; and the value set for sex would be male, female.

(3) Postulate. Every property value set contains at least two values:

Ω (Undefined, not relevant)
Θ (Missing, relevant but not known)

This Postulate is included for logical consistency as well as to provide a means of distinguishing undefined or missing values from zero. To illustrate these values, consider the payroll example. If the payroll files carried additional personnel data, the property "draft status" would be undefined and not relevant (Ω) for female employees, while if its value were missing for a male employee, it would be relevant but not known (Θ). In actual practice, symbols (such as 99999 or 88888) which can be distinguished from relevant, known values have often been assigned by convention to represent these possibilities in the value set.

(4) Postulate. Every entity has one and only one value assigned to it from each property value set.

For example, a certain employee may have the values (721, male, Don, $15.25) associated with him; i.e., employee number is 721, sex is male, etc. (Note that [721, male, Don, $15.25] is not a property value set, but a list of particular values of the various properties chosen to represent an entity.)

(5) Definition. A *Coordinate Set (Q)* is a finite set of distinct properties.

Thus $Q = (q_1, q_2) =$ (employee number, hourly payrate) is a coordinate set.

(6) Definition. The *Null Coordinate Set* contains no properties.

(7) Definition. Two coordinate sets are *equivalent* iff they contain exactly the same properties. (Iff means *if and only if.*)

Thus, the coordinate set (employee number, hourly payrate) is equivalent to the coordinate set (hourly payrate, employee number). The set operations: union, intersection, and complementation apply to coordinate sets. Definitions (5), (6), and (7) are subsidiary concepts leading to the definition of a property space.

(8) Definition. The *property space* (P) of a coordinate set (Q) is the cartesian product

$$P = V_1 \times V_2 \times V_3 \times \cdots \times V_n,$$

where V_i is the property value set assigned to the ith property of Q. Each point (p) of the property space will be represented by an n-tuple $p = (a_1, a_2, a_3, \ldots, a_n)$ where a_i is some value from V_i. The ordering of properties in the set is for convenience only. If $n = 1$, then $P = V_1$. If $n = 0$, then P is the Null Space.

The property space is a conceptual way of representing the set of all possible combinations of the values of all the properties. The "×" used does not denote multiplication, but rather the selection of one element (value) from each of the components (property value sets). Thus, each point p represents one particular value from property q_1, one particular value from property q_2, one particular value from property q_3, and so on for all the properties in the coordinate set. For example, the property space (P) of the coordinate set (employee number, hourly payrate) contains all points obtained by taking one coordinate as an employee number from the value set of employee number and one coordinate as an hourly payrate from its value set. If the value set for Employee Number contains 1,000,000 values and there are 15 payrates in its value set, then P contains 15,000,000 points. One of these points might be $p = (721, \$1.25)$.

(9) Definition. The *Datum Point* (d) of an entity (e) in a property space (P) is a point of P such that if $d = (a_1, a_2, a_3, \ldots, a_n)$, then a_i is the value assigned to e from the ith property value set of P for $i = 1, 2, 3, \ldots, n$. (d) is the *representation* of (e) in (P).

In our example, if the company has 14,000 employees, then these should be represented by 14,000 datum points in the property space P.

(10) Theorem. Every entity has exactly one datum point in a given property space.

This follows immediately from II(4) and II(9). Note that several entities may be represented by the same datum point. Thus, the property space P of the coordinate set (employee name, hourly payrate) may contain the datum point (John Doe, 2.34). If there are several people (i.e. entities) with this name and hourly payrate, then each of these entities will be represented by this single datum point.

A data processing system must be able to handle any entity that can possibly occur. The definition of a property space takes full cognizance of this necessity. Thus, there might not be anyone making $2.34 per hour when the system is installed, but there certainly could be at a later time. In practice, of course, most points in a property space will never become datum points.

(11) Definition. A *discriminatory property space* for a set of entities is a property space in which no datum point represents more than one entity. The corresponding coordinate set will be called a *discriminatory coordinate set.*

The property space of (employee number, hourly payrate) is discriminatory since the personnel department is careful to assign a unique employee number to each employee. However, the property space (sex, hourly payrate) is not discriminatory, because it only contains 30 points and there are 14,000 entities represented by a subset of these 30 points. A single point in this property space might be the datum point for 2,000 entities having the same sex and payrate. A property space may be discriminatory for one set of entities but become nondiscriminatory for another set. For example, the property space of (employee name, hourly payrate) will be discriminatory for any set of employees in which there are no two people with the same name making the same salary.

(12) Definition. If (a) Q and Q' are coordinate sets, (b) Q is discriminatory, and (c) for all Q' contained in Q, Q' is nondiscriminatory, then, Q is a *basic discriminatory coordinate set*.

The coordinate set that consists of (department number, departmental-employee number) is a discriminatory coordinate set since no two employees in the same department are assigned the same departmental-employee number. It is also basic since neither of the two subsets is discriminatory.

III. Lines and Functions of Lines

For the remainder of this paper we will postulate a single property space P. To distinguish datum points from other points does not further the development; therefore, no distinction will be made. It should be observed, however, that whatever is true for points in general is also true for datum points.

Preceding sections dealt with the method of definition of a property space and the relation among entities, properties, and values. Once a coordinate set has been chosen, we need only be concerned with operations within this property space. For a particular data processing application a property space must be defined which encompasses all the required information for the application—input, intermediate results, and output.

(1) Definition. A *line* (L) is an ordered set of points chosen from P. The *span* (n) of the line is the number of points comprising the line.

(2) Notation. A line L of span n is written as

$$L = (p_1, p_2, \ldots, p_n).$$

The term *line* is introduced to provide a generic term for a set of points which are related. In a payroll application the datum points might be individual records for each person for five working days of the week. These records might contain employee number, number of hours worked, payrate, and the day of the week. For each employee we can define a line such that the first point is the record for Monday, the second Tuesday, etc. The span of this line is five.

(3) Definition. A *Function of Lines* (FOL) is a mapping that assigns one and only one value to each line in P. The set of distinct values assigned by an FOL is the *value set of the FOL*.

The value set of an FOL may—but need not—be contained in the property value set of any one of the properties of P.

(4) Notation. It is convenient to write an FOL in the function form $f(X)$, where f is the FOL and X is the line.

An FOL represents a mapping of each line in the property space into a set of values. This is the same as saying that an FOL assigns a value to each ordered set of points in the property space. Thus lines are used to combine related points while FOL's provide the necessary means for operating on this set of related points.

Consider the above example, which defined a line as the five points with each day's work record as a point. An FOL for such a line would be the value of the product:

(hours worked) * (payrate),

summed over five days of the week (i.e. summed over the points of the line). The value set of this FOL would be weekly gross pay. Its values lie mainly outside of the range of values assigned to either hours worked or rate of pay. Another FOL for the same line would be the maximum number of hours worked in a day. Its value set would be the same as the value set for hours worked.

(5) Definition. Two FOL's are *equal* iff they assign the same value to the same line, for all lines in P.

(6) Definition. An FOL is *independent of order* if it assigns the same value to all lines that contain exactly the same points, regardless of the order of the points on the line.

In the payroll example, the sum of

(hours worked) * (payrate)

over the points of the line is independent of the order in which the sum is formed. Thus, this FOL is independent of order.

(7) Definition. The *Null FOL* is one whose value set contains only Ω.

The definition of a null FOL is included for logical consistency.

(8) Definition. An FOL is of *span n* if it is not a null FOL and if it assigns the value Ω to all lines whose span is not n.

This definition makes it possible to distinguish FOL's that are defined only for lines with a specific number of points. The value of such FOL's is undefined (Ω) for all lines with either more points or fewer points. (Note that the FOL is not necessarily defined for *all* lines of span n.) As an example of an FOL of span 2, consider the set of lines defined by the pairs of points (p_1, p_2), where the value of the property sex for p_1 is male, and that for p_2 is female. Let the FOL be the number of children that are born to this pair; clearly, this FOL is of span 2.

(9) Definition. An *Ordinal FOL* (OFOL) is an FOL whose value set is contained within some defining set for which there exists a relational operator ⓡ which is irreflexive, asymmetric, and transitive. Furthermore, the following defining sets with the indicated relational operator have been assigned specific names:

Defining set	Name
$\Omega < \Theta < -\infty <$ set of real numbers $< +\infty$	Real (operator, ⓡ is "less than")
$\Omega < \Theta <$ set of negative integers $< 0 <$ set of positive integers	Integer (operator, ⓡ is "less than")
Ω ⓡ Θ ⓡ False (F) ⓡ True (T)	Boolean
False (F) ⓡ True (T)	Selection

An ordinal FOL will be said to be of the same "type" as its defining set.

A defining set is any set of values which has the proper ordering relation. For example, the set of integers, the set of even integers, the set of all letters between M and R are three possible examples of defining sets for an FOL.

The requirement that the relational operator is irreflexive (i.e., a ⓡ a is not true), asymmetric (i.e., a ⓡ b and b ⓡ a cannot both be true), and transitive (i.e., if a ⓡ b and b ⓡ c then a ⓡ c) means that the defining set is simply ordered. Note that the integers and the real numbers are simply ordered under both $<$ and $>$.

Note that an OFOL can be of more than one type, since certain defining sets are subsets of others. For example, an Integer OFOL is also a Real OFOL, but not conversely. A Selection ordinal FOL is also a Boolean ordinal FOL.

(10) Definition. An FOL which is not an ordinal FOL will be called a *non-ordinal* FOL.

IV. Areas and Functions of Areas

Heretofore it has been useful to consider, as a whole, the total information representable within a data processing system. To be a model for data systems, the Information Algebra must concern itself with portions or subsets of this total information. These portions are termed areas and are analogous to files and portions of files.

(1) Definition. An *area* is any subset of P.

The property space P represents the totality of all possible information in a data processing system. An area A represents only a portion or subset of this information. Having defined subsets of P, we can now apply the operations of set theory. Note that subsets may be joined, intersected, complemented, etc.

Consider the following example:
P is defined over the properties (Part Number, Balance, Quantity Ordered, Quantity Received, Dollar Value of Quantity Ordered). The value sets of each property are defined over the positive and negative integers. Area A_1 (corresponding to the Balance File) of P is the set of datum points of P representing physically existing parts; area A_2 (corresponding to the Transaction File) is the set of datum points of P which represent events called "transactions."

At this stage in the development of the Information Algebra, we have shown how to identify the parts of a property space which correspond to files. We shall see shortly how to express the relationships between these files.

Note that both lines and areas are made up of sets of points; the primary difference is that the points of a line are ordered.

(2) Definition. A *Function of Areas* (FOA) is a mapping that assigns one and only one value to each area. The set of distinct values assigned by an FOA is defined to be the *value set of the* FOA.

(3) Notation. It is convenient to write an FOA in the functional form $f(X)$, where f is the FOA and X is the area.

The definition of FOA's serves several useful purposes. One of the things it allows us to do is to name areas, by using the value defined by the FOA as a name. Other more elaborate uses of FOA's can be made by performing arithmetic or relational operations on the points of the area. In the previous example, we can define an FOA by summing all the values of the property Dollar Value of Quantity Ordered for each point in A_2.

(4) Theorem. If (a) B is an area of P, and (b) M is a line which contains all points in B and no others, then for each FOA f there exists an FOL f' that is independent of order such that $f(B) = f'(M)$ for all lines M which satisfy (b).

Proof. If we can define an f' with the required characteristics then we have proved that it exists.

By definition, f assigns a unique value, say v_B to B. Consider a line M composed of all the points in B and no others. Define a mapping g with the property that $g(B) = v_B$. Then by the definitions of FOL and FOA, g is an FOL. If M'

is a line which differs from M only in the ordering of points, then define $g(M') = v_B$. Then $f' = g$ is the required FOL.

V. Bundles and Functions of Bundles

In the data processing applications, related data from several sources must be grouped. Various types of functions can be applied to these data to create new information. For example, files are merged to create a new file. Methods of joining and evaluating functions of the information from several subsets of a property space are defined in this section.

(1) Definition. An *Area Set* $\aleph$ of order n is an ordered n-tuple of areas $(A_1, A_2, \ldots, A_n) = \aleph$.

Area Set is a generic term for a collection of areas in which the order is significant. $\aleph$ consists of one or more areas which are considered simultaneously. The files "Transaction File" and "Master File" form an Area Set of order 2.

(2) Definition. The *Bundle* $B = B(b, \aleph)$ of an area set $\aleph$ for a Selection OFOL b is the set of all lines L such that if (a) $\aleph = (A_1, A_2, \ldots, A_n)$, and (b) $L = (p_1, p_2, \ldots, p_n)$ where p_i is a point of A_i for $i = 1, 2, \ldots, n$, then $b(L) = $ True. The function b will be called the *Bundling Function* for B. The Bundle will be said to be of the same order as the area set.

This definition states that individual points from various areas of the property space can be grouped into lines by a Selection OFOL. A bundle thus consists of a set of lines $\{L_i\}$, each of order n, where n is the order of the area set (i.e., the number of areas in $\aleph$). Note that each line in the bundle must contain one and only one point from each area.

The concept of bundles gives us a method of conceptually linking points from different areas so these points may be considered jointly.

As an example, consider two areas with the names "Master File" and "Transaction File," each containing points with the property Part Number. Allow the Selection OFOL to apply equality to the value set for this property, i.e., determine the truth of the statement: Part Number in Master File equals Part Number in Transaction File. Each line of the Bundle over the area set (Master File, Transaction File) defined this way consists of a pair of records (points) which have equal part numbers.

(3) Definition. A *Function of a Bundle* (FOB) is a mapping that assigns an area to a bundle such that

(a) there is a many-to-one correspondence between the lines in the bundle and the points in the area;
(b) the value of each property of each point in the area is defined by an FOL of the corresponding line in the bundle;
(c) the value set of such an FOL must be a subset of the corresponding property value set.

This function allows the formation of a new area (file) which consists of the points linked by the lines over which the Bundling Function is defined. Thus, each line of the bundle maps into a point (not necessarily unique), the values of whose coordinates are determined by a set of FOL's. Another way of expressing this concept is to say that each point of the new area is a function of some or all the points (records) in each of the areas (files) in Area set $\aleph$. The Bundling Function selects information from each of the areas (files) for simultaneous consideration and the FOB states how the value of each property of a desired area (output file) is obtained. There will be a point in the FOB (output file) for each line in the Bundle.

In a Status of Inventory example, consider the areas Master File (containing points with the non-null properties: Part Number and Balance), and the Transaction File (containing points with the non-null properties: Part Number and Amount Ordered). A Bundle can be defined which consists of those lines joining points of each area which have equal part numbers. The FOL which defines the formation of the area New Master File uses the mapping:

New Part Number
$\quad$ = Old Part Number

New Balance
$\quad$ = Old Balance − Amount Ordered

Note that this mapping is applied only to points with equal part numbers. Other problems will involve Boolean as well as algebraic rules to determine the co-ordinates of the points in a new area.

(4) Definition. The *null FOB* is one which assigns the area containing the single point $(\Omega, \Omega, \ldots, \Omega)$ to all bundles.

The definition of the null FOB is included for logical consistency.

(5) Notation. An FOB may be expressed in three equivalent ways

(a) $F \equiv (f_1, f_2, \ldots, f_k, \ldots, f_m)$

(b) $F \equiv (q_1' = f_1, \ldots, q_k'$
$\qquad = f_k, \ldots, q_m' = f_m)$

(c) $F \equiv \begin{cases} q_1' = f_1 \\ q_2' = f_2 \\ \vdots \quad \vdots \\ q_k' = f_k \\ \vdots \quad \vdots \\ q_m' = f_m \end{cases}$

where F is the FOB, q_k' is the kth property for points in the area assigned by the FOB, m is the number of properties, and f_k is a function of the q_{ij} for $i = 1, 2, \ldots, n$ and $j = 1, 2, \ldots, m$, where q_{ij} is the jth property of the ith area of the area set; n is the span of the lines comprising the bundle.

In the form (a) the f corresponding to every q' must be specified.

In forms (b) and (c) any expression $q_k' = f_k$ may be missing, in which case $q_k' = q_{nk}$ will be understood, where q_{nk} denotes the value of the property q_k for points in the nth area, i.e., the area which contains the last point of each line. If none of the f's is specified, i.e., $F \equiv (\ )$, then $q_k' = q_{nk}$ for $k = 1, 2, \ldots, m$.

This notation allows three ways of writing the Function of a Bundle (FOB). The coordinates (properties) of each point in the area defined by the FOB are expressed in terms of the q_k's. Note that in the double subscript notation the second subscript specifies the particular property, while the first subscript indicates from which area of the area set the property value is to be taken for calculating the function. To save writing, it is assumed that if the value of some property in the generated area is not specified by an FOL, then it is equal to the value (for all lines) of the corresponding property in the last (or nth) area included in the bundle. It is further assumed that, if none of the q-primes is explicitly defined, each is equal to its corresponding property in the nth area. This convention thus permits the FOB to be used as a "match-select" operator, since those points in the last area of the bundle which fail to satisfy the requirements are excluded from the area defined by the FOB.

An example may help to clarify some of the concepts associated with the Function of a Bundle.

Consider a property space defined by the four properties: Type, Unit Cost, Unit Price, Unit Net. The value sets for the properties are as follows:

q_1 = Type (A, B, ... , Z)
q_2 = Unit Cost (.01, .02, ... , .50)
q_3 = Unit Price (.51, .52, ... , 1.00)
q_4 = Unit Net (.01, .02, ... , .99)

A Cost File corresponds to an area A_1 for which Type and Unit Cost have non-null values while Unit Price and Unit Net are not relevant.

A Price File corresponds to an area A_2 for which Type and Unit Price are relevant while the other properties have null values.

Areas A_1 and A_2 form an area set $\aleph = \{A_1, A_2\}$.

$B = F(b, A_1, A_2)$ is a bundle for which the bundling function b is $q_{11} = q_{21}$. The bundle consists of all those lines defined by the pairs of points, one taken from each area, for which the values of the property Type are equal.

The lines are of span 2. To compute the value of Unit Net for each item, an FOL may be applied:

$$f = q_{23} - q_{12}.$$

This FOL subtracts the Unit Cost in A_1 from the Unit Price in A_2.

An FOB defines a new area A_3 for which Type and Unit Net have non-null values. A_3 is defined by the relationship:

$$A_3 = F(q_{11} = q_{12}, A_1, A_2)$$

$$F \equiv \begin{cases} q_1' = q_{21} \\ q_2' = \Omega \\ q_3' = \Omega \\ q_4' = q_{23} - q_{12} \end{cases}$$

Using the convention of Notation (5), F may be written as

$$F \equiv \begin{cases} q_3' = \Omega \\ q_4' = q_{23} - q_{12} \end{cases}$$

The values of the unspecified properties in A_3, q_1' and q_2', are assumed to be carried over from A_2, the last area of the bundle.

To recapitulate, an FOB defines an area wherein each point corresponds to a line in the bundle; more than one line may correspond to the same point. Each coordinate of a point in this defined area is a function of the coordinates of the points comprising a line. Notation (5) indicates three ways in which the functional relationships may be expressed.

(6) Notation. The area A assigned by an FOB F to the bundle B may be expressed as either:

$$A = F(B)$$
$$A = F(b, \aleph)$$
$$A = F(b, A_1, A_2, \ldots, A_n)$$

(7) Notation. $A(s)$ will stand for an area that is determined from

$$A(s) = F(s, P), \quad \text{where } F \equiv (\ ).$$

s will be called the *selection function* for A.

Notation (7) is an application of the last rule in Notation (5), which permits an FOB to function as a "match-select" operator. An area $A(s)$ contains points chosen from an entire property space, as determined by a selection criterion s.

VI. Glumps and Functions of Glumps

Having provided the basis for discussion of data processing operations involving several subsets (files) of the property space, we now provide the basis for discussion of operations within P or its subsets (files). Actions such as summarizing within a file will be permitted through the concepts developed in this section.

(1) Definition. If A is an area and g is an FOL defined over lines of span 1 in A, then a *Glump* $G = G(g, A)$ of an area A for an FOL g is a partition of A by

g such that an element of this partition consists of all points of A that have identical values for g. The function g will be called the *Glumping Function* for G. The concept of a glump does not imply an ordering either of points within an element, or of elements within a glump.

The FOL g is defined over lines of span 1 (i.e., is defined for all points) in A. The points in A are divided into subsets based on the values into which the points are mapped by g. If two points are mapped by g into the same value, then they are in the same element of the partition; otherwise, they are in different elements. The elements of the glump are actually subsets of A.

(2) Notation. The glump element which is the subset of A defined by points whose value for g is c will be denoted by $G(g, A)|_{g=c}$.

One of the distinctions which may be made between a bundle and a glump is that a glumping function partitions strictly on the basis of its assigned values for points, whereas discrimination by bundling functions is contingent on the truth of some statement. (Clearly, they coincide at times.) A second distinction is that bundles can involve many areas whereas glumps are concerned with only one. A third difference is that the elements of bundles are lines (which means that the points are ordered) whereas the elements of glumps are sets of unordered points.

As an example, let the Back Order File consist of points with the 3 non-null properties: Part Number, Quantity Ordered, and Date. Define a glumping function to be simply Part Number. Each element of this glump consists of all orders for the same part. Note that different elements of the glump will not necessarily contain the same number of points.

(3) Definition. A *Function of a Glump* (FOG) is a mapping that assigns an area to a glump such that

(a) there is a many-to-one correspondence between the elements of the glump and the points of the area,
(b) the value of each property of a point in the assigned area is defined by an FOA of the corresponding element in the glump,
(c) the value set of the FOA must be a subset of the corresponding property value set.

This definition is analogous to that of Function of Bundles, except that here we have an FOA instead of an FOL.

As an example, let us again consider the Back Order File, and include Unit Price with each Part Number. Now we can define an area that is an FOG of the back order file, in which each point has three non-null properties—Part Number, total back order quantity for each part, and total cost of this quantity. Note that a simple FOA (addition of the values of the new area) gives the total cost for one complete set of Stock Items.

(4) Notation. An FOG may be expressed in three equivalent ways

(a) $H \equiv (f_1, f_2, f_3, \ldots, f_k)$
(b) $H \equiv (q_1' = f_1, q_2' = f_2, \ldots, q_k' = f_k)$
(c) $H \equiv \begin{cases} q_1' = f_1 \\ q_2' = f_2 \\ \vdots \quad \vdots \\ q_k' = f_k \end{cases}$

where H is the FOG, f_i is an FOA, q_i' is the ith property for points in the area, k is the number of properties. In form (a), the f corresponding to every q' must be specified. In forms (b) and (c), any expression $q_i' = f_i$ may be missing, in which case $q_i' = \Omega$ will be understood.

Notation (3) indicates the similarity in the writing of a function of a Glump and a function of a Bundle. However, while the f's of an FOB are FOL's, the f's of an FOG are FOA's. A double subscript notation is not used in the FOG since only a single area is involved in the functional relationship H. Observe also that in the FOG, any property not explicitly defined is considered Ω, while in the notation of writing FOB's, a missing property is based on the last point on the line.

(5) Notation. The area A assigned by an FOG, H, to the glump $H(g, A_1)$ may be expressed as either:

$$A = H(G) \quad \text{or} \quad A = H(g, A_1).$$

VII. Ordering of Areas

Introducing areas gave us a concept analogous to that of files; bundling gave us functions similar to matching; and glumping gave us a way of handling operations such as summarizing within one file. In all three of these concepts, the notion of "order" was deliberately ignored since it was not necessary. However, in certain practical applications its introduction is required.

(1) Definition. An *Ordering* $O(f, A)$ of an area A by an OFOL f of span 1 is a set of points $(p_1, p_2, \ldots, p_n)$ chosen from A such that the set exhausts A and either $f(p_i) < f(p_{i+1})$ or $f(p_i) = f(p_{i+1})$.

An ordered area is analogous to a data processing situation in which a file is sorted.

(2) Definition. An ordering is said to be *simpler* for a given OFOL iff there exist no points such that $f(p_i) = f(p_{i+1})$. An ordering is said to be *partial* for a given OFOL iff there exists at least one pair of points (p_i, p_{i+1}) for which $f(p_i) = f(p_{i+1})$.

If successive points are not mapped by an OFOL into the same value, then the area is simply ordered under the OFOL. For example, define two OFOL's for ordering a deck of playing cards. The first OFOL orders each suit and the cards within each suit, to give strict inequality; the second OFOL merely orders by face value, yielding equality for four cards with the same face value. The former ordering is simple, the latter ordering is partial. In the data processing field, any sorted file which has duplicate keys is partially ordered; if all the keys are unique the file is simply ordered.

(3) Theorem. If f is an OFOL of span 1 defined over an area A, and $G(f, A) = (G_1, G_2, \ldots, G_m)$ is a glump of A by f, where the number of points in G_i is n_i, $i = 1, 2, \ldots, m$, then there are exactly $n_1! \, n_2! \cdots n_m!$ orderings of A by f.

Proof. By definition, $G(f, A)$ is a glump of A by f such that each glump element contains points for which $f(p_i) = f(p_{i+1})$ for all points in each element. If a given glump element G_k contains n_k points, then the number of ways of ordering such points by f is equal to the number of permutations of n_k things taken all at a time, or $n_k!$ (n_k-factorial). The total number of orderings of A by f, then, is the product of the number of orderings of each glump element, or $n_1! \, n_2! \cdots n_m!$.

(4) Corollary. There exists at most one simple ordering of an area A by an OFOL f.

Proof. If the ordering is simple, it follows from Definition VII(2) that there exist no points such that $f(p_i) = f(p_{i+1})$. Hence, $G(f, A)$ partitions A into areas each containing exactly one point, so that $n_1 = n_2 = \cdots = n_k = 1$. Therefore, $n_1! \, n_2! \cdots n_m! = 1$.

VIII. Operations of FOL's

Several types of FOL's have been used in the examples in previous Sections. The present Section deals with operations on FOL's. Since an FOL may be defined as a single property, operations on FOL's may be operations on properties. However, the definition of an FOL is recursive in the sense that an FOL may operate upon values, upon functions of values, upon functions of functions of values, and so forth.

A single example will be used throughout the present Section. The entity: an employee, uniquely identified by his name and address, who may work Monday through Friday, and is paid on the basis of hours worked and pieces produced.

The coordinate set of the space designed for this entity is (name, address, day of week, hours worked, pieces produced). Consider lines of span 5 selected from this property space. A line consists of the five points representing a given employee's record for each day of the week. The following FOL's are defined for illustrative purposes:

f_1 = name
f_2 = address
f_3 = sum of hours worked during week
f_4 = sum of hours worked in week with daily maximum of 8 hours
f_5 = sum of pieces produced during week
f_6 = bonus hours (to be paid at .20 per hour) = $f_3 - f_4$
f_7 = bonus indicator = $(f_5 / f_3 < 13)$
f_8 = productivity incentive rate
   = $(.20 \leftarrow f_7 \rightarrow .25)$
f_9 = overtime pay indication = $(f_3 = f_4)$

The meaning and use of functions f_1 through f_9 will be developed below.

(1) Definition. Concatenation (¢) is a binary operator with properties that

(a) if $f_1 \; ¢ \; f_2 = f_3$, then, for any L, $f_3(L)$ is the couple $[f_1(L), f_2(L)]$;
(b) if $f_1 \; ¢ \; (f_2 \; ¢ \; f_3) = f_4$, then $f_4(L)$ is the triple $[f_1(L), f_2(L), f_3(L)]$;
(c) $f_1 \; ¢ \; (f_2 \; ¢ \; f_3) = (f_1 \; ¢ \; f_2) \; ¢ \; f_3 = f_1 \; ¢ \; f_2 \; ¢ \; f_3$;
(d) in general, $f_1 \; ¢ \; f_2 \neq f_2 \; ¢ \; f_1$.

The concatenation function is used for adjoining values of distinct properties. In the example, the property space is defined over the two properties (name, address), among others, and it is desired to match points in two or more areas on the combined name and address, the concatenation $f_1 \; ¢ \; f_2$, "name and address," produces the required "linked" values.

Concatenation with a function derived through concatenation adjoins an element to the concatenated function in the indicated order. The concatenation function is associative but not commutative.

(2) Definition. Addition (+) is a binary operator such that $f_1 + f_2$ is a real ordinal FOL whose values are determined by the following table:

f_1 \ f_2	Ω	Θ	R_2	A_2
Ω	Ω	Ω	Ω	Ω
Θ	Ω	Θ	Θ	Ω
R_1	Ω	Θ	$R_1 + R_2$	Ω
A_1	Ω	Ω	Ω	Ω

where R_1 and R_2 are any real numbers, and A_1 and A_2 are any other values.

If f_1 and f_2 are both real, they are added algebraically. The value of the sum of an unknown (Θ) with another unknown or with a real number is unknown. The value of addition in all other cases is undefined (Ω). In the example, $f_3 + (-f_4)$ = bonus hours.

(3) Definition. Multiplication (∗) is a binary operator such that $f_1 * f_2$ is a real ordinal FOL whose values are determined by the following table:

f_1 \ f_2	Ω	Θ	R_2	A_2
Ω	Ω	Ω	Ω	Ω
Θ	Ω	Θ	Θ	Ω
R_1	Ω	Θ	$R_1 * R_2$	Ω
A_1	Ω	Ω	Ω	Ω

where R_1 and R_2 are real values, and A_1 and A_2 are any other values.

If f_1 and f_2 are both real, algebraic multiplication applies. If one operand is real and the other is Θ, or if both operands are Θ, the result is Θ. The operation yields Ω for all other values. In the above example, $f_8 * f_5 + .2 * f_6 =$ weekly pay.

(4) *Definition.* *Division* (/) is a binary operator such that f_1/f_2 is a real ordinal FOL whose values are determined by the following table:

f_1 \ f_2	Ω	Θ	0	R_2	A_2
Ω	Ω	Ω	Ω	Ω	Ω
Θ	Ω	Θ	Ω	Θ	Ω
0	Ω	Θ	Ω	0	Ω
R_1	Ω	Θ	Ω	(R_1/R_2)	Ω
A	Ω	Ω	Ω	Ω	Ω

where R_1 and R_2 are any real numbers $\neq 0$, and A_1 and A_2 are any other values.

In the example, f_5/f_3 yields a productivity index used to determine a bonus indicator, or figure of merit, for each employee.

(5) *Definition.* *Negation* (−) is a unary operator such that $-f$ is a real ordinal FOL whose values are determined by the following table:

f	Ω	Θ	0	R	A
$-f$	Ω	Θ	0	$-R$	Ω

where R is any real number $\neq 0$, and A is any other value.

If f is real, negation inverts its algebraic sign. Negation does not alter the values Ω and Θ. In the example, if f_5 equaled 577 for some employee then $-f_5 = -577$.

(6) *Definition.* OR ($\oplus$) is a binary operator such that $f_1 \oplus f_2$ is a boolean FOL whose values are determined by the following table:

f_1 \ f_2	Ω	F	Θ	T	A_2
Ω	Ω	Ω	Ω	Ω	Ω
F	Ω	F	Θ	T	Ω
Θ	Ω	Θ	Θ	T	Ω
T	Ω	T	T	T	Ω
A_1	Ω	Ω	Ω	Ω	Ω

where F and T are values "false" and "true," respectively, and A_1 and A_2 are any other values.

f_1 and f_2 are boolean variables which may assume the value "false" and "true." Note that $F \oplus \Theta = \Theta$ whereas $T \oplus \Theta = T$. If, in the example, $f_7 = F$ and $f_9 = T$, then $\oplus f_9 = T$.

(7) *Definition.* AND @ is a binary operator, such that $f_1 @ f_2$ is a boolean FOL whose values are determined by the following table:

f_1 \ f_2	Ω	F	Θ	T	A_2
Ω	Ω	Ω	Ω	Ω	Ω
F	Ω	F	F	F	Ω
Θ	Ω	F	Θ	Θ	Ω
T	Ω	F	Θ	T	Ω
A_1	Ω	Ω	Ω	Ω	Ω

where F and T are the values "false" and "true," respectively, and A_1 and A_2 are any other values.

Again, if $f_7 = F$ and $f_9 = T$, then $f_7 @ f_9 = F$; also note that $F @ \Theta = F$, $T @ \Theta = \Theta$.

(8) *Definition.* NOT ($\neg$) is a unary operator such that $\neg f$ is a boolean FOL whose values are determined by the following table:

f	Ω	F	Θ	T	A
$\neg f$	Ω	T	Θ	F	Ω

where F and T are the values "false" and "true," respectively, A is any other value.

If f has the values of True or False, NOT inverts its true value. In the example, if $f_7 = F$, $\neg f_7 = T$.

(9) Definition. *Equals* ($=$) is a binary operator such that $f_1 = f_2$ is a selection FOL whose values are determined by the following table:

	$f_1 = f_2$
$v_1 = v_2$	T
$v_1 \neq v_2$	F

where v_1 and v_2 are the values of f_1 and f_2, respectively.

This FOL was used in the example to define f_9.

(10) Definition. *Less than* ($<$) is a binary operator such that $f_1 < f_2$ is a selection FOL whose values are determined by the following table:

	$f_1 < f_2$
$v_1 < v_2$	T
Any other condition	F

where v_1 and v_2 are the values of f_1 and f_2, respectively.

This FOL was used in the example to define f_7.

(11) Definition. *If–otherwise* ($\leftarrow$, $\rightarrow$) is a ternary operator such that, $f_1 \leftarrow f_2 \rightarrow f_3$ is an FOL whose values are determined by the following table:

f_2	$f_1 \leftarrow f_2 \rightarrow f_3$
T	v^1
F	v_2
Θ	Θ
Any other	Ω

where v_1 is the value of f_1, v_3 is the value of f_3.

In the example, f_8 is defined as .20 $\leftarrow f_7 \rightarrow$ .25. That is, if f_7 is true, the value of f_8 is .20; if f_7 is false, the value of f_8 is .25.

IX. Constructs and Extensions of the Algebra

This Section will illustrate the use of some of the basic concepts of the Information Algebra and relate them to more familiar notions. In particular, the discussion below of files makes it clear that the representation of the information of a file in a property space is not a unique process. Exploration of this fact suggests certain extensions of the algebra, some of which are discussed below. The introduction of certain expressions derived from the basic definitions of bundles, glumps, or both will be called bundle, glump, or bundle-glump *constructs*.

(1) Definition. A *bundle argument* A_i is one of the areas in the area set $\aleph$ over which the bundle B

$$B = B(b, \aleph)$$

is defined.

(2) Definition. If F is a function of a bundle and L is some line of the bundle

$$p_F(L)$$

is the point assigned to L by F.

(3) Definition. The *intersection* $I_i(B, A_i) = B_n A_i$, of a bundle B with a bundle argument A_i, is the set of all points of A_i each of which is on some line L of the bundle B.

These definitions are introduced to permit easy reference to the concepts described.

(4) Theorem. For a given bundle B over an area set $\aleph$ and a given bundle

argument A_i, there exists an FOB F_i such that $I_i(B; A_i) = F_i(B)$.

Proof. Let $F_i \equiv \{q_j' = q_{ij}, j = 1, 2, \ldots, m\}$. Then F_i assigns to any line L of the bundle B, the point $p_F(L)$ which is the intersection of L with A_i. Q.E.D.

(5) *Theorem.* For any area set $\aleph$ and any area A_i of the set, there exists a bundle B over $\aleph$ and an FOB F_i, such that $A_i = F_i(B)$.

Proof. Choose b as the bundling function which has the value True for every possible line. Every point of each A_i is contained in some line of this bundle. Choose the function F_i defined by

$$F_i \equiv \{q_j' = q_{ij}, j = 1, 2, \ldots, m\}.$$

Then

$$I_i(B; A_i) = A_i = F_i(B)$$

Q.E.D.

(6) *Definition.* The *Area of a Bundle* (*AOB*) is written as $A(B)$ and is defined to be the set of all points each of which is contained in some line of the bundle.

From this definition $A(B)$ appears to be a new primitive. Actually $A(B)$ is the join of the intersections, I_i, of B with each A_i, and each I_i is defined by an FOB, $F_i(B)$. Thus the area of a bundle is a bundle construct:

$$A(B) = F_1(B) \cup F_2(B) \cup \cdots \cup F_n(B)$$

(7) *Theorem.* For any area set $\aleph = (A_1, \ldots, A_n)$ there exists a bundle B such that $A(B) = A_1 \cup A_2 \cup \cdots \cup A_n$.

Proof. Let the bundle over $\aleph$ be the one which consists of all possible lines, and the FOB's be the $\{F_i\}$ defined in the proof of Theorem 4. Then $F_i(B) = A_i$, where $i = 1, 2, \ldots, n$; and the theorem is proved.

(8) *Notation.* The set of points of A_i not in I_i is the complement of I_i with respect to A_i and will be designated by $I_i'(A_i)$. In contrast the complement of I_i, with respect to the whole property space will be designated by I_i'.

It follows that $I_i'(A_i) = A_i \cap (I_i')$.

Application of Bundles to File Updating. Consider the notion of updating a file in its simplest form. This amounts to adding some records to a master file, deleting other records from the master file, modifying some records of the master file using information obtained by matching records from other files, and preserving the rest of the records of the master file intact.

In the notation of the Information Algebra, let A_n be the area whose points represent the records of the master file; let A_1 represent new records; let $A_2, A_3, \ldots, A_{n-1}$ represent files of transaction records to be matched with master file records for deletion and modification. Finally, let A_n^{new} be the area resulting from "updating" A_n, i.e., the updated master file. The following expression for the bundle construct U_p will now be derived:

$$A_n^{\text{new}} = U_p(b; F_1; A_1, \ldots, A_n),$$

where the A's were discussed above, and b and F_1 are discussed below.

In deriving U_p, note first that b is a bundling function defined so as to match records from $A_2, A_3, \ldots, A_{n-1}$ with records of A_n. This creates a bundle $B(b, A_2, A_3, \ldots, A_n)$. The processing of the transactions in $A_2, \ldots, A_{n-1}$ against the matched master file records of A_n is accomplished by an FOB, F_1, and results in modified records which constitute an area $M = F_1(B)$. The FOB, F_1, will be different for each specific updating operation. It defines the rules by which the records on the master file (A_n) are modified by the records on the transaction files ($A_2, \ldots, A_{n-1}$). For example, in a payroll file a simple updating would normally involve changes of pay rate, addresses, exemptions, modifications of year-to-date gross pay, etc. In each case, the FOB will associate the null point with certain lines thereby "deleting" certain records.

The records of A_n, which b selects for modification must not be incorporated in A_n^{new}. They constitute the intersection $I_n(B; A_n)$ of the bundle B with A_n. Removing the records $I_n(B; A_n)$ and incorporating the unmodified records of A_n is accomplished by forming the area C, the complement of $I_n(B; A_n)$ with respect to A_n: $C = I_n'(A_n)$.

The area A_n^{new} is thus the join of C, A_1, and M:

$$A_n^{\text{new}} = A_1 \cup M \cup C$$
$$= U_p(b; F_1; A_1, \ldots, A_n)$$
$$= A_1 \cup F_1(B) \cup I_n'(A_n)$$

An Application of Glumps to File Structure. In order to relate the notions to be presented in this paragraph to more familiar notions, consider the organization of information commonly used with magnetic tapes.

Consider first a simple file, File-1, which consists of a set of records (representing entities) R, each of which has certain fields (properties) $f_1, \ldots, f_{nR}$.

The first of these fields f_1 might be a sort key and the records R might be recorded on tape in order by the sort key f_1. In the notation of the Information Algebra, $f_1, \ldots, f_{nR}$ would be chosen as property coordinates of a property space P and the set of all records of File-1 would be an area in P.

File-1 is simple, in that it contains information about only one kind of entity. The entity might be a collision insurance policy and the recorded properties might be policy number, vehicle engine number, other properties describing the vehicle, effective date of policy, etc.

Consider next a more complicated file, File-2, whose records have headers R and trailers S and the following form:

$$R$$
$$f_1(R)$$
$$f_2(R)$$
$$\vdots$$
$$f_{nR}(R)$$
$$S \; f_1(R)$$
$$f_1(S)$$
$$f_2(S)$$
$$\vdots$$
$$f_{nS}(S)$$

In this file, R may be considered one entity type with properties $f_1(R), f_2(R), \ldots, f_{nR}(R)$. R might be a collision insurance policy header, $f_1(R)$ might be the policy number, $f_2(R)$ the effective date of the policy, etc.

Each R may also contain any number of groups S representing a second type of entry, whose property $f_1(R)$ denotes the particular R that it belongs to, and whose properties $f_1(S), f_2(S), \ldots, f_{nS}(S)$ are fields which describe S. S might be a vehicle insured under the collision policy and $f_1(S)$ the vehicle engine number, etc. A fleet of vehicles of any size can be insured under one policy. Thus, belonging to any R there may be many S entities.

In order to handle this file with the Information Algebra, a property space is chosen with properties

$$f_1(R), f_2(R), \ldots, f_{nR}(R),$$
$$f_1(S), f_2(S), \ldots, f_{nS}(S).$$

An R-header is considered as an entity represented by a datum point whose properties $f_1(R), f_2(R), \ldots, f_{nR}(R)$ are defined and whose properties $f_1(S), f_2(S), \ldots, f_{nS}(S)$ are Ω. Similarly an S-group may be considered as an entity represented by a datum point whose properties $f_2(R), \ldots, f_{nR}(R)$ are Ω, and whose properties $f_1(R), f_1(S), \ldots, f_{nS}(S)$ are defined. The set of datum points representing R-header entities constitute an area A_R. The set of datum points representing entities of type S constitute an area A_S. All the datum points representing S's belonging to a given R constitute a glump element of a glump of A_S by constant $f_1(R)$. The set of all datum points representing entities of both types constitute an area A_{RS} which is the join of A_R and A_S. The set of all datum points of type R or S belonging to a given R (i.e., for which $f_1(R)$ is constant) constitute a glump element of a glump of A_{RS} by $f_1(R) = $ constant. This glump element is the *whole* R record (e.g., a complete collision insurance policy with all the insured vehicles).

The information in File-2 can be represented in the property space in other ways by entirely different datum points by emphasizing different entities. For example, an entity C could be chosen which is an S supplemented by the R to which it belongs. All the properties of the property space are defined for this entity C. The collection of datum points representing C-entities constitute a file, File-3, having the form:

    File-3
    C
       $f_1(R)$
       $f_2(R)$
       $\vdots$
       $f_{nR}(R)$
       $f_1(S)$
       $\vdots$
       $f_{nS}(S)$

Note that this File-3 is a simple file like File-1 having only one kind of entity. In terms of a tape machine a price is paid for this greater logical simplicity: all the policy information is repeated for each vehicle.

The set of all datum points in File-3 constitute an area A_C. There is a one-to-one correspondence between points of A_C of File-3 and the points of A_S of File-2. Since A_C can be glumped by $f_1(R) = $ constant, there is a one-to-one correspondence between the glump elements of the glumps of A_C and A_{RS}. Although "information content" has not been defined, it is intuitively clear that corresponding glump elements contain identical information but there is one more point (an R-type point) in a glump element of File-2; there is a one-to-one correspondence between the S-type points of a glump element of File-2 and the C-type points of a corresponding glump element of File-3.

It is possible to use functions of a bundle to transform File-2 to File-3 and conversely:

Let
$$q_1 = f_1(R), \ldots, q_{nR} = f_{nR}(R),$$
$$q_{nR+1} = f_1(S), \ldots, q_{nR+nS} = f_{nS}(S).$$
$$A_C = F_1(R) = F_1(b_1; A_R, A_S),$$

where b_1 is True if $q_{1.1} = q_{2.1}$ and

$$F \equiv \begin{cases} q_1' = q_{1.1} \\ \vdots \\ q_{nR}' = q_{1.nR} \\ q_{nR+1}' = q_{2.nR+1} \\ \vdots \\ q_{nR+nS}' = q_{2.nR+nS} \end{cases}$$

$$A_R = F_2(b_2, A_C),$$

where b_2 is True for all lines and

$$F_2 \equiv \begin{cases} q_1' = q_1 \\ \vdots \\ q_{nR}' = q_{nR} \\ q_{nR+1}' = \Omega \\ \vdots \\ q_{nR+nS}' = \Omega \end{cases}$$

$$A_S = F_3(b_3, A_C),$$

where b_3 is True for all lines and

$$F_3 \equiv \begin{cases} q_1' = q_1 \\ q_2' = \Omega \\ \vdots \\ q_{nR}' = \Omega \\ q_{nR+1}' = q_{nR+1} \\ \vdots \\ q_{nR+nS}' = q_{nR+nS} \end{cases}$$

The above demonstration has shown that these two different ways of looking at a file are in some sense "equivalent." In general two area sets are intuitively equivalent in "information content" if each area of the first set is in some way expressible in terms of the areas of the other area set and vice versa. More specifically:

(9) Definition. Two area sets $\aleph_1 = (A_{1,1} \cdots A_{1,n})$ and $\aleph_2 = (A_{2,1} \cdots A_{2,m})$ each of whose areas is expressible as an FOB of the other area set are said to be *bundle-equivalent*.

In File-2 and File-3, sets $\aleph_1 = (R, S)$ and $\aleph_2 = (C)$ are *bundle-equivalent*.

Use of Bundles and Glumps to Extend the Update Construct to 2-level Files. The bundle update construct defined above suffices to update area $A_n = A_{RS}$ in File-2 if the entities R are independent of the associated S-entities. However, it frequently occurs that although an R-group has certain properties whose values are independent of the associated S's and other properties whose values are functions of all or some of the S-groups contained in it, the handling of such a file requires the use of both bundles and glumps and more entities.

Let the entities D and E be introduced as follows: the totality of properties of D and E is the totality of properties of R. D has properties $q_1, q_2, \ldots, q_m$ defined, and all others Ω; m is less than n_R, and is defined by arranging the q_i's

so that q_1 through q_m are those properties independent of S, such as policy number, name of insured, etc. E has properties $q_1, q_{m+1}, \ldots, q_{nR}$ defined, and all others Ω; q_{m+1} through q_{nR} are the properties which depend only on S-entities, such as number of vehicles insured, total amount of insurance, etc.

The updated area A_{RS}^{new} is a bundle-glump construct depending on several bundling and glumping functions, FOG's, FOB's, and areas, and is defined below in easy stages.

The updating of area A_S is handled by the Update construct described earlier. More precisely, given some new S-entities making up an area A_{SN} and transactions which constitute an area A_{ST}, there exist a bundling function b_1 and an FOB F_1 such that

$$A_S^{\text{new}} = U_p(b_1; F_1; A_{SN}, A_{ST}, A_S).$$

Similarly A_D^{new} is defined in terms of an area of new D-entities A_{DN}, transactions A_{DT}, a bundling function b_2, and FOB F_2 such that

$$A_D^{\text{new}} = U_p(b_2; F_2; A_{DN}, A_{DT}, A_D).$$

Since the properties of an E-entity depend only on associated S-entities, A_E is defined by the use of glumps. More precisely, there is a glumping function $g_1 = q_1$ and an FOG H_1 over the area A_S such that

$$A_E^{\text{new}} = H_1(g_1, A_S^{\text{new}}).$$

An R-entity has properties some of which are D-entity properties and some of which are E-entity properties. Consequently, A_R can be defined in terms of A_D and A_g through the use of an FOB. More precisely, given A_D^{new} and A_E^{new} there is a bundling function b_3 and an FOB F_1 such that $A_R^{\text{new}} = F_3(b_3, A_D^{\text{new}}, A_E^{\text{new}})$. Finally, A_{RS}^{new} is the join of A_R^{new} and A_S^{new}: $A_{RS}^{\text{new}} = A_R^{\text{new}} \cup A_S^{\text{new}}$.

The introduction of several entities A, B, D, E instead of a single entity C permitted the arrangement of this file on tape in such a way as to minimize storage space. The result is a more complicated update computer program which is reflected in the Information Algebra by a more complex update construct involving glumps as well as bundles.

X. Sample Problem (Payroll)

The problem is to create a New Pay File from the information given in an Old Pay File, a Daily Work File, and a New Employee File. The Daily Work File contains the daily hours-worked records for each employee for the week. The Old and New Pay Files contain other information about the employees including the year-to-date totals of salary earned. The New Employee File provides rate-of-pay information about new employees for incorporation in the New Pay File. The solution is based on collecting and summarizing each employee's daily work records and on the matching of records in the Old Pay File, the summarized Work File, and the New Employee File to create the up-to-date records for the New Pay File for both old and new employees.

A property space sufficient to contain all the information is constructed on nine properties. Four areas corresponding to the four files are defined in this space. The properties, their value sets, and the relevance of each property to each area shown in Table 1, where an "X" denotes relevant information and a "Ω" denotes nonrelevant data for that file.

The solution to this Payroll Problem is fully expressed by the following relationships. The New Pay File is expressed as the union of two areas, one derived for the old employees and the other for the new employees. Each of these areas in turn is a function of a bundle of two areas.

The first area of each bundle is the same and is a function of a glump of the daily work file. The second area of one bundle is the Old Pay File and the second area of the other bundle is the New Employee File.

$$\text{NP} = F_1[q_{12} = q_{22}, H(q_2, \text{DW}), \text{OP}] \quad (1)$$
$$\cup\ F_2[q_{12} - q_{22}, H(q_2, \text{DW}), \text{NE}]$$

$$H \equiv \begin{cases} q_2' = q_2 \\ q_5' = \Sigma\ [q_5 \leftarrow (q_5 < 8) \\ \qquad\qquad \to 1.5 * q_5 - 4] + f_1 \\ f_1 = 0 \leftarrow (f_2 < 40) \\ \qquad\qquad \to 0.5 * f_2 - 20 \end{cases}$$

$$F_1 \equiv \begin{cases} q_7' = q_{27} + q_{15} * q_{24} \\ q_8' = q_{28} + 1 \\ q_9' = q_{15} * q_{24} \end{cases}$$

$$F_2 \equiv \begin{cases} q_1' = PF \\ q_7' = q_{15} * q_{24} \\ q_8' = q_{28} + 1 \\ q_9' = q_{15} * q_{24} \end{cases}$$

The terms of equation (1) are identified in Figure 1. Let us examine each element of this equation more closely. Consider first the expression $H(q_2, \text{WD})$. The Daily Work File (DW) consists of the daily hours-worked entries for all employees. Using q_2(Man ID) as a glumping function, each glump element contains the entries for a single employee. $H(q_2, \text{DW})$ is a function of a glump which calculates the values of certain properties for a set of points where each point corresponds to a single glump element. These values are defined by

$$H \equiv \begin{array}{l} q_2' = q_2 \\ q_5' = \Sigma\ [q_5 \leftarrow (q_5 < 8) \to \\ \qquad\qquad 1.5 * q_5 - 4] + f_1 \\ f_2 = \Sigma\ ]q_5 \leftarrow (q_5 < 8) \to 8] \\ f_1 = 0 \leftarrow (f_2 < 40) \to 0.5 \\ \qquad\qquad * f_2 - 20. \end{array}$$

In the definition of H, f_1 and f_2 are functions of areas. Note that $f_2 = 40$ unless $q_5 < 8$, in which case $f_2 = \Sigma\ q_5$. From

TABLE 1. Payroll Problem—System Information

Properties	Value Set	Areas			
		Old Pay File OP	Daily Work File DW	New Employee File NE	New Pay File NP
q_1 = File ID	PF, DW, NE	X (always PF)	X (always DW)	X (always NE)	X (always PF)
q_2 = Man ID	00000 ··· 99999	X	X	X	X
q_3 = Name	20 alphabetic characters	X	Ω	X	X
q_4 = Rate	00.00 ··· 99.99	X	Ω	X	X
q_5 = Hours	00 ··· 24	Ω	X	Ω	Ω
q_6 = Day number	0 ··· 7	Ω	X	Ω	Ω
q_7 = Total salary	00000.00 ··· 99999.99	X	Ω	Ω	X
q_8 = Pay period number	00 ··· 52	X	Ω	X	X
q_9 = Salary	000.00 ··· 999.00	X	Ω	Ω	X

this, it follows that $f_1 = 0$ if $f_2 < 0$; and $f_1 = f_2/2 - 20 =$ half of excess time over 40, otherwise. The overtime rule used in q_5' states that a man is paid time and a half for each day's excess over 8, and also is paid time and a half for the excess over 40.

All other properties are undefined for this function of a glump.

The set of points obtained from $H(q_2, \text{DW})$ constitutes an area which is the first area of the area set $[H(q_2, \text{DW}), \text{OP}]$. A bundle is defined over this area set. The bundling function for this bundle is a match on the Man ID's in each area, i.e., $q_{12} = q_{22}$. q_{12} is the property *Man ID* in the first area of the bundle; q_{22} is the corresponding property in the second area of the bundle.

$F_1[q_{12} = q_{22}, H(q_2, \text{DW}), \text{OP}]$ is a function of a bundle which maps each line of the bundle into a single point for each employee. The definition of F_1 is:

$$F_1 = \begin{cases} q_7' = q_{27} + q_{15} * q_{24} \\ q_8' = q_{28} + 1 \\ q_9' = q_{15} * q_{24} \end{cases}$$

The subscripts of the q-primes denote the property in the new area which is being defined by the equation. The first subscript of each unprimed q indicates the area of the bundle and the second subscript identifies the particular property. Thus, q_{27} is the property *Total Salary* for points in the *second* area of the bundle.

Each of the nonstated q-primes is understood to have the same value as the corresponding q in the last (in this case, the second) area of the bundle. Thus,

$q_1' = \text{PF}$ $q_3' = \text{Name}$ $q_5' = \Omega$
$q_2' = \text{Man ID}$ $q_4' = \text{Rate}$ $q_6' = \Omega$

while the remaining q-primes are defined by F_1.

Similarly, a second function of a bundle is defined for the area set $[H(q_2, \text{DW}), \text{NE}]$:

$F_2[q_{12} = q_{22}, H(q_2, \text{DW}), \text{NE}]$

where F_2 is defined by $F_2 \equiv$
$\begin{aligned} q_1' &= \text{PF} \\ q_7' &= q_{15} * q_{24} \\ q_8' &= q_{28} + 1 \\ q_9' &= q_{15} * q_{24} \end{aligned}$

and the remaining q-primes are understood to be taken from the New Employee File:

$q_2' = \text{Man ID}$ $q_4' = \text{Rate}$ $q_6' = \Omega$
$q_3' = \text{Name}$ $q_5' = \Omega$

Now two areas of up-to-date pay information have been derived. All that remains is to join them by the set operation *union* to form the New Pay File.

INDEX OF TERMS AND SYMBOLS

The following list of terms defined in this paper and of symbols used in a particular way throughout will facilitate reference to the definitions and to those places in the text where symbol conventions are introduced. The double asterisk (**) indicates operation on FOL's.

	Section
a (particular value)	II(8)
A (area)	IV(1)
A_i (Bundle argument)	IX(1)
$A(B)$	IX(6)
**ADDITION (+)	VIII(2)
**AND (@)	VIII(7)
**@ (AND)	VIII(7)
Area (A)	IV(1)
Area of a Bundle $A(B)$	IX(6)
Area Set ($\aleph$)	V(1)
$\aleph$ (area Set)	V(1)
b (bundling function)	V(2)
B (bundle)	V(2)
Bundle (B)	V(2)
Bundle Argument (A_i)	IX(1)

AN INFORMATION ALGEBRA

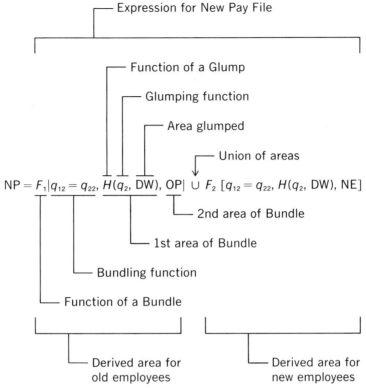

FIGURE 1 Identification of terms of equation (1).

	Section		Section
Bundle-equivalent bundling function (b)	IX(9) V(2)	FOB (Function of a Bundle)	V(3)
Complement of I_i (I_i')	IX(8)	FOG (Function of a Glump)	VI(3)
**CONCATENATION ($\cent$)	VIII(1)		
**$\cent$ (CONCATENATION)	VIII(1)	FOL (Function of a Line)	III(3)
Coordinate Set (Q)	II(5)		
d (Datum Point)	II(9)	Function of a Bundle (FOB)	V(3)
Datum Point (d)	II(9)		
discriminatory	II(11)	Function of an Area (FOA)	IV(2)
**DIVISION (/)	VIII(4)		
e	II(1)	Function of a Glump (FOG)	VI(3)
element of a glump	VI(1)		
Entity	II(1)	Function of a Line (FOL)	III(3)
**EQUALS (=)	VIII(9)		
f	III(4), IV(2)	g (glumping function)	VI(1)
F	V(5)	G	VI(1)
F and T	The Boolean values False and True	glump element glumping function (g)	VI(2) VI(1)
	IV(2)	H	VI(4)
FOA (Function of an Area)		iff	means "if and only if"

	Section
**IF – OTHERWISE ($\leftarrow, \rightarrow$)	VIII(11)
(I_i) Intersection of Bundle with Bundle Argument	IX(3)
I_i' (complement of I_i)	IX(8)
Intersection of Bundle with Bundle Argument (I_i)	IX(3)
line (L)	III(1)
L (Line)	III(1)
**LESS THAN ($<$)	VIII(10)
missing (Θ)	II(3)
Θ (missing)	II(3)
**MULTIPLICATION ($*$)	VIII(3)
**NEGATION ($-$)	VIII(5)
**NOT ($\neg$)	VIII(8)
**OR ($+$)	VIII(6)
O (Ordering)	VII(1)
OFOL (Ordinal FOL)	III(9)
Ordering (O)	VII(1)
Ordinal FOL (OFOL)	III(9)
p (point)	II(8)
P (Property Space)	II(8)
p_F (Point assigned to L by F)	IX(2)
point (p)	II(8)
Property	II(1)
Property Space (P)	II(8)
Property Value Set (V)	II(2)
q	II(1)
Q (Coordinate Set)	II(5)
R	Designates a real number
relational operator (Ⓡ)	III(9)
s (selection function)	V(7)
Selection	III(9), V(7)
selection function (s)	V(7)
Selection OFOL	III(9)
Span of an FOL	III(8)

	Section
Span of a line	III(1)
T and F	The Boolean values True and False
Undefined (Ω)	II(3)
Ω (Undefined)	II(3)
v	II(1)
V (Property Value Set)	II(2)
Value	II(1)
Value set of an FOA	IV(2)
Value set of an FOL	III(3)
Value set of a Property (V)	II(2)

SYMBOLS NOT CONTAINED WITHIN A WELL-KNOWN DEFINING SET

(All but the last four are Operations on FOL's.)

	Section
$+$ (ADDITION)	VIII(2)
@ (AND)	VIII(7)
¢ (CONCATENATION)	VIII(1)
/ (DIVISION)	VIII(4)
$=$ (EQUALS)	VIII(9)
$\leftarrow, \rightarrow$ (IF – OTHERWISE)	VIII(11)
$<$ (LESS THAN)	VIII(10)
$*$ (MULTIPLICATION)	VIII(3)

	Section
$-$ (NEGATION)	VIII(5)
$\neg$ (NOT)	VIII(8)
$\oplus$ (OR)	VIII(6)
ℵ (Area Set)	V(1)
Θ (missing)	II(3)
Ⓡ (relational operator)	III(9)
Ω (undefined)	II(3)

ABSTRACT FORMULATION OF DATA PROCESSING PROBLEMS

JOHN W. YOUNG, JR.
HENRY K. KENT

There are three stages in the application of high speed digital computers to data processing problems:

1. Systems Analysis—the task of determining what is to be done
2. Programming—a statement of how it is to be done
3. Coding—a translation of this statement into machine language

A wide variety of automatic coding techniques has been developed. This paper presents a first step in the direction of automatic programming as well as a tool which should be useful in systems analysis.

Since we may be called upon to evaluate different computers or to find alternate ways of organizing current systems, it is necessary to have some means of precisely stating a data processing problem independently of mechanization.

The notation presented here provides such a precise and abstract way of specifying the informational and time characteristics of a data processing problem and should enable the analyst to organize the problem around any piece of hardware. This notation could be used in the input to a new type of automatic programming system in which a problem is stated not only independently of machine, but without specifying the structure of files or sequence of operations where such specification is not needed for logical correctness. Further, the application of a graphical version of the notational system will show the relationships among information in input and output so that redundancies can be eliminated and alternative ways of processing studied.

The context of our analysis is that the objectives of the data processing system have been stated in terms of the required outputs; these outputs are not considered as subject to revision. On the other hand, although the inputs may

---

SOURCE: Young, J. W., H. K. Kent, "Abstract Formulation of Data Processing Problems," *Journal of Industrial Engineering*, Nov.–Dec., 1958, pp. 471–479.

be organized in any desired fashion, it appears necessary or at least convenient to state one of the possible input organizations from which any equivalent one can be derived. It should be noted that the input may supply any one of a number of equivalent pieces of information, e.g., either customer name to be copied directly onto an output or an identification number from which the name can be looked up.

BASIC COMPONENTS OF DATA PROCESSING PROBLEMS

A data processing problem can be described in terms of four kinds of basic components:

1. Information sets
2. Documents
3. Relationships
4. Operational requirements

Information Sets

An information set, which we call P_i, is a list of all possible items belonging to the same class. From the items in these lists are drawn the data which will flow through the system. The members of an information set may be considered as all the possible entries which could be made in a specified blank on a document. Examples of information sets are customers' names and addresses, part numbers, unit prices, dates, invoice numbers, etc. Whether an information set or a collection of related information sets exists in the system as a file will depend on the ultimate mechanization of the problem. As far as an abstract statement of this phase of problem is concerned, it is necessary to state only the information sets and their relationships. Governed by these relationships, the analyst is then free to choose whatever file structure will make the overall system most efficient.

Documents

A document, which we call D_j, is a collection of related information items, D_{jk}, where k is the number of the items on the jth document. Documents are either inputs or outputs; examples are shipping notice, invoice, daily sales report, etc.

Relationships

The third element which must be described is the set of relationships among the information sets, the documents, and the items on the documents. A relationship between information sets specifies the correspondence between the items in one set and the items in some other set. For example, relationships exist between the corresponding items of the set of customers' names and customers' numbers, salesman's names and customers, day-month-year's and dates, etc.

The relationship which shows how an item on an output document is derived, or from where it is taken if it is simply copied from another document, is called the defining relationship. In addition, there are producing relationships which define under what stimuli the documents or certain items on the documents are produced, and conditional relationships which specify under what conditions any of the foregoing apply.

Operational Requirements

The final element in a data processing problem is the set of operational requirements of the system. These are the requirements which are not related to the logic of the problem, such as the volume of input documents per day, the number of copies of outputs, the size or color of the documents, the elapsed time between the receipt of an input and the production of an output, etc.

ABSTRACT STATEMENT OF A PROBLEM

An abstract statement of a problem can be made by preparing two lists: one of the information sets and the other of the documents.

The list of information sets presents the name of each set along with certain data about the number of items in each set, the number of characters required for each item, and the qualitative and quantitative relationships among sets. The list of documents presents for each document the items on the document and the information set each item belongs to. It also lists the defining and producing relationships for each item on the documents as well as the operational requirements associated with the documents.

Thus the output documents are completely specified in terms of the transformations which are applied to the inputs. At the same time, the relationships among the information sets enable logical substitutions to be made in the input to achieve the same output.

In order to describe the notation and its properties, we will proceed by means of an illustrative example.

VERBAL STATEMENT OF PROBLEM

Suppose that we have a manufacturer who maintains stock at several scattered warehouses. Sales are made and reported to the warehouses, which in turn make the shipments and at the same time send information to the central office for billing purposes. The central office sends invoices to the customers, receives payments, furnishes monthly statements, and prepares a daily cumulative sales report for management purposes. Each customer is served by a given salesman who in turn works out of a given warehouse.

There are five documents in this system, two inputs (shipping notice, customer payment) and three outputs (invoice, monthly statement, daily cumulative sales report).

Three hundred shipping notices are received daily and each has, on the average, five line items. An invoice is prepared for each shipping notice. After determining the unit price and extended price for each line item, the total price per shipment is calculated. The invoice must be sent within two days after receipt of the shipping notice.

About two hundred payments are received daily, each covering on the average one invoice. The customer includes the invoice number with the payment.

Monthly statements are sent out between the tenth and the fifteenth of the month to all customers with an open balance as of the tenth of the month. The invoices which were dated after the tenth of the previous month and which have not yet been paid are itemized on the statement. Old unpaid invoices are consolidated and shown only as an old balance.

The daily sales report is produced within two days after the close of business. It includes the gross sales for the date and a breakdown by salesman of daily sales and cumulative sales for the past month.

An abstract statement of this problem is presented in Table 1 (List of Information Sets) and Table 2 (List of Documents). A graphical representation of the information sets is presented in Figure 1 and a combined graphical representation of both the information sets and documents is presented in Figure 2. Graphical notation is explained in Figure 3. Symbols are explained in Table 3.

The following format is used for the information sets:

P_i	Name	Number of items	Number of characters (Numeric, N; Alphabetic, A; Alphanumeric, A/N)	Relationships
P_{10}	Customer name and address	2000	50 A/N	$P_{10} \approx P_2$
⋮	⋮	⋮	⋮	⋮
P_{18}	Salesman name	50	15A	$P_4 \approx P_{18} \sim P_{11}$

In this problem, the customer's name is always linked with his billing address and P_{10} is simply a name for the collection of the two thousand names and addresses in the system. The fifty alphanumeric characters required represent the maximum number for any one name and address. We could, if desired, also list the mean and standard deviation, the range, or any other statistic which is applicable. The other information sets are obtained by examining all of the data flowing through the system and simply listing the sets as they are found. In this problem, there are eighteen such information sets.

RELATIONSHIPS AMONG INFORMATION SETS

One relationship that frequently exists between sets of information items is a one-to-one correspondence. Whenever there is this one-to-one correspondence

TABLE 1 Information Sets

		n	L	Relationships
P_1	Date	—	6N	$P_1 = P_7 \times P_8 \times P_9$
P_2	Customer Identification No.	2000	5N	$P_2 \approx P_{10}, P_2 \sim P_4$
P_3	Ship to code	9	1N	$P_{14} \approx P_2 \times P_3$
P_4	Salesman No.	50	2N	$P_2 \sim P_4 \sim P_{11}$
P_5	Model No.	150	5A/N	$P_5 \approx P_{12} \times P_{13}$
P_6	Quantity ordered	—	2N	
P_7	Day	31	2N	$P_1 = P_7 \times P_8 \times P_9$
P_8	Month	12	2A	$P_1 = P_7 \times P_8 \times P_9$
P_9	Year	10	2N	$P_1 = P_7 \times P_8 \times P_9$
P_{10}	Customer N / A	2000	50A/N	$P_{10} \approx P_2$
P_{11}	Warehouse name	10	12A	$P_4 \sim P_{11} \sim P_{15}$
P_{12}	Part No.	800	3A/N	$P_5 \approx P_{12} \times P_{13}$
P_{13}	Color	20	2A	$P_5 \approx P_{12} \times P_{13}$
P_{14}	Ship to address	6000	50A/N	$P_{14} \approx P_2 \times P_3$
P_{15}	Pricing area	8	1A	$P_5 \times P_{15} \sim P_{17}$
P_{16}	Invoice No. (Shipping Notice No.)	—	5N	$P_{16} \approx D_2$
P_{17}	Unit price	—	5N	$P_5 \times P_{15} \sim P_{17}$
P_{18}	Salesman name	50	15A	$P_{18} \approx P_4$

n = number of elements in set
L = number of characters (numeric—N, alphabetic—A, or alphanumeric A / N) in each element.

TABLE 2 Document Descriptions

Shipping Notice—D1—Input

Items	Verbal Description	Information Set	Defining Relationship
D_{1-1}	Date	P_1	
D_{1-2}	Shipping Notice No.	P_{16}	
D_{1-3}	Customer Identification No.	P_2	
D_{1-4}	Ship to Code	P_3	
D_{1-5}	Salesman No.	P_4	
$[D_{1-6}]$	Quantity of Order	P_6	
D_{1-7}	Model No.	P_5	
D_{1-8}	Line Item		$D_{1-6,7}$

Volume: $\Box\, D_1/P_{7E} = 300$
$\Box\, D_{1-8}/D_1 = 5$

Invoice—D2—Output

Items	Verbal Description	Information Set	Defining Relationship
D_{2-1}	Date	P_1	$t_E(D_2)$
D_{2-2}	Invoice No.	P_{16}	
D_{2-3}	Customer Identification No.	P_2	
D_{2-4}	Customer Name and Address	P_{10}	
D_{2-5}	Ship to Address	P_{14}	
D_{2-6}	Warehouse shipped from	P_{11}	
$[D_{2-7}]$	Quantity of order	P_6	
D_{2-8}	Model No.	P_5	
D_{2-9}	Unit Price	P_{17}	
$D_{2-10}]$	Extended Price		$P_{17}(D_{2-8} \times P_{15})$
D_{2-11}	Total Price		$D_{2-7} \cdot D_{2-9}$
D_{2-12}	Line Item		$\Sigma\, D_{2-10}$
			$D_{2-7 \text{ to } 10}$

Producing Relationship: $D_1 \to D_2$
Operational Requirement
Volume: $\Box\, D_2/P_{7E} = 300$
Time: $t_E(D_2) - t_E(D_1) < 2$ days

TABLE 2 Document Descriptions (*continued*)

Customer Payment—D3—Input

Items	Verbal Description	Information Set	Defining Relationship
D_{3-1}	Date	P_1	
$[D_{3-2}$	Invoice No.	P_{16}	
D_{3-3}	Amount		
D_{3-4}	Line Item		$D_{3-2,3}$

Volume: $\Box\ D_3/D_{7E} = 200$
$\phantom{\text{Volume: }}\Box\ D_{3-4}/D_3 = 1.5$

Monthly Statement—D4—Output

Items	Verbal Description	Information Set	Defining Relationship
D_{4-1}	Customer Name and Address	P_{10}	
D_{4-2}	Date	P_1	$10/P_{8E}(D_4)/P_{9E}(D_4)$
			Statements are to be dated the 10th of the month.
D_{4-3}	Customer's (old) balance		$D_{4-8}(D_{4-1}, P_8 - 1) - \Sigma\ \bar{C}_{4-1}D_{3-3}(D_{4-1})$
$[D_{4-4}$	Invoice No.	P_{16}	D_{2-2}
D_{4-5}	Date of Invoice	P_1	D_{2-1}
$D_{4-6}]$	Amount of Invoice		D_{2-11}
D_{4-7}	Line Item		$D_{4-4,5,6}$
D_{4-8}	New Balance		$D_{4-3} + \Sigma\ D_{4-6}$

Producing Relationship: $P_2 \times P_8 \to D_4 | D_{4-8} \neq 0$

(statements are produced each month for each customer with a non-zero balance)

$$D_2 \to D_{4-7} | C_{4-1} \wedge C_{4-2}$$

(an invoice is included in the statement if both condition C_{4-1} and C_{4-2} are true)

Special Conditions: C_{4-1}: $[P_8(D_2) = P_8(D_4) \wedge P_7(D_2) < 10] \vee [P_8(D_4) - 1 = P_8(D_2) \wedge P_7(D_2) > 10]$

(the invoice was dated after the 10th of the preceding month but before the 10th of this month.)

$$C_{4-2}: \bar{\exists}\ D_3\ [D_{3-2}(D_{2-2})]$$

(a payment has not been received for the invoice)

Operational Requirements:

Volume: $\Box\, D_4/P_{8E} = 500$
(the average number of statements issued per month is 500)
$\Box\, D_{4-7}/D_4 = 4$
(the average number of invoices (line items) itemized per statement is 4)

Time: $10 < P_{7E}(D_4) < 15$
(statements are to be produced between the 10th and the 15th of the month)

Daily Cumulative Sales Report—D5—Output

Items	Verbal Description	Information Set	Defining Relationship
D_{5-1}	Date	P_1	
$[D_{5-2}$	Salesman No.	P_4	
D_{5-3}	Salesman Name	P_{15}	
D_{5-4}	Sales this date		$\Sigma\, D_{2-11}(D_{5-1}, D_{5-2})$
$D_{5-5}]$	Cumulative sales this month		$D_{5-4} + C_{5-1} \cdot D_{5-5}(D_{5-1} - 1)$
D_{5-6}	Line Item		$D_{5-2\cdot 3\cdot 4,5}$
D_{5-7}	Total gross sales this date		$\Sigma\, D_{5-4}$

Producing Relationships: $P_1 \to D_5$
$P_4 \to D_{5-6}$

Conditions: $C_{5-1}: P_7(D_{5-1}) \neq 1$

Operational Requirements:

Volume: $\Box\, D_{5-6}/D_5 = 50$

Time: $t_E(D_5) - t_1(D_5) < 2$ days

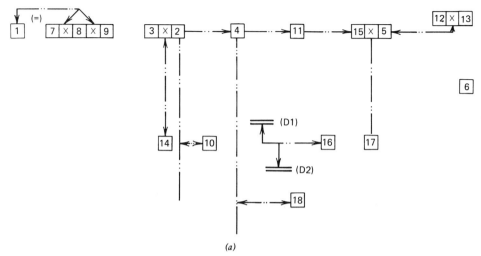

FIGURE 1 Graphical representation of information sets.

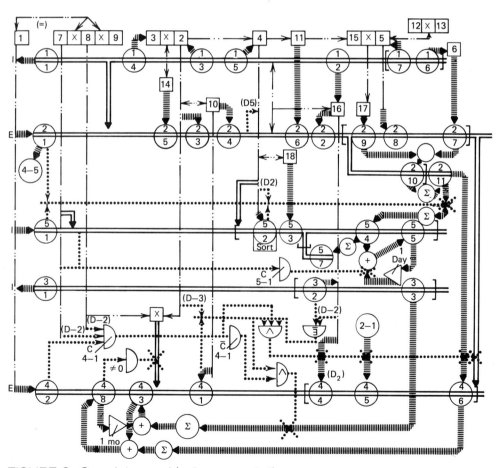

FIGURE 2 Complete graphical representation.

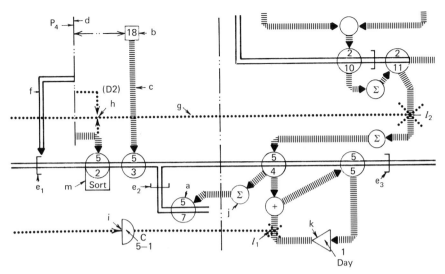

FIGURE 3 Explanation of graphical notation.

(a) This circle represents an information item on a document, viz., D_{5-7}, with the document number above the line, and the item number within the document, below. The heavy double line (generally horizontal and possibly with branches) connects all the items on one document.

(b) Each information set, P_i, is shown as a square, in this case P_{18}.

(c) An item on a document which is an element of some P_i is connected to it; the arrow runs out of the document for an input, into it for an output (i.e., one may think of the flow of information to and from the system as following the arrows).

(d) Isomorphism (or homomorphism) between information sets is shown by a double (or single) headed arrow. In this case we are saying that $P_{18} \simeq P_4$, with the square standing for P_4 having been extended downward by the broken vertical line.

(e) The small square brackets enclose those items making up a line item of a document, in this case $D_{5-2,3,4,\text{and }5}$. Note that D_{5-7} is not a line item and is excluded by the bracket e_2.

(f) The producing relationship is shown by a double heavy line with an arrow pointing to the document or line item produced, e.g., P_4 produces a line item on D_5. The line runs from the extension (the vertical broken line) of the square box representing P_4 to the bracket representing a line item on D_5.

(g) The dotted line is used to connect various elements of a condition.

(h) The condition at h is that D_{5-2} is equal to P_4 (D_2), represented by condition lines from D_{5-2} and P_4 with arrows to the condition line g. That D_2 is involved is shown by writing in (D_2).

(i) This is the standard symbol for a gate and states the condition involved, viz., C_{5-1}. The input to the gate (the flat surface) is connected to the item or items involved in the condition. The output goes to the operation affected by the condition.

(j) The summation sign indicates that D_{5-7} is equal to the sum of D_{5-4}.

(k) The triangle represents a delay equal to the time indicated within it. Thus yesterday's D_{5-5} is added to D_{5-4} to produce today's D_{5-5}.

(l_1 & l_2) The dotted cross through an information transfer line (l_1) or a sum (l_2) (or any other operation or production) indicates that this transfer or operation is performed if the condition is true.

(m) The sort notation implies that the document or line item must be sorted by the item indicated.

TABLE 3 List of Symbols

Symbol	Description
P_i	A list of all possible information belonging to the same class
p_i	A specific member of the class
D_j	A document
d_j	A specific document belonging to the D_j class
D_{jk}	A collection of entries on the document D_j
[ ]	Line item
$\simeq$	Isomorphic (one to one correspondence)
$\sim$	Homomorphic (many to one correspondence)
$\times$	Cartesian product, e.g., $P_j \times P_k$ means a pair of p_j and p_k
$\subset$	Contained in
$\leftarrow$	Produces
E	Extrinsic time (real time)
I	Intrinsic time, e.g., date written on document
( )	Function
C_{m-n}	nth Condition relating to the mth Document If
$\overline{C}_{m-n}$	Negation of C_{m-n}, i.e., $\overline{C}_{m-n}$ is true if and only C_{m-n} is false
$\sqsubset$	Number of
$\overline{\sqsubset}$	Average number of
$\exists$	There exists
$\overline{\exists}$	There does not exist

between the items in one set and the items in another set, we shall, borrowing a term from abstract algebra, call this relationship isomorphism. This relationship holds, for example, between P_{10} (customers' names and addresses) and the information set that consists of customer identification numbers, P_2. That is, to each specific customer name and address corresponds exactly one customer identification number and vice versa. This relationship is read P_{10} isomorphic to ($\simeq$) P_2.

If the correspondence between the items in the information sets is many-to-one rather than one-to-one, we shall call the relationship homomorphism. For example, there are many salesmen who work out of a given warehouse, so that we may say that P_{18} (salesman name) is homomorphic to P_{11} (warehouse name). Thus given the salesman, we can find the warehouse from which he works, but not vice versa. In other words, if P_j is homomorphic to P_k then given an element of P_j we can find the corresponding element of P_k. However, unlike isomorphism, the relationship is not symmetric; given an element P_k there is a whole set (with one or more members) of elements of P_j which correspond to it.

Since both isomorphism and homomorphism are transitive,[1] we can derive relationships which are not explicitly given. For example, if P_{10} (customer name and address) is homomorphic to P_{18} (salesman name) which in turn is homomorphic to P_{11} (warehouse name), we can then say P_{10} is homomorphic to P_{11}.

Our graphical notation, shown in Figure 1, provides the simplest way to derive all of the relationships among the information sets.

The double-ended arrow indicates an isomorphism between sets and the single arrow, homomorphism. The relationships hold only in the direction of the arrows so that P_{10} which is isomorphic to P_2 is also homomorphic to P_4, P_{11} and P_{15}. For convenience in drawing, the arrows can be connected to a line extending from the squares.

An item of information is sometimes the result of combining several other items; e.g., the model number is determined by the combination of part number and color, or expressed symbolically $P_5 \simeq P_{12} \times P_{13}$; this is read P_5 isomorphic to P_{12} cross P_{13}. Care must be taken to distinguish between this so-called cartesian or cross product, and arithmetic

---

[1] If a, b, and c are three elements and R is a relationship, then R is said to be transitive if aRb and bRc together imply aRc.

multiplication, which we always denote by $a \cdot b$ or just ab. The above says that there is a 1-to-1 relationship between the elements of P_5 and all pairs (p_{12}, p_{13}) where p_{12} is a member of P_{12} (i.e., a specific part number) and p_{13} is a member of P_{13} (i.e., a particular color).

Again, the customer's actual shipping address (P_{14}) (store, warehouse, etc.) is determined by the combination of his identification number (P_2) and the ship-to code, i.e. $P_{14} \simeq P_2 \times P_3$.

In the case of date (P_1), we say that P_1 is equal to (rather than isomorphic to) $P_7 \times P_8 \times P_9$ (day, month and year) since an element of P_1 actually consists of the day, month and year taken from P_7, P_8 and P_9. Since certain triplets of day, month and year do not exist, it would be more correct to say that P_1 is contained in (symbolized by $\subset$) $P_7 \times P_8 \times P_9$. Operationally, however, this should be no problem since dates such as Feb. 31, 1958, should not enter the system.

P_{16}, the information set consisting of invoice numbers and shipping notice numbers (in this problem the same number is used for both) must be handled in a slightly different manner. Since these numbers are created at the same time as the documents they identify, they bear a definite relationship to these documents. In fact, there is a one-to-one correspondence between the numbers and the documents so that we may say that

$$P_{16} \simeq D_2 \quad \text{and}$$
$$P_{16} \simeq D_1$$

This implies a relationship between P_{16} and any of the information sets to which the items on the document belong, viz:

$$P_{16} \sim P_j$$

where P_j is any information set to which an item on the document belongs. In other words, given an invoice number we can find the specific entries on the document.

The following format is used for the documents:

Name of Document		Number (D_j)	Input/Output
Items	Verbal Description	Information Set	Defining Relationship
⋮	⋮	⋮	⋮
D_{jk}	—	P_i	—

Producing Relationships:
Special Conditions:
Operational Requirements
   Volume:
   Time:

For each document we record its name, number, and whether it is an input or output document. The items D_{jk} are entered along with their verbal description, the information set to which they belong, and, for output documents, the defining relationship, which tells from where the item is taken or how it is derived. If no relationship is shown, the item is derived directly or indirectly from the producing document. Whatever producing relationships and special conditions apply are listed at the bottom together with the volume and time requirements.

One characteristic of a document is that it contains not only information that appears only once but also information, known as line items, which may be repeated any number of times. The line item for the shipping notice (D_1) is D_{1-6} and D_{1-7}, quantity ordered and model number, and is distinguished by placing square brackets around it. A particular shipment may consist of seven different models so that the line item would be repeated seven times, each containing a different quantity ordered and model number. In order to refer to a line item as a whole without the necessity of repeating its elements, an item, D_{1-8}, is added and used whenever reference is made to an operation which pertains to each of the line items. For convenience, the items contained in a line item are listed in the defining relationship column. Thus D_{1-8} is shown to consist of D_{1-6} and D_{1-7}.

We distinguish between two kinds of time, extrinsic or E time and intrinsic or I time. Extrinsic time is the time at which an event occurs, e.g., the production of an output document; and intrinsic time is the time which is an information item on a document, e.g., the date written on the document.

Using this symbolism we can now express elapsed time relationships. For example, if the invoice must be produced within two days after the receipt of the shipping notice by the system, we would write:

$$t_E(D_2) - t_E(D_1) < 2 \text{ days}.$$

Absolute instead of relative time can also be expressed, e.g.,

$$10 < P_7[t_E(D_4)] < 15$$

implying that the day (P_7) of production of D_4 must be between the 10th and 15th (of the month). Expressions of this kind will often be condensed from $P_7[t_E(D_4)]$ to just $P_{7E}(D_4)$, since P_7 is an information set consisting of time units. We can apply arithmetic operations to dates as in the specification of D_{5-5} in Table 2 where $D_{5-5}\ [(D_{5-1}) - 1]$ means that we take D_{5-1} (date), subtract 1 from it, and look up D_{5-5} from the D_5 with this date.

The volume statements on the shipping notice state that the average number (indicated by the square C with the bar over it) of shipping notices (D_1) per day (P_{7E}) is 300 and the average number of line items (D_{1-8}) per shipping notice is 5. The special symbols are used with an eye toward future automatic programming input.

A graphical representation of the shipping notice and the invoice together with the information sets is shown in Figure 4.

The items on a document are connected by a (generally horizontal) double line and the item numbers are entered in a circle. By convention the circles on input documents are connected to their corresponding information sets by a line with an arrow pointing toward the information set. The line items are enclosed by square brackets. The graphical representation makes it easy to pick up redundant information. For example, D_{1-3} is a member of P_2 which in turn is homomorphic to P_4. D_{1-5}, which is a member of P_4, could therefore be derived from D_{1-3} by a table look-up. In other words, it was not necessary to input the salesman's number as it could have been derived from a knowledge of the customer number. Note also that it was not necessary to know what meaning was given to D_{1-3}, D_{1-5}, P_2, and P_4 in order to determine this redundancy.

PRODUCING RELATIONSHIPS

We have discussed certain relationships among the information sets; other kinds of relationship may exist among docu-

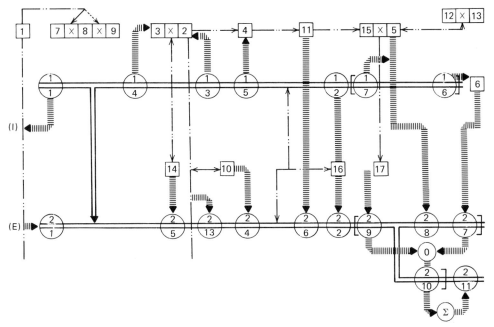

FIGURE 4 Graphical representation on D_1, shipping notice and D_2, invoice.

ments, and between documents and information sets. One of these relationships is that of "produces." In general, a document will be produced in one of two ways:

1. Periodically, once every day, week, etc.
2. Irregularly, once for each occurrence of an input document or of a condition dependent on input data.

For example, since an invoice (D_2) is created for every shipping notice (D_1) that is received, we say that

$$D_1 \rightarrow D_2 \text{ (read } D_1 \text{ produces } D_2\text{)}$$

The producing relationship is concerned with the existence rather than the content of the document involved. The daily sales report is produced each day, so we say the date produces the report; however, this does not tell us what is on the sales report or from where its information is taken. In general, though, the items on an output document will be derived from the document which produced it, unless the defining relationship specifies some other way to derive these items.

DEFINING RELATIONSHIPS

The defining relationship tells us the content of the output items. The exact specification of how this content is to be generated is left to the subsequent mechanization of the problem. Where no defining relationship is given, it is understood that the items are to be copied or derived from the input document that produces the output. These items are copied directly from the input when both input and output items belong to the same information set. They are derived indirectly from the input when the input and output items belong to different but related information sets. For example, the customer name and address (D_{2-4}) on the invoice (D_2) belongs to P_{10} (the information set consisting of customer names and addresses). No defining relationship is given opposite D_{2-4} in Table 2; thus for the derivation of D_{2-4} we must look to the input document (D_1) which produces D_2. However, since no item in D_1 is a member of P_{10} (that is, the customer name and address does not appear on the shipping notice), we must find some item on D_1 whose information set is related to P_{10}. Referring to Figure 1, we note that P_2 is isomorphic to P_{10} and that D_{1-3} is a member of P_2. Thus we know that D_{2-4} can be derived from D_{1-3} via the relationship between P_{10} and P_2. In other words, given the customer identification number on an invoice we can find the customer name and address for the corresponding invoice. Note that nothing need be said as to how this is done. It may be desirable to maintain a file within the system and find the customer's names by a table look-up or this may be done manually outside of the system. We have merely stated the requirement that it be done.

One way of symbolizing that an item in one information set is to be derived from an item in another set is by means of the function notation of mathematics. For example, the unit price depends upon both the model number and the pricing area in which the item is sold so that D_{2-9}, the unit price, (which belongs to P_{17}) is specified in Table 2 as P_{17} ($D_{2-8} \times P_{15}$). This is read as P_{17} of D_{2-8} cross P_{15}, and means that the system is to look at the combination of a model number (D_{2-8}) and a particular pricing area (belonging to P_{15}) and put the corresponding member of P_{17} (unit price) in the space designated as D_{2-9} on the invoice.

Note that the pricing area (P_{15}) may be derived in a number of ways, e.g., from a member of P_{11} (via $P_{11} \sim P_{15}$), from a member of P_4 (via $P_4 \sim P_{15}$), or from a member of P_2 (via $P_2 \sim P_{15}$). That is, the pricing area can be found from either the warehouse name, the customer identification number, or the salesman number. Again, the particular choice of deriving the pricing area is left to the subsequent mechanization of the problem.

The defining relationship may be a simple mathematical relationship between various items; e.g., the extended price (D_{2-10}) is found by multiplying the quantity ordered (D_{2-7}) by the unit price (D_{2-9}). Thus, D_{2-10} is defined as $D_{2-7} \cdot D_{2-9}$ (read D_{2-7} times D_{2-9}). In the graphical notation, Figure 4, this is indicated by drawing information flow lines from D_{2-7} and D_{2-9} to a small circle which contains the arithmetic operation to be performed. The output of this operation is then shown going to D_{2-10}. The fact that D_1 produces D_2 is shown by the vertical double line with the arrow pointing from D_1 to D_2.

The defining relationships may become very complex. They may contain conditional statements or may involve arithmetic operations on time units.

CONDITIONS

Decision-making is another important part of any data processing problem; here we have available the help of symbolic logic. For example, the condition C_{4-1}, the first condition on the fourth document, is of the form (p and q) or (r and s) where p, q, r, and s are equalities or inequalities on certain dates; $\wedge$ is used for "and," and "$\vee$" for "or." This condition states that the invoice was dated after the 10th of the preceding month but before the 10th of this month. C_{4-1} is written as:

$$[P_8(D_2) = P_8(D_4) \wedge P_7(D_1) < 10] \\ \vee [P_8(D_4) - 1 \\ = P_8(D_2) \wedge P_7(D_2) > 10]$$

A bar over a condition indicates the negation of that condition, i.e., $\bar{C}_{m-n}$ is true (or false) if C_{m-n} is false (or true).

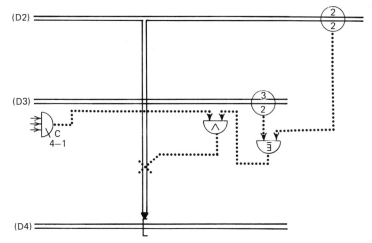

FIGURE 5 Graphical notation of a conditional relationship.

The statement in a defining relationship of a calculation involving a condition is often simplified by interpreting C_{m-n} as a number which is 1 if the condition is true and 0 if it is false. For example, one entry on the monthly statement (D_4) is the customer's old balance (D_{4-3}). This is defined as last month's new balance for this customer minus payments against invoices dated prior to the period covered by this statement. D_{4-3} is then defined as:

$$D_{4-8}(D_{4-1}, P_8-1) - \Sigma \bar{C}_{4-1} D_{3-3}(D_{4-1})$$

Thus the system is to find the new balance, D_{4-8}, on last month's, P_8-1, statement for this customer, D_{4-1}, and subtract from it the sum of this customer's payments, $D_{3-3}(D_{4-1})$ which do not meet the condition that the payments were made during the time period covered by this statement, $\bar{C}_{4-1}$, where C_{4-1} is the condition which specified the period.

The second condition, C_{4-2}, on the monthly statement states that a payment has not been received for the invoice; here "∃" is used for "there exists" and "$\bar{\exists}$" for "there does not exist" so that C_{4-2}, $\bar{\exists}\ D_3[D_{3-2}(D_{2-2})]$, implies that the system started with D_{2-2}, and invoice number, and found that there did not exist in the system a customer payment, D_3, with the same invoice number (D_{3-2}).

Conditional statements may be involved in the producing relationship. In describing the conditions to be met before some action is taken, we use a vertical line to symbolize "if," e.g., in saying

$$D_2 \rightarrow D_{4-7} \mid C_{4-1} \wedge C_{4-2}$$

or an invoice (D_2) produces a line item on the statement (D_{4-7}), if C_{4-1} and C_{4-2} are true.

Figure 5 illustrates this producing relationship. D_2 is shown producing the line item D_{4-7} by having the producing arrow run to the square bracket indicating the line item. The condition on the producing relationship is indicated by the X on the producing line leading to a gate. The input to this gate are the conditions C_{4-1} and C_{4-2}. The condition C_{4-2} is diagrammed completely showing that the input to the gate containing the "there does not exist" symbol is D_{3-2} and D_{2-2}.

If the condition is simple, it may follow the vertical line itself as in

$$P_{10} \times P_8 \rightarrow D_4 \mid D_{4-8} \neq 0$$

or a statement, D_4, is produced each month (P_8) for every customer (P_{10}) who has an open balance ($D_{4-8} \neq 0$).

CONCLUSIONS

We have shown how a simplified problem can be stated in pseudomathematical terms. Such a tool has many advantages for a systems analyst. In reviewing an existing problem it provides the ability to insure that all of the input entries are used to produce outputs. By presenting all of the alternative methods of deriving information, it facilitates making decisions on the best organization of the inputs. It also helps in determining the cost of producing outputs, one criterion for their inclusion. For programming purposes, it provides an unambiguous statement of the problem to the programmer. The maximum number of files, record lengths, file densities, volumes, amount and type of computation required, etc., can be easily determined. The graphical presentation, which can be modified to suit the needs of the user (e.g., by including descriptive labels), should be helpful in determining the best organization of files and subroutines and in providing a check on redundant and superfluous information.

At first glance it may seem that we have made the problem statement more, rather than less, complicated. However, the statement of a problem in these terms clearly cannot be more complex than the problem itself. The cost of current data processing no longer permits incomplete or partial solutions; and ultimately, we would like the computer to solve the problem of its own best use. The first step toward this goal must be a statement of the problem which the computer can understand.

REFERENCE

1. Ackerman, S. A., "Symbolic Logic," *Journal of Industrial Engineering*, Vol. 8, No. 5, 1957, pp. 293–299.

SYSTEMATICS—
A NON-PROGRAMMING LANGUAGE FOR DESIGNING AND SPECIFYING COMMERCIAL SYSTEMS FOR COMPUTERS

C. B. B. GRINDLEY

In the early days of commercial data processing by computer, there were no systems analysts as we know them today. The job of getting work on to the computer was seen to be only one of programming. True, before he could write and test his programs, the programmer had to get the facts from the user. In the event, this task turned out to be one of major difficulty. How often did the programmer complain that the user did not know what he wanted? Today it is not too difficult to see why this was so. No one person knew all the facts. The clerks did not know how their separate tasks fitted into the broad system. The managers did not know the detailed work that went on within these separate tasks. More important, many of the rules that the programmers were seeking did not exist. Managers, supervisors, and clerks frequently exercised their discretion when deciding on a course of action. Instead of saying what rules they followed in each case, they said it was a matter of experience and judgement.

To cap it all, after a while it began to be seen generally that simply to repeat the existing system was not good use of a computer. The old system was designed to suit the capacity of human beings to do work. The capacity of a computer to do work is in many ways greater than that of human beings, in some ways less. But whether greater or less, it is essentially different. On the face of it, it is unlikely that the old system will do. A new system should be designed to exploit the different capacity of the computer. Gradually the analysis of users' requirements, and the design and specification of an information system to meet these requirements, emerged as a prior and separate task from programming.

SYSTEMS ANALYSIS

Although the systems analyst now claims recognition as an expert in his own right,

SOURCE: Grindley, C. B. B., "SYSTEMATICS— A Nonprogramming Language for Designing and Specifying Commercial Systems for Computers," *Computer Journal*, August 1966, pp. 124–128.

his title has not been clearly stated. The overall job of getting work on to a computer is still something of a relay race, where the analyst passes the baton to the programmer just about when he feels he's had enough. In some cases the analyst prepares flow charts which consider quite detailed computer problems, and from which the programmer codes his program. In other cases he simply identifies broadly the output required without attempting to provide all the detailed rules showing how that output is derived. In a series of studies that Urwick Diebold undertook about two years ago it was found that programmers spent less than half their time actually writing and testing programs. The rest of their time was taken up mainly with delays caused by something called "systems queries." On further analysis, systems queries were found to be due to:

1. Changes made by the analyst to the system as originally specified.
2. Misunderstandings between the analyst and the programmer over the specification.
3. Errors in the specification.
4. Omissions in the specification.

Two things emerged quite clearly. The first was that within the overall job of transferring work on to the machine there are two distinct jobs involved. These jobs are distinct since they serve different needs. Job I considers the user's needs and is concerned with designing an information system. Job II considers the computer's needs and is concerned with designing a programming system which will satisfy these information requirements using a computer. The second thing was that Job II is not just a detailed extension of Job I. Both jobs explore, *in final detail*, the separate problems with which they are concerned. Before the programming function begins to plan *how* a particular job is to be performed efficiently on a certain computer, it must be furnished with a specification showing, for all cases that can arise, *what* is required.*

DESIGNING A SYSTEM

If the systems man is to provide rules for every case that can arise, then he is, in fact, constructing a model of the information system he is designing. Even if such a model were not required before programming could properly be undertaken, it would still be valuable to construct it, because it is only by constructing such a model that the designer is forced to face all the implications of his system. It is a way of trying out his ideas. What are the requirements for model building? They are twofold:

Firstly: to break down the problem into its component parts.
Secondly: to describe precisely the relationship of all the parts with one another.

The scientist frequently has to construct models. He has developed techniques and languages for this purpose, for example, mathematics. The need of the systems analyst is clear. To perform Job I he needs the techniques and languages to construct models of information systems.

Commercial programming languages do not at present meet this need. Let us examine why this is so. It is interesting to compare them with scientific programming languages, for example, ALGOL, which do appear to satisfy the needs of the scientist. This is because his needs were different. True, he needed to construct models. But long before the advent

---

*A definition of the separate tasks of the systems analyst and of the programmer is offered in Appendix 1 to this paper.

of the computer, mathematics was already developed to the point where it was rich in concepts and techniques for this purpose. Scientific programming languages have been able to make use of existing mathematical notation. The systems analyst had no such language for designing information systems. English existed, but to the extent that programming languages have imitated English they have missed the point. English is entirely unsuitable for model building. It is imprecise and open to interpretation. To illustrate this it is only necessary to look at the version of English used in statutes in an attempt to cover a situation completely, and then to see the lengths to which courts have to go in practice in order to interpret these attempts. The commercial programming language should ideally have provided for the analyst his equivalent of mathematics. But, no matter how great is the claim of any commercial programming languages to be problem orientated, they are, without exception, designed to suit the way a computer goes about processing data. There is good reason for this. It is the aim of such languages to be automatically translatable into a machine's language. Commercial data processing has had to make efficient use of computers. The difficulties of producing compilers to translate efficiently into a machine's language have acted as a restraint upon the free development of concepts useful to the analyst.

SYSTEMATICS

A new language has been developed to meet this need. It is called *systematics*. This language is solely concerned with techniques and concepts useful to systems analysts in designing information models to meet user's requirements. This has been achieved largely because no attempt has been made to provide it with a compiler. It is thus completely computer-independent. It is interesting to compare this feature with one of ALGOL's most important uses. The 1958 Zurich Conference set for ALGOL three objectives. One was that it could be used to describe computational processes in publications. This objective has been successfully achieved to the extent that the language is now frequently used to describe processes which it is never intended to perform on a computer. In such descriptions ALGOL expressions are used for which no compiler exists. Essentially, systematics is a tool for specifying solutions to information systems problems. More important, it is also a tool for developing such solutions. Like most tools for scientific analysis, it points to deficiencies in proposed solutions. It shows the analyst where more information is required, where certain circumstances have not been covered, where rules suggested are inconsistent. The models constructed may be of large total systems or of detailed parts of such systems. The system may eventually be performed on a computer, or partly on a computer and partly using other processing methods, or not on a computer at all.

How does systematics work? Whilst not attempting a full description of the language in this paper, the three most important features are described. In addition, a detailed example of systematics applied to a simplified payroll is given in Appendix 2. Before discussing these three features, it should be said that the overall objective of each of them is to give precision to the building of information systems' models comparable to that supplied by mathematics. Why then not use mathematics? In fact, mathematical concepts and notation are used freely where appropriate. Indeed, any existing, precise, and standard notation may be used within systematics if found to be useful in the context of the information system being designed, e.g., statistical and Boolean notation. Systematics provides a framework within which these existing concepts, together with a small but growing number of concepts peculiar to systematics itself, may be used. But this framework is designed to suit the problems encountered in information systems design. These problems are different from mathematical problems. They differ principally in two ways. Firstly, in information systems, the relationship between a given result and the information from which it is derived is usually relatively simple. The complexity of the problem arises from the very high number of alternative relationships which may apply depending on the values and states of other items of information. In mathematics the relationship between results and the information from which they are derived is generally far more complex, but the number of alternative relationships is relatively small. Secondly, relationships in information systems depend not only upon quantities but also upon other qualities which we are not accustomed to measure and to which the assignment of numerical values appears inappropriate, e.g., sex, behavior, location, etc.

THREE PRINCIPAL FEATURES OF SYSTEMATICS

Alternative Conditions

The first feature is concerned with the problem of fully exploring all of the high number of alternative conditions which can arise in information systems. The language thus recognizes two entirely different sorts of statements.

  (i) Statements of condition — Connectives
  (ii) Statements of derivation — Expressions

The statement of condition is contained within a Boolean AND/OR matrix. It treats all combinations of the possible states of relevant items of information and shows what action follows each of them.

$$
\begin{array}{c|cccc}
 & \multicolumn{4}{c}{\text{OR}} \\
I_1 & p & & q & \\
I_2 & \checkmark & x & \checkmark & x \\
\text{AND} & & & & \\
 & E_1 & E_2 & E_3 & E_4
\end{array}
$$

Essentially it says if item 1 is state p AND item 2 is state $\checkmark$ do expression 1, OF if item 1 is state p AND item 2 is state x do expression 2, OR etc. The statement of derivation or expression simply shows how a derived piece of information is related to the pieces of information from which it is derived, e.g.,

Derivative = state of item 1
+ state of item 2.

Definition of Qualities

The second feature is concerned with giving precision to qualities other than quantity. It provides for two things. Firstly, for each item of information, the way in which its particular state can vary significantly from case to case is shown. Secondly, a name or reference for each of these states is given. For example:

Item   *Variability*
Area   Europe: Africa: Rest

Variability should not be confused with "range," frequently given for each item of information by systems analysts. Range is for checking the validity of each item as read by the computer. Variability is for the purpose of exploring all significantly different states of the items within the connectives just described. For example, if "area" were one of the items in the Boolean matrix, then what to do in the case where its state was "Africa," or "Europe," or "Rest" would have to be shown. And the names Africa, Europe, and Rest would have to be used. The quality of area may thus be talked about with precision. It is defined as being one of these three states.

Classification of Information

The third main feature is the classification of each item of information according to the part it plays within the model. One method of classification is according to permanence. Four conditions are recognized. Does the state of the item remain unchanged during the operation of the model or does it change? If it changes, then is it up-dated, originated, or destroyed? Another method is by generation. For those items that change their state, it is shown whether they have yet changed—and if so, how many times. A further major classification of the items of information is into a hierarchy of classes, sub-classes, sub-sub-classes, etc. It follows here principles well-established in logic, scientific method, and in language construction.

The example given in Appendix 2, together with the notes at the foot of the Appendix, should illustrate in more detail the principles of systematics. The language is no more than a specialized branch of mathematics. It provides additional tools which are particularly valuable to the design of information systems. It lays the foundation for a body of knowledge appropriate to this new field of activity. The word "new" is used advisedly since it is only the recent use of the computer for processing information that has made it necessary to design and construct information systems with such precision. Until the appropriate body of knowledge and techniques are reasonably well formulated, it is not intended to inhibit the development of systematics by providing it with a compiler.

APPENDIX 1

Definition of Systems Analysis and of Programming

Systems Analysis (ignoring work other than that to be performed by computer)— Expressing the relationship between the data fed into a computer system and the information to be produced by it.

Programming—Expressing the relationship between the data fed into a computer system and the information to be produced by it in a manner which is efficient in terms of the capabilities of that system and in a manner which can be interpreted by that system.

Implications of definitions

1. The *analyst* requires no computer knowledge. True, before constructing a detailed model of an information system, satisfaction that it is feasible to perform the work on the computer should be obtained, as far as is possible. Also, when the analyst is concerned with the interface between user and machine (i.e., design of input and output documents), computer considerations are involved. Both these activities will require liaison with the programming function. But the statement remains broadly true. The analyst is concerned with identifying the user's needs and constructing an information system to meet them.

2. The *programmer* is solely concerned with efficiency problems involved in processing the information system on a given computer, and with translating his solution of these problems into the machine's language. Translation from systematics could, of course, be done automatically. The programmer's basic job therefore is concerned with computer efficiency. If, and only if, computers are made where processing efficiency no longer matters, or efficiency problems can be solved automatically, will his job disappear.

APPENDIX 2

Example of Systematics Applied to a Simplified Payroll

LEVEL 1
THE MODEL

E¹ Calculate Pay

Dictionary

Reference	Description	O	I	R	Variability
E¹ P01 E	number	✔	✔	✔	NNNN
P02	name	✔		✔	A→
P03	sex			✔	M : F
P04	birth			✔	NN/NN/NN
P05	reduced N.I.		✔	✔ : x	
P06	pension scheme			✔	O.K.: No N.I.: No G.P.
P07 (a)	department		✔	✔	NN
P08	marital status			✔	Single: Married: Widow
P09	rate of pay			✔	0d.—25/0d.
P10	date		✔		NN/NN/NN
P11	holiday		✔		0—15
O01	gross pay	✔			N→
O02	net pay	✔			N→
P12 (a)	hours worked		✔		0—80 (¼)
P13	sports			✔	6d. : 1/-
P14	hospital			✔	0d. : 3d.
O03	national insurance		✔		N→
U01 (a)	hours to date			✔	N(¼)
U02	gross pay to date		✔	✔	N→
O04	tax	✔			N→
U03	tax to date	✔			N→
P15	tax code			✔	0—999
O05	other deductions		✔		N→
O06	maximum hours		✔		N→
etc.					

LEVEL 2
ROUTINES
(only National Insurance shown)

Characteristics statement

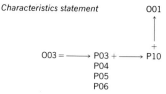

Conditional statement

Connective statement

Intermediates
- A — Age
- B — 40 hr earnings
- C — Flat rate
- D — Graduated pension

Connectives

Flowchart

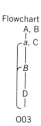

a: C = ——→ A, B, P03, P05.
b: P06, O01.

LEVEL 3
EXPRESSIONS

A = ——→ P10 − P04

$$B = \longrightarrow \frac{O01}{\sum_a P12(a)} \times 40$$

$$D = \longrightarrow 4\tfrac{1}{4}\% \times O01 \; r > .5d. \;\; \overset{1d.}{}$$

O03 = ——→ C + D

CONNECTIVES

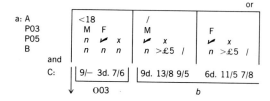

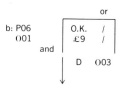

N.B. (i) Assumed for simplicity that hours worked not zero.
(ii) Intermediates not calculated; assumed to be zero.

Key to symbols used:

$=\rightarrow$	derived from
$+\rightarrow$	together with
$\rightarrow$	followed by
n	not applicable
/	the remainder of the range within straight brackets
x	round y and above to nearest
ry	x.

Notes: Systematics describes the information model at three levels of detail.

Level 1. The model itself. A model is restricted to a major and separate operation performed upon one class only. At this level a broad statement of the purpose of the model is given identifying the information class concerned. This is followed by the dictionary. The dictionary lists and classifies every item of information used in the model and defines its variability. Permanence is indicated by P = permanent, U = up-dated, O = originated. Subscripted "a" indicates a sub-class. On its own it indicates the principal of that sub-class, i.e. the item of information that uniquely identifies a particular member. In brackets it indicates a subordinate of that sub-class. O, I, and R show whether the item is output, input, or held as a record, or what combination of these apply.

Level 2. Routines. A separate routine is provided for each up-dated item and for each originated item. It states which other items are required to derive the item concerned and under what conditions the routine is performed.

It also states the "intermediate" steps taken on the way to the final derivation, and how these are related to each other through the connectives. References are provided for each expression (capital letters) and each connective (small letters).

Level 3. Expressions and Connectives. The derivation for each routine is stated in final detail at this level.

THE USE OF DECISION TABLES WITHIN SYSTEMATICS

C. B. B. GRINDLEY

A range of techniques under the general title of Systematics is being developed for designing and describing information systems. The main feature of these techniques is that they allow the user to concentrate on the design and description of the information system without having to consider the computer strategy problems concerned with how the system is to be implemented. Some of these have been described elsewhere (Grindley, 1966). One of the basic techniques in this earlier paper was called an AND/OR Matrix. As King (1967a) has shown, this AND/OR Matrix is a form of decision table. During the field work which has since taken place, this particular technique has undergone some modification.

SOURCE: Grindley, C. B. B., "The Use of Decision Tables within Systematics," *Computer Journal*, August 1968, pp. 128–133.

The present form of the decision table and the way in which it is used is described in this paper.

GENERAL DECISION TABLES

A full description of decision table techniques is given in King (1967b). The general decision table takes the form shown in Fig. 1.

In its extended entry form the condition stub contains the names of items whose state governs the actions to be taken. In the condition entry columns the different combinations of the states of these items are explored. Within the action stub, the names of the items to be derived are shown. In the action entry columns the different actions required to perform the derivations are given under the appropriate combination of states. See Fig. 2.

This means, when sex is male and length of service is less than 5 years, pension contribution is 2s. and holiday entitlement is 16 days, etc.

Condition Stub	Entry
Action Stub	Entry

FIGURE 1

| Sex | Male | Female .... etc. |
Length of Service	<5 years	<5 years ....
Pension Contribution	2s.	1s. 6d. ....
Holiday Entitlement	16 days	15 days ....

FIGURE 2

| Sex | Male | Female .... etc. |
Length of Service	<5 years	<5 years ....
Pension Contribution	2s.	1s. 6d. ....
Holiday Entitlement	16 days	15 days ....
Go to	Table 14	Table 15 ....

FIGURE 3

| | Rule 1 | Rule 2 .... etc. |
| Sex | Male | Female .... |
Length of Service	<5 years	<5 years ....
Pension Contribution	2s.	1s. 6d. ....
Holiday Entitlement	16 days	15 days ....
Go to	Table 14	Table 15 ....

FIGURE 4

Furthermore, the next step to be taken for each set of conditions may be indicated as in Fig. 3.

Each set of conditions, together with the appropriate action is called a *rule*. See Fig. 4.

DECISION TABLES IN SYSTEMATICS

In Systematics, the basic derivation statement is called an *element*. The element is in fact a special case of decision table in that it displays three special features. The form of the decision table just illustrated will be modified in order to illustrate each of these features.

The first feature of the element is that it is confined to providing the rules for the derivation of one derivative only. For example, length of service may influence pension contribution and holiday entitlement. A general decision table

| | Pension contribution | |
| Sex | Male | Female .... etc. |
Length of service	<5 years	<5 years ....
Go to	2s. Table 15	1s. 6d. .... Table 16 ....
	Holiday entitlement	
Sex	Male	Female .... etc.
Length of Service	<5 years	<5 years ....
Go to	16 days Table 15	15 days .... Table 16 ....

FIGURE 5

might explore all relevant states of length of service. The element explores only those which affect one derivation, say, pension contribution. A separate element explores those which affect holiday entitlement. If some states affect both, then these states appear in both elements. All the rules for the derivation of the derivative are given in the one element. We thus identify a shift of emphasis from *fully exploring a set of conditions*, to *fully exploring the derivation of a single item*. There is, therefore, one separate element for each derivative. The element is named after its derivative, as in Fig. 5.

The second feature is to introduce a fresh set of conditions known as *primary conditions*. These conditions determine when the element is performed. They are distinct from the normal decision table conditions (called *secondary conditions*) which determine the particular derivation to apply. Primary conditions trigger the element as a whole. Secondary conditions govern the individual rules. For example, the element holiday entitlement may be performed for each person on the payroll. Entitlement in a particular case may depend upon sex and length of service. A change of employee's number would therefore be the primary condition, sex and length of service would be the secondary conditions. This example illustrates an interesting difference usually found between primary and

	Holiday entitlement		
Primary conditions	Employee's number	For each	
Secondary conditions	Sex	Male	Female ..... etc.
	Length of service	<5 years	<5 years ....
	Go to	16 days Table 15	15 days .... Table 16 ....

FIGURE 6

secondary conditions. Secondary conditions explore different states of an item, e.g., less than five years, five to ten years, over ten years, etc. Primary conditions usually explore the mere change of state of the item, irrespective of what its new state is, e.g., each employee's number. Thus we now have two sets of conditions as shown in Fig. 6.

The third feature is to substitute a "use" list for the "go to" information. "Go to" provides a processing sequence. Systematics is not concerned with sequence outside that needed to derive one isolated derivative. To go further is to be concerned with computer processing strategy which is deliberately outside the scope of Systematics. On the other hand, it is extremely useful for the designer of the information system to know what part the derivative plays within the total system. The names of any other elements which use the derivative are therefore given, as in Fig. 7.

It is difficult to produce a complete statement of use manually. A computer can perform this function quite simply.

Holiday entitlement			
Employee's number	For each		
Sex	Male	Female ... etc.	
Length of service	5 years	5 years	
	16 days	15 days	

Use: Deductions, holiday pay.

FIGURE 7

ADVANTAGES

What are the main advantages expected to be obtained from using this special form of decision table?

More Manageable

In the first place, the tables themselves become very much more manageable. The various combinations of conditions can often amount to a large number of entries in the table. If these entries are limited to those giving rise to the derivation of one item only, then the table becomes easier to construct.

Output Orientated

This "one element per item" approach enables the design method to be entirely output orientated. This was one of the early objectives of Systematics—that the designer could work back step by step from the output wanted, and thus discover what rules and supporting information were required as he went along. For example, let us consider a possible approach to information systems design. In this approach an existing system is not presumed; the analyst sets out to design one. He identifies a basic output requirement, say an amount owing or "invoice total." He then asks, is this item given to the system or derived by it? If given, he merely records the fact. If derived, he attempts to analyze its derivation using the element technique described. He then lists all the items

DECISION TABLES WITHIN SYSTEMATICS

DERIVATIONS		ORIGINS		
		Item	Given	Derived
Level 1	Invoice total	Item totals		✓
Level 2	Item totals	Quantity	✓	
		Price	✓	
		Discount		✓
Level 3	Discount	Customer type	✓	
		Quantity ordered this year		✓
Level 4	Quantity ordered this year	Quantity	✓	
		Customer number	✓	
		Date	✓	

FIGURE 8

referred to in this element and classifies them "given" or "derived." Again the "givens" are merely recorded, but the "derived" items are each further explored in a similar manner. The designer thus develops a cascade until he reaches a level at which all items are given. See Fig. 8.

If an element were not confined to the derivation of one item, this approach would be inhibited. For example, considering level 3, it is possible that certain customer types affect the customer's credit limit as well as, or instead of, discount. But to explore credit limit at this time would be to divert attention from the scheme of thought being developed.

Avoids Processing Sequence

It is important to distinguish between two sequence problems involved in computer systems. Firstly, there are those inherent in the derivation. For example, where

Deductions = Tax + Graduated Pension

it is essential to calculate Tax and Graduated Pension before calculating Deductions. Secondly, there is the superimposed sequence enforced by the "one job at a time" attitude of the computer, e.g., the sequence shown in Fig. 9.

It is to be noted that the sequence shown in Fig. 10 is an equally valid

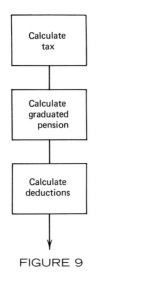

FIGURE 9

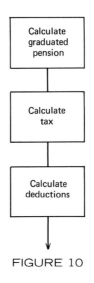

FIGURE 10

Deductions		
Employee's number (given)	For each	(primary conditions)
Nil		(secondary conditions)
	Tax (derived) + Graduated pension (derived)	

FIGURE 11

solution to the problem. It may, for some reason, be a less efficient solution in terms of computer processing, however. The strategy adopted within the computer system is of vital importance. But it is confusing to have to consider it at the same time as the information derivation statements are being designed and specified. It also makes such specifications unnecessarily complicated to read. Furthermore, it is premature to consider computer strategy until the requirements have been specified.

The removal of the "go to" feature allows the analyst to avoid nonessential statements of sequence. Sequence essential to the solution of the derivation problem is catered for in two ways:

1. The element is self-triggering. That is, its primary conditions state when it should be performed.
2. Elements which should have been performed previously are named. See Fig. 11 for an example.

Being self-triggering, the element does not automatically follow a previous element (processing sequence) but is done for each employee, i.e., each time employee's number changes. The elements for any derived items referred to, however (tax and graduated pension) must be performed first.

Localizes Attention

Perhaps the chief advantage when designing a system, and certainly when amending it later, will be the facility to localize attention to any degree defined by the analyst or the problem, and to ignore the rest of the system. Consider the elements which might be involved in the cascade in Fig. 8. These are shown in Fig. 12.

Any one of the four elements could be considered separately. This is because we know that the whole story of a particular derivation and also its place in the overall system is described in the one element. Taking element 3, we can be sure that no alternative method of deriving discount is described in another element. We also know that it is to be derived for each product ordered and that the "quantity ordered this year" is to be derived first.

SIMILARITY TO NERVOUS SYSTEM

It is of interest to note that the form of decision table described bears some relationship to the neuron. The animal nervous system appears to rely upon signals traversing a network of neuron cells largely of the type shown in Fig. 13.

Part I, the dendrites, consists of devices for interpreting signals from other neurons. Part II, the body, generates the output signal. The strength* of the out-

---

*Signals within the nervous system are electrical. Their strength varies according to frequency of pulse rather than voltage of each pulse. This distinction is ignored here.

Invoice Total (1)

Order no.	For each
	Sum of item totals

Item Totals (2)

Product no.	For each
	Quantity (price − discount)

Discount (3)

Product no.	For each		
Customer type	Home	Home	Overseas
Quantity ordered this year	>100	Else	—
	$\frac{1}{10} \times$ Price	0	$\frac{1}{5} \times$ Price

Quantity ordered this year (4)

Product no., Customer no., Year	For each
	Sum of quantity

FIGURE 12

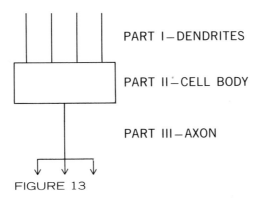

FIGURE 13

put signal appears to depend upon the interpretation, by the dendrites, of the strengths of the input signals and upon the physical properties of the particular neuron body. Part III, the axon, carries the output signal to the input areas (dendrites) of other neurons. For a readable further description of the nervous system, see Wooldridge (1963).

The element is the basic building brick within Systematics. The information model designed consists of a number of

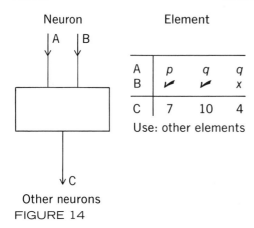

FIGURE 14

these elements. Their similarity to nature's basic building bricks is illustrated in the simple example shown in Fig. 14.

The neuron illustrated receives signals from the axons of other neurons A and B. Various combinations of strengths of these signals are interpreted by the dendrites; these interpretations stimulate the generation of signals of various strengths in the body of the cell which, in turn, form the output C. C is distributed by the axon as a stimulant to other neurons.

The element shown similarly receives signals from the output of other elements A and B. Various combinations of strengths of these signals are explored. The strength of the output C varies according to these combinations; e.g., if $A = p$, and $B = \checkmark$, then $C = 7$, etc. The value of C may then stimulate action in other elements.

SYNTACTIC DESCRIPTION

A syntactic description of the Systematics language is given in Fig. 15. This description is largely in Backus-Naur form. A difficulty arises, however, since Backus-Naur form provides a linear description, whereas Systematics takes essentially a two-dimensional or tabular form. Rather than attempt to produce an exact description therefore, those separators and terminators which are implied by the tabular form have been left out.

It will be noted that a GIVEN item may have many SUBMISSIONS. For example:

⟨NAME⟩ = ⟨SUBMISSIONS⟩
Quantity Ordered = 7, 14, 83, 6,

being the different quantities order for different products or by different customers. The question arises as to which submission is relevant to a particular calculation. The rule is:

TERM, the latest submission at the time the element is triggered.
SERIES, all submissions since the element was last triggered.

e.g., Fig. 12, element 2, quantity is the amount submitted for a particular value of product number.

Fig. 12, element 4, quantities are all those submitted for a particular product, customer, and year.

The element is a statement in two-dimensional form. In the example in Fig. 7, the element fits into the linear Backus–Naur description as shown in Fig. 15.

Notes

(1) Special Symbols:
under ⟨FUNCTION⟩   S = Sum
                    F = Frequency
under ⟨RELATIONAL
       OPERATOR⟩   Ex = Except
under ⟨VALUE⟩   — = Not
                    applicable

(2) The equals sign is assumed in the tabular form.
(3) Where more than one TERM or SERIES is present under primary

```
⟨MODEL⟩ ::=              ⟨GIVENS⟩, ⟨DERIVEDS⟩
  ⟨GIVENS⟩ ::=           ⟨GIVEN⟩ | ⟨GIVENS⟩, ⟨GIVEN⟩
  ⟨DERIVEDS⟩ ::=         ⟨DERIVED⟩ | ⟨DERIVEDS⟩, ⟨DERIVED⟩
    ⟨GIVEN⟩ ::=          ⟨NAME⟩ = ⟨SUBMISSIONS⟩
      ⟨SUBMISSIONS⟩ ::=  ⟨SUBMISSION⟩ | ⟨SUBMISSIONS⟩, ⟨SUBMISSION⟩
        ⟨SUBMISSION⟩ ::= ⟨IDENTIFIER⟩ | ⟨NUMBER⟩ | ⟨LOGIC⟩
      Where:  IDENTIFIER  is an alphanumeric string which attempts to identify
                          uniquely a member of a class, e.g. a customer's number.
      and:    NUMBER      is a number subject to the laws of arithmetic (in any spe-
                          fied radix), e.g. the value of age or gross pay.
      and:    LOGIC       is an alphanumeric string identifying an attribute of class
                          members, e.g. a member's sex or tax code.
      and:    NAME        is an alphanumeric string uniquely identifying each GIVEN
                          and each DERIVED item, e.g. Holiday entitlement.
    ⟨DERIVED⟩ ::=        ⟨NAME⟩ = ⟨ELEMENT⟩
    ⟨ELEMENT⟩ ::=        ⟨TRIGGERS⟩ ⟨DERIVATIONS⟩
                       | ⟨TRIGGERS⟩ ⟨ALT DERIVATIONS⟩
                       | ⟨TRIGGERS⟩ ⟨LTD DERIVATION⟩
      ⟨DERIVATION⟩ ::=   ⟨TERM⟩ | ⟨SERIES⟩ | ⟨DERIVATION⟩ ⟨OPERATOR⟩
                                                          ⟨DERIVATION⟩
        ⟨TERM}  ::=      ⟨NAME⟩ | ⟨LITERAL⟩
  Where: LITERAL         is a number or an alphanumeric string.
        ⟨SERIES⟩ ::=     ⟨FUNCTION⟩ | ⟨NAME⟩
          ⟨FUNCTION⟩ ::= S | MAX | MIN | AV | LIST | F
        ⟨OPERATOR⟩ ::= + | − | × | /
      ⟨TRIGGERS⟩ ::= ⟨TRIGGER⟩ | ⟨TRIGGERS⟩, ⟨TRIGGER⟩
        ⟨TRIGGER⟩ ::= ⟨DERIVATIONS⟩ ⟨EVENT⟩
          ⟨DERIVATIONS⟩ ::= ⟨DERIVATION⟩ | ⟨DERIVATIONS⟩, ⟨DERIVATION⟩
          ⟨EVENT⟩ ::= ⟨STATES⟩ | FOR EACH
            ⟨STATES⟩ ::= ⟨STATE⟩ | ⟨STATES⟩, ⟨STATE⟩
              ⟨STATE⟩ ::= ⟨RELATIONAL OPERATOR⟩ ⟨DERIVATION⟩
                ⟨RELATIONAL OPERATOR⟩ ::= ⟩ | ⟨ | ≤ | EX | =
    ⟨ALT DERIVATIONS⟩ ::=  ⟨ALT DERIVATION⟩ ⟨ALT DERIVATION⟩
                         | ⟨ALT DERIVATIONS⟩, ⟨ALT DERIVATION⟩
      ⟨ALT DERIVATION⟩ ::= ⟨CONDITIONS⟩ ⟨DERIVATION⟩
        ⟨CONDITIONS⟩ ::= ⟨CONDITION⟩ | ⟨CONDITIONS⟩, ⟨CONDITION⟩
          ⟨CONDITION⟩ ::= ⟨DERIVATION⟩ ⟨VALUE⟩
            ⟨VALUE⟩ ::= ⟨STATES⟩ |−| ELSE
    ⟨LTD DERIVATION⟩ ::= ⟨CONDITIONS⟩ ⟨SERIES⟩ | ⟨CONDITIONS⟩
                          ⟨DERIVATION⟩ ⟨OPERATOR⟩ ⟨SERIES⟩
```

FIGURE 15

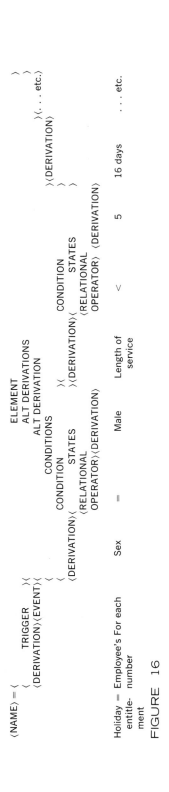

FIGURE 16

or secondary conditions, OR logic applies to a horizontal list and AND logic to a vertical list; e.g.,

Customer No., Year	For each

means for each change in Customer No. OR in Year and

| Sex | Male .... |
| Length of Service | <5 years .... |

means where Sex is Male AND where Length of service is <5 years. The order in which the TERMS or SERIES are written is arbitrary and of no consequence.

REFERENCES

Grindley, C. B. B. (1966). Systematics—a non-programming language for designing and specifying commercial systems for computers, *Computer Journal*, Vol. 9, p. 124.

King, P. J. H. (1967a). Some Comments on Systematics, *Computer Journal*, Vol. 10, p. 116.

King, P. J. H. (1967b). Decision tables, *Computer Journal*, Vol. 10, p. 135.

Wooldridge, D. E. (1963). *The Machinery of the Brain*, McGraw-Hill Book Company.

SOME APPROACHES TO THE THEORY OF INFORMATION SYSTEMS

B. LANGEFORS

1. ORGANIZATION OF INFORMATION

1.1 Analysis of Information Systems

Information systems serve to provide the different functions within an organization with information necessary for rational decisions.

Thus the information system has to be designed to handle such functions as collecting, storing, processing, and displaying of data. It is therefore also natural that such information systems normally grow in a way that is to a great extent dependent on the development of automatic data processing machinery.

This, however, involves great risks; for a system may grow up which will process large masses of data that are not used for decisions (and yet ignore important data) and process them in a way which is far from optimal and on an equipment which is not designed to make the system efficient for its purpose.

What is sorely needed in this area is a systematic (i.e., scientific or mathematical) technique for establishing the real needs for information within an organization. Thus it has to define the information needed, its volume, the time intervals at which it is required, and that at which it is available, the data from which it can be produced, and the process — or alternative processes — needed for its production, and the form for presentation of the results.

It is important to note that this analysis is not called for merely to enable automatic data processing to be installed efficiently but is basic to any rational design of an information system — that is, for any efficient administration.

An analysis of an information system must therefore, in the first place, be hardware independent. It has to provide an abstract definition of the system itself.

SOURCE: Langefors, B., "Some Approaches to the Theory of Information Systems," *BIT* 3, 1963, pp. 229–254. © Scandinavian Computer Science Foundation.

Only after such an analysis has been made is it appropriate to consider hardware implementation. In this stage, however, specific hardware should not yet be considered but rather the common characteristics of hardware systems in general. Only when this analysis has been made, with respect to the general properties of all (or most) data processing systems, should one evaluate the suitability of different specific hardware systems and their effect upon the overall organization of the system of information.

At this phase of the analysis it is appropriate to consider the impact of the organization of information on the total workload and on the data processing system and try to find an optimal, or at least efficient, solution.

There are some basic propositions made here, in connection with the systematic approach advocated, which appear to be in contradiction to present practices or assumptions. One is the hypothesis that in most cases it is possible to isolate and define the relevant organization functions in a separate operation to be performed before the actual design of the system is attempted. It is thus assumed that these functions are defined from the basic goals of the organization and therefore will not need to await the detailed construction of the system.

The other hypothesis is that it is possible to define all input information necessary to produce a desired output. The basic assumption here is that actually any information can be defined only in terms of more elementary information, which will then occur as input parameters. Therefore, once a class of information is defined, then it is known what input information is required for its production. The point here is that it should not be necessary to work out formulas, or even computer programs, before it can be determined what input data are needed. In fact, it is quite possible to work out formulas or programs for an entity where important variables are missing; so starting by programming is no safeguard against ignoring important data.

1.2. Organization of Data for Minimum Transport

One problem of organizing data is concerned with the data transport needed to bring data from the storage place to the processor. Another is to organize data in a way that is meaningful for people. In the context of the first of these problems, it seems appropriate to consider such data as are stored in the main memory (or any memory in which they are available in one basic word-time of the processor) to be used for processing, as being available without transport work. The problem of data organization then is how to store data in mass storage so as to minimize the work needed to transport these data to the main store when needed, or to make them available within the time required.

In business information systems it is common that data can be retrieved from files during a simple scanning of the file (and the batch of transactions), after sorting of transactions. When this is not possible, random access is often said to be required. Contrary to this, the situation in engineering data processing is often one where more complicated magnetic tape handling is required. This has also been found to be the case in many applications such as production scheduling.

Insofar as the so-called random-access memories require a general access time that is much longer than the word-time of the processor, while having smaller access time to data placed in neighboring positions than to those in other positions, the data transport minimization is not a problem except for so-called serial access memories. In fact these two classes of memories are merely different instances of one and the same kind of pseudo-random access systems. It will be shown later how many data handling routines that can not be run as a simple scanning of the master file involved can be classified as a more general handling type called rectangular handling. Obviously the type of handling of files required must be considered when searching for minimum transport system solutions. Likewise, it is obvious that the processing period (the time interval between two processing runs of the same file) is of importance for the transport work.

2. ANALYSIS OF THE INFORMATION NEEDS OF THE FIRM

We make the basic assumption about the firm that a set of "functions" can be defined which must be in action in order for the firm to be able to fulfill its objectives. In fact, it seems probable that the only way to define the objectives of a firm is to define its "functions."

Our approach to the system analysis will be to start by defining the basic functions of the firm, and then go on to find the relation of each function to the set of information classes available.

2.1. The Basic Functions of the Firm

We take as the basic functions of the firm those operations the firm has to perform in order to fulfill its objectives, and also those operations which are not directly concerned with the objectives but are indirectly necessary in order for the firm to be able to perform the directly necessary operations.

Thus a firm which has as its main objective the production of a set of objects has, of course, the directly necessary function of running the production. Another directly necessary function is buying the raw materials out of which the products are made. Indirect functions are, for example, the paying of salaries to the employees or the cost accounting needed to serve the directly necessary function of setting prices on products and controlling the efficiency of the firm.

In defining the functions it is important to isolate each single function on the basis of its proper reasons for existence, without being too much influenced by its present organization. To take a simple example: the function of paying salaries is concerned with the computation of the salary for each individual person at each salary period; that is, one must compute the amount payable after deductions, and associate this amount with an identification of the person concerned.

It is not within the scope of this function, however, to define which different products or customer orders are to be charged with the amount payed. The fact that this is normally computed in connection with payroll processing corresponds to a common solution to the problem of economic data transport and is not to be mistaken for an indication that it belongs to the function of salary pay.

2.2. The Information Needs of Firm Functions

For every function of the firm, a set of different information classes has to be provided. This information will be needed either to monitor the performance of a function or as a basis for a decision which has to be made by a human in order to control a function. An example of the first kind is the salary pay where the computed amount per person can directly be used to trigger a pay action. Instead, a function of sales forecasting will clearly be used only as one basis for human decision-making.

The first step in the analysis of information need is to define for each function of the firm the different classes of information needed. Further, for each such class we also have to determine the requirement of information as a function of time. Thus, for instance, the salary pay function calls for the information specifying for each person concerned his identification and the amount to be payed to him. In addition, it will also, in most cases, be required to list all deductions made for each person.

This information will be required at each pay period — each week, for instance. In this way, for each function, a set of required information sets is defined; and for each of these, another set of information sets is required for its production.

The definition of the basic functions does not seem to be computable by some routine procedure at present. Rather it has to be defined by careful analysis of the goals of the firm. This seems difficult and will certainly be regarded so by most firms. In fact, it is believed to be in many cases a hindrance for taking an analysis of this kind as a first step to automating information processing. However, it should be obvious on a second thought that no firm can be assumed to work in a rational way if, at least, it has not clearly defined its different functions and the information needed for each of them. This is not a problem that can be solved in one day or two. It is well worth much effort and even long research. This is to say that a firm which really knows its objectives and how to fulfil them should find no difficulty in defining the basic functions and its information requirements.

Although no procedure for routine or automatic solution of this basic problem exists, it is obvious that in principle this task can be made fairly easy by listing all the functions and their direct information needs for firms of different types. Thus to specify these entities for any firm would mainly call for consulting such a list or set of lists for firms of the type in question, and deciding which functions to accept and which to omit.

Such lists may not be available today. It will not be long after the introduction of an approach like ours, however, until they will become at hand, and in a continuously improving set. Let us now assume that the task of listing the functions and their information needs has been accomplished. Let us assume, as a simple example, that the firm is so simple that only two functions are to be served. Let us call them A and B.

Further, for these two functions we have found the information sets a, b and c, d to be directly required.

We now have the problem to define for a, as well as for b, c, and d, which information is needed for its production. Again this might be done by means of lists if available. In any case it has to be done by considering the definition of the information a in terms of more elementary concepts. No programming or formula working should be done to this end, however; but it may be necessary to make some calculations in order to determine whether some information which is logically motivated for a does actually play a significant numerical role.

Suppose we find in this way that to obtain a we need four information sets e, f, g, h, say. One of these may represent a set of computer programs.

How do we go about designing a processing system for this information need? Before we tackle this problem, we shall introduce some tools for such problems in general.

2.3. Definition of Precedence of Information Sets

We shall say that when a set of information sets (b, c, d, for instance) is needed in order to produce the information set

a, then b, c, d constitute the precedence set for a. We introduce the notation

(2.31) $\mathscr{P}(a) = b,c,d;$

a is called the succedence set of $\mathscr{P}(a)$.

To indicate that b is a precedent of a we use the notation

(2.32) $b < a.$

Then the definition of $\mathscr{P}(a)$ can be written as

(2.33) $\mathscr{P}(a) = \{x; x < a\}$

where we use the common set definition notation.

In information processing it is common to produce from some precedents, one of which is x (say), a new, updated version of x. We shall use the notation x' to denote the updated version of x. Obviously

$$x < x';$$

If we have the pair of precedence relations

(2.31a) $\begin{cases} \mathscr{P}(a) = b, c, d; \\ \mathscr{P}(c') = c, d; \end{cases}$

then we shall always suppose that c' is first produced and then a is produced after c has been replaced by c', whenever this is possible. More generally we shall use the rule (when no specific ordering is prescribed):

(2.35)   whenever $x < y \land y \not< x'$ then $x' < y$

or in words: whenever x precedes y, and y does not precede x', then we take x' to precede y, instead of x.

When a set a contains only elements which are also contained in another set q, then we denote this fact by writing

(2.36) $q \supseteq a$ or $a \subseteq q$

and a is said to be a *subset* of q in case we do not know whether $q = a$ may be true or not. If $q \supseteq a$ and $q \neq a$, then we write

(2.36a) $q \supset a$ or $a \subset q$

and call a a *true subset* of q. It is obvious that it is of interest to know which precedence sets are subsets of other precedence sets. Therefore when we list a set of precedence relations, we shall try to group them by subset relations.

We shall have occasion to use the concept of *set intersection* $x = y \cap z$, which will be defined as the set x such that if $q \in x$ then $q \in y$ and $q \in z$ (or if q is a member of x then it is a member of both y and z).

For example, let

$$y = a1, a2, a3, a4$$
$$z = a2, a4, a6, a8$$

then $x = y \cap z = a2, a4$.

Obviously, if $z \subset y$, then $y \cap z = z$.

The precedence relations such as (2.31) can also be specified by a precedence matrix P^{00}, where the precedence relation $\mathscr{P}(a) = b, c, d;$ is represented by a column labelled a.

To show this we shall take the simple example brought up in section 2.2 which we give some extension here.

(2.37)

	a	b	c	d	e'	i'	j'	l'
$P^{00} = a$		1		1		1		1
c				1				1
e	1	1		1	1	1		1
f	1			1				
g	1		1	1				
h	1			1				
i		1	1			1	1	
j			1				1	
k			1				1	
l				1				1

The columns of P^{00} correspond to the precedence relations $\mathscr{P}(a) = e, f, g, h$; $\mathscr{P}(b) = a, e, i$; $\mathscr{P}(c) = g, i, j, k$; $\mathscr{P}(d) = a, c, e, h, l$; $\mathscr{P}(e') = e, f, g$; $\mathscr{P}(i') = a, e, i$; $\mathscr{P}(j') = i, j, k$; $\mathscr{P}(l') = a, c, e, l$.

The precedents are listed to the left of the matrix, labelling the associated rows, whereas the succedents are listed above their associated columns. The rows of P^{00} indicate the succedents of the set corresponding to the row. Thus g is seen to have the succedents a, c, e'. According to (2.35) we need not have e', i', j', l' among the rows. A set which is indicated to have i, for instance, as a precedent will in most cases use i' instead, following the rule (2.35).

From inspection of P^{00}, (2.37) we find

(2.38)
$$\mathscr{P}(e') \subset \mathscr{P}(a)$$
$$\mathscr{P}(i') = \mathscr{P}(b)$$
$$\mathscr{P}(j') \subset \mathscr{P}(c)$$
$$\mathscr{P}(l') \subset \mathscr{P}(d)$$

Further we see that b and d have no succedents. In fact their rows are empty and have been deleted from P^{00}. They are *terminal* sets. Similarly e, f, g, h, i, j, k, l have no precedents. They are thus *initial* sets.

The rest have an intermediate position. They can be ordered by levels. These levels can be found by matrix multiplications involving P^{00} [2]. In our example, however, we can obtain the ordering by inspection. We want the ordering for instance in order to find out how to draw the precedence graph [2] associated with P^{00} in a proper way. Thereby we want to draw sets, which are grouped by the subset relations, as neighbors. We look for ordering relations for the groups.

We want to place the terminal sets to the right. Both b and d are terminals. However b is grouped with i' and d with l'. Further

$$a < i' < j' < c < l'.$$

Hence we obtain the order [cf. also (2.38)]

$$(a, e') < (b, i') < (c, j') < (d, l').$$

We rewrite P^{00} in this order

(2.39)

	a	e'	b	i'	c	j'	d	l'
$P^{00} = a$			1	1			1	1
c							1	1
e	1	1	1	1			1	1
f	1	1						
g	1	1			1			
h	1						1	
i			1	1	1	1		
j					1	1		
k					1	1		
l							1	1

In Fig. 2.31 we have drawn the graph corresponding to P^{00}. If the reader will do this for himself, he will find that it is much easier if he follows the order indicated by (2.39) than if he used (2.37).

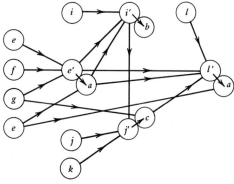

FIGURE 2.31

In drawing Fig. 2.31 the rule (2.35) has been followed. Further, in the drawing we have introduced the simplification that one circle touching another one will have all the precedents of that one and in addition those which are indicated by lines flowing into itself. (Thus a has precedents e, f, g and also h.) Arrows have been drawn at the points of contact of two circles to indicate which of them is to have also the precedents of the other.

2.4. Files and Processes

We have seen that as a rule a certain kind a of information is obtained by some combination out of a set $P(a)$ (= the *precedence* set of a) of other kinds of information. We shall use the word computation to denote the set of all operations which uses $\mathscr{P}(a)$ and produces a. In general it is wise to assume the possibility of producing different versions of a from $\mathscr{P}(a)$ in different ways, using different procedures. For instance, one may know a procedure which will give an approximation to a while one is still searching for another procedure which would produce "better values of a." We therefore use the word *computation* for a for the set of all feasible procedures for getting some approximation to a, and write for short Comp(a). It is important to try to define $\mathscr{P}(a)$ with regard to the computation of a and not be satisfied with a subset of it which might be sufficient for one of the feasible procedures of Comp(a). We may also find it convenient to refer to the precedence set $\mathscr{P}(a)$ of a as the *precedence set* of Comp(a), denoted as $\mathscr{P}$ Comp(a). Also a may be regarded as a member of the succedence set S Comp(a) of Comp(a) and also as a member of the succedence set $S[P(a)]$ of the precedence set $\mathscr{P}(a)$ of a.

The information in the succedence set a of a process as well as that in the precedence sets $\mathscr{P}(a) = b, c, d$; will consist of a set of values of one information set, often called the *key*, and, for each value of the key, a group of corresponding values of the information classes contained in a set. One package consisting of one key value and the corresponding values within the set will be called a *record*. Thus for any value k of the key there may (or may not) be a record a_k of the output information set and a record, $b_k, c_k, \ldots$ of each input set. A set of records for a set of k-values will be called a file, and we shall in most cases consider $a, b, c, \ldots$ to be the files of all $a_k, b_k, c_k, \ldots$.

In some cases it will be important to make a distinction between "standing files" (or files proper), on one hand, and files consisting of temporary data such as input data, output data, and intermediary results, on the other hand. We shall then use the word "*transfile*" for the files of temporary data.

To each precedence relation there is associated a computation. The actual procedure, however, may be designed in a variety of ways. Thus whereas $\boldsymbol{P}^{00}$ or its graph is a unique representation of the system of computations there may be different systems of actual procedures for the implementation.

Among the computations will also be sorting operations for a file, necessary between two computations proper.

The number of entries $\neq 0$ in any row of $\boldsymbol{P}^{00}$ equals the number of computations for which the corresponding data set is an input. Thus, for instance, $\boldsymbol{P}^{00}$ of (2.39) shows that e (or e') is used as input in six different computations. If the computations are all taken as separate computer runs, we thus will have a multiplicity of data input transport for all files. This multiplicity may be even higher than that obtained by computing unit entries in (2.39) because some computations may call for multiple scanning of some files. We can indicate this by replacing the units in $\boldsymbol{P}^{00}$ by the numbers giving the multiplicity of the corresponding input scan. Note that then it is not possible to indicate in $\boldsymbol{P}^{00}$ also a multiple scan of an output file. If, however, the precedence matrix P^{00} for the graph (Fig. 2.31 for instance) is replaced by the incidence matrix P^{10}, then it is possible to indicate multiplicity of both input and output. We return to this point later.

If we indicate also the (relative) volume or scan time for each file along with P^{00}, it will be possible to see how much transport is saved by taking together two or more computations into the same process.

We introduce some assumed relative volumes and some assumed multiplicities of input scans in our example:

Volumes (relative)

(2.41)

a	b	c	d	e	f	g
1	1	2	1	10	2	1

h	i	j	k	l	
1	10	20	2	10	Total = 61

Considering also e', i', j', and l' we get 111.

(2.42)

	a	e'	b	i'	c	j'	d	i'
$P^{00} =$ a			1	1			1	1
c							5	1
e	1	1	1	1			1	1
f	1	1						
g	1	1			1			
h	1						1	
i			1	1	1	1		
j					1	1		
k					1	1		
l							1	1

The system as described by (2.42) may be called the basic *topological system*.

We obtain from P^{00} the total transports of each input file in the topological system. For instance for c we get $(5 + 1)2 = 12$.

Volume (input)

a	c	e	f	g	h
4	12	60	4	3	2

i	j	k	l
40	40	4	20

Total = 189 for input.

Total output volume = 55 for (a,b,c, d,e,i',j',l'); thus total volume = 244.

Hence we find that in our example the computations system implies a (relative) transport volume for input of 189 against the volume 59 ($= 61 - b - d$) for single input. This gives a *topological transport factor* of about 3.3.

The multiple transport can be reduced by taking some computations into the same computer process. For this to be possible the computer memory must be sufficient to store the programs for these computations, or else there will be data transport for shuffling programs between records.

Grouping computations corresponds to grouping columns of P^{00}. It is easy to see from P^{00} how much transport is saved by grouping any computation with any other. Thus from (2.42) we find that grouping a and e' saves one scan of e, f and g (i.e., it saves a transport volume = 13).

No other computation saves as much when grouped with a. We list all the pairs which save most by being grouped.

a, e saves	13
b, i	21
c, j	32
d, l	23
Total	89

The other possible combinations, such as e, b, do not have to be considered since whenever $P(e) \subset P(a)$ then Vol $[P(e) \cap P(b)] <$ Vol $[P(a) \cap P(b)]$. [Vol(x) is used to denote the transport volume of x.]

Thus with this grouping (which happens to correspond to grouping by subsets) the total transport goes down to

100, corresponding to the much improved transport factor of $100/59 = 1.6$.

The grouping of computation may call for shorter data blocks in the files in order to leave memory space for the group programs. This will then lead to an increase of transport volume which will also have to be considered (cf. Example in section 2.7).

In general it will be desirable (and possible) to group more than two computations. Let us assume that in our example the memory space permits four programs simultaneously and that all programs have equal space requirements. Let us test which further grouping would then be best.

We find

a,e grouped with b,i saves		10
a,e	c,j	1
a,e	d,l	10
b,i	c,j	10
b,i	d,l	11
c,j	d,l	0

Thus if we group b,i and d,l we save 11, and are then able to group also a,e and c,j, which saves 1; so this solution saves an additional 12, bringing the total down to $100 - 12 = 88$, a factor of about 1.3.

Instead if we choose to group a,e and d,l plus b,i and c,j, we save 20; and the groupings a,e and b,i plus c,j and d,l, finally saves 10. Thus the best solution is the grouping

a, e, d, l
b, i, c, j
Total transport $= 80$.

In the general case one will have to test for different combinations of computations where the memory space and the space requirements for the different programs have to be considered.

We have seen that the subset relations [such as $P(e) \subset P(a)$] give valuable aid in suggesting efficient computation grouping and reducing the number of possible combinations.

2.5 Reducing the Number of Transput Equipment

The precedence matrix P^{00} also gives information about the input and output (or *transput*) requirements. Thus the column for a shows that in addition to an output unit for a, one needs input units for e, f, g and h. Thus, in all, five units would be needed for this computation. If it is grouped with that for e', there will have to be a unit for e' as well, thus increasing the number to six units altogether.

The group a, e, d, l similarly is seen to call for twelve transput units.

In general, the amount of transput equipment with a computer has to be small for cost reasons. One then will have to put together (or consolidate) several of the *elementary files* we have considered so far, into larger files. A consolidation of files corresponds to grouping of rows of P^{00}.

Consolidation of files may cause excessive data transport by causing data to be input to a process where they are not used. Grouping of computations on the other hand may eliminate such excessive transport. Hence, grouping of computation in general will have a still greater beneficial effect than we have seen above.

As a simple illustration we may consider what happens if the files g and h would be consolidated. We write the corresponding rows of P^{00}.

	a	e'	b	i'	c	j'	d	l'
g:	1	1	0	0	1	0	0	0
h:	1	0	0	0	0	0	1	0

We see that in two of the three inputs of g we will have to transport h as a *deadweight*. Similarly during one of the two transports of h we will have g as a deadweight. Thus a deadweight transport of $2h + 1g = 3$ is induced. Grouping of a and e will eliminate one of these h-transports, and grouping a,e with d,l will eliminate also the excessive g-transport, reducing the total deadweight transport caused by consolidating g and h to 1.

As further examples we note that consolidation of k and j would imply no deadweight transport; but consolidation of i with j causes one excess transport of j and one of j' so that the resulting deadweight transport would be 40. Grouping of b, i, c, j, on the other hand, would enable i, j, k to be consolidated without any deadweight transport at all. This is easily seen from P^{00}.

When we consider consolidation of files, the picture of excessive data transport becomes somewhat complicated, being composed of duplicated input of some files to several processes, and of deadweight transport of some data in these files which are not utilized as input in some of the processes to which they are fed together with the true input data for those processes.

However, the total excessive transport is very easily detected, being simply the number of passes of that file minus one. Thus we need not keep record of the different kinds of excessive transport.

2.6. Incidence Matrix of Process and Transput

We have seen that the precedence relations and the precedence matrix P^{00} are suitable to describe some of the relations of our study. However, some facts could not be described in that way. Thus the number of file passes in a process could not be specified in P^{00} for both input and output files. Further, P^{00} does not offer a convenient description of how computations are grouped to composite processes. Both these deficiencies are remedied if we introduce the concept of incidence between process and input and output. In this way the precedence matrix is replaced by the *incidence matrix* P^{10}. The incidence concept considers not only precedence relations among data sets but shows how any process takes some data sets as precedents (or input) and other data sets as succedents (or output). In the incidence matrix P^{10} one row is taken for each process (or computation) and one column for each data set. The number of scans that a file is treated by a process is used as the *incidence number between the file and the process*. Incidence numbers are given different signs for input and output. Here we use minus-sign to indicate output.

To start with a simple illustration let us take the computations

(2.61)   comp(a): $P(a) = e, f, g, h$;
            comp(e): $P(e') = e, f, g$;

illustrated by the first two columns in (2.42) and write the corresponding incidence matrix P^{10}

(2.61a)

	a	b	c	d	e	e'	f	g	h	i	i'	j	j'	k	l	l'
P^{10} = comp (a)	−1				1		1	1	1							
comp (e')					1	−1	1	1								

Note that we have now written numbers for both input and output so that by replacing the units by any appropriate number we are in P^{10} able to indicate any number of input and output scans.

Note further that in P^{10}, as well as in P^{00}, we can easily see the fact that $P(e') \subset P(a)$ and also that consolidating, for instance, g and h would cause a deadweight transport of h in comp(e').

Also, a row in P^{10} indicates clearly the number of input and output sets for the associated process.

Now suppose that we decide to group comp(a) and comp(e') into one process pr(a, e'). This decision could have been made on inspection of P^{00} or P^{10}. It can be performed by replacing the rows comp(a) and comp(e') by one single row pr(a, e'):

(2.61b)

$P^{10} = $ pr(a, e')

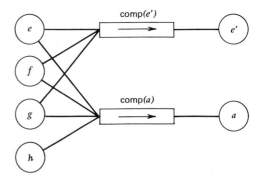

FIGURE 2.61

We may also introduce a change in the graphical representation which is analogous to the change in passing from P^{00} to P^{10}. Thus we introduce also in the graph a representation of the process (or computation); see Fig. 2.61.

We have used a rectangular element (or a line element) to represent the computations or processes whereas the data sets are represented as points. This is because processes have two ends, the input and output end, like a line has. Thus data sets are represented as 0-dimensional entities and processes as 1-dimensional ones. It is thus that P^{00}, being a relation between 0-dimensional entities, has two superscripts of 0 while in P^{10} the first superscript (the one associated with the rows of the matrix) is 1 to indicate that the matrix rows are associated with the 1-dimensional objects called processes.

When the two computations are grouped into pr(a,e') this is illustrated by Fig. 2.62.

Now we can write down the incidence matrix P^{10} corresponding to P^{00} of (2.42). Thereby we assume that the output set d in pr(d,l') has to be scanned the same number of times as its input c, i.e., five times. This fact could not have been indicated in P^{00}. We make the assumption, in writing down P^{10}, that we have decided to group all computations which have input sets that are subsets of input sets of other computations, with these

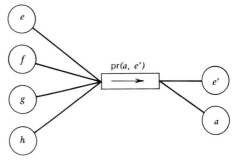

FIGURE 2.62

latter computations. Thus pr(e') is grouped with pr(a) giving pr(a,e'), and similarly we obtain pr(b,i') pr(c,j'), and pr(d,l').

(2.62)

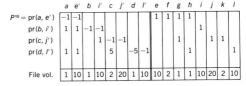

Total file volume = 111

Note:

In (2.62) we have indicated e' as input instead of e in all processes except pr(a,e'), as it should be (likewise for i', j', l').

In (2.62) we have also added a row (the last one) to indicate the assumed file volumes.

In Fig. 2.63 we have drawn the graph corresponding to P^{10} of (2.62).

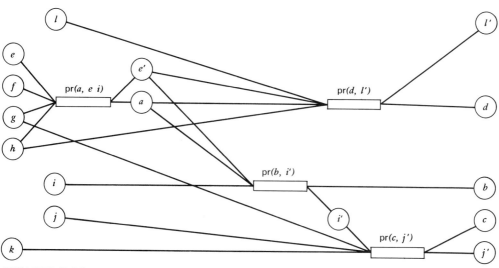

FIGURE 2.63

2.7. Adaptation to Hardware System

In designing the information system we have to consider limitations imposed by hardware design or hardware cost.

Thus for all practical considerations we can take as an axiom that standing files and transput files are stored in a storage which is much slower than the main memory of the computer. Therefore we say that the filed data are *transported* to and from the main memory. In addition to this general property of using auxiliary file storage we have to consider mainly two hardware limitations. The first is that limited memory space may make it uneconomic to group too many computations together, because all necessary programs cannot be kept in the main memory simultaneously. Therefore programs may have to be transported to and from the memory. This may be a larger transport volume than the data transport it saves. The second limitation is that the number of file storage (or data transport) units cannot be chosen large enough to transport all data sets separately. This limitation may again be a reason for not grouping some computations although this might have saved data transport. As an example note that comp(b) needs three input data sets and one output; i.e., it would need four transput units (see 2.42). Instead it is seen from (2.62) that pr(b,i') needs three inputs and two outputs or five transput units in total. Alternatively, the number of transput units may be reduced by consolidating several files into one file (see section 2.5), giving rise to deadweight data transport and extra runs for merging files.

Finally, it must be born in mind that in using memory space to group computations, and thereby reduce the number of file scans, may necessitate shorter data blocks which will increase data transport by increasing the number of interblock gaps, which are also a source of deadweight transport.

In addition to all this we also will have to consider the time spent in replacing files (for instance tape reels) between processes.

The optimum design of the data structure for a system will have to consider all the above mentioned effects upon data transport and try to minimize resulting total transport. This problem is to establish different feasible solutions (by which will here be meant system lay-outs that are compatible with memory and transput equipment limitations) and to find one among these which corresponds to minimum transport, in other words to find an optimum solution. This is easily seen to be a complicated problem, and the possibility that the computation program complications associated with grouping of computations may add one further difficulty makes the problem a very difficult one, indeed.

We do not take up here the problem of constructing an efficient algorithm for finding an optimum solution. Instead we give some brief study to the much more modest problem of finding a feasible solution.

We introduce the simplification of assuming that first a system of grouping of computations has been decided on. Thereafter we look for a way to define a consolidation of the files which will reduce the number of transput units to a prescribed value.

We now study the problem of finding a feasible solution by assuming that we have already decided on the grouping of computations of our previous example as given by (2.62). We assume that we have to comply with the limitation of having only four transput units (in this specific example).

In trying to find a suitable file consolidation (or a feasible solution) we might start at the initial end, that is, with pr(a,e'). We may illustrate the consolidation construction by listing the input-output relations and use parentheses to enclose consolidated sets.

(2.71) $\quad (e)(f,g,h) \rightarrow (a)(e')$
$\quad\quad\quad\quad (a)(e',i) \rightarrow (b)(i')$
$\quad\quad\quad\quad (i',g)(j,k) \rightarrow (c)(j')$
$\quad\quad\quad\quad (a)(e')hlc \rightarrow (d)(l')$
$\quad\quad\quad\quad\quad\quad\quad 5 \quad\quad 5$

Considering which consolidation to choose for the first row in (2.71) [that is, $(2.71)_1$], we notice that a is much smaller than e'. Hence, since a is an output that will have to be printed in a separate process, it may be wise to keep a and e' separate. Similarly it may seem wise to keep e separate from f, g, h. This leaves the consolidation of $(2.71)_1$, as shown.

For $(2.71)_2$ separation of b is natural for the same reason as given for a in $(2.71)_1$.

Obvious arguments lead to $(2.71)_3$. When we come to $(2.71)_4$ we see that no solution can be obtained in a straightforward way. We will be forced to go back and start all over again. It is obvious that we may have to iterate this scheme several times. In the vastly more complicated cases of real applications this may be a rather complex operation. *In the normal way of handling this problem, one will most often have done a lot of programming for the earlier procedures before one reaches a stage where it becomes clear that a restart will be necessary.*

Therefore in actual practice, the manner of analyzing for feasible solutions using P^{10} or the graph may save man-years of programming.

In most cases a more straight-forward solution procedure may result if we start at the terminal sets rather than at the initial sets.

We now try to start at the terminal end, i.e., at the process giving d.

First we do some simple calculations. We observe that c and d are passed five times in pr(d,l'). Hence it is reasonable to try to keep them separate from other files. If we do this, we are forced to consolidate input data such as h with standing files such as l and e (we assume all files to be updated in some process; that is, e, i, j, and l, and only those are standing files). This means that, for instance, the operation pr(h) of copying $\hat{h}$ from punched tape (say) to h on magnetic tape [pr(h): $\hat{h} \to h$] will instead have to be performed as pr(h): $\hat{h}, l, e \to h, l, e$, thus causing two extra passes of l and e. ($\hat{h}$ is used to denote the h-data when in punched form.) This extra transport has a volume of $2[\text{vol}(l) + \text{vol}(e)] = 40$. Instead, if we would have to consolidate, for instance, as much as a, f, g, h, with c, it means an extra transport of 5 vol(a,f,g,h) = 25 which is far better. On the other hand when punched data are read in, the accompanying file storage transports may be without consequence. This is mostly true when data transport is either buffered or handled by time-sharing. Therefore it may often be sufficient to consider the transports studied here. Data transport buffering or time-sharing may often permit us to count only input transput.*

$$
\begin{aligned}
(2.72)\quad & (h,c)(a,e',l) \xrightarrow{d,l'} (d)(a,e',l') \\
& (i',g,h)(j,k) \xrightarrow{c,j'} (h,c)(j',k) \\
& (a,e',l)(i) \xrightarrow{b,i'} (b)(i',g) \\
& (a,e,l)(i,g,h,f) \xrightarrow{a,e'} (a,e',l)
\end{aligned}
$$

Now we rearrange to original sequence and at the same time introduce the modifications made in files in later stages in all stages (thus for instance f is to go with i, g, h in a, e', therefore also in b, i').

$$
\begin{aligned}
(2.73)\quad & (e,l)(i,g,h,f) \xrightarrow{a,e'} (a,e'l) \\
& (a,e',l)(i,g,h,f) \xrightarrow{b,i'} (i',g,h)(b) \\
& (i',g,h)(j,k) \xrightarrow{c,j'} (h,c)(j',k) \\
& (h,c)(a,e',l) \xrightarrow{d,l'} (d)(e',l')
\end{aligned}
$$

We have now also cancelled during updating of a file the data that are not used any more. For instance, f is not used after pr(a,e'), which can be seen easily in (2.72) because we underlined those data which are used in the process. Therefore when (i,g,h,f) is updated during pr(b,i'), we leave f out. Here, for instance, h' is used instead of h although $h' = h$ here. This is necessary since h and h' occur in different files in the same program, and similarly for g', k', e'' and l'' in other files.

In Fig. 2.71 the graph of the feasible solution (2.73) is shown.

In order to estimate the quality of our solution we compare the resulting data transport volume with that of the system (2.62) without transput limitation and with the basic topological one as well as with the theoretical minimum.

	a	b	c	d	e	f	g	h	i	j	k	l
(2.62)	3	6	1	5	4	1	2	2	3	2	1	2
(2.73)	3	6	1	5	5	2	4	8	4	2	2	5
Vol.	1	1	2	1	10	2	1	1	10	20	2	10

Total transport volume (2.62) = 154
(2.73) = 216

Total file volume, counting old and updated files, is 111 which constitutes the theoretical minimum transport.

Thus in our example the excess transport due to limited process grouping is 43 or about 35% of file volume. The additional transport caused by limitation to 4 transput units is 65 or about 60%.

We also compare our total transport volume with that associated with the basic topological scheme of (2.42). Thereby we have to add 4 times vol(d) because in section 2.4 we did not consider 5 scans of d at the output as we have done here. We therefore obtain the basic transport volume = 249.

We have mentioned that whereas the grouping of some processes or computations into one larger process may reduce the volume of data input transport, this saving may be partly lost by the necessity

---
*We have written the produced results above the arrows. Thus in (2.72)$_1$ d, l' over the arrow indicate that in the output set d, a, e' l' only d and l' are produced, a and e' being only copied.

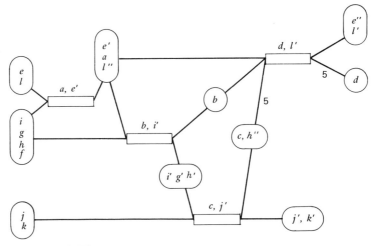

FIGURE 2.71

to use shorter blocks. This may be necessary for storing all programs needed in the grouped computations. In order to show that this is actually a problem worth considering when small memories are used, let us take up an example which, although much simplified, uses fairly realistic data volumes. In the above studies topological structures rather than numerical relations were of primary importance. Our present problem depends largely on size relations between input data, programs, and memory space. In order to be of interest, the example has to use fairly realistic values or relations.

Example

We assume that we have two processes, pr(b,i') and pr(c,j'), and we want to see what can be saved by grouping them into one single process, pr(b,i',c,j'). Let us assume that the number of elementary input files is 10 for each of the processes, pr(b,i') and pr(c,j'). Further assume that the number of inputs to pr(b,i',c,j') is 15 rather than 20 so that grouping the two processes together saves a transport volume corresponding to 5.

We further assume that each record in an elementary file contains 30 characters (or an equivalent amount of information). This would mean that the total input set for one processing would be $10 \times 30 = 300$ characters for pr(b,i') or pr(c,j') and $15 \times 30 = 450$ characters for pr(b,i',c,j'). These are seen to be fairly normal sizes. For each of the four computations (for b, i', c, and j', respectively) there will be one program; and in addition, one program is needed for the overall organization. Let us assume 2000 characters for each program, which means $5 \times 2000 = 10000$ for the grouped process, a normal figure.

We now assume that the elementary files are consolidated into three tape files with five elementary files in each. We thus obtain three tape files, 1, 2 and 3, say. For pr(b,i') we read in 1 and 2, for pr(c,j') 2 and 3, and for pr(b,i',c,j') (assuming three input units) 1, 2, and 3. Each consolidated file record will then contain 150 characters.

We also have to consider the interrecord gaps on the tapes and the memory size. We assume the interrecord gap to correspond to 150 characters and we assume the memory to be just sufficient to accept the grouped process pr(b,i',c,j').

Under these assumptions we must use tape blocks with only one record. Thus the blocklength will be 150 characters and the input volume for each process step will be $3 \times 150 = 450$ characters for the three file records; and in addition the same amount in record gaps will be transported. The resulting volume is seen to correspond to 900 characters.

Now if we use the separate processes pr(b,i') and pr(c,j'), each run has the extra available memory space of the programs for the two computations left out. Thus in addition to the earlier input data space of 450 characters, we now get $2 \times 2000 = 4000$ characters. This leaves 4450 characters for two input blocks in each run. Hence we can use blocks of 2200 characters, say. In this case the interrecord gaps represent only about 7% of the blocks. The volume of the input transport in this case will be two times that of the 300 characters for each run of pr(b,i') and pr(c,j'), or 600 characters for computing one set of b, i', c, and j'; that is about 640 with allowance for gaps. Thus in our example the grouped process will lead to an increase in input transport from 640 to 900, rather than the saving which might have been expected.

It is also of interest to note that we have assumed three input tape units in the grouped process but only two in the separate processes. If we make the more "fair" assumption of permitting only two input tapes also for the grouped process, we may assume 225 characters per record in each file to obtain the same total of 450. Rather than increasing the transport volume this will reduce the volume because there will now be only two interrecord gaps (of 150 characters) against three gaps above. In this way the transport volume is reduced from 900 to 750.

This is still more than the 600 needed for two separate processes.

It is easy to see that if the memory size would be increased, say by the amount of 6000 characters, then the gaps would have negligible influence and the full saving of grouping the processes would be obtained.

It may be of interest to see what memory size corresponds to our assumptions. We had 5×2000 characters for program (including output data) and 450 characters for input data, or 10450 characters in all, a not too rare size; but it is more common to have double the sizes for both total data volume and total program space, calling, of course, for twice the above memory size.

This example is another illustration of the fact, already encountered in (2.71), that the design of the information handling system is inherently of an iterative character. Thus in a real situation similar to that of the example, we might have started with an assumed blocklength and on this basis have found it suitable to group the computations into one single process. After having done this, we may find that we have to use a smaller block which then makes the grouping uneconomic; and a new solution, with two separate processes, may instead have to be chosen. It is also clearly exhibited by the example that a larger memory size does not only increase computation speed and reduce data transport volume. It also serves to reduce the systems design work, for instance by removing the need for some iterations on the design structure.

3. RECORD LAY-OUTS

It was shown in section 2.2 how information requirements were traced backwards in the precedence graph until it was defined in terms of what we shall now name *elementary files*.

An elementary file will consist of data from a minimum of two information classes so that each record in an elementary file (each *elementary* record) contains a minimum of two *terms*. One of these is the sorting key, the other the value of some function of this key. Thus an elementary file might be defined by the pair: identification number, person name. Another might be: identification number, salary. In many cases, however, it is not appropriate to use such small elementary files. Then the elementary file will contain the key and two or more information classes in a package that is in no case partitioned further. As an example we may find that in order to obtain a unique definition of all person identification numbers we must associate these with both the name and the address of the respective persons. The elementary record would thus, in this case, at least contain the three terms identification number, name, and address.

When elementary files are consolidated (as in Fig. 2.71), then the elementary records are assembled into *file records* and these file records obtain a certain structure by the way the consolidation is done. It is in this way that the relatively complicated records of business data files come into being. It is seen that one might expect that the record formats are defined already at this stage of the analysis. Thus we have another kind of work that should be done well in advance of the actual, detailed programming, contrary to present usage.

Let us take the consolidated file $(i,g,h,f) = F_1$ (say) of Fig. 2.71. From (2.62) we find

i goes into $\mathrm{pr}(b,i')$
g goes into $\mathrm{pr}(a,e')$
h goes into $\mathrm{pr}(a,e')$
f goes into $\mathrm{pr}(a,e')$

It will be convenient to be able to move that part of the record $F_1(i,g,h,f)$ which is to be copied only during a process. If we look at Fig. 2.71 we see that in process (b,i') i', g is to be moved from F_1 to the output file $(i'g)$ in the cases when a record from F_1 is not changed. Thus we find that the organization of the records of F_1 should be as shown in Fig. 3.11. We have introduced the name F_{11} for (i',g) and F_{12} for (h,f), as is seen in Fig. 3.11.

We assume that the data sets i, g, h, f consist of elementary items in the following way:

$$i = i_1, i_2, i_3;$$
$$g = g_1, g_2;$$
$$h = h_1, h_2, h_3, h_4;$$
$$f = f_1, f_2, f_3,;$$

where $i_1 = g_1 = h_1 = f_1 =$ the common sorting key, which is also the key for the whole file F_1 and therefore will be denoted by F_{10}.

With this information we now have the record lay-out for F_1 as shown in Fig. 3.12.

This record organization can be described by the COBOL way of using level numbers with 01 for the record name, 02 for the data items or data sets on the highest level within the record, and so forth.

```
01   F₁
  02   F₁₀
  02   F₁₁
    03   i
      04   i₂
      04   i₃
    03   g₂
  02   F₁₂
    03   h
      04   h₂
      04   h₃
      04   h₄
    03   f
      04   f₂
      04   f₃
```

THEORY OF INFORMATION SYSTEMS

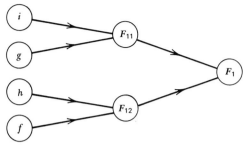

FIGURE 3.11

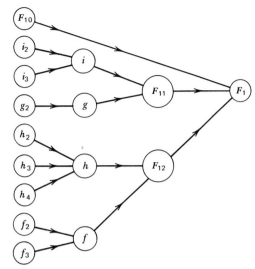

FIGURE 3.12

We have, so far, seen how an organizational structure is imposed upon a record by the overall structure of the information system. This is not all that is needed to determine the record format, however. Thus, often the way records are stored on tape, for instance, affects data transport time and processing time in opposite ways. Thus the choice will depend on whether the file will be used in processing where transport time is dominating or processing dominates or neither. The decision in this respect may have to wait until programming is being done, although for some files the situation already at the time of data structure analysis will be decisive.

Also, the length of tape blocks will have to be decided at the same time; and for some firms the coding form will be open to special decision.

REFERENCES

1. B. Langefors, *Information Retrieval in File Processing.* BIT, Bind 1, Hefte 1, p. 54 and Hefte 2, p. 103.
2. B. Langefors, *Computation of Parts Requirements For Production Scheduling.* BIT, Bind 2, Hefte 2, p. 91.

PROBLEM STATEMENT LANGUAGES IN MIS

DANIEL TEICHROEW

INTRODUCTION

The term "problem statement language" is used here to mean a language which is intended to be a means of expressing the requirements for a data processing task rather than a procedure for accomplishing the task. The term nonprocedural is sometimes used though it is hard to define, Young (1965). The desirability of nonprocedural problem statement languages has long been recognized; see for example, Chapin (1958, 1960) and Ladd (1957). A number of papers proposing and discussing such languages have been published. In this section, the literature will be summarized and some of the major concepts that have appeared will be identified. Previous surveys of this literature (and of other related languages) are given by Young (1965), Shaw (1965), and Air Force Systems Command (1965).

Six major approaches have been selected for detailed examination. The earliest is the work by Young and Kent (1958) [YK]; Information Algebra [IA] is the result of work by the CODASYL Development Committee (1962); Lombardi [LO] has published a number of papers from 1958 to 1964; Langefors [LA] has published a series of papers in BIT; Accurately Defined Systems (ADS) has recently been released by the National Cash Register Co. (1967); and TAG, Meyers (1962), though not documented in the public literature, has some interesting features. The material used in this discussion is based on these publications.

All six approaches are concerned with the problem definition phase of IPS building.

> IA: "The goal of this work is to arrive at a proper structure for a machine-independent problem-defining language at the systems level of data processing."

SOURCE: Teichroew, D., "Problem Statement Languages in MIS," *Proceedings, International Symposium of BIFOA*, "Management Information Systems—A Challenge to Scientific Research," Cologne, July, 1970, pp. 253–270.

LA: "A formal method for performing systems analysis of information systems in business and elsewhere is needed in order to save systems work and programming and to obtain better systems."

YK: "There are three stages in the application of high speed digital computers to data processing problems:
1. Systems Analysis—the task of determining what is to be done.
2. Programming—a statement of how it is done.
3. Coding—a translation of this statement into machine language.

This paper presents a first step in the direction of automatic programming as well as a tool which should be useful in systems analysis."

LO: "[The language] relies exclusively on non-procedural representation of processes as sets (tables) of relations between data and results (there are no control statements such as GO TO, etc.) instead of procedure descriptions (which are one-to-one translations of flow charts)."

ADS: Is specifically intended for complete specification of problem requirements: "The completion of the ADS gives the definer a well documented application that includes all of the information requirements of the problem."

TAG: "The Time Automated Grid (T.A.G.) technique is a computer tool for use in systems definition, analysis, design and program definition.

A more complete list of techniques related to problem statement is given in Figure 1. (The meaning of the column headings is given at the end of the subsection.) The language developed by Bosak (1967) is not included because it is a file processing language rather than a problem statement language. The output decompositions method [ODM] (Grosz, 1963) and SCERT (Herman, 1964) are not included because they are primarily design techniques. The large block programming languages [Basic Functions, (Jones, 1961; Procter and Gamble, 1965), MAST (Boeing, 1967), BEST (National Cash Register, 1965)] are primarily programming techniques and do not contribute directly to non-procedural problem statements. General purpose programming languages, such as COBOL and PL/1, require problem definition in terms of the data manipulation procedures. This is also true of the other programming languages such as Iverson (1962), BCL (Hendry, 1967) and Dataless programming (Balzer, 1967).

The six selected approaches are classified in Figure 2 according to which of the seven main phases of the life cycle of a system they cover. The objectives explicitly stated in the papers are summarized in Figure 3.

YK, ADS, and TAG are problem statement techniques that use a practical, straight-forward approach with very little or no attempt to develop a theory of data processing. They consist of a systematic way of recording the information that an analyst would gather in any case, and any experienced analyst could use either with very little instruction. IA is more concerned with developing a theory. It uses a terminology and develops a notation that is not at all natural to most analysts, and it is not at all obvious how it would be used in any particular case. LO is more of a programming language

Technique	PROBLEM STATEMENT							Provides aids to the design process
	Outputs and inputs	Data definition	Processing requirements				Other data	
			VARIABLES	LARGER GROUPS				
Young and Kent (YK)	OUTPUT DOCUMENTS INPUT DOCUMENTS	INFORMATION SETS	DEFINING RELATIONSHIPS	PRODUCING RELATIONSHIPS		TIME VOLUMES	No
Information Algebra (IA)	INPUT AREAS OUTPUT AREAS	PROPERTY SPACE	COORDINATE DEFINITIONS	MAPPING		None	No
Lombardi (LO)	ORDERED INPUT ORDERED OUTPUT FILES	FIELD, RECORD FILE, BUNDLE	FIELD DECLARATIONS	CONTROL PREDICATES		None	Yes
Langefors (LA)	INITIAL INFORMATION SETS TERMINAL INFORMATION SETS	INFORMATION SETS	Not mentioned	PRECEDENCE RELATIONSHIPS		VOLUMES	Yes
TAG (Meyer) (TAG)	INPUT OUTPUT FILE	FIELD	SEVERAL TYPES OF VARIABLES	None		VOLUME FREQUENCY	Yes
ADS (ADS)	INPUT REPORT	FORM FORMAT	COMPUTATION FORM	SELECTION RULES		VOLUMES CYCLES	No
Bosak	ENTRY SET PROPERTY LISTS	PROPERTY	PROPERTY FUNCTION	FILE FUNCTION		None	No
Grosz - ODM	SOURCE GENERATED ON FILE	FIELDS	STATEMENT	None		VOLUME FREQUENCY SEQUENCE	Yes
Basic Functions (Jones)	FILES	FIELD, RECORD	TASK SPECIFICATION	None		None	Yes
Basic Functions (P&G)	FILE	FIELD, RECORD	SUBFUNCTION	FUNCTION		None	Yes
MAST	TRANSACTION RECORD REPORT	DATA SET	STATEMENTS	None		VOLUME FREQUENCY	Yes
BEST	RECORD REPORT	FIELD	STATEMENT	Macros		None	Yes
SCERT	←————— STARTS WITH SPECIFICATION OF RUNS —————→							Yes
BCL	Not separately treated	DATA STRUCTURE JUXTA POSITION	Data structure alternation and repetition of arithmetic expression			None	No
Dateless Programming	PL/1 notation and convention	Hierarchical canonical data collection	FUNCTION	Group operators or statements		None	No
Programming	←————————— Processing Procedures —————————→							No

FIGURE 1 Comparison of selected techniques for use after requirements are determined.

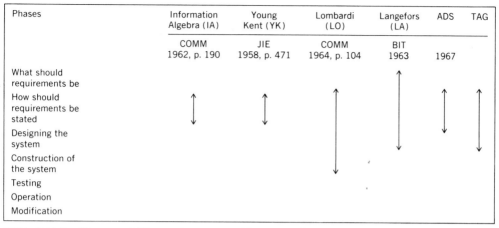

FIGURE 2 Phase of life cycle covered by technique.*

Objectives	IA	LA	YK	LO	ADS	TAG
Non-procedural	x				x	x
Abstract formulation		x	x			
Reduce work of systems analyst	x					x
Machine independent	x	x	x		x	x
Toward overall systems optimizing	x		x			
Manipulation of PS	x		x			
Machine readable PS				x		x
Maximize use of creative personnel						x
Standardization and documentation						x

FIGURE 3 Objectives of techniques.

than a problem statement technique. In order to use it one must have completed the system design, i.e., laid out the file processing runs that are needed. (The language as described in the literature applies to batch processing only.) The technique, however, is relevant because it presents a nonprocedural technique for stating processing requirements once the runs are determined. LA starts with a precedence relationship among information sets (files), but he does not indicate how these are obtained. This technique therefore is more relevant to the analysis of a problem statement and to the design of a system. However, it does suggest some desirable features of a problem statement technique. These six approaches, on the surface, appear to be very different; but upon detailed examination, they have some major similarities. Figure 4 compares them as they relate to the second phase of the life cycle, "how should requirements be stated," in five categories: (i) form of the problem, (ii) data, (iii) relationships, (iv) other information, and (v) method of presentation. This table summarizes the comparison that follows below.

The discussion in this section is based on the published descriptions. To the best of our knowledge only ADS is in

---

* All these papers are included in this book except Lombardi.

PROBLEM FORM		Information Algebra	Young & Kent	Lombardi	Langefors	ADS	TAG
INPUT		INPUT AREAS	INPUT DOCUMENTS	ORDERED INPUT FILES	INITIAL INFORMATION SETS	INPUT	INPUT DOCUMENTS AND FILES
		OUTPUT AREAS	OUTPUT DOCUMENTS	ORDERED OUTPUT FILES	TERMINAL INFORMATION SETS	REPORT	OUTPUT DOCUMENTS
DATA		ENTITY PROPERTY PROPERTY VALUE PROPERTY VALUE SET COORDINATE SET DATUM POINT PROPERTY SPACE LINES AREAS (BUNDLE, GLUMP)	Not used ITEM Not named INFORMATION SET	Not used FIELD Not named Not used RECORDS FILES BUNDLE	Not used Not used Not used Not used INFORMATION SET	VARIABLE VALIDATION RULES	DATA NAME
PROCESSING REQUIREMENTS		MAPPINGS COORDINATE DEFINITION	PRODUCING RELATIONSHIPS FOR DOCUMENTS DEFINING RELATIONSHIPS FOR OUTPUT ITEMS	CONTROL PREDICATES FIELD DECLARATIONS	PRECEDENCE RELATIONSHIPS AMONG INFORMATION SETS Not used	LOGIC COMPUTATION LOGIC	PERIOD and PRIORITY Not specifically included
OTHER INFORMATION		NONE	OPERATIONAL REQUIREMENTS VOLUMES ELAPSED TIME		SIZE OF FILES	VOLUMES	VOLUMES
PRESENTATION			GRAPHICAL NOTATION		PRECEDENCE GRAPH	FIVE FORMS	INPUT/OUTPUT ANALYSIS FORM

FIGURE 4 Comparison of four proposed problem statement techniques.

current use and only TAG has a computer program to process the problem statement.

FORM OF THE PROBLEM

The first category for comparing the six approaches deals with how they define the overall problem. In this comparison, it is desirable to quote the authors on what they say about data processing systems and how they should be designed.

IA: "An information system deals with objects and events in the real world that are of interest. These real objects and events, called 'entities' are represented in the system by data. The data processing system contains information from which the desired outputs can be extracted through processing. Information about a particular entity is in the form of 'values' which describe quantitatively or qualitatively a set of attributes or 'properties' that have significance in the system. Data processing is the activity of maintaining and processing data to accomplish certain objectives."

LA: "There are some basic propositions made here in connection with the systematic approach advocated, which appear to be in contradiction to present practices or assumptions. One is the hypothesis that in most cases it is possible to isolate and define the relevant organization functions in a separate operation to be performed before the actual design of the system is attempted. It is thus assumed that these functions are defined from the basic goals of the organization and therefore will not need to await the detailed construction of the system. The other hypothesis is that it is possible to define all input information necessary to produce a desired output. The basic assumption here is that actually any information can only be defined in terms of more elementary information, which will then occur as input parameters. Therefore, once a class of information is defined then it is known what input information is required for its production. The point here is that it should not be necessary to work out formulas, or even computer programs, before it can be determined what input data are needed. In fact, it is well possible to work out formulas or programs for an entity, where important variables are missing, so that starting by programming is no safeguard against ignoring important data."

YK: "The content of our analysis is that the objectives of the data processing system have been stated in terms of the required outputs; these outputs are not considered as subject to revision. On the other hand, although the inputs may be organized in any desired fashion, it appears necessary or at least convenient, to state one of the possible input organizations from which any equivalent one can be derived. It should be noted that the input may supply any one of a number of equivalent pieces of information, e.g., either customer's name to be copied directly onto an output or an identification number from which the name can be looked up."

LO: "The common denominator of file processes is the production of output files as functions of input files."

ADS: "The starting point is the definition of the reports—what output information is required. Once the reports are defined, the next step is to find out what information is immediately available. This is followed by laying out the information system in between the output and input. The origin of all information needs to be specified. The outputs of this system are always looked at in terms of inputs."

TAG: "The technique requires initially only output requirements of a present or future system. These requirements are analyzed automatically [by a computer program] and a definition is provided of what inputs are required at the data level."

All six approaches assume that the problem statement starts at output. What is required output? (IA: ". . . from which the desired outputs . . ."; LA: ". . . then it is known what input . . ."; YK: ". . . in terms of desired outputs . . ."). Therefore, a necessary part of the problem statement should be the description of the desired output. As will be seen later, this requirement is implied by LA in the definition of TERMINAL SEETS* and in YK by the definition of output DOCUMENTS. IA does not mention required output as such and, in fact, in the example given in the paper says, "The problem is to create a new pay file from. . . ."

*Words which have particular meanings in any of the languages are typed in capital letters.

INFORMATION ABOUT DATA

A problem statement must have some description about the data that will be processed. The most extensive data description facility is the one used by IA. This starts with the concept of an entity which has a connotation of a physical entity in the real world such as an employee, a paycheck, or an order. Each ENTITY has PROPERTIES which describe that entity; e.g., an employee has an employee number, hourly rate, etc. For any given ENTITY there is a value for each property. The PROPERTY VALUE SET is the set of all possible VALUES that a PROPERTY can have in the problem. The COORDINATE SET is the list of all PROPERTIES that appear in the problem. A DATUM POINT is a set of values, one for each PROPERTY in the COORDINATE SET, for a particular ENTITY. The PROPERTY SPACE is the set of all DATUM POINTS, i.e., all possible points obtained by taking the cartesian product of all possible PROPERTIES. Once this PROPERTY SPACE has been defined, further definitions deal with subsets of this space. A LINE is a subset which is roughly equivalent to a record, and an AREA is a subset roughly equivalent to a file. Other subsets of the PROPERTY SPACE are BUNDLES and CLUMPS. The basic reason for this choice of data description is to use the concepts of a set theory as the formulation for a theory of data processing. (The authors of IA rejected data description by arrays as being too limited.)

In YK, the basic units of data are called ITEMS and they correspond to a PROPERTY in IA. The term INFORMATION SET is used for the set of all possible values of a particular item and is, therefore, equivalent to the PROPERTY VALUE SET in IA. The information that can be provided for each

INFORMATION SET are: (i) the number of possible values, (ii) the number of characters or digits, and (iii) relationships. The following relationships are defined:

Relationship	Description	Symbol	Graphic Symbol
Isomorphism	One-to-one correspondence	–	←——→
Homomorphic	Many-to-one correspondence	–	——→
Cartesian product	$P_j X P_k$ means a pair of P_i and P_k	X	
	Contained in	C	
Equal to		=	

The relationships may be used to make statements such as: there is one employee number for each employee name and address. YK did not want to make any statements about the file structure; and, hence, there are not terms that do respond to records or files. YK also provides a graphical notation for showing relationships.

In LO, the definition of data is more conventional, including FIELD which corresponds to PROPERTY, RECORDS, FILES, etc. The word BUNDLE is used to denote a set of files that are merged on a single input or output unit.

In LA, there is no definition of data corresponding to data items. The problem definition starts with collections of data which are called INFORMATION SETS. This corresponds roughly to the notion of a file in common terminology. LA introduces the concept of an elementary file in which each record contains a data value and enough "keys" to identify it uniquely.

ADS provides three forms on which data are described: reports, inputs, and history. Each of these forms provides space for some information describing the particular report or input: name, media, volume, and sequence, and space for each variable. For each variable the forms provide space for name, how the value of the variable is obtained (input, computation, a history) and a cross-reference, how often the variable appears, and size (number of characters).

TAG provides one form which contains space for data describing the document (or file) and space for each variable. Figure 5 shows a comparison of the data required by YK, TAG, and ADS.

PROCESSING REQUIREMENTS

In the standard programming languages, each program or subprogram includes statements which produce output, statements which test the conditions under which the output is produced, and statements which compute the values of the variables that appear in the output. It has been argued by Lombardi, in particular, that a nonprocedural language must separate the statement of what output is to be produced when, from the statement of the procedure for producing the value of the variables that appear in the output. All six approaches follow this concept.

In IA the basic operation that the problem definer can use to state his processing requirements is a mapping of one subset of the PROPERTY SPACE into another subset. Two kinds of mappings are defined. One corresponds to operations within a given file. For example, suppose a tape contains time cards, sorted in order by employee number, one for each day of the week. A mapping could be defined which would take the set of (five) POINTS for each employee into one new POINT which would contain the total for the week.

FOR WHOLE DOCUMENT	YOUNG & KENT	TAG	ADS
1. NAME	✓	✓	✓
2. TYPES OF DOCUMENTS	INPUT, OUTPUT	INPUT OUTPUT FILE	INPUT REPORT HISTORY
3. WHEN PRODUCED	PRODUCING RELATIONSHIP	PERIOD (S, MI, H, D, W, MO, Q, Y) PRIORITY (NUMERIC) FREQUENCY	SELECTION RULES
4. MEDIA	NOT MENTIONED	NOT MENTIONED	

5. SEQUENCING CONTROL
  MAJOR
  INTERMEDIATE
  MINOR
	NOT MENTIONED	NOT MENTIONED	FOR INPUT
			SEQUENCES
			  MAJOR
			  INTERMEDIATE
			  MINOR

6. VOLUME
  AVERAGE (A)
  MINIMUM (M)
  PEAK (P)
	
	✓
	✓
	✓
		
		✓
		✓
		✓
			EXPECTED VOLUME
			FOR HISTORY, INPUT
			AND REPORT

7. DESIGNED FOR P, A, M	NOT MENTIONED	DESIGNER'S CHOICE	NOT MENTIONED
8. OTHER DATA		REFERENCE AUDIT	
FOR EACH VARIABLE			
1. NAME	✓	✓	✓

2. HOW USED		FI – FIXED; INFORMATIONAL
		FF – FIXED; FUNCTIONAL
		VF – VARIABLE; FACTOR
		VR – VARIABLE; RESULT
			MODIFIED BY
			  – FORMULA
			  – PARTICULAR VARIABLE
			HOW OFTEN?
			  – NEVER
			  – PARTICULAR CYCLE
			  – FIXED TIME
			  – LOGICAL CONDITION

3. COMPUTATION FORMULAS	DEFINING RELATIONSHIPS FOR VARIABLES ON OUTPUT REPORTS	NOT INCLUDED (MAY BE ADDED AS COMMENTS)	COMPUTATION FORM LOGIC FORM
4. SEQUENCING ROLE	NOT MENTIONED	✓	✓
5. VALIDATION RULE	NOT MENTIONED	NOT MENTIONED	✓

6. FORMAT
  A or N
  SIZE (NO. of CHAR.)
  FOR OUTPUT
	IN INFORMATION SET TABLE	
		✓
		✓
		Ordering number of P for presence.
		RATIO
			
			✓
			✓

7. NO. OF TIMES PER DOC.
  MINIMUM
  AVERAGE
  MAXIMUM
	✓		✓

FIGURE 5 Description of documents. (A checkmark denotes provision for including the information listed in the left-hand column.)

The second type of mapping corresponds to the usual file maintenance operations in which POINTS from a number of input files are processed to produce new output files. These two types of mappings are called GLUMPING and BUNDLING, respectively. The actual computation of the PROPERTY VALUES of the new POINTS produced by a mapping is specified by a COORDINATE DEFINITION which must contain a computational formula for each PROPERTY in the COORDINATE SET.

In YK the major unit of processing is a PRODUCING RELATIONSHIP; there must be one PRODUCING RELATIONSHIP for each type of output document. This PRODUCING RELATIONSHIP gives the conditions under which a document will be produced. This statement may contain

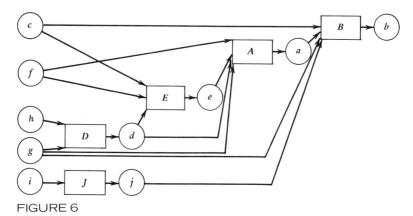
FIGURE 6

conditions (Boolean expressions) that depend on values of data ITEMS or on time. For example, a PRODUCING RELATIONSHIP might be "a monthly statement is produced for a customer each month for all customers with a non-zero balance." A PRODUCING RELATIONSHIP may also state that Document D_2 is produced for each input D_1. The values of the data ITEMS which appear in the output documents are calculated using a DEFINING RELATIONSHIP. There must be one defining relationship for each data ITEM which appears on an output document.

In LO the statements which control whether or not an output record is produced are called CONTROL PREDICATES. There must be one CONTROL PREDICATE for each record for each type of output file. The CONTROL PREDICATES, in general, are Boolean expressions which may involve the use of INDICATORS. The values of the variables which appear on the output records are produced by FIELD DECLARATIONS. The CONTROL PREDICATES are separated from the FIELD DECLARATIONS and the order within groups is immaterial. The CONTROL PREDICATES and FIELD DECLARATIONS are evaluated at the end of each PULSE in PHASE.

In LA the relationships are given for production of information sets and, hence, correspond to PRODUCING RELATIONSHIPS. However, they are stated only as precedence relationships; e.g., information sets a, b, and c are necessary to produce d. No functional relations are given. A problem statement may be represented by a graph as shown in Figure 6.

In ADS some basic information is specified about when reports are to be produced. However, in many cases this is supplied by written notes. This information may be regarded as analogous to the PRODUCING RELATIONSHIPS in YK. ADS requires that each variable be identified as coming from input, computation, or history. A form is provided for specifying the computations; this specification is somewhat limited. Another form is used to state logical conditions and these may modify computations or input.

TAG provides for stating how often output will be produced by specifying a period. The available codes are: second, minute, hour, daily, weekly, monthly, quarterly, and yearly. A priority can be assigned to distinguish a sequence ordering between two documents with the same period. TAG does provide a means for stating which data elements are to be computed, but it does not provide for stating the formula for the computation. (The formula can be included in the "Comments" section of the form, but it will not be analyzed by the program.)

OTHER INFORMATION

IA and LO do not specify any additional information; LA assumes that the relative size of files is available. YK and ADS both provide for specifying time and volume requirements.

YK define two kinds of time: extrinsic (when an event occurs) and intrinsic (the time written on a document). Volumes of documents may be expressed in terms of averages over some time period. The operational requirements consist of a volume for each document (input and output) and a time statement for each output document.

ADS permits specification average and maximum volume in input, report, and history forms. In addition each variable in HISTORY is characterized by how long it is to be retained; this may be a fixed number or may depend on a computation.

TAG provides for volume information for documents, size information on data elements, and repetition information on data elements within documents.

PRESENTATION

Both ADS and TAG have well structured forms for recording the problem statement. They differ in that ADS has a few very structured forms while TAG has a single form.

ANALYSIS, SUMMARY, AND CONCLUSION

Only ADS is in current use and only TAG is implemented in the sense of having computer programs to process the problem statement. One might question why these techniques have not been used. One reason may be that preparing a rigorous and complete problem statement requires (or at least seems to) more time than the present procedure in which problem statement, systems analysis, and programming are collapsed into one indistinguishable process. A second reason may be that there has not been any immediate advantage to an analyst or programmer to invest additional time in a more systematic problem statement. Such advantage could come from either or both of (1) facilities to manipulate the problem statement symbolically and (2) a computer processing of the problem statement itself.

IA represents an attempt to develop a problem statement notation that might be manipulated symbolically. The use of set notation and the usual set operations appear a reasonable start for a language in which data processing problems can be expressed. To our knowledge IA has only been used once [McGee and Katz (1965)], and therefore the practical usefulness of IA remains to be demonstrated. It is also not clear how one uses a problem statement expressed in IA in the system design. Both of these questions (the usefulness of IA for problem statement and the derivation of a design from such a statement) provide promising areas for research.

The above discussion has indicated a number of concepts that should be included in a problem statement language in order to facilitate both symbolic manipulation and computer processing. They are summarized here in order to point to desirable characteristics of future problem statement languages.

Form of the Problem

There can be no question that the basic purpose of an IPS is to produce outputs. However, it is not clear that limiting the statement of data processing requirements to outputs only is desirable. Conceptually, one does not want to prejudice the systems design by stating inputs that may not be needed. Frequently, however, certain inputs must be accepted by the IPS and then the problem statement might as well include the facility for specifying them. Also, the conditions that must be stated (the PRODUCING RELATIONSHIPS in YK) in the absence of specification of inputs can become very complicated. The statement of problems will probably be simplified if the problem definer can state his requirements either in terms of "events" which require action in the IPS or in terms of outputs required; whichever is convenient for him, i.e., either in terms of input or of outputs. Providing this convenience may complicate the analysis of the problem statement by the computer, but the additional processing time is probably worth it.

Data Description

IA is the only approach that associates data with the real world through the use of the ENTITY concept. It should be noted, however, that the IA language in itself does not depend on how the PROPERTY SPACE is obtained, i.e., whether it is derived from real ENTITIES or from a set of abstract variables. It is desirable to give the problem definer as much help as possible in defining his data, and the analogy to the real world through entities is the best method available. Hence, it might as well be part of a problem statement language as long as it does not restrict the language in defining data abstractly.

It is important to distinguish between two possible uses of VALUE SETS. (A VALUE SET consists of all possible values of a variable.) The first use is for variables in which only one value will be in the machine at any one time, and the second use is for variables which may have many values at any one time. For example, the variable "warehouse number" may have many values in the memory at one time whereas the "quantity" of a particular part number at a particular warehouse will have only a single value at any particular time.

In the first case, the VALUE SETS may be used for validation of input data. ADS, for example, permits validation rules to be given for each variable. In practice, validation is a complex process depending on combination of variables rather than on single variables; and such rules are difficult to state on the ADS forms. It may be more desirable to specify validation reports as outputs of the system; these then can include any processing specification permitted by the language for specifying variables on output reports.

The second use of VALUE SETS will be in providing information about how much memory space will be required. The basic question is how the problem definer states the role the variables play. In COBOL the definition is through the structure definition in the DATA DIVISION and the use of OF and IN; in PL/1 nested qualifiers separated by periods are used. In YK the relationships among INFORMATION SETS are used to present this information. In future problem statement languages it would be desirable to infer much of the qualifier-identifier relationship of variables from the processing statements themselves and only ask information that is not included there. It may be possible to obtain all needed information from VALUE SETS and the processing requirements.

The information in a problem statement must be sufficient to infer what variables will have to be stored in the auxiliary memory and in the main memory. A variable must be stored in the memory if:

1. It appears in an update statement, e.g., of the form $X(,,) = X(,,) + Y$, where X might be "gross pay to date" and Y "the pay this week."
2. It is used in a statement without its value being supplied by the input being processed, e.g., "number of exemptions" in a payroll problem. This variable would appear as input on a new-hire transaction and in a change transaction, and would be used in pay computations.

ADS permits the problem definer to specify variables to be available in HISTORY. These may be either intermediate variables that are used in a number of places or variables whose values the problem definer believes will have to be stored.

It is immediately clear from the preceding paragraph that one cannot determine what variables fall into these categories unless the problem statement contains information about the time at which processing requirements occur. In the first case (1), there must be some way of stating that payroll is computed weekly and the "gross pay to date" is cleared (set to zero) at the end of each year. Similarly in the second use (2) it must be clear that a new-hire transaction occurs only once while the pay computations occur regularly.

Other Information

None of the problem statement languages have a well developed procedure for describing the time aspects of requirements, though YK, ADS, and TAG provide some capability. Some help in developing an acceptable "time" language might be obtained by studying the master time routines in simulation languages such as SIMSCRIPT or the executive systems for real-time systems.

It may be noted that time specifications are required not only for determining which variables will be stored but also for determining feasible and optimal storage organizations. The criteria used to determine optimality includes both memory space and processing time. One important factor to be considered is organization of data to reduce memory space, by such techniques as header-trailer organization as used in hierarchical files and IDS. In order to do this, one must be able to infer the header-trailer relationships from such information as qualifiers. Another important factor is the question of what data should be stored semi-permanently and which need be held only temporarily. Again the analogy to simulation may be useful: SIMSCRIPT, for example, distinguishes between PERMANENT and TEMPORARY ENTITIES.

The second part of the criterion is to reduce processing time. This can be done by reducing the number of accesses to external memory. Since a number of different types of processing requirements must be accomplished, the problem statement must contain both the values of each type and the time periods over which they occur so that accesses to auxiliary memories can be grouped whenever possible.

Presentation

Graphical techniques are extremely useful in many areas of stating specifications, e.g., blueprints for construction specification and flowcharts for algorithms. A graphical technique for the problem statement was given by YK and this has since been extended by Young (1967) under the acronym GRIST. The problem statement proposed by LA is equivalent to a directed graph. At the present state of development of problem statement languages, it appears unlikely that graphical techniques other than flowcharts and graphs will be very useful. Some experimental work with the proposed techniques including GRIST, appears justified.

Future problem statement languages will undoubtedly depend on forms, probably somewhere between these two extremes. Good forms can be extremely useful in acting as questionnaires.

REMARKS

How much information about a problem should be collected and when? In current practice the analyst will normally start with the general overall and summary data, and gradually he will become more specific until he has enough details to be able to write the programs himself. In contrast, ADS attempts to have the analyst specify all the details of the problem statement at one time.

The best procedure may be compromise between current practice and the ADS approach. Description of data, for example, could be divided into two levels:

- Composition — how the date is made up of smaller units of data
- Representation — hardware related items such as number of bits, precision, etc.

The composition information clearly is needed as part of the problem statement. Representation information, on the other hand, may not really be needed until program construction begins. A similar categorization could be made for processing requirements. Ideally a problem statement would require specification of necessary data (composition information) for data, for example, and make optimal the statement of information which is not needed until later. This is because sometimes it is easier to record all relevant data at one time.

REFERENCES

Abrahams, D. W., M. F. Lipp, and J. Harlow, "Quantitative Methods of Information Processing System Evaluation," AD 433.220, *ITT*, Paramus, New Jersey, October 1963, 99 + VII pages.

Air Force Systems Command, "Advanced Programming Developments: A Survey," AD 614 704, ESD TR-65-171, *Electronic Systems Division* in cooperation with Computer Associates, Inc., Bedford, Mass, Feb. 1965, 101 p.

Arnold, R. F., H. L. Garner, R. M. Karp, and E. L. Lawler, "Mathematical Models of Information Systems, AD 483-281, *Dept. of Elec. Eng., U. of Michigan,* April 1966, 70 + VIII pages.

Am. Standards Assoc., Comm. X3.6, "Report on Proposed American Standard Flowchart Symbols for Information Processing." *Communications ACM* Vol. 6, No. 10, October 1963, p. 599–604.

Auerbach Corp., "Interim Report on Computer Performance Effectiveness Evaluation Study," 1243-TR-I, Philadelphia, PA., Feb. 1965.

Baum, C. and L. Gorsuch, Editors, "Proceedings of the Second Symposium on Computer-Centered Data Base Systems," *Systems Development Corporation*, Santa Monica, California, 1965.

Berul, L., "Information Storage and Retrieval, A State-of-the-Art Report," AD 630 089, *Auerbach Corporation*, Philadelphia, Pa., September 1964.

Brandon, D. H. and F. Kirch, "The Case for Data Processing Standards," *Computers and Automation,* November, 1963, p. 28–31.

Briggs, R. B., *A Mathematical Model for the Design of Information Management Systems*, (M.S. Thesis, Div. of Natural Sci.), U. of Pittsburgh, 1966, 88 p.

Buchholz, W., "File Organization and Addressing," *IBM Systems Journal*, June, 1963, p. 80–111.

Butler, D. D., O. T. Gatto, "Event-Chain Flow Charting Autosate: A New Version," AD 622-744, *The Rand Corp.*, Santa Monica, California, October, 1965.

Cainer, D., "Symob," Data Proc., March-April, 1964, p. 106-8, *Data & Control,* March, 1964, p. 24-5.

Calingeart, P., "System Performance Evaluation: Survey and Appraisal," *Communications ACM*, 10, I. January 1967, p. 120-8.

Chambers, J. C., R. C. Hain, *The Design of a Management Information System*, Corning Glass Works, Corning, New York, 1966, 32 p.

Chandler, A. R., "AGIL II: A General Input Language for On-Line Information Control," Vols, I, II, AD 489 421, AD 489 422, *The Mitre Corp.*, Bedford, Mass., August 1966, 291 pages.

Chang, W., "A Queuing Model for a Simple Case of Time Sharing," *IBM Systems Journal*, Vol. 5, No. 2, 1966, p. 115-125.

CODASYL Development Committee, "An Information Algebra Phase I Report," *Communications ACM*, 5, 4, April 1962, p. 190-204.

Control Data Corporation, "3400/3600 INFOL General Information Manual," No. 601324000, Palo Alto, California, June 1965.

Connors, T. L., "ADAM—Generalized Data Management System," *Proc., AFIPS 1966, SJCC Proc*, p. 193-203.

Dantzig, G. B., and G. H. Reynolds, "Optimal Assignment of Computer Storage by Chain Decomposition of Partially Ordered Sets," Operations Research Center, College of Engineering, *University of California, Berkeley*, ORC 66-6, March, 1966, 8 p.

Day, R. H., "On Optimal Extracting From a Multiple File Data Storage System: An Application of Integer Programming: *J. of the Oper. Res. Soc. of Amer.* 13, 3, May-June, 1965, p. 482-494.

Dearden, J., "How to Organize Information Systems," *Harvard Business Review*, 42, 2, March-April 1965.

Defense Intelligence Agency, Washington, D.C., "IDHS 1410 Formatted File System: Programming Maintenance Manual," AD 648 020, 1966.

D'Imperio, M., "Data Structure and Their Representation in Storage," Parts I and II, *NASA Technical Journal*, 9, 3-4, 1964, p. 59-81, p. 9-54.

Dixon, P., "Decision Tables and their Applications," *Comp. and Auto*, 13, 4, April 1964, p. 14-19.

Dixon, P. J., and J. D. Sable, "DM-I, A Generalized Data Management System," *Proc. AFIPS 1967 SJCC conf.*, Vol. 30, p. 185-198.

Dzubak, B. J., C. R. Warburton, "The Organization of Structured Files," *Communications ACM 8*, 7, July 1965, p. 446-452.

Farr, L., V. Labolle, and N. E. Willmorth, "Planning Guide for Computer Program Development," AD 465 228, *System Development Corporation*, TM 2314/000/00, May 1965, 179 pages.

Fisher, D. L., "Data, Documentation and Decision Tables," *Communications ACM*, 9, I, January 1966, p. 26-31.

Franks, E. W., "A Data Management System for Time-Shared File Processing Using a Cross-Index File and Self-Defining Entries," *Proc. ALFIPS 1966 SJCC*, p. 79-86.

Gagliardi, U. O., C. G. Ying, and L. G. Hall, "Mathematical Programming Techniques for Information System Design," AD 605 826, *ESD*—TDR 64-530, Bedford, Mass., July 1964, 56 pages + appendices.

Garner, H. L., E. L. Lawler and T. F. Piatkowski, "Mathematical Models of Information Systems," AD 648 823, Sys. Engrg., Lab, *U. of Michigan*, January 1967, 153 p.

Gatto, O. T., "Autosate," *Communication ACM*, 7, 7, July 1964, p. 425-432.

Gaver, "Probability Models for Multiprogramming Computer-Systems," AD 640 706, *Carnegie Inst. of Tech.*, September 1966, 34 p.

General Electric, IDS/COBOL, GE-400 Series, Reference Manual, Program Number CD400HI.002, *General Electric*, Information Systems Division, 83 p.

Gibbons, G. D. and F. B. Thompson, "DEACON Breadboard Processing," AD 612 165, *General Electric TEMPO*, Santa Barbara, Calif., RM 64TMP-12, September, 1964, 84 pages.

Ginsberg, A. S., H. M. Markowitz and P. M. Oldfather, "Programming by Questionnaire," *Rand Corporation*, Santa Monica, California, April 1965, 42 p. See also Proc. AFIPS, 1967 SJCC, Vol. 30, p. 441-446.

Gosden, J. A., and R. L. Sisson, "Standardized comparisons of Computer Performance," *Proc. IFIPS Congress*, 1962, p. 57-61.

Gotlieb, C. C., "General Purpose Programming for Business Applications," in *Advances in Computers*, F. L. Alt ed., I, Academic Press, New York, London, 1960, p. 1-42.

Greenberger, C. B., "The Automatic Design of a Data Processing System," *Proc. IFIPS Congress* 1965, I, p. 227-282.

Gregory, R., and R. L. Van Horn, *Automatic Data Processing Systems*, Wadsworth Publishing Co., Belmont, Calif., 1965, p. 218-238.

Grems, Mandalay, Pracniques: "Terms Frequently Combined in Problem Description," *Communications ACM*, 6, I, January 1963, p. 31.

Grosz, M. H., "Systems Generation Output Decomposition Method," *Standard Oil Company of New Jersey*, July 1963.

Hain, G., and K. Hain, "Automatic Flow Chart Design," *ACM 20th National Conference*, 1965, p. 513-523.

Henderson, P. B., Jr., "A Theory of Data Systems for Economic Decisions," Ph.D. Thesis, *Massachusetts Institute of Technology*, June 1960, 22 p.

Herman, D. J. and F. C. Ihrer, "The Use of a Computer to Evaluate Computers," *Proc. AFIPS, 1964 SJCC*, Vol. 25, p. 383-395.

Homer, E. D., "A Generalized Model for Analyzing Management Information Systems," *Mgmt. Sci.*, 8, 4, July 1962, p. 500 ff.

Hutchinson, G. K., "A Computer Center Simulation Project," *Communication ACM*, 8, 9, September 1965, p. 559-568.

Hutchinson, G. K., and J. Maguire, "Computer Systems Design and Analysis Through Simulation," *Proc., AFIPS 1965 FJCC*, Pt I, 161-167.

IBM, "Study Organization Plan Documentation Techniques, C20-8075-0, 1961, 60 pages.

IBM, "SHARE 7090 9PAC, Part I: Introduction and General Principals," 7090 Programming Systems, Systems Reference Library, File 7090-28, Form J28-6166-1, 1961, 32 pages.

IBM, "Generalized Information System Application Description," *IBM Application Program*, 1965, E20-0149-0.

Informatics, Inc., "Auxiliary Programming for Intips System," *Informatics Inc.*, Bethesda, Maryland, October 1966., Tech. Rpt. RADC-TR-66-550, 226 p.

Iverson, K. E., *A Programming Language*, John Wiley & Sons, New York, London, 1962, 286 p.

Iverson, K. E., Programming Notation in Systems Design, *IBM Sys. J.*, 2, 1963, p. 117-128.

Johnson, L. R., "On Operand Structure, Representation, Storage, and Search," *IBM*, Yorkton Hts., New York, December 1961, 63 p. and appendices.

Jones, R. T., "Basic Commercial Data Processing Functions Generalized Approach to Analysis, Programming and Implementation," *IBM Systems Research Institute*, New York, California, 1964, 38 p. and appendices.

Kleinrock, L., "Time-Shared Systems: A Theoretical Treatment," *J. ACM* 14, 2, April 1967, p. 242-261.

Kriebel, C. H., "A Resume of Mathematical Research on Information Systems," Graduate School of Ind. Administration, *Carnegie Inst. of Technology*, AD 616 113, February 1965, 13. p.

Kriebel, C. H., "Operations Research in the Design of Management Information Systems," *Carnegie Institute of Technology*, April 1966, 18 p. and bibliography.

Kozmetsky, G. and P. Kircher, *Electronic Computers and Management Control*,

McGraw-Hill Book Co., New York, 1956, p. 275-289.

Laden, H. N. and R. R. Gildersleeve, *Systems Design for Computer Application*, Wiley, 1963.

Langefors, B., "Some Approaches to the Theory of Information Systems," *BIT* 3, 1963, p. 229-254.

Langefors, B., "Information Systems Design Computations Using Generalized Matrix Algebra," *BIT*, 5, 1965, p. 96-121.

Lieberman, I. J., "A Mathematical Model for Integrated Business Systems," *Mgmt. Sci.*, 2, 4, July 1956, p. 327-336.

Lombardi, Lionello, "Theory of Files," *Proc. Eastern Joint Comput.* Conf. 1960, p. 137-141.

Lombardi, Lionello A., "A General Business-Oriented Language based on Decision Expressions," *Communications ACM*, 7, 2, Feb. 1964, p. 104-111.

Lynch, W. C., "Description of a High Capacity, Fast Turnaround University Computing Center," *Communications ACM*, 9, 2, Feb. 1966, p. 117-123.

Markel, G. A., "Towards a General Methodology for Systems Evaluation," AD 619-373, *HRB Singer, State College*, Pa., July 1965, 73 p.

McGee, R. C. and H. Tellier, "A Re-evaluation on Generalization," *Datamation*, July-August, 1960, p. 25-29.

McGee, W. C., "The Formulation of Data Processing Problems for Computers," *Advances in Computers*, F. L. Alt. ed., 4, 1964, p. 1-52.

Miller, J. C., and R. M. Gordon, "Management Information Systems: Lifeblood of Management," *Ideas for Management*, Systems & Proc. Assoc., 1962, p. 386-414.

Montalbano, M. S., "Tables, Flow Charts, and Program Logic," *IBM Systems Journal*, Sept. 1962, p. 51-63.

Montillon, G. D., "A Computer User-Oriented System," *Communications ACM*, 8, 2, Feb. 1965, p. 117-124.

Moravec, A. F., "Designing the Fundamental Information System: a Guide to Productive Data Processing," *Institute of Management Sciences*, Febr. 16-19, 1966, Dallas, Texas, 26 p.

Myers, D. H., "A Time-Grid Technique for the Design of Information Systems," *IBM Systems Research Institute*, New York, New York, 1962, 50 p.

National Cash Register, BEST General Information Manual, Dayton, Ohio.

Nee, D., "Application of Queuing Theory to Information Systems Design," ESD-TDR-64-428, *Air Force Systems Command*, Bedford, Mass., June 1964, 105 p.

Nelson, D. B., R. A. Pick, and K. B. Andrews, "GIM-I, a Generalized Information Management Language and Computer System," *Proc. AFIPS, 1967 SJCC,* Vol. 30, p. 169-173.

Nielsen, N., "An Analysis of General Purpose Computer Time-Sharing Systems," Ph.D. dissertation, *Stanford University*, 1966, 374 p.

Pollack, S. L., "DETAB-X: An Improved Business-Oriented Computer Language," *The Rand Corporation*, Memo RM-3273-PR, August 1962, 18 p.

Procter & Gamble, Co., *Basic Functions Manual*, Cincinnati, Ohio, March 1967.

Rosin, R. F., "Determining a Computing Center Environment," *Communications of the ACM*, 8, 7, July, 1965, p. 463-488.

Salton, G., "Manipulation of Trees in Information Retrieval," *Communication ACM*, 5, 2, Feb. 1962, p. 103-114.

Shaw, C. J., "Theory, Practice, and Trend in Business Programming," AD625-003, *Systems Development Corporation*, July 1965, 18 p.

Spiegal, J. J. K. Summers, and E. M. Bennet, "Aesop: A General Purpose Approach to Real-Time, Direct Access Management Information Systems," *The Mitre Corporation*, Bedford, Mass., June 1966, 31 p.

Spitzer, J. F., J. Robertson, and D. H. Neuse, "The COLINGO system Design Philosophy," *Proc. 2nd Congress Info. Sys. Sci.*, 1965, p. 33-47.

Statland, N., R. Proctor, and J. Zelznick, "An Approach to Computer Installation Performance Effectiveness Evaluation," AD 617 613 *Air Force Systems Command*, Bedford, Mass., June 1965, 155 p.

Steil, G. P., "File Management on a Small Computer, the CIO System," *Proc. AFIPS,*

1967 SJCC, Vol. 30, p. 199–212.

Steinberg, A. W., "Some notes on the similarity of three management science models and their analysis by connectivity matrix techniques," *Mgmt. Sci.*, 9, 1963, p. 341–343.

Taylor, A., J. R. Hillegass, N. Statland, "Quantitative Methods for Information Processing Systems Evaluation," AD 435 557, *Auerbach Corp.*, ESP-TDR-64-194, January 1964.

Teichroew, D., "Information Systems in Information Display," *Information Display*, 2, 6, November/December 1965, p. 33–39.

Thiess, H. E., "Mathematical Programming Techniques for Optimal Computer Use," *Proc. ACM 20th Ntl. Conf.* 1965, p. 501–512.

Turnburke, V. P. Jr., "Sequential Data Processing Design," *IBM Systems Journal*, March 1963.

Ver Hoef, E. W., "Design of a Multi-Level File Management System," *Proc. ACM National Conference*, 1966, p. 75–86.

Warshall, Stephen, "An Algebraic Language for Flow Charts," *Armed Services Technical Information Agency*, Arlington Hall Station, Arlington 12, Va., Technical Operations Incorporated, Burlington, Mass., Report no. TO-B-20-33, August 1960, 34 p.

Williams, W. D., and P. R. Bartrum, "Compose/Produce: A User-Oriented Report Generator for a Time-Shared System," *Proc. AFISPS, 1967 SJCC*, Vol. 30, pp. 635–640.

Youchak, M. I., D. D. Rudie, and E. J. Johnson, "The Data Processing System Simulator (DPSS)," *Proc. AFIPS, 1964 FJCC*, Vol. 26, ptl. pp. 251–276.

Young, J. W., Jr., "Non-Procedural Language: A Tutorial," 7th *Annual Tech. Meeting, South Calif. Chapter ACM.*, March 23, 1965, 24 p.

Young, J. W. and H. Kent, "Abstract Formulation of Data Processing Problems, *J. of Ind. Engr.*, Nov.-Dec. 1958. p. 471–479. See Also *Ideas for Management*, Internat. Systems-Procedures Assoc., 1959.

Ziegler, J. R., "A Modular Approach to Business EDP Problem Solving," *Proc. ACM 20th Ntl. Conf.*, 1965, p. 476–484.

BEYOND PROGRAMMING: PRACTICAL STEPS TOWARDS THE AUTOMATION OF DP SYSTEM CREATION

J. RHODES

Commercial DP is under fire. Users are becoming increasingly dissatisfied with the service which their DP departments are producing. Manpower costs are rising, but productivity is often static or actually falling; lead times are lengthening; quality has not improved dramatically over the last five years. Yet there is normally very little that the DP department can do. Short-term pressures are overwhelming. The best resources are fighting fires; there is no time for methods improvement. Maintenance absorbs more and more effort. The manufacturer always has a reason for recommending a technical change, and each change has a conversion overhead. In most cases,

user complaints can be narrowed down to three specifics: cost, lead time, and reliability. And, in most cases, the DP department has little opportunity for improvement because of the impact of change.

Many systems are built on the erroneous assumption that they are basically static. In fact, most systems are highly dynamic: even during their initial development they are subject to continual modification, and once installed they are constantly being tuned, improved, or converted. One typical installation with 80 systems of various types has undertaken 3,000 significant modifications in five years, an average of eight per system per year. If, as is frequently the case, 50% of an installation's manpower is tied up in maintenance, then the original cost of a project is doubled over the life of the system.

How can the basic effectiveness of the DP department be improved without major changes? Solutions which require radical revisions of current practice nearly always fail, because of inertia, the Not-Invented-Here syndrome, lack of training, or often simply because of lack of time to get new ideas properly planned and introduced. It might take

---

SOURCE: Rhodes, J., "Beyond Programming: Practical Steps Towards the Automation of D.P. System Creation," Copyright J. Rhodes (Sept. 1972).

several years for the methods used in analysis to be revised. But although the major activities—analysis, design, programming, etc.—are difficult or risky to change, the interfaces between the activities are not. And since the interfaces are often the points at which most bad practice occurs, they are correspondingly valuable points at which to introduce improvements.

Three kinds of improvement are possible at the interface points: improved quality control, better documentation, and better presentation of information for the next activity. At the critical interface point between analysts and programmers, within the loosely defined activity of computer system design, there is ample opportunity for all three forms of improvement. In this instance, improved quality control requires an ability to check the output of the design activity for completeness and efficiency, within the overall design constraints of the system; better documentation involves an exact expression of the required computer system components; and better presentation of information demands the organization of the system description material in a form directly transferrable into source language form. And, bearing in mind the dynamic nature of systems discussed earlier, the documentation on the interface must be amenable to extensive modification.

It is important that any methods improvement program should result in a coherent set of techniques and facilities, rather than a disjointed set of tools. One such coherent technique, designed to meet the priorities and principles described, is based upon a set of specification matrices. These matrices describe the key relationships which exist with a computer system. Figure 1 describes the basic elements within a computer system. Programs are sets of processing operations, each of which completes a major activity—raw data editing, for example. Files are sets of data, each of which holds data in a particular state—sorted on a particular key, for example. Programs can be further defined as sets of actions to be taken, dependent upon sets of conditions. Files consist of records, which contain particular sets of data—the data for one particular customer, for example.

Thus, the system may be described in terms of programs and files, the programs in terms of conditions and actions, and the files in terms of records and data

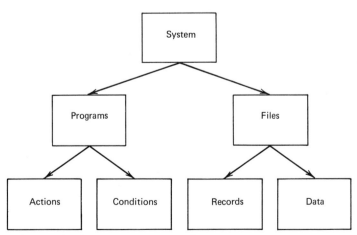

FIGURE 1 System components.

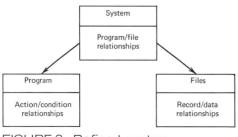

FIGURE 2 Refined system components.

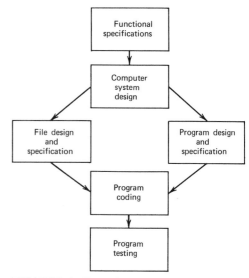

FIGURE 4 Development stages.

elements (see Figure 2). These sets of relationships can most conveniently be described as two-dimensional matrices. Figure 3 shows such a matrix.

This technique results in a clear set of documentation, which spans the stages of computer system design, and program and file specification. These stages and their conventional output are described in Figure 4. In a conventional environment, it would now become the responsibility of the programmer to code the appropriate programs. This process counts for much of the cost and some of the opportunity for error and misunderstanding in the system development process. The automatic translation of the specification into COBOL code obviously offers considerable advantages, in terms of time, cost, accuracy, and standardization. A "preprocessor" is a program which generates another source-level program, which is then ready for conventional compilation. Figure 5 illustrates this process. The use of pre-processors to automatically translate system specification matrices into COBOL allows program elements to be built up automatically and then consolidated into programs. Figure 6 shows the revised stages of development as described in Figure 4.

Putting these concepts to the test in a practical example, Figure 7 shows the flow of a simple commercial application. Program EDIT reads data cards containing details of sales. These cards are validated, correct input being placed on the sequential file TRANS, and errors listed on report ERRORS. These transactions will be manually corrected and re-input via file AMENDS. Program UPDATE updates direct access master file MASTER, which holds all transactions aggregated by point of sale. A summary report lists the total number of updates made and the transactions for which there was no corresponding point of sale record on the MASTER: these will be resubmitted via AMENDS at a subsequent date. Finally, UPDATE puts information relating to successful updatings of MASTER onto output file

	Files			
Programs	A	B	C	D
W	✓			
X	✓	✓		
Y	✓			✓
Z				✓

FIGURE 3 Matrix defining a system.

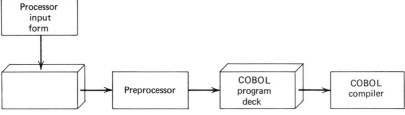

FIGURE 5 Use of a preprocessor.

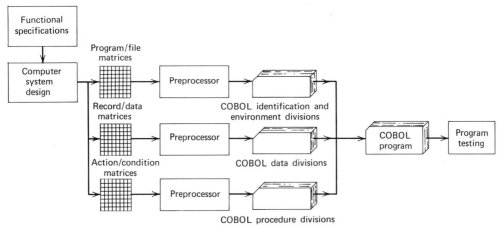

FIGURE 6 Development stages using automatic specification programming.

UPDRECS, which is formatted and output by program PRINT.

Figure 8 shows the matrix representation of this system. In the table, the symbol P indicates a program; the symbols I, O, and U indicate input, output, and updated (both input and out) files, respectively. This gives virtually the same information as the paragraph above, or the chart in Figure 7, and by showing the program/file relationships in the system, completely defines the system flow.

Two additional matrices are required to describe the files: one to state the keys by which the files may be accessed and one to state which record types exist in which file. These matrices are not shown for this example. They would also contain such information as file organization (e.g., index-sequential). Taken together, the matrices contain sufficient information for the programmer to write the identification and environment divisions for all COBOL programs in the system, together with the FD's with record layout COPY statements in the data division. These divisions represent the "envelope" within which the programmer must write his working storage requirements. This envelope can be generated automatically by a COBOL generating processor directly from the matrices. Thus the system documentation is directly input to the programming function and completed automatically without need for manual intervention.

The next major component of the system is the record specifications. In the example, which is typical of many commercial applications, there is, fundamentally, only one set of information. Although it is subjected to a number of

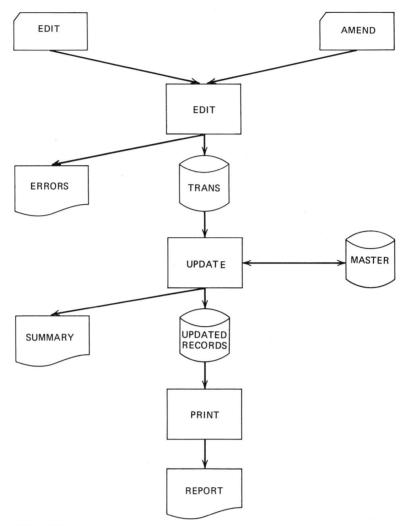

FIGURE 7 The example system.

transformations, the basic information about a sale is represented in every file. Figure 9 illustrates the data content of the "basic record."

To this basic information is added other information particular to a file: for example, file ERRORS would need a diagnostic indicating which edit check had been filed; file MASTER would omit the date of sale but would have many sets of item volumes and prices. It is fundamental to the success of any commercial system that the basic record specifications are standardized. An increasing number of installations, even those which prefer to give programmers the widest creative latitude, are turning to centrally-controlled COPY libraries. In these circumstances, it is highly desirable to generate the COBOL record descriptions directly from the system specifications. It is also highly desirable

P			EDIT	
	P		UPDATE	
		P	PRINT	
I			CARDS	
I			AMENDS	
Ø			ERRORS	
Ø	I		TRANS	
	U		MASTER	
	Ø	I	UPRECS	
	Ø	Ø	SUMMARY	
		Ø	REPORT	

FIGURE 8 Program/file specification.

1				TRANS	−A
	1			MISMATCH	−B
		1		UPRECS	−C
			1	MASTER	−D
2	2	2	2	POINT-OF-SALE	X(20)
2	2			DATE-OF-SALE	9(6)
		2	2	DETAILS OCCURS 20	
2	2	3	3	ITEM-SOLD	X(4)
2	2	3	3	VOLUME	9(5)
	2	3	3	PRICE-PER-UNIT	9(5)
	2			ERROR-INDIC	X

FIGURE 10 The record/data specification in program "update."

to have a single document showing all the records in the system, and the data descriptions which are common to more than one record. Figure 10 shows such a document, again arranged as a matrix.

Figure 10 shows, for simplicity, just the data items used by the UPDATE program. The numbers on the left are arranged in columns. Each column represents one record. The numbers themselves indicate the levels at which the items occur. Record TRANS contains only the basic information from the input cards. This information is repeated in MISMATCH, which also shows the price-per-unit information obtained from the master file and indicator to show the cause of the mismatch. UPRECS shows the MASTER records which have been updated. These two files therefore have

POINT-OF-SALE
DATE-OF-SALE
ITEM-SOLD
VOLUME
PRICE-PER-UNIT

FIGURE 9 Basic record.

identical record descriptions. Unlike the two previous records, they do not contain DATE OF SALE; however, they have a number of occurrences of ITEM SOLD, VOLUME, and PRICE PER UNIT.

As in the case of the previous matrices, this table can be input directly to a processor which generates record descriptions, which can be held on a library and copied down into the FD's of the programs by the previous processor described. One technicality remains. It is necessary the data names be unique in COBOL: this can be achieved by using the processor to automatically suffix each name in each record as the characters A, B, C, or D as shown on the extreme right of Figure 10. Figure 11 shows one of the record descriptions which would be generated from the matrix.

The remaining element is the procedure itself. Once again, the objective is to describe system requirements in a form of documentation which can be

```
01  UPRECS-A.
  02  POINT-OF-SALE-A PICTURE
      X(20).
  02  DETAILS-A OCCURS 20.
      03  ITEM-SOLD-A PICTURE X(4).
      03  VOLUME-A PICTURE 9(5).
      03  PRICE-PER-UNIT-A 9(5)
```

FIGURE 11 Generated COBOL record.

immediately transformed into a COBOL source code. Again, the matrix shape is preferred. Decision tables have had an undistinguished history in commercial data processing; the benefits seldom worth the effort. Decision tables are ideal when the programmer or analyst is faced with a complex set of conditions: sufficient complexity seldom occurs and when it does, the tabular notation has to be painstakingly converted into procedure.

However, the tabular notation has one great advantage: it frees the writer from having to understand all the procedural relationships. He can simply state the problem and let a processor write the procedure out for him. And, in addition, a processor can both check logical consistency and optimize the resultant code. Figure 12 shows a simple part of a specification, which one might expect to encounter in the update program. The first part of the Figure states the processing required. The flowchart illustrates a procedural statement of the problem; the table illustrates the problem in tabular form.

The flowchart is a competent programmer's correct solution to the problem. However, the table illustrates the problem without the necessity for procedure. Moreover, the table is suitable for direct input to a COBOL-generating processor, whereas the flowchart is not; the table can be easily incorporated in the source program as procedural comment, and it can be understood easily by analyst, user, and, indeed, manager alike. The flowchart demands that one understand the solution to the problem; the table only that one understand the problem itself.

In the table, the analyst or programmer may simply list all the conditions he finds, then all the actions. Then using a simple "Y = Yes, N = No, X = Do It" notation, he can work through the problem stating one rule per column. The middle column, for example, represents the first sentence in the specification above. And, having thus tamed the problem, he can let a processor generate the necessary COBOL.

In summary, this approach provides a step-by-step development process which has been found to meet the objectives discussed earlier in this article. It helps to standardize and defines the system designer/programmer interface; it ensures that the system documentation always reflects the true state of the system, because the documentation is part of the code generation process; it enables the impact of systems modifications to be measured and the changes implemented easily.

The concepts described in this article have been put into practice by Hoskyns Systems Research, the New York City DP consulting organization. The specification matrices for programs and files, records and data, and actions and consultations, are input to three proprietary preprocessors as shown in Figure 6. These are known as the Program File Processor (PFP), Record Data Processor (RDP), and TABLEMASTER, respectively.

Introduced into the United States in 1972, the input to the processors is termed Hoskyns Systems Language/ 1 (HSL/1). HSL/1 is employed in a number of large IBM users, where initial results are very promising. In each case, a large percentage of COBOL

"If A equals B and B is not greater than C, make Z equal to X.
If A is not equal to B and B is not greater than C, make W equal to Z and Z equal to X.
If A is equal to B, B is greater than C and D is less than E, make U equal to V and Z equal to X."

(a)

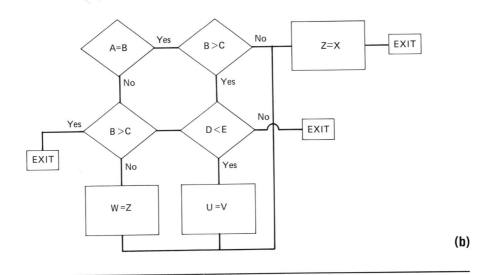

(b)

Conditions $\begin{cases} Y & Y & N & A = B \\ Y & N & N & B > C \\ Y & & & D < E \end{cases}$

Actions $\begin{cases} & & X & W = Z \\ X & & & U = V \\ X & X & X & Z = X \end{cases}$

(c)

FIGURE 12 Procedure specifications: (a) English problem statement, (b) procedural statement, (c) tabular statement.

code is being generated directly from the specifications; and a number of the benefits outlined at the beginning of this article are beginning to be realized.

If commercial DP is to satisfy its users, to build systems more inexpensively and more quickly, and to be able to respond to changes more flexibly, then increasingly the production and amendment process will be automated; and the commercial DP industry will move beyond programming.

PROBLEM STATEMENT ANALYSIS: REQUIREMENTS FOR THE PROBLEM STATEMENT ANALYZER (PSA)

D. TEICHROEW

1.1 MAJOR FUNCTIONS

This section describes the structure of the Problem Statement Analyzer in terms of seven major functions. It is conjectured that these must be carried out in the order listed since the output of any one is required before the functions following it can be done. Each of these will be described and implemented as separate modules. PSA will eventually be used to determine what logical precedence relationships exist among these functions.

1. Processing control
2. Lexical syntactic and semantic analysis
3. Data definition analysis
4. Network (static) analysis
5. Dynamic (timing) analysis
6. Volume analysis
7. Analysis of additional requirements

The discussion in this section covers the following for each of the seven functions:

1. Purpose and general description
2. The information that is needed in the analysis and consequently must be provided in the Problem Statement
3. The algorithms or processing performed
4. Files produced for use for the function (and for later functions)
5. The outputs produced for each of the five groups of users of output from PSA. Emphasis here will be placed on the contents of these outputs and particularly on their value to the users.
6. Questions remaining to be answered before the function can be completely specified

As with any information processing system, it is the output requirements which dictate the functions to be incorporated in the system. The output is of two major types.

SOURCE: Teichroew, D., Problem Statement Analysis: Requirements for the Problem Statement Analyzer (PSA), ISDOS Working Paper No. 43, Dept. of Industrial Engineering, University of Michigan, Ann Arbor, 1971, pp. 20–53.

First, a tabular, coded representation of the problem statement is required for subsequent analysis by other parts of the ISDOS system. This information will be coded if possible, for ease of manipulation and saving in storage space.

The second output from the PSA — and at this stage of the implementation of ISDOS it is the most important one — is the set of analyses, validity checks, etc., to be prepared for the problem definer, data and function administrator, and physical systems designers. This output is designed to provide both analysis of the problem statement, containing information that would not normally be easily available, and restatements of critical sections of the problem statement, for use as a check. Analyses should be produced for each Problem Statement Unit (PSU) and consolidated analysis produced for the whole problem statement or any part designated by the Problem Definer.

The structure of the Problem Statement Analyzer is shown in Table 1.1. This table summarizes in one column for each function the information in topics 2, 3, 4, and 5.

1.2 PROCESSING CONTROL

PSA, as in any other IPS, will accept inputs and produce outputs. Most IPS's today operate under an operating system directed by statements in a job control language. In the case of PSA, the conception of an operating system will be generalized to a set of hierarchical levels each with its own set of PSU's which are defined in PSL by Problem Definition Management.

Considerations in developing the processing control include the following:

1. The inputs to the system (PSA) should be accepted in a form as convenient to the user as possible. There must be methods to accept the input in various modes on various devices. These should be utility routines that perform simple checks on the input so that trivial errors do not completely invalidate large runs.
2. The users should be able to specify in a convenient manner how much, and what, analysis they desire. They should not be deluged with unwanted output.
3. The processing procedures must be organized with a view to efficient use of the computer. The cost of PSA would be prohibitive if every little change in a Problem Statement required a complete rerun of the whole problem.

1.3 LEXICAL, SYNTACTIC, AND SEMANTIC ANALYSIS

1.3.1 Purpose and General Description

The syntactic and semantic analysis function will convert selected parts of the source PSL Problem Statement into some coded machine-readable form. There are three basic steps involved in performing this function: (1) token conversion, (2) syntax analysis, (3) semantic analysis.

Token* conversion, sometimes called lexical analysis, is the process of translating the source PSL characters into some internal coded machine representation which are then stored for any future reference. This conversion is carried out for savings in file space. All reserved words, simple names, and numerics will be recorded in a symbol table; and any further reference to them will be in terms of their individual table index

---

*PSL tokens are: (1) Simple Names, (2) Qualified Names, (3) Reserved Words, (4) Quoted Literal, (5) Numeric symbols, (6) Specified Symbols.

TABLE 1.1 Summary of Problem Statement Analyzer

Function	Processing control	Lexical, syntactic, semantics analysis	Data definition
2. Additional data needed from Problem Definition beyond what was needed for the previous functions			
Identification		Problem Statement in PSL	
Requirements	Requirements for processing control specified by Problem Definition Management		
Define			All statements
Compute			All statements
3. Algorithms used or processing performed	1. Selection of requested analysis and output 2. Updating	1. Conversion of character strings into tokens 2. Syntax analysis 3. Semantic analysis	Construction of Data Structure matrix
4. Files produced for later stages		1. Coded source file of updated problem statement 2. Symbol table	Data Structure matrix
5. Outputs provided for			
Problem Definition Management	As requested		
Problem Definer		Source PS with error comments PS structure	Problem Statement Directory
System Definer Data and Function Administrator Physical systems designers and constructors			

Function	Network analysis	Dynamic analysis	Volume analysis	Additional processing
2. Additional data needed from Problem Definition beyond what was needed for the previous functions				Systems Data
Identification				
Requirements		Occurrence Generate Perform	Volume statement	
Define			Replication statements	
Compute				
3. Algorithms used or processing performed	Construction of the network			
4. Files produced for later stages	Network; incidence matrix	Timing data added to network	Volume data added to network	
5. Outputs provided for				
Problem Definition Management				
Problem Definer	Network analysis reports	Timing analysis	Volume analysis reports	
System Definer			Variables appearing as volume parameter	
Data and Function Administrator				
Physical systems designers and constructors				

numbers. Other files will be built, based on efficiency considerations, in later analysis functions. Some files built at this point could be: (1) a problem structure file, (2) a synonyms reference file, (3) a COPY statement file.

Part of the processing control function (Section 1.2) can be performed in this step. That is, only those parts of the Problem Statement that the problem definer wishes analyzed need be converted. This facility will reduce the amount of output produced and reduce computation time subsequent to this function. The token conversion or lexical analysis is the first step that must always take place before any other analysis can be accomplished. Errors at this step would be illegal PSL characters, illegal formation of simple names, reserved words, etc., and illegal formation of PSL comment delimiters.

The next step in this function is a syntax analysis. There are several alternative methods available. Some of them are: (1) keyword (reserved) word-driven analysis, (2) combination keyword and table driven analysis, (3) completely table driven analysis. The methods increase in generality and flexibility, as well as difficulty of implementation, in the order in which they are listed. All methods must provide a mechanism for checking "combinations of tokens" or statements. The PSL language has defined the form or appearance of statements. Before any meaning (or semantic intent) can be attached to the statements in a PS, their form must be verified against the syntactic rules of the PSL language. The errors to be detected at this step will be ones related to improper use of the PSL syntax rules or models.

Only relatively simple semantic analysis can actually be performed in this function as compared with the semantic analysis performed in later functions.

For example, semantic errors detected at this step would be dual definition of simple names (e.g., using a simple name for both an identifier and as a PSU name), improper opening and closing of problem units (illegal nesting of SUBPROBLEMs), etc.

1.3.2 Input

The input to this function is the problem statement expressed in PSL.

1.3.3 Algorithms and Processing Performed

The lexical analysis converts source character strings and stores them in a coded file. A syntactic analysis of each statement is performed as it is being converted. The semantic analysis that can be performed at this point is done during the statement conversion and syntactic analysis. Those statements that reference the whole problem statement or another PSU (i.e., COPY, SYNONYM) should be noted in appropriate files.

1.3.4 Files Produced

A machine-readable form of the problem statement must be produced, along with the symbol table which is a directory to all simple names, reserved words, etc., in the Problem Statement.

1.3.5 Outputs Produced

The external outputs produced by this function will include all diagnostics on syntax and semantics that can be performed at this point.
1. To the problem definer
    a. The user information output will be a SOURCE LISTING of the user selected portion of problem statement with error comments. The listing will be a copy of the input, line-for-line,

without any editing* performed. Any discrepancies with PSL syntax that would cause ambiguity or error in later phases must be identified at this point. A list of difficulties that must eventually be corrected should also be produced. Error comments will immediately follow any source line errors. More than one error marker may appear against a line as a single source line may contain more than one statement and/or a statement may contain more than one error. PSA will, as much as possible, identify all errors in a statement and not just give up after encountering the first error.

  b. If the input is a modification or addition to an existing PS, the complete PS with diagnostics will be produced. (A convenient, economical method for doing this must be developed as part of the Processing Control function.)
2. To the problem definition manager (PDM).
  Probably no output should be produced automatically, but the PDM should at any time be able to call for a copy of the current status of all the problem statements under his jurisdiction.
3. To the data administrator (DA). Same as to PDM.
4. To the Physical Systems Designer and Constructor.
  By the time the tokens are coded, much of the work of translating into object code is already done.

Therefore, if the hardware on which the problem will be executed is known, it *may* be possible and desirable to carry out the translation and compute (i) the amount of memory space required to store the PROCESSES object code and (ii) the amount of computing time required to execute the PROCESS.

1.4 DATA DEFINITION ANALYSIS

1.4.1 Purpose and General Description

Introduction. The definition and analysis of data is the fundamental problem in stating requirements for information processing systems. Consequently it is necessary to discuss the process by which this is done and the problems that have to be solved, in considerable detail at this point.

The overall problem definition procedure begins with a recognition of a situation which requires outputs from an IPS. This situation may arise because one or more managers require some information which they are not now getting, or it may arise because of a desire to implement a business model. For example, a production department may have developed a new production control procedure; this procedure may require a number of new outputs from the IPS. The situation, however it arises, in turn may require a number of new inputs. The need for a new IPS or a modification to an existing one may also arise from the need to act on new inputs.

However the need to produce outputs or to accept inputs arises, the PD will normally begin with a list of one or more outputs that the system is to produce. The PD starts by stating things about each output as a unit: in particular, the

---

*It may be desirable to perform some editing to make the listing easier to read, without forcing the problem definer to do the necessary spacing on input. This will also be necessary when detailed information (e.g., PICTURES) is required.

conditions under which it is to be produced and its structure in terms of subgroups of data. The PD must then describe the structure of each of the subgroups in terms of other groups until all the subgroups are exploded and consist only of single-valued data elements. For each output data element the problem definer must also give a way of determining the value.

PSL and PSA Objectives in Data Definition. PSL and PSA have been designed with the following objectives in mind:

  i. To simplify the work of the problem definer as much as possible by giving him definitional and clerical aids in PSL, and computer aids in PSA. For example, the local synonym statement in PSL makes it unnecessary to use long simple names in a PSU. A short unique abbreviation is used to reduce the writing effort.
  ii. To facilitate the problem of identifying particular cases when more than one instance can occur, both to help problem definers and to help SODA.
  iii. To help the problem definer make a correct statement of his data requirements.
  iv. To help the data and function administrator in his task of ensuring consistent data definition and correct use of data elements. To achieve these objectives a number of features have been incorporated in PSL and PSA. Table 1.4 shows these features and how they contribute to the objective.

Identification of Data Elements. In PSL terminology a DATA ELEMENT is the smallest unit of data which has a unique name and a value; e.g., a DATA ELEMENT might be QTY which may have any value from 000 to 999. ("Smallest unit" means that the PD cannot refer to a part of the element value—say, to only the last two digits of QTY.)

Two cases can arise for any data element:

  i. The data element has a single value in the system at any one time; i.e., the data element named PI has only one value π, i.e., 3.14. . . .
  ii. The data element has more than one value in the system at one time, i.e., QTY, PART_NUMBER.

In the information processing systems that are of interest, most data elements will have more than one value; therefore case ii will be treated as the normal case.

PSL permits (and PSA must analyze) three methods for locating particular instances of data that have more than one value:

- Grouping
- Qualification
- Identification

Each of these is described below.

GROUPING. The PD in describing his requirements does so in terms of names, e.g., DATA ELEMENT NAMES. He may group DATA ELEMENT NAMES and give the group a name. Such a name is referred to in PSL as a GROUP NAME. It is important to clearly understand the distinction among: GROUP NAME, which is the name or tag of a group of DATA ELEMENTS (or DATA GROUPS); ELEMENT NAME, which is the name or tag of a data element; and VALUE, which is the magnitude or state of a particular data element.

The grouping described above, and available in PSL through the Group Definition statement, is a grouping of data names (either element or group names); and at each node there may be

TABLE 1.4 How Features of PSL Contribute to Objectives

Features of PSL	Features			
	Simplify work of problem definer	Identify particular cases	Help in correct problem statement	Help data and function administrator
Replication clause	Realistic situations can be described simply			Gives maximum size of data sets
"Value are" clause	Helps PD state validation criteria		Two data elements with same values may be two names for the same element	
Synonyms	Reduce writing/ punching effort, make corrections for simple spelling errors		Corrects two, problem definers use of different name for same value	Can make two different names refer to the same value
Qualification	Reduces writing/ punching effort	Indicates sequence of identifiers		
Identification		Provides a way of identifying particular case desired		Points out the important keys in the Data Base
COPY	Reduces writing effort			Enables the use of certain data structures (templates) in many places
Other attributes of data	Indicates attributes which must be specified			

replications or multiple instances (e.g., PART_NUMBER). Consequently there must be some way to refer to an individual instance. The PSL model for this situation is as follows:

1. The PD may, in each group replication, state a variable which uniquely identifies a particular instance; e.g., in a customer order, line items may be replicated and identified by line item numbers.
2. If the PD does not designate an identifier, PSA will automatically add an identifier name. This serial number will be given a unique name obtained by adding some suffix to the node name which is assumed to be always present. The created identifier must then be checked (in the symbol table) for uniqueness. Other suffixes may be necessary to obtain a unique identifier name.

QUALIFICATION. In assigning names to data elements the objective is to use basic names as much as possible and distinguish among different usages by qualification. For example, QTY may represent quantity; and the different cases should be qualified, e.g.,

ORDER_QTY

ON_HAND_QTY

This applies also to time qualification, e.g.,

GROSS_PAY_TO_DATE

EMPLOYEE_GROSS_PAY_THIS PERIOD

PSA will include in the Problem Statement Directory a list of qualifiers for each data name. It will also provide a list of variables qualified by a particular variable.

IDENTIFICATION. In addition to the grouping described above there is another kind of grouping. For example, the problem definer may be interested in all valid values of the data element PART_NUMBER. This could be represented as a tree.

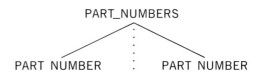

The set of all PART_NUMBER is called PART_NUMBERS and described in PSL by

PART_NUMBERS IS
(PART_NUMBER REPLICATED NOPN_TIMES)

PART_NUMBERS IDENTIFIED BY PART_NUMBER,

For each Data Set the problem definer specifies α and ρ Value Sets.* A data set can have only one identifier; if a different one is desired, a new DS must be defined; e.g., the Data Set "Customers_Orders" may be identified by "Order_Number." To distinguish those customer orders not yet processed requires another identifier. Hereafter the term identifier means any Boolean (logical) and arithmetic function of elementary identifiers; i.e., it is a combination of keys that uniquely identify a data set from the data base. The following apply to identifiers in PSL and PSA:

1. An identifier may identify more than one data set.
2. The identifier relationship is not transitive; i.e., if A identifies B, B may or may not identify A because if a value of B is given (say 5), there are maybe two instances of A that both have the B value of

---

*The α Value Set contains values actually in the system. The ρ (for potential) Value Set contains all values which have not been excluded by statements made by the Problem Definer.

5. The inverse relation, of course, does divide A into subsets.
3. All the definitions imply an instant of time, and special identifiers are required to deal with time. (This could be eliminated by stating that each identifier must include one that involves time.)
4. If the relationship between two identifiers is isomorphic, either could be chosen as the identifier.
5. For each identifier, the PD will have to indicate the α and ρ Value Sets and also a mechanism for changing the α Value Set during execution.

Classification of Data Elements by Type. A PD may wish to specify that a data element belongs to a class which has certain properties. For example, in a real time system the PD may be concerned with all the variables which represent states of the system. He could define S to be the state and have it identified by STATE NUMBER. However, this may not be as convenient as using a mnemonic name. A general facility for this will be provided in PSL and PSA and will be used, in particular, for time variables. All variables which deal with time will be so designated. This will mean that any computational statement in which it appears can be checked for dimensional consistency.

1.4.2 Input

The inputs to this function are all the Statements in the Structure and Computation Section in the set of PSU's to be analyzed. All the data in the structure statements are required including data PICTURES.

1.4.3 Algorithms and Processing Operations

The purpose of this function is to apply semantic interpretation to the syntactically correct statements made in the structure/computation section of a PSU. This analysis must be done for each PSU. First, a consistency check is made of the definitions to assure that no loops or circular definitions occur in any Principal Data Set. This is accomplished, at least conceptually, by constructing a Data Structure Matrix, DSM.

The Data Structure Matrix is the vehicle for recording information necessary for consistency check as well as being a tabular representation of the definitions. The use of matrix techniques allows for calculation of data level numbers,* determining uniqueness of identifiers, and relationships among and between identifiers.

The Data Structure Matrix should be available for use by the problem definer for error correction if he so desires. If a consistent DSM is available, it should also be possible to generate a COBOL Data Division Structure for the user. This reformatted output of the Principal Data Set might also shed additional light on definition problems. If document format and pictures have been provided, sufficient information is available to construct the graphic representation of input/output documents.

Problem Statement Directory. The first step in generating a data structure matrix is to construct a directory of user defined simple names. The directory consists of an entry for each simple name that has been defined in the problem being analyzed. For each entry the dictionary gives the following:

1. Entry number—entries are numbered consecutively.
2. User-defined name (in alphabetical order). Qualifiers are included if used.

*Data elements are level zero. Data Sets containing only data items are level one, etc.

Data set at level	Data sets at level	0		1	2	3	4	5
		Identifiers	Data elements					
0	Identifiers							
	Data elements							
1								
2								
3								

CODE: I: Data Set J is an identifier for Data Set I
 1: Data Set J appears once and only once in Data Set I
 N: Data Set J appears n times in Data Set I
 V: Data Set J appears a variable number of times in Data Set I
 F: Data Set J appears as a factor in calculating of Data Set I
 C: Data Set J is calculated by a function (appears in diagonal only)
 H: Data Set J is in history (appears in Diagonal only)

FIGURE 1 Structure matrix showing how data sets J whose names appear as column headings are related to data sets whose names appear as row headings (I).

3. Type — possible values are:
 Problem name
 Subproblem name
 PSU name
 PDS name
 Group name
 Element name
 Undefined name
 Synonym
4. Usage — possible values are:
 Input
 Output
 History
 Inquiry
 Policy
 Calculated — entry appears in LHS of a function statement
 Factor — entry appears in RHS of a function statement
 None — for PDS, Problem, Sub-problem name
5. PSU number
6. Line number. In "Source Listing" Report
7. Miscellaneous

Following each entry may appear one or more diagnostics:

- Data output but not input or calculated or accessed.
- Stored data not updated or accessed
- Data input but not output or factor
- Invalid type-usage combination

The Data Structure Matrix. The data structure matrix has the form shown in Figure 1. It will be obtained by sorting the data names alphabetically within level and inserting the additional data from the PS in the correct place.* Data sets at level 0 (e.g., data elements) are divided into two groups, those that are

*An entry in DSM (i,j) will be made if data set j is subordinate to data set i.

identifiers and those that are not. Notes on DSM:

1. Since the data sets are grouped by ascending level number, the matrix will be blank above the diagonal.
2. If a function of more than one data set is required to identify another, the function must be named and specified.
3. The identifiers must include time (see section on *Problem Statement Directory*).

The data structure matrix will permit the problem definer to inspect visually the structure of the data sets and hence ensure that he has defined them correctly. Since a PSU has one (the Principal Data Set), all the data names in a PSU can be determined by examining the appropriate PDS name row.

The data structure matrix will be the fundamental tool used by SODA to determine the range over which computations are to be performed.

Analysis of the Identifier Section. This part of the data structure matrix could also be expanded to show relationships among sets of identifiers.

COBOL Data Division Structure. The structure of each terminal data set (i.e., a data set which is not a member of any other data set) can be presented in the same form as the COBOL Data division by assigning it a level 01, then an 02 level to its constituents, and so on.

1.4.4 Files Produced

The DSM will be one file produced by this function.

1.4.5 Output

The DSM should be available to the problem definer, and Data and Function Administrator, in an appropriate format.

1.5 NETWORK (STATIC) ANALYSIS

1.5.1 Purpose and General Description

The purpose of network analysis is to

i. Develop a description of the static relationships among the data sets and functions,
ii. Identify errors or omissions in the problem statement,
iii. Provide inputs necessary for the dynamic (timing) analysis function.

This phase of PSA attempts to construct a representation of the static relationships among the data sets and processes. In PSL, an output is identified and expressed as a Principal Data Set. The PDS is composed of data sets at level 0 (data elements) and other data sets, each of which may be composed of data sets at level 0 and other data sets. For each PDS, it is necessary to check that there exist PROCESSES (or relationships) which show how each of the data sets at level 0 in an output PDS can be derived. The data elements which can be used in these derivations are any which appear on input data sets, or have been derived, previous to the time when they are needed in the data set now under consideration.

1.5.2 Input

The input to this phase is the collection of processes identified in the Lexical Function (Section 1.3) and the Data Structure Matrix constructed in the previous function (Section 1.4). The processes (derivations) are recorded during the lexical analysis of the source Problem Statement from the actions specified by computation statements, decision tables, implied computation in reports, and data control statements.

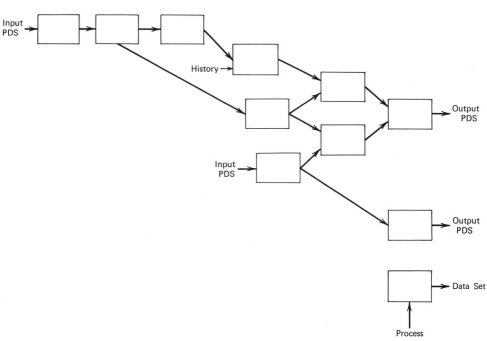

FIGURE 2 Conceptual diagram of network of problem statement.

1.5.3 Algorithms or Processing Performed

Definition of a Network. The static description of the problem statement can be represented conceptually as a (usually very large) network such as shown in Figure 2. The network consists of:

i. Nodes (each node is called PROCESS)
ii. Arcs (each arc is a DATA SET)

Each PROCESS has one output (a data set) which may be an input to one or more other PROCESSES. A PROCESS may be either

i. A function statement,
ii. A grouping of data names into a higher level DATA SET, or
iii. Assignment statements that are not named.

The network is constructed by starting with output PDS and working backwards until each input to a PROCESS is either part of an input PDS or comes from HISTORY.

Types of Processes. The purpose of an IPS is to produce outputs; outputs are expressed as PRINCIPAL DATA SETS. These outputs are triggered by:

- The occurrence of external events which require action which, in turn, may result in one or more outputs.
- Outputs prescheduled to occur at certain times or when the values of one or more data elements satisfies certain conditions.
- Combinations of the two above triggering conditions.

As a consequence of the assumptions about the type of IPS that are of interest:

- Production of output requires execution of one or more program modules.
- Data may have to be "stored." Consequently the values of stored data are a function of time.

Any output needed must be "produced." Hence a statement specifying when this production is to occur must be given for each output. This will be discussed under Dynamic Analysis (Section 1.6).

A statement of what the output contains must also be given. The problem definer specifies each data group that is not itself an element in another data group, i.e., terminal group or a PDS in an output PSU. (A terminal data set is one with blank columns in the DSM, Figure 1.)

- PD states conditions under which it is "produced." Unless otherwise stated this implies all instances are produced; e.g., if PAYCHECK is produced weekly and is identified by EMPLOYEE, then each week paychecks for all employees would be produced.
- This in turn triggers production of all data sets belonging to the PDS until data sets at level zero are reached. DATA SETS at level zero are defined by assignment statements (functions). These productions apply to the whole range of the applicable identifier.

Therefore, the PROCESSES must be specified for

A. Output data sets
B. Nonterminal data of level 1 and above
C. Data sets at level 0
D. Input data sets

Specification of PROCESSES

PRODUCTION OF TERMINAL DATA SETS (i.e., Output PSU's)
- The structure of the Terminal Data Set has been defined by the PD in the structure section of the PSU.
- The IDENTIFIERS have been given by the PD or been deduced by PSA.

PRODUCTION OF NONTERMINAL DATA SETS
- The production of nonterminal data sets is completely determined by the production of terminal data sets and the structure of the nonterminal data sets as specified in the structure section of the PSU.
- Since GROUP definitions are local to a PSU, except for PDS definitions, the definitions of GROUPS will be consistent in this analysis if they were found consistent in the PSU analysis (see Section 1.4).

PRODUCTION OF DATA SETS AT LEVEL 0
- The PD has specified for each DS what the identifiers are (or this has been deduced by PSA).
- From PROCESS of the next higher level DS has come the specification of the identifier for this data set, including time identifiers.
- Two cases can arise: the identifiers are the same or the identifiers are not the same
- The basic PROCESS has the form

DN1=ASSIGNMENT STATEMENT OF(DN2, DN3, . . . , DNN),

where (a) the assignment statement contains only

$+, -, *, /, **$ operations,

and (b) any DN which is not a DS is treated as a constant.
- Each of the DS has an identifier. Two cases arise:
  (a) The same identifiers apply to each DS. In this case (as proposed by Grindley) DN1 is to be evaluated for each value of the identifier.
  (b) The identifiers for the different DN's are not the same. Three possibilities then are:
   i. The PD made an error.
   ii. The PD intended an operation to "collapse" an identifier; e.g., HR_WORKED

(EN, DAY) was intended to be summed for all days in the week to give HR_WORKED (EN, WEEK).

iii. The PD intended to select a particular case of the DN which has the "excess" identification; e.g., TOTAL_ SALES (SALESMAN_NO, WEEK) = UNIT SALES (SALESMAN_NO, PRODUCT, WEEK) * PRICE. Here he may have meant to select a particular product and forgot to note it.

PRODUCTION OF INPUT DATA SETS

- Input DATA SETS can contain structure and computation statements. Therefore the analysis will proceed in the same way as with other DATA SETS.

Matrix Representation. The network may be presented by an incidence matrix I_{ij} where

$I_{ij} = 1$ if Data Set i is an input to PROCESS j

$= -1$ if Data Set i is an output from PROCESS j

$= 0$ otherwise

As before, it is assumed that each PROCESS has a single Data Set as an output, and the output of a PROCESS may be a Data Set at any level. The incidence matrix can be used to develop the network, and either the incidence matrix or the network can be used to detect the errors in the problem statement that are described below in Section 1.5.5.

Remarks

1. The network represents the static structure of the problem in the same way that the wiring diagram of a computer represents the static structure of the computer.
2. The analysis in this section is based on the work of Langefors and Briggs and the implementation of SODA by Nunamaker.
3. The first network analysis step will be to determine whether it can be subdivided into subnetworks. If any part of the network is completely separate (disjoint) from the rest, it can be treated separately. Since the amount of time required for analysis increases rapidly as the size of the network increases, it will be advantageous to treat each completely disjoint network separately.
4. It will also be desirable to decrease the size of the problem as much as possible by combining PROCESSES which form self-contained units. The precise definition of self-contained unit remains to be worked out, but a simple example is the following: Suppose A, B, C, D, and E have the same identifiers, and two processes in which these elements appear are:

(1) $\qquad A = B + C$;

(2) $\qquad D = A + E$;

then $D = B + C + E$

This reduces the number of nodes in the network by one, if A is not used elsewhere in the network. (Later SODA, of course, may wish to treat the original PROCESSES separately.)

1.5.4 Files Produced

The result of the analysis described in the previous section is a file which contains a coded representation of the network.

1.5.5 Outputs Produced

The basic output of this phase is the Network Analysis Report. It is a presentation of the network (Figure 2) together with additional relevant information and diagnostics. The report consists of a description of each node and its preceding nodes. Also, as part of this report or in a separate one, will be the list of succeeding nodes. The report therefore contains one entry for each node. The nodes will be in alphabetical order. The following information will be given for each node:

(1) Entry number—assigned consecutively
(2) Name
(3) Type of Node—possibilities are:
    Assignment statement
    Named function
    Output PDS
    Group DATA SET
    Data element
(4) Identifiers
(5) Miscellaneous

For each entry the subnodes that are appropriate to the type of node will be listed. For each subnode the following information will be given:

(1) Subnode number—assigned consecutively
(2) Name
(3) Type—same as Type of node
(4) Identifiers
(5) Source

For each entry, or each subnode, diagnostics will be given as appropriate:

1. Source missing—if the subnode required in the definition of the node has not been defined.
2. Inconsistent identifiers—if the identifiers of the node and subnodes are not consistent with the type of node.
3. Circuits or loops in the network.
4. Inputs not used. (If input is not included in the problem statement, and the network analysis shows that input is needed.)
5. Outputs produced that are not input to any node.
6. History Data Sets that have never been updated.

1.6 DYNAMIC (TIMING) ANALYSIS

1.6.1 Purpose and General Description

The network analysis gives a representation of the static relationship among the DATA SETS and PROCESSES defined by the problem definer. The dynamic analysis is concerned with developing a representation of the timing or dynamic relationships among the inputs and outputs. Clearly this analysis can be carried out only after the network analysis is complete, and must come before the volume analysis because volume analysis is only meaningful when it is related to specified time intervals.

The major objectives of the dynamic analysis, which are each described in more detail in the subsection shown, are:

1. To describe the dynamic relationships among inputs and outputs as specified by the problem definer and to indicate any inconsistencies and missing data.
2. To check the time consistency of PROCESSES.
3. To determine what data have to be stored over what periods of time.
4. To determine times over which target system must operate.

1.6.2 Input Needed

The dynamic analysis is based on the network prepared in the network analysis phase described in the previous section. The information from the problem statement that is needed for this analysis comes from

- the OCCURRENCE statement
- other statements such as GENERATE, PERFORM which trigger outputs.

1.6.3 Algorithms or Processing Performed

The operations necessary in dynamic analysis are described in the following subsections as outlined in Section 1.6.1.

Dynamic Relationships among Inputs and Outputs. The target IPS, in general, must produce outputs and must accept inputs. It will always have some manual override; so it will be assumed that an operator places the system in one of a number of possible states, e.g., OFF, AVAILABLE, MAINTENANCE, SUPERVISORY, etc. The AVAILABLE state is the one for which the problem definer describes the requirements, and the dynamic analysis described here is concerned only with this state.

ACCEPTING INPUTS. When an input arrives, it must be checked for "acceptance." The input must carry with it, usually in the first few symbols, a code which is recognized by the IPS. One of two situations can occur:

(i) The IPS has previously received the instructions about what is to be done with the input.
(ii) The input carries with it the instructions about what is to be done.

(This dichotomy of course does not preclude, under (i), the possibility of parameters in the input controlling or altering what is actually done. Therefore, (i) really means that the total set of what might be done under all possible cases has been previously given to the IPS. If (ii) occurs, the IPS may have to first carry out analysis, design, and construction of program modules and files of the type performed by PSA, SODA, and the rest of ISDOS for the type (i) inputs. This implies that these programs must be one of the tasks defined for the target IPS.

In either case, (i) or (ii), the processing that the IPS undertakes is the result of two types of analysis:

(i) Definition, or *what*? For example, if input A causes the output B, the structure of B and the formulas for computing the values that appear on it must be described. The definitions are static since they do not depend on when B is required.
(ii) Dynamic, or *when*? How soon after event A must B be produced?

A basic premise in the ISDOS approach is that the description of these two aspects (static and dynamic) should be separated so that one can be changed without affecting the other. Furthermore, these descriptions must be separated in order to develop automatic, optimum procedures for accomplishing the necessary processing. In general, it is feasible to separate the two types of description explicitly. One possible difficulty that arises is in the case where two descriptions are interrelated so that they cannot be separately stated; e.g., if output B is different in structure when it is produced within one second after event A than when it is produced one hour after A. It is not yet clear whether such situations can or cannot conveniently be described in PSL.

PRODUCING OUTPUTS. Outputs that the system must produce must also be described by both "what" and "when" types of information. The dynamic description of outputs is concerned with *when* outputs are produced. Three cases can arise:

  (i) The output must be produced at a fixed time measured in some specified calendar; e.g., report C is required at 6:00 p.m., March 23, 1971.
  (ii) The output must be produced regularly at a fixed time measured in some specified calendar; e.g., report D is required daily (on each working day) at 6:00 p.m.
  (iii) The output must be produced when certain situations occur. These conditions can depend on:
  - Time only; e.g., issue a notice when a part has not been requisitioned for two years.
  - The value of some data element; e.g., issue a reminder notice to a customer when his balance exceeds a given amount.
  - Some combination of time and value of a data element; e.g., issue a reminder notice to a customer when his balance has exceeded a given amount for more than a certain length of time.

ANALYSIS OF CONDITIONS. The value of a data element will not be changed in the IPS files except as the result of some input or as the result of some processing which has been carried out either to accept inputs or to produce outputs. Therefore, to properly accommodate all the conditions outlined above requires that two kinds of records be kept in the target IPS.

The first type of data is the time associated with events; e.g., the first case under (iii) above would require that the latest time associated with requisitioning each part be kept. The appropriate action could then be initiated in several different ways: a suspense file could be kept, a periodic search of the latest times could be made, etc.

The second type of data is a list of the inputs or other PSU's which change the value of a particular data element. In the second case under (iii) above it would be necessary to keep a record of all PSU's that cause a customer's balance to be changed. The appropriate action could then be initiated when a prespecified condition is met.

The recording of these types of information is consistent with the concept of "tagging" in a data base. A "tag" is attached to a data element containing various kinds of information such as who has accessed the data, the time it is to surface to trigger an event, etc. Such a tag is useful for several purposes. In Restart and Recovery procedures it is necessary to know what programs have accessed the data. Similarly, such information is necessary for security and audit-trails. It is also useful for monitoring efficiency of storage structures.

DYNAMIC ANALYSIS REPORTS. The analysis of ways in which outputs in the target system can be triggered leads to the classification of such outputs into three types.

  TYPE I.   Outputs required at points defined by time only.
  TYPE II. Outputs required as a result of inputs only.
  TYPE III. Outputs required as a result of time and input events.

The analysis carried out in this phase of PSA will result in reports for each of the three types of outputs. The reports are described in Section 1.6.5.

Time Consistency of PROCESSES. One of the identifiers of data elements is time. The various ways in which time can be used must be specified in such a way that relationships can be analyzed, e.g., between hourly and daily, to-date, etc. A monthly report produced by the second working day of each month may contain the results of all the transactions received during the preceding month. Therefore, there must be some way of stating that requirement. This is a general case of the need for being able to define for any data element that it belongs to a particular set, e.g., all parts that are green and weigh over 1000 pounds.

The timing analyzer should be able to check that only "time" data items enter into time computations. Functional statements can be analyzed for correct use of the time dimension; e.g., $A = B + C$ is incorrect if time is an identifier of A but not of B. (This is, in fact, only a special case of verifying that the dimensional units of PROCESSES must be consistent.) There are two units that have to be checked: (1) units of Data Sets involved; (2) verification of the identifiers, which involves not only the units of the identifier but also their appropriateness; e.g., summing over weeks does not give day.

The two requirements described in this section will require designation of several types of time:

(i) "Real" time which will be recorded by the system and attached to the data sets if necessary.
(ii) Time stated on documents.

Determination of Data to be Stored. The purpose of this aspect of timing analysis is to determine what data elements have to be stored. There are only two reasons why values are stored:

(i) If they are available at one time but needed at some other time.
(ii) It is easier to store than to recompute their values.*

*Some of this analysis must be delayed until later because only SODA can decide on the tradeoff of storage versus computational cost.

It is assumed that if the PD wishes values stored, which he may want to see at some time in the future, he must list them in a PSU which could be, say, a report produced every n years. In this way the cost of storing the values will be identified and the PD can determine whether having the values stored is worth it to him. The problem definer may list data elements whose values he wishes to have stored in a HISTORY PSU. This will permit the PD to refer to HISTORY as source of the value of the data element. In line with general PSA philosophy the fact that a data element is included in HISTORY will be checked and a diagnostic will be printed if PSA analysis indicates the value is not needed.

It will probably be necessary to distinguish between three types of data items:

(i) Data items that must be stored but do not change over time except when they are replaced by a value from an external transaction, e.g., marital status of an employee.
(ii) Data items that change over time because they may be modified by any appropriate transaction, e.g.,

$$QOH = QOH - QS \text{ or}$$
$$QOH = QOH + QR$$

(iii) Data items that specifically include time in their definition, e.g.,

$$GPTY = GPTY + GPTP$$

In this case, the timing analyzer should be able to verify that the timing relationships are correct and insure that the variables are automatically initialized at the right time.

Times during which the Target IPS Must Operate. If inputs to the target system arrive randomly, it is necessary to state the time period during which they will

	TYPE II Outputs	TYPE III Outputs
Inputs		
Data elements		

FIGURE 3 Dependence of TYPE II and III outputs on inputs and data elements.

arrive. In PSL these times are given by the EARLIEST and LATEST statements.

1.6.4 Files Produced

This phase will produce:

(i) Information which will be added to the network file. This information will include for each node of the network the appropriate time information as described above.
(ii) Other information necessary for the physical systems design and construction phase.

1.6.5 Outputs

The outputs produced during the dynamic analysis phase are outlined here in the order in which the analysis was described in Section 1.6.3.

Dynamic Relationships among Inputs and Outputs. The first report will be a listing of all TYPE I outputs alphabetically within time frequency required; e.g., hourly reports will be listed before daily reports. (SODA will probably cause processing cycles for each cycle represented in these requirements.)

The reports of TYPE II and III outputs will show which inputs and data elements are involved in determining when outputs are to be produced. Conceptually the information may be represented in a matrix as shown in Figure 3.

Another report will be "a calendar of events." This is a list of the identifiable events within the Problem Statement arranged in relative chronological order. Where it is possible to ascribe absolute times to an event, this will be done. (For example, it may be possible to ascribe the absolute time of "not before 5:30" to the event of arrival of input.)

Diagnostics on these reports will show inconsistencies of input and output relationships.

Time Consistency of PROCESSES. The report will also show the time requirements of execution for each PROCESS or network of PROCESSES. SODA now does this by dividing PSU's into three cases. TAG requires the PD to give priority and time, and it assigns a cycle number to each document. This does not mean that the processes must be executed together; rather, it means that they can be treated as a group for timing and volume calculations. For example, customer orders may enter the system randomly during the day, and each one must be processed within two minutes of arrival. For a multi-programmed, multi-tasked system, it can be the basis for designing priorities. For timing and volume calculation they can be considered as x orders per day.

Data to be Stored. A report will list the data items that have to be "stored," including their names, length of time they have to be stored, etc. A basic distinction would be made between temporary and permanent files in order to permit calculation of storage requirements in the volume analysis phase.

Times during which Target IPS Must Operate. This report will show the earliest and latest times specified on inputs that arrive randomly.

Diagnostics. Diagnostics, in addition to those mentioned above, will include input elements that are needed but not defined. Give time information about when they are needed.

1.71 VOLUME ANALYSIS

1.7.1 Purpose and General Description

The network analysis shows the structural relationships among data sets and processes; and the timing analysis shows the time relationships, i.e., when terminal data sets must be available. The volume analysis is concerned with another aspect of the problem statement: how many times processes must be carried out within given time periods as developed in the time analysis phase, how many characters of data must be transferred, and how much storage is required. It will develop estimates* of upper or lower bounds of information (number of characters) transferred in and out of primary storage, and of how much primary storage is needed.

1.7.2 Input

The volume analysis will require all the input in the problem statement used up to this point. In addition, the volume

---
*Probably probabilistic.

statements, detailed data on number of characters per data item (PICTURES), number of IDENTIFIER Data Elements (and Data Sets), and number of replications in structures will be required for the more detailed calculations.

1.7.3 Algorithms or Processing Performed

1. Determining how many times PROCESSES must be executed. The timing analysis phase has produced a list of PROCESSES grouped into classes in such a way that, for each PROCESS, the number of times it must be carried out within that time interval is listed. (To do this completely requires a knowledge of the storage structure so that the necessary search algorithm can be timed. This function will carry the analysis as far as it can.)

   The output of this algorithm would be
   (i) Diagnostics on inconsistency or ambiguity of identifiers of the various data sets associated with a given PROCESSES.
   (ii) The number of times each PROCESS must be executed in the chosen time interval if this can be determined (in cases where volumes or replications are stated in terms of symbolic names).

2. Volume of data transfer in and out of the machine. If PICTURES and output formats are given, this algorithm would calculate the number of characters which have to be read in and written out of the machine by multiplying the number of times the PROCESSES are carried out as computed in 1 above.

3. Number of characters of storage required. This would be computed by multiplying the number of characters required for each of the items that must be stored as determined in the timing analysis phase. This would be an upper bound since it would assume that all the identifiers for each data item would be stored with it. Therefore, it would be desirable to show how much storage is required for data and how much for identifiers.
4. Transport volume to and from auxiliary memory. The transport volume could be computed for all PROCESS on the assumption that each DATA SET must be read in or written out for the range over which the PROCESS is executed. Again, this is an upper bound since in practice some of the PROCESSES would be grouped into program modules and executed in core together and consequently the transport volume would be reduced.

1.7.4 FILES PRODUCED

Outputs: Volume Analysis Report. Volume of input and outputs are specified in the *Volume* report. These volumes should be given in terms of symbolic names to which values can be assigned by the System Definer(s). One report should give a list of variables used in volume specifiers in alphabetical order, showing, for each one, a list of PSU's where it is used. This report gives the systems definer the list of variables for which he will have to supply values.

The next report adds to the network analysis for each process the number of operations required in terms of number of symbols.

This forms the basis for the volume computations.

1.8 ADDITIONAL REQUIREMENTS

1.8.1 Purpose and General Description

The three previous functions have developed and analyzed three aspects of the problem statement:

(i) Network analysis of PROCESSES and DATA SETS has established the structural (static) relationships among them.
(ii) Timing analysis has established the relationships among the times when outputs are required and when inputs are available.
(iii) Volume analysis has established the number of times the PROCESSES must be carried out within the requirements of (i) and (ii).

The purpose of the processing procedure analysis is to combine these three aspects of the total requirements, to analyze the combination for inconsistencies, ambiguities, etc., and finally to generate alternative processing procedures as far as is possible (or worthwhile) before an explicit statement of hardware characteristics is available.

In this analysis, additional processes may have to be added because of "systems" level requirements stated by the systems definer. Therefore, a distinction will be made between PROCESSES as defined by the Problem Definer and S PROCESSES, which will be processes defined by ISDOS in order to accomplish the required processing.

These may include building new data sets, for example, for holding transactions until the processing has been completely validated; and sorting, merging, etc.

```
PROBLEM IS PSA;
    SUBPROBLEM IS PROBLEM_DEFINITION_MANAGEMENT;
    ...
    END SUBPROBLEM PROBLEM_DEFINITION_MANAGEMENT;
    SUBPROBLEM IS PROBLEM_DEFINITION;
    ...
    END SUBPROBLEM PROBLEM_DEFINITION;
    SUBPROBLEM IS SYSTEM_DEFINITION;
    ...
    END SUBPROBLEM SYSTEM_DEFINITION;
    SUBPROBLEM IS DATA_AND_FUNCTION_ADMINISTRATION;
    ...
    END SUBPROBLEM DATA_AND_FUNCTION_ADMINISTRATION;
    SUBPROBLEM IS PHYSICAL_SYSTEMS_DESIGN_AND_CONSTRUCTION_
    INTERFACE;
    ...
    END SUBPROBLEM PHYSICAL_SYSTEMS_DESIGN_AND_CONSTRUCTION_
    INTERFACE;
END PROBLEM PSA;
```
EXHIBIT 1   PSA subproblem in PSL format.

PROCESSING CONTROL — Sequence of processing, level, and amount of analysis

LEXICAL ANALYSIS, SYNTACTIC AND SEMANTIC ANALYSIS:

    LEXICAL ANALYSIS — Identification of groups of symbols which will be recognized in PSA input

    SYNTACTIC ANALYSIS — Detection and identification of violation of the syntax in the input

    SEMANTIC ANALYSIS — Detection and identification of violation in the semantics in the input

DATA DEFINITION ANALYSIS — Preparation of data and function directory

NETWORK ANALYSIS — Development of static structure of data and function definition

DYNAMIC (TIMING) ANALYSIS — Development of the dynamic structure of the problem, determining the consistency of timing requirements, and determining the data that has to be stored in the target information processing system

VOLUME ANALYSIS — Detailed analysis of amount of computation and data transfer required

ADDITIONAL REQUIREMENTS ANALYSIS — Addition of requirements that have to be added to meet systems level requirements

EXHIBIT 2   Major processing functions in PSA.

A METHODOLOGY FOR THE DESIGN AND OPTIMIZATION OF INFORMATION PROCESSING SYSTEMS

J. F. NUNAMAKER, JR.

INTRODUCTION

The design of an Information Processing System (IPS) can be divided into two major problems:

1. What are the requirements of an information system; e.g., what outputs should be produced?
2. What is the best way to produce the required outputs, on time, given the requirements developed in 1?

This paper is concerned with the second problem. However, the methodology developed to solve the second problem provides cost information that can be used by the problem definer to assist him

---

SOURCE: Nunamaker, J. F., "A Methodology for the Design and Optimization of Information Processing Systems," *AFIPS Conference Proceedings*, Vol. 38, 1971 Spring Joint Computer Conference, May, 1971, pp. 283–293.

with the determination of the requirements. One approach to determining the requirements of an information system involves comparing the information value of a report with the cost of producing the report.

Considerable time and money are expended in system design and programming when a firm acquires, leases, or in any way uses a computer. With each new system the task of getting the system operational seems to take longer and becomes more costly than the time before. A methodology for analyzing and designing Information Processing Systems is needed if we are to keep from getting further behind.

The general purpose of this paper is to continue the formalization of the process for designing Information Processing Systems and to improve it by increased application of operations research techniques and by more use of the computer itself.

SODA (Systems Optimization and Design Algorithm) is presented as a methodology for automating the system design functions. The objective of SODA is to generate a complete systems design starting from a statement of the processing requirements.

INFORMATION PROCESSING SYSTEMS

An Information Processing System is here defined as a set of personnel, hardware, software packages, computer programs, and procedures that have been assembled and structured so that the whole set accomplishes some given data processing requirements in accordance with some given performance criterion.

An important aspect of this definition is that it includes an explicit statement of the "performance criterion" by which performance of the system is measured. A consequence of including performance measures is that the emphasis is focused on the overall performance of the system rather than on any one part. The study of large scale IPS is in essence a study of the performance of the total system: hardware, software, and other procedures.

One characteristic of an IPS is that data files are stored on auxiliary memories, and it takes a number of interrelated computer programs to meet the specific requirements of the problem definer. The large number of interrelated programs distinguishes the problem that is described here from that aspect of Computer Science which is concerned with individual programs. These systems almost always depend on a large amount of data, now frequently called a data base. A duality exists between the programs and the data, and the structure of each is quite important.[1,2]

The selection of expensive hardware for a given set of requirements is frequently involved, and often the expenses increase significantly since the requirements of the system are continuously changing.

What is needed is a flexible systems design process that can accommodate changing requirements.

THE IPS DESIGN PROCESS

The IPS design process has a number of similarities to any physical design process, such as a production plant or a bridge. In each case there must be an initial recognition of a need. Next, preliminary studies are conducted in which major alternatives are considered, the technical feasibility determined, and costs of alternatives estimated. If a decision to proceed is made, the requirements must be stated in sufficient detail for designing the system. The design phase consists of preparing a set of specifications (blueprints) which are detailed enough for the construction phase.

The major functional activities and decision points in the design process of IPS are shown in Figure 1. The design process is initiated through the statement of requirements from "Problem Definers." After the requirements have been documented, the systems analysts consider the equipment available and any constraints (such as the existing system) on the design activity. The design phase consists of producing the specifications for the four major parts of the system:

- Hardware and software packages that will be used
- Programs to be written
- System Scheduler, schedule for sequencing the running of the programs
- Data Organization, specifications of file structure and how the files will be stored in hardware memories

Beneath the surface similarities between the design process of physical structures and IPS there are some differences in emphasis. Typically, more attention is given to the planning and generating of alternative designs in developing a pro-

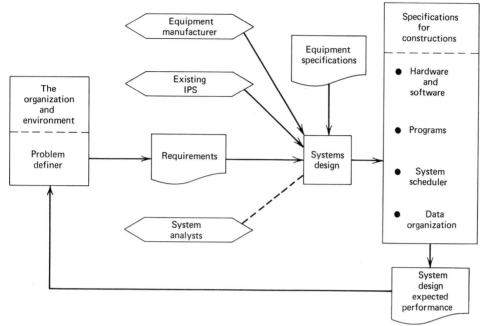

FIGURE 1 The systems design process.

duction facility than in development of an IPS. This is probably true because there are more external constraints associated with the design of a new facility, e.g., architects, contractors, equipment suppliers, and governmental zoning commissions. In the design and implementation of the Information Processing System, the requirements for the formalization of the design functions are not so apparent; hence there is a tendency to do some of the design work concurrently with construction of the system. This practice often leads to problems.

External constraints, similar to those involved in the physical design process, can be created to formalize the requirements of the Information Processing Department (IPD) through the use of information budgets.[3] The information budget will directly involve management in the operation of the IPD and force more attention to be given to the IPS design.

CURRENT PRACTICE IN SYSTEMS DESIGN

While some formal techniques have been proposed and the computer is sometimes used for calculating estimated processing time, most of the systems design is done in an *ad hoc* basis. The need for formal analysis techniques, of course, has long been recognized.[1,2,4,5]

Information Processing Department managers generally recognize a distinction between systems analysts and programmers. The systems analyst is usually responsible for systems design. In most cases he has had no formal education or training for systems design and has obtained his knowledge by experience. He uses little in the way of tools other than graphical communication devices such as flowcharts and decision tables.

Complicating the problems of inexperience and lack of training of the systems analyst is the fact that IPS problems must be subdivided to reduce the task for a systems analyst to a reasonable size, and this introduces the problem of coordination of the many systems analysts on the project. In addition it involves the coordination of problem definers, systems analysts, and programmers.

The systems design is carried out by one or more analysts who obtain the statement of requirements from those who specify what processing is to be done and what output is needed. The analysts specify programs and file design in sufficient detail so that programmers can write the programs and the files can be constructed.

There is considerable difference of opinion on how much detail of the systems design should be documented and how much other communication between systems analysts and programmers should be allowed. Despite all these problems, information processing systems are being designed.

There are, however, some major undesirable features of the present procedures for systems design. These are:

1. The performance criteria and the requirements of the IPS are not stated explicitly.
2. Programs become the only up-to-date documentation.
3. Accommodation of changes to the IPS is expensive.
4. The design process takes too long.
5. Construction of the system frequently starts before the system is completely designed in an effort to save time.
6. Few alternatives are examined in the design phase.
7. The systems do not work correctly.
8. They are costly to design.
9. Procedures become inefficient as changes in the IPS occur.

Optimization has not been completely overlooked in the design and implementation of Information Processing Systems, but any such effort has been applied to the evaluation and selection of equipment usually for a specified application. This should not be surprising since the commitment for computing system hardware represents a sizable outlay. Since a contract is to be considered, it represents a decision point in which management becomes directly involved. Because of this, care is given to the consideration of alternatives to assure the "best decision" is made—perhaps too much care, in relationship to the return that can be expected from review focused only on this decision point.

The difficulties both of time lag and problem expression summarized above are familiar to anyone who has been involved in management of a large-scale Information Processing System department. In addition to these existing difficulties, the design problem will become even more difficult in the future. The hardware is already able to accommodate more powerful software systems than are available. Moreover, attempts to develop sophisticated software systems have been very expensive.

The first three points listed above are related to the deficiencies in problem definition and documentation. Points four and five are concerned with the time required to carry out the systems design, and the sixth point relates to the lack of optimal use of resources. Points seven, eight, and nine are concerned with the realization that the systems often do not work properly. They are very costly and become inefficient over time.

It is becoming more and more apparent that the drawbacks listed in the previous section are likely to become more of a problem in the future. The design of IPS to handle more complex requirements cannot be achieved without unrealistic expenditures of effort. The number of analysts, designers, and programmers required to handle these more complex requirements are not likely to be available.

There are basically four ways to improve the situation:

1. Education to increase the number of personnel and improve their quality.
2. Improvement of manual system design tools, techniques, and procedures.
3. Use of generalized rather than tailor-made software.
4. Automation of the system design process.

Most attention has been devoted to the first three approaches. The SODA methodology concentrates on the fourth approach and rests on the premise that specifications for the IPS can be generated directly from a statement of user requirements for a limited class of processing requirements. The specifications must be detailed enough to verify feasibility and to evaluate the performance of the proposed system, but not more detailed than enough to specify construction because producing "too detailed specifications" is costly; and they may have to be changed in any case. Also, "too detailed specifications" are embedded in processing procedures which tend to bind the design unnecessarily, at too early a stage and with negative payoff.

Since the purpose of the IPS is to produce outputs, it must respond to changing inputs. One of the design problems is to decide what changes should be accommodated with what degree of ease. Certain components are easier to change than others; e.g., it is easier to change a program than to change the operation performed in the hardware. It is easier to change the data in a file than it is to change the structure of the file.

The systems design decision hierarchy consists of certain decisions which constrain later activities. For example:

1. Selecting a hardware configuration constrains everything that follows to what can be accomplished with the selected hardware.
2. Selecting the size of the main memory, number and type of input and output units, and auxiliary memory constrains still further what can be accomplished.
3. Selecting an operating system determines what processing organization is possible.
4. Selecting compilers and utility programs constrains what the programmer can do for himself. The programmer, when he writes a particular program module, is constrained by the data organization, the input and output formats, and the software and hardware on which the program is run.

The purpose of our discussion of design is to identify the design decisions that are made at any point, to enumerate the decisions that are possible, and to develop methods for:

1. Determining when (in the design process) the decisions should be made.
2. What is the optimal or suboptimal decision in a particular case.

Currently each organization tends to develop its own procedures with very little evaluation of methods developed elsewhere. IPS design can benefit by the synthesis of the available knowledge and practical know-how, and by the development of more powerful analytical methods to replace the current *ad hoc* methods.

In the course of the evaluation of computers and their use as information processing devices a variety of tools has been developed to aid the analyst in making the necessary design decisions and in facilitating the construction of these systems. Steiger[6] describes some of the steps in systems design; and it is clear that as these aids have become more sophisticated, the computer is being used more and more extensively in the design process. For example, there are commercial computer systems simulation packages[7,8,9] available for use in the evaluation of computer system performance.

However, most of the use of the computer has been in the construction phase, i.e., in the generation of computer code from source language statements. These source language statements could only be prepared once the system had been designed, i.e., once the analyst had decided what hardware would be used, how data would be organized and stored in the hardware, how the processing would be combined into programs, and how these programs would be organized (in what order the programs would be run to accomplish the total processing requirements). It is exactly these decisions which have a major influence on the performance of the system, i.e., on how much computer hardware and how much time are needed to satisfy a given set of processing requirements.

SODA

It is with respect to systems design decisions that much work needs to be done, and SODA is presented as a methodology for the design and optimization of IPS. The existing systems simulators[7,8,9] assume as given a systems design, i.e., a description of each program, schedule for a set of runs, and structure of the data files. SODA is intended to specify a systems design from a statement of the requirements and to generate the set of programs and data files.

SODA consists of a number of submodels that are solved using mathematical programming, graph theory, and heuristic procedures. Since the overall design problem is very large, it is convenient to view the algorithm as a multilevel decision model with the decision variable of one level becoming a constraint at the next level and so on. The partitioning of the problem into a multilevel structure implies that a different set of decision variables is required for each level of the algorithm.

The decision-making structure of SODA is described by (1) specification of the inputs and outputs, (2) specification of decision variables and determination of feasible alternatives, (3) selection of an objective function, (4) expression of objectives as a function of decision variables, (5) explicit statement of constraints which limit the value of the decision variable, and (6) solution, i.e., determination of the values of the decision variables.

SODA is a set of computer programs which begins with the initial statement of requirements (i.e., what the system is to do) and proceeds through the design and specification of the system. SODA is not concerned with the determination of which requirements are to be stated. The assumption is made that the problem definer (PD) can accurately identify his requirements. The major components of SODA are:

Problem Statement Language (PSL)

SODA/PSL is a technique for stating the requirements of the IPS independent of processing procedures. It also provides the capability for easily handling changes in requirements.

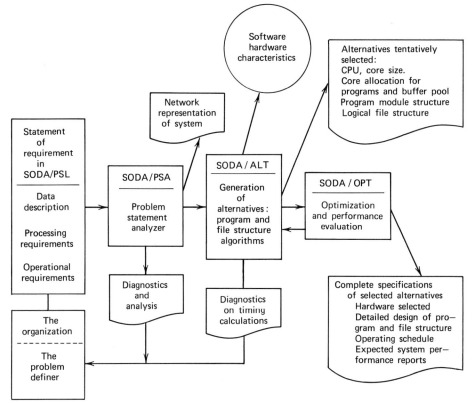

FIGURE 2 Soda: Systems Optimization and Design Algorithm.

Problem Statement Analyzer (PSA)

SODA/PSA is a program for analyzing the statement of the problem and organizing the information required in SODA/ALT and SODA/OPT. This program also provides feedback information to the problem definer to assist him in achieving a better problem statement.

Generation of Alternative Designs (ALT)

SODA/ALT is a procedure for the selection of a CPU and core size and the specification of alternative designs of program structure and file structure.

Optimization and Performance Evaluation (OPT)

SODA/OPT is a procedure for the selection of auxiliary memory devices and the optimization and performance evaluation of alternative designs.

Refer to Figure 2 for an overview of SODA.

The output of SODA is (1) a list specifying which of the available computing resources will be used, (2) specifications of the programs generated, (3) specifications of the file structure and the devices on which they will be stored, and (4) a schedule of the sequence in which the programs must be run to accomplish all the requirements.

SODA selects a set of hardware and generates a set of programs and files that satisfy timing requirements, core memory, and storage constraints, such that the hardware cost of the system is minimized.

SODA is limited to the design of uni-programmed batch systems, sequential auxiliary storage organization, the specification of linear data structures, and the selection of a single CPU. The model is deterministic. The problem statement technique is intended to handle "report oriented" data processing systems. Refer to Figure 3 for the interaction of the levels of SODA.

The overall structure of SODA/ALT and SODA/OPT is given in Figure 4 which describes the decision variables, objective function, alternatives, constraints, and solution techniques for each level in a summary form.

PROBLEM STATEMENT LANGUAGE

It is assumed that someone called a Problem Definer (PD) is familiar with the operation of the organization and has the necessary training to describe the processing requirements of the organization. The Problem Definer states his data processing requirements in a problem statement (PS) according to SODA/PSL, and the requirements are input to SODA/PSA in the form of a subset of a PS called a Problem Statement Unit (PSU). A PSU consists of three major categories: the data description, processing requirements, and operational requirements. The data description is defined by Elementary Data Sets (eds) and Data Sets (ds). The processing requirements consist essentially of a set of formulas called Processes (pr). The operational requirements consist of information on volumes, frequency of output, and timing of input and output.

An eds consists of a Data Name (dn), Data Value (dv), Descriptor Name (sn) and Descriptor Value (sv).

The sales of model X in the north region is an example of an eds.

There are four types of Data Sets and SODA makes use of the Data Set type in the file structure algorithm. An Input Data Set is any input data to the IPS. A Storage Data Set is that data which is stored in the IPS. A Terminal Data Set consists of output reports or forms and is not retained in permanent storage in the IPS. A Computed Data Set is the output of a Process that is neither a Storage nor a Terminal Data Set.

A Data Set may have up to three Descriptor Pairs to uniquely identify a Data Value. The set of Descriptor Pairs is

Data name		Descriptor value	Descriptor name		Data value
SALES_MODEL_X	(in the)	NORTH	REGION	(is)	500

A Data Set is the set of all eds with the same dn. The sales of model X in all regions of the country is an example of a Data Set.

Data name		Descriptor value	Descriptor name		Data value
SALES_MODEL_X	(in the)	NORTH	REGION	(is)	500
SALES_MODEL_X	(in the)	SOUTH	REGION	(is)	600
SALES_MODEL_X	(in the)	EAST	REGION	(is)	300
SALES_MODEL_X	(in the)	WEST	REGION	(is)	400

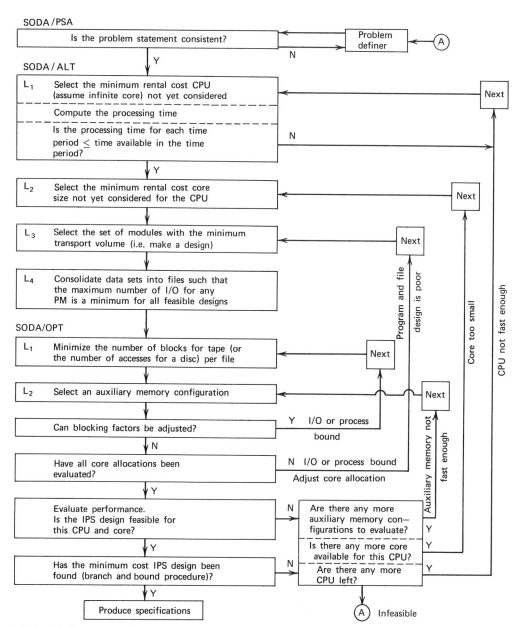

FIGURE 3 Interaction of the levels of SODA.

Level	Decision variables	Objective	Alternatives	Constraints	Optimization techniques
			SODA/ALT		
1	CPU	Select the minimum rental cost CPU class	List of available CPU classes	Time available for processing	Search for ordered CPU classes
2	Core size	Select the minimum rental cost core size	List of available core sizes	–Times available for processing –CPU class	Search of ordered core sizes
3	Program module	Select the set of modules with the minimum transport volume	Feasible grouping arranged in tree structure	–Core Size available for modules	Network analysis and branch and bound search over feasible alternatives
4	Data structures	Select Files such that the maximum number of I/O for any module is a minimum	Feasible grouping arranged in tree structure	–Program Modules	Network analysis and branch and bound search over feasible alternatives
			SODA/OPT		
1	Storage structure	Minimize the number of inter block gaps for tape or the number of accesses for a disk	1 char/block to upper limit	–Program Modules –Data structure	Nonlinear programming model
1	Number and type of auxiliary memories	Minimize the variable reading and writing time	List of available devices	–CPU and Core Size –Program Modules –Storage structures –Data structures	Integer programming model

FIGURE 4. Decision levels of SODA/ALT and SODA/OPT.

referred to as an Identifier, and a Descriptor Pair consists of a Descriptor Name and Descriptor Value.

Let id_k be the kth Identifier where $dn[id]$ is a shorthand notation for writing $\langle dn, sn \rangle$, $\langle sv, dv \rangle$.

Example: The GROSS_PAY of five employees each identified by the Identifier consisting of a single Descriptor EMPLOYEE_NUMBER is written as

$$\langle GROSS\_PAY\ [id_1] \rangle$$

and is equivalent to:

$$\langle GROSS\_PAY, sn_1 \rangle,$$

$$\langle sv_1, sv_2 \ldots sv_5, dv_1, dv_2 \ldots dv_5 \rangle$$

where:

sn_1 = EMPLOYEE_NUMBER
$sv_1 = 17$ $dv_1 = \$263$
$\vdots$ $\vdots$
$sv_5 = 21$ $dv_5 = \$300$

A Process is the smallest unit of processing requirement that may be grouped, but never subdivided, by SODA. A Process produces a Data Set and is a well defined assignment-type statement. These are four types of Processes defined in SODA/PSL. They are: (1) COMPUTE, (2) SUM, (3) IF, and (4) GROUP.

The Compute Process is a variation of the familiar assignment statement found in FORTRAN, ALGOL, or COBOL.

An example of the COMPUTE Process is:

NET_PAY $[id_1]$
= COMPUTE GROSS_PAT $[id_1]$
− DEDUCTIONS $[id_1]$

This expression consists of three Data Sets, NET_PAY $[id_1]$, GROSS_PAY $[id_1]$, DEDUCTIONS $[id_1]$ and all are identified by id_1 (EMPLOYEE_NUMBER). It is implied that whenever the statement is to be executed it is executed for all valid Descriptor pairs. The number of Descriptor pairs and the number of each type of arithmetic operation (e.g., multiplication, addition) is used by SODA/ALT for the purpose of estimating running times. The number of each type of arithmetic operation is obtained from the list of Processes.

The SUM Process is an expression that makes it convenient to sum over two or three Descriptor Names and the associated Descriptor Values. For example, one could sum labor cost by employee numbers and department numbers. The IF Process is a conditional expression. The GROUP Process groups Data Sets that are required to represent output reports. The four types of Processes enable SODA/PSA to construct precedence graphs of the Processes and Data Sets that are necessary for the Program and File Structure Algorithms.

The time and volume characteristics of the IPS are also described in the problem statement. Time requirements are specified by stating absolute time deadlines; i.e., paychecks must be produced by 4:00 PM on Friday. The statement of the time requirements for the output reports of the IPS is expressed by a Need Vector. The Data Set volumes are computed from the volumes stated for each eds. All time and volume information is expressed in units specified by the problem definer.

The structure of a Problem Statement in SODA/PSL consists of:

Problem Statement Name
   List of Identifiers
   List of Descriptors
      Descriptor Name
      Number of Descriptor Values for each Descriptor Name
   List of Data Sets
      Data Name
      Volume of Data Set
      Type of Data Set

    List of PSU
        Contents of each PSU
        PSU Number
        PSU Name
        Need Vector
        List of Processes
        END of PSU

END OF PROBLEM STATEMENT

SODA/PSL is a nonprocedural Problem Statement Language in the sense that the PD writes a PSU without imposing any procedural ordering on the Processes. The precedence relationships of the Processes are inferred by SODA/PSA.

The PS must contain sufficient detail so that systems analysts and programmers could use it (if necessary) to design and implement the Information Processing System with no additional information.

PROBLEM STATEMENT ANALYZER

The problem statement analyzer (SODA/PSA) accepts the requirements stated in SODA/PSL, analyzes them, and provides the problem definer diagnostics for debugging his problem statements and reports. SODA/PSA also produces a number of networks which record the interrelationships of Processes and data, and passes the networks on to SODA/ALT.

Each type of input and output is specified in terms of the data involved, the transformation needed to produce output from input, and stored data. Time and volume requirements are also stated. SODA/PSA analyzes the statement of the problem to determine whether the required output can be produced from the available inputs. The PS stored in machine-readable form is processed by SODA/PSA which:

1. Checks for consistency in the PS and checks syntax in accordance with SODA/PSL; i.e., verifies that the PS satisfies SODA/PSL rules and is consistent, unambiguous, and complete.
2. Prepares summary analyses and error comments to aid the problem definer in correcting, modifying, and extending his PS.
3. Prepares data to pass the PS on to SODA/ALT.
4. Prepares a number of matrices that express the interrelationship of Processes and Data Sets.

There are a number of papers that discuss the use of graphs and their associated matrices for the analysis of program and data structure.[1,2,10,11,12,13]

SODA/PSA follows the papers of Börje Langefors[1,2] and Raymond B. Briggs.[10] Langefors discusses the use of matrix algebra and graph theory to represent the processing units and data units in an IPS. Langefors' work differs from others using graph theory for this purpose in that it includes a performance criterion to be optimized.

Briggs added to the matrix definitions of Langefors and provided the necessary structure to develop a Program and File Structure Algorithm.

The problem statement is defined in SODA/PSA as the set of Processes required, the set of Data Sets needed by each Process, and the precedence relationships of the Processes (pr) and Data Sets (ds).

SODA/PSA generates the P, P^*, and E matrices for each PSU and for the entire IPS.

P — Precedence Matrix: Data Sets
  $p_{ij} = 1$ if ds_i is a direct precedent of ds_j,
  $p_{ij} = 0$ otherwise.

P^* — Precedence Matrix: Processes.
  $p^*_{ij} = 1$ if pr_i is a direct precedent of pr_j,
  $p^*_{ij} = 0$ otherwise.

The precedence matrices are checked for consistency using Marimont's procedure.[14]

E – Incidence Matrix: Processes and Data Sets

$e_{ij} = 1$ if ds_j is an input to pr_i
$e_{ij} = -1$ if ds_j is an output of pr_i
$e_{ij} = 0$ if there is no incidence between ds_j and pr_i

Let v_j be the volume of ds_j, l_i be the number of inputs and outputs for each Process, and m_j be the multiplicity of Data Set transport for ds_j. Let m_j represent the number of times ds_j is an input or output of a set of Processes.

$$l_i = \sum_{j=1}^{k} |e_{ij}|; \; i = 1, 2, \ldots n.$$

$$m_j = \sum_{i=1}^{n} |e_{ij}|; \; j = 1, 2, \ldots k.$$

Transport volume for ds_j is

$$tv_j = m_j \cdot v_j.$$

The transport volume for the set of Data Sets is

$$TV = \sum_{j=1}^{k} m_j \cdot v_j.$$

Transport volume is used as a criterion to evaluate alternative program and data designs and is discussed in the next section.

Let ds_j be represented by a ○ and pr_i be represented by a □. An example of an incidence graph and the associated incidence matrix is given in Figure 5. The R, R^*, and M matrices are generated for the entire set of Processes.†

R – Reachability Matrix: Processes.

The R matrix is used to check precedence violations in the grouping procedure of SODA/ALT.

$$R = (P^*)V(P^*)^2 V \ldots V(P^*)^{q-1},$$

where q is the index of the nilpotent matrix P^*.

†A procedure is discussed briefly in the next section for partitioning the entire set of Processes into smaller groups when the number of Processes is very large.

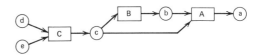

The associated incidence matrix is:

		\multicolumn{5}{c}{Data sets}	
		a	b	c	d	e	l_i
Processes	A	−1	1	1	0	0	3
	B	0	−1	1	0	0	2
	C	0	0	−1	1	1	3
	m_j	1	2	3	1	1	
	v_j	20	20	20	20	20	
	tv_j	20	40	60	20	20	

The transport volume for the Data Sets (TV) in this example is 160 units.

FIGURE 5 Incidence graph and matrix.

$r_{ij} = 1$ if pr_i has any precedence relationship with pr_j,
$r_{ij} = 0$ otherwise.

R^* – Partial Reachability Matrix: Processes.

The R^* matrix is used to calculate the M matrix.

$$R^* = (P^*)^2 V(P^*)^3 V \ldots V(P^*)^{q-1}.$$

$r^*_{ij} = 1$ if pr_i has a higher (2 or more) order precedence with pr_j,
$r^*_{ij} = 0$ otherwise.

It was shown by Briggs[10] that by using a theorem proved by Warshall[15] R and R^* can be constructed without first computing successive powers of P^*.

M – Feasible Process Grouping Matrix: Processes.

If $m_{ij} = -1$ there exist higher (2 or more) order relationships between pr_i and pr_j and pr_i cannot be combined with pr_j. If $m_{ij} = 0$ there is no precedence ordering and pr_i can be combined with pr_j. This indicates a feasible but not necessarily profitable grouping. If $m_{ij} = 1$ there is a direct precedent relationship and pr_i can and should be combined with pr_j since this indicates a feasible and profitable grouping. If $mi_{ij} = 2$ there is an immediate reduction in logical input/output requirements when pr_i and pr_j are grouped.

$m_{ij} = -1$ if r^*_{ij} or $r^*_{ji} = 1$.
$m_{ij} = 0$ if $r^*_{ij} = 0$ and $r^*_{ji} = 0$ and $p^*_{ij} = 0$ and $p^*_{ji} = 0$; except when $p^*_{il} = 1$ and $p^*_{jl} = 1$ or $p^*_{li} = 1$ and $p^*_{lj} = 1$.
$m_{ij} = 1$ if $r^*_{ij} = 0$ and $r^*_{ji} = 0$ and $p^*_{ij} = 1$ or $p^*_{ji} = 1$.
$m_{ij} = 2$ if $r^*_{ij} = 0$ and $r^*_{ji} = 0$ and $p^*_{il} = 1$ and $p^*_{jl} = 1$ or $p^*_{li} = 1$ and $p^*_{lj} = 1$.

pr_l has a first order precedence or succedent relationship with pr_i and pr_j.

A list of all feasible pairs for grouping of Processes is constructed from the M Matrix and passed to SODA/ALT.

GENERATION OF ALTERNATIVES

The information system design phase begins after the requirements have been stated, verified, and analyzed in SODA/PSA. SODA/ALT accepts as input, the output of SODA/PSA and a statement of the available computing resources, hardware, and utility programs. The hardware alternatives are ordered in a tree structure as shown in Figure 6.

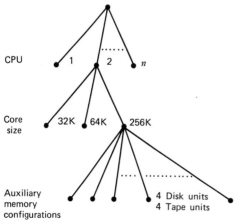

FIGURE 6 Hardware alternatives.

A feasible CPU and core size are specified using a heuristic timing procedure. An ordered search of different CPU's is made in an attempt to find the minimum CPU assuming an infinite core memory with no auxiliary memory and all Processes and Data Sets in real core. The premise is that if a CPU cannot perform adequately under these "ideal" conditions it cannot possibly be adequate with limited core constraints.

The processing time for each time period (i.e., a week) is computed. If this is less than the actual time available to do this, then a lower bound of CPU capability is found. If not, the next CPU is tried. If it appears that some shifting of load from one or more time periods (i.e., week 1) to other time periods (i.e., week 3) could solve the problem, then the problem definer is advised about it and given a change to "level" the requirements.

Having found a CPU that will perform adequately under infinite core assumptions, an ordered search (starting with the smallest core size) is made of available core sizes for this CPU.

Using as constraints CPU and core size, a graph theoretic model generates alternative Program Module and File Designs. A Program Module (pm) is a set of Processes grouped together by SODA/ALT. A file (f) is one or more Data Sets that are grouped together. From the Matrix of feasible groupings (M) a list of feasible (profitable) groupings of Processes is obtained. The list of feasible groupings is partitioned into three cases to reduce the number of alternative program designs that must be evaluated. The cases separate output reports into classifications of due dates for reports. The cases are then divided into subcases. (A subcase is a group of Processes that has no Process precedence link to other subcases.) For each subcase, feasible Program Modules of size 3, 4, ... , N are generated, where N is the number of Processes in the partitioned list (or subcase).

It is known that by grouping Processes into a composite process called a Program Module, the multiple input and output of Data Sets can be reduced. Such grouping of processes, however, requires additional main memory for the Program Modules. In generating an efficient design, it is necessary to decrease the transport volume (total number of characters read in and written out of main memory) in order to reduce the processing time. If Data Set volumes remain constant, in order to decrease the transport volume, the multiplicity (the number of times a Data Set is input and output) of Data Set transport must be decreased. After the Program Modules are specified, the Data Sets are consolidated into Files for the purpose of reducing the number of input/output Files required and for better utilization of storage in auxiliary memory. Process grouping is shown to correspond to a grouping of rows of the incidence matrix, and data set consolidation is shown to correspond to a grouping of columns.

Program Module and File Design are concerned with the reduction of processing time and can be summarized by the two methods by which the processing time can be reduced. SODA/ALT determines:

1. Which operations (Processes) will be grouped into Program Modules, the objective being to reduce total transport volume and thus total processing time.
   a. Group Processes which will eliminate the writing out and reading in of a Data Set. For instance Figure 7a.
   b. Group Processes which require the same Data Sets. For instance Figure 7b.

(a)

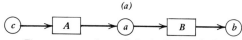

The transport volume of ds_a is eliminated when pr_A and pr_B are grouped.

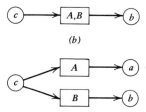

(b)

The transport volume is reduced when pr_A and pr_B are grouped since ds_c is read only once.

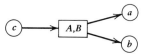

and it may be profitable to also group ds_a and ds_b.

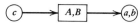

FIGURE 7 Methods for reducing processing time.

2. Which Data Sets will be grouped into Files (f):
   a. Organize the data structure of the Files so that data which are needed together are close together in order to reduce searching time.
   b. Organize the data structure of the Files such that fewer logical input/output devices are needed.

The following matrices are used in SODA/ALT to generate alternative Program Module and File Designs.

S — Program Module Selection Matrix: Program Modules and Processes.

The S matrix represents the alternative grouping of Processes.

$s_{ij} = 1$ if pr_j is a member of pm_i,
$s_{ij} = 0$ otherwise.

E' — Incidence Matrix: Program Modules and Data Sets.

The S matrix is multiplied by the E matrix to produce the new incidence matrix E', where $E' = S \Lambda E$.

The Boolean matrix operators "Λ" and "V" for Process grouping follow the rules of Boolean algebra, with the following exceptions for the Boolean addition operator:

$0 \vee -1 = -1$

$1 \vee -1 = 0$ if the output ds_j is used only by the pm_i with which it is grouped and ds_j is not a Terminal ds or Storage ds.

$1 \vee -1 = 1$ if the output ds_j is required in a Process that is a member of another Program Module or is a Terminal ds or Storage ds.

D — Feasible Data Set Grouping Matrix: Data Sets.

$d_{ij} = 1$ if $e'_{ij\alpha}$ and $e'_{ij\beta} = 1$ or $e'_{ij\alpha} = -1$ and $e'_{ij\beta} = -1$ and $ds_{j\alpha}$ and $ds_{j\beta}$ are the same data types and have a common descriptor or if ds_i and ds_j are input Data Sets and have a common descriptor.

$d_{ij} = 0$ otherwise.

$ds_{j\alpha}$ and $ds_{j\beta}$ are two data sets required by pm_i. The same data type for $ds_{j\alpha}$ and $ds_{j\beta}$ refers to the classification of Input ds, Computed ds, Storage ds, and Terminal ds. The test for a common Descriptor Name is intended to link Data Sets together that have a relationship other than common input or common output of a Program Module.

G — File Selection Matrix: Data Sets and Files.

The G matrix represents the alternative grouping of Data Sets.

$g_{ij} = 1$ if ds_i is a member of f_j,
$g_{ij} = 0$ otherwise.

E'' — Incidence Matrix: Program Modules and Files.

The E' matrix is multiplied (as described earlier) by the G matrix to produce the incidence matrix E'' of Program Modules and Files, where

$$E'' = E' \Lambda G.$$

The selection procedure for program design is organized as a tree structure with all feasible alternatives ordered in terms of core memory requirements and transport volume. The procedure for File design is organized by descriptors (keys) and the number of input/outputs Files required for each Program Module.

If software modules such as sort modules are required to process Files they are inserted in the IPS Design.

The next step is to look for design improvements and to select a specific number and type of auxiliary memory units.

OPTIMIZATION AND PERFORMANCE EVALUATION

The optimization and performance evaluation phase generates a storage structure and scheduler, selects auxiliary memory devices, and searches for ways to improve the IPS design. SODA/OPT may return control to SODA/ALT to select another CPU, core size, or to select another set of Program Modules and Files.

SODA/OPT selects the minimum cost hardware configuration that is capable of processing the stated requirements in the time available. This phase consists of a number of mathematical programming models and timing routines that are used to (1) optimize the blocking factors for all Files, (2) evaluate alternative designs, i.e., specify the number and type of auxiliary memory devices, (3) assign Files to memory devices, and (4) generate an operating schedule for running program Modules. These submodels follow the work of Schneidewind[16] and Thiess.[17] Refer to McCuskey[18] for another approach to the design of data organization.

In SODA/OPT the performance criterion is optimized within the constraint set by the capability of the hardware and by the processing requirements. SODA/OPT produces a report describing the system and stating its predicted performance. On the basis of this, the Problem Definer may decide to change his PS, or accept the design; SODA/OPT then provides detailed specifications for the construction of the system. The output of SODA/OPT is:

1. A list specifying which of the available computing resources will be used.
2. A list of the Program Modules specifying the input, output, and computations to be performed in each.
3. A list of Files to be maintained, specifying their format and manner in which they will be stored. Assignment of Files to memory devices.
4. A statement of the sequence and manner in which the Program Modules must be run to accomplish all the requirements.

IMPLEMENTATION

SODA has been written for the Univac 1108 in FORTRAN at Case Western Reserve University,[19] the program has also been implemented on the IBM 360/67 at the University of Michigan by Professor Daniel Teichroew. The program is currently being modified and rewritten for the CDC 6500 at Purdue University.[20]

The Case version of SODA has the following drawback: if there are two feasible solutions,

(1) Small CPU, large core
(2) Larger CPU, smaller core

such that (2) is cheaper, then the algorithm would not find (2). The assumption is made (the Case version) on the ordering of the hardware tree that this situation (2) would not occur. This was done in order to simplify the search procedure and to reduce the size of the large combinatorial problem involved. A branch and bound procedure is being implemented in the Purdue version so that alternative (2) would be found. The Case version also does not automatically evaluate all Program Module and File Designs. A partial set of Program Modules and File designs ($\sim 50,000$) is generated. A smaller number of designs (~ 200 for the Company Y example) must be selected manually and then input to SODA to be evaluated. Procedures have been developed[20] to eliminate the need for man-machine interaction in SODA/ALT, and the procedures are being implemented in the Purdue version.

The program has been run using an example problem called Company Y. The example consists of 117 Processes and 180 Data Sets. Approximately 30 runs were required to debug the problem statement for Company Y.

A single run for SODA/PSA takes about 120 seconds of execution on the UNIVAC 1108. The total time required for the Company Y example is difficult to estimate since the SODA program was not run from beginning to end at one time; many of the submodels were run, then a data file passed to the next submodel, and so on.

A series of hypothetical computers is described in the hardware file. The hardware file consists of three CPU's with five core options for each CPU. The auxiliary memory option consists of two types of tape drives and two types of disk units.

REFERENCES

1. B. Langefors
   Some approaches to the theory of information systems
   BIT 34 1963.
2. B. Langefors
   Information system design computations using generalized matrix algebra
   BIT 52 1965.
3. J. F. Nunamaker Jr, A. B. Whinston
   Computing center as a profit center
   Computer Sciences Department TR52 Purdue University. Lafayette, Indiana. January, 1971.
4. L. Lombardi
   Theory of files
   Proceedings of the Eastern JCC. 1960.
5. D. Teichroew
   ISDOS—A research project to develop methodology for the automatic design and construction of information processing systems
   ISDOS Working Paper Number 1 Case Institute of Technology, August, 1967.
6. W. Steiger
   Survey of basic processing function: Literature and related topics and (2) Derivation of primitives for data processing
   ISDOS Working Paper Number 11 Case Institute of Technology, May, 1968.
7. D. J. Herman, F. H. Ihrer
   The use of a computer to evaluate computers
   AFIPS Conference Proceedings Spring Joint Computer Conference Volume 25 pp. 383–395. 1964.
8. L. R. Huesmann, R. P. Goldberg
   Evaluating computer systems through simulation
   The Computer Journal Volume 10 Number 2, August, 1967.
9. *Digest of the second conference on applications of simulation*
   Sponsored by SHARE/ACM/IEEE/SCI. New York, December, 1968.
10. R. B. Briggs
    A mathematical model for the design of information management systems
    MS Thesis Division of Natural Science University of Pittsburgh, 1966.
11. T. C. Lowe
    Analysis of boolean program models for time shared paged environments
    Communications of the ACM Volume 12 Number 4, April, 1969.
12. T. C. Lowe
    Automatic segmentation of cyclic program structures based on connectivity and processor timing
    Communications of the ACM Volume 12 Number 1, January, 1970.
13. C. V. Ramamoorthy
    Analysis of graphs by connectivity considerations
    Journal of the Association for Computing Machinery Volume 13 Number 2, April, 1966.
14. R. B. Marimont
    A new method of checking the consistency of precedence matrices
    Journal of the ACM Volume 6 Number 2, April, 1959.
15. S. Warshall
    A theorem on boolean matrices
    Journal of the ACM Volume 9 Number 1, January, 1962.
16. N. F. Schneidewind
    Analytical model for the design and selection of electronic digital computing systems
    D B A Dissertation University of Southern California, 1965.
17. H. E. Thiess
    Mathematical programming techniques for optimal computer use
    Proceedings ACM 20th National Conference, 1965.
18. W. A. McCuskey
    On automatic design of data organization
    AFIPS Conference Proceedings Fall Joint Computer Conference Volume 37, 1970.
19. J. F. Nunamaker Jr.
    On the design and optimization of information processing systems
    PhD Dissertation Case Western Reserve University, 1969.
20. J. F. Nunamaker Jr.
    SODA: Systems Optimization and Design Algorithm
    Computer Sciences Department TR51 Purdue University, Lafayette, Indiana, January, 1971.

Introduction to SECTION IV

Fourth Generation System Analysis Techniques

This section consists of only one paper, "Automation of System Building," by Teichroew and Sayani. The paper describes the ISDOS project at the University of Michigan; ISDOS is an acronym for Information System Design and Optimization System. Just as computer *applications* are being integrated, *techniques* for each phase of system development are being integrated. ISDOS has the objective of specifying a system design from a statement of the requirements and automatically generating the set of programs and data files. The concept is far enough along in implementation (as shown by the papers on PSL/PSA and SODA in the previous section) to prove the validity of the approach.

ISDOS integrates Phases II through VII of the system development cycle. The only phase excluded is Phase I, the System Survey. For most organizations the System Survey is conducted only once. However, for firms which have a policy of acquisition, or which set up new divisions for new product lines, updating of the System Survey, using the SOP approach, is feasible. The deficiency in SOP is its lack of interaction with the Planning Model. The objectives of the firm and the resource levels required to undergird those objectives have significant impact on the computer-based management information system. Although ISDOS provides a means to optimize system development for a major system, the integration of major systems with the MIS and the priority of computerization is ascertained from SOP, the basis for the MIS for the entire organization. Therefore, SOP must be updated and must be interactive with the Planning Model.

The fourth generation system development approach will provide for such interaction.

AUTOMATION OF SYSTEM BUILDING

DANIEL TEICHROEW
HASAN SAYANI

The building of computer-based information systems to serve the management and operation of organizations has become a large and visible activity. Furthermore, one need only note some of the recent news items regarding the system building process to be convinced that it does not always lead to satisfactory results.[1] In the past, the emphasis on improvement of techniques has been on methods to help the programmer. Programming is certainly an essential step in the process, but it is only one of the steps. The attention and emphasis it has received is completely out of proportion to its role in the building of systems, and this has resulted in insufficient attention paid to the improvement of other steps. It is becoming more generally recognized that the other steps in the system building process must also be improved.

What we call an "information system" consists of two subsystems: a management system and an information processing system. The management system consists of the organization, its objectives, the individuals or groups in it, and the rules and procedures under which they work. The information processing system is the subsystem which consists of hardware, programs, noncomputerized procedures, etc., that accomplish the storage, processing, and communication of information necessary for the functioning of the management system. An essential element in this view is that the information processing system (IPS) must, or at least should, be designed to serve the management system. This characterization of information systems

---

[1] See for example: "Burroughs Sued by Trans World for $70 Million," **Datamation**, Dec. 1, 1970, p. 47; "Bell's BIS: Bottomless Well," **Datamation**, July 15, 1970, p. 35; "Chrysler's Private Hard Times," Fortune, April 1970; "Computer Classic," The Economist, Oct. 24, 1970, p. 94.

SOURCE: Teichroew, D., H. Sayani, "Automation of System Building," *Datamation*, Aug. 15, 1971, pp. 25–30.

is particularly relevant to management information systems.[2]

SYSTEM BUILDING

Organizations normally go through a number of phases in building information systems. Initially, the potential use of the computer is treated as a one-time task for a few programmers. Soon it becomes obvious that the task is much bigger than first suspected, and during the second phase more manpower is assigned to the project. The third phase begins when it is recognized that a series of systems will have to be built and that a procedure will have to be developed. This results in the establishment of a systems department. A fourth stage is reached when it becomes apparent that the systems being built have many features in common and considerable reduction in effort might result from using standard building blocks.

Most medium and large organizations have reached at least the third phase; they recognize that information processing system building will be a continuous activity and have established systems departments. System building therefore usually involves three groups: top management, users, and system builders. The users, in general, are the functional divisions such as manufacturing, finance, personnel, etc.; however, a system may frequently be designed to serve more than one function. The system builders are centralized in a systems department. Top management is involved because it must define the responsibility of the user and the builders and adjudicate differences. It must also allocate resources and assign priorities where resources are scarce.

Frequently, one of the activities that the system departments undertake early in their existence is the development of a set of procedures and standards to be followed in the building process. The number of papers and books describing such procedures and standards has increased very rapidly in the last few years. Most organizations, however, prefer to develop their own. Review of the published methods and the manuals developed by organizations for their own use indicates that the procedures are basically similar though they may differ in details. Here a brief outline of the major steps will be given in order to illustrate the need for, and potential scope of, automation.

The first step in the process is a request which indicates the need for a new system or the modification of an existing one. This request, ideally, is originated by the user, though it may come from the systems department if many users will be served. The request should contain sufficient information to initiate the next step.

Feasibility, or impact analysis, is the second major step. It consists of estimating the potential benefits of a system to satisfy the expressed request. A proposed system must be developed in sufficient detail to estimate the costs. The impact on the organization and the existing systems is evaluated since this may affect both benefits and costs.

While the analysis in the second step may frequently be extensive and time-consuming, neither the statement of user requirements nor the description of the proposed system is in sufficient detail to proceed with the construction of the proposed system. The third step is to determine the user needs in full detail and to describe them in a form which the users can agree to, and which

---

[2]Emphasis in this paper is on the use of the information processing system which serves management because this is the most important and the most difficult type of system to construct today. The techniques discussed, however, are applicable with only minor changes to other types: routine business data processing, command and control, information storage and retrieval, message switching, and process control.

is also suitable for the design and construction phases that follow. In some cases extensive analysis is required to verify that the detailed requirements, as stated, do in fact satisfy the more general needs stated in the first phase. It may in fact be necessary to simulate the management system. This phase is sometimes referred to as the "functional specifications" or "logical system design" phase.

The fourth phase—the physical systems design—is concerned with developing the specifications for the proposed computer-based information processing system that will accomplish the logical requirements detailed in the previous phase. Ideally, this is an elaboration of the proposed system used in the feasibility phase to estimate potential costs; if not, the feasibility results may have to be modified. The design of the proposed system consists of selecting, within whatever constraints that may exist, the processing organization (real-time, batch, etc.) and hardware, and designing the programs and data base. The result of this phase is a set of specifications.

These specifications are used in the system construction phase to build the actual target system. The new hardware requirements, if any, go to the hardware acquisition group for procurement. The program specifications go to the programming department which writes and tests the various programs. The data base specifications go to the group which has the responsibility for constructing the data base that will be needed. Normally, all these activities are the responsibility of the systems department. Other specifications may go to the personnel department for training and educational requirements.

In the sixth phase, all the components, already tested individually, are brought together and tested as a system. Errors discovered in this phase must be corrected; this may require going back several phases.

During the seventh phase, the system is in operation. Since requirements may change or errors may be discovered, a change control procedure must be established to ensure that changes are appropriately recorded. The performance of the system must be compared with the estimates made in the feasibility phase.

The amount of attention paid to the steps in the process depend, of course, on the size of the system. If the system can be built by one person in a short period of time, he can usually build it satisfactorily without explicitly following the procedure. As the number of individuals involved increases, the formal procedure and complete documentation become essential. Unfortunately, documentation is usually neglected because the system building process is essentially manual. Formal techniques are not widely used; the steps are carried out by individuals using pencil and paper, and the documentation consists of descriptions in English supplemented by flowcharts, tables, etc. Formal, computer-aided techniques are used only in that part of the construction process in which higher level statements are compiled into object code.

Another consequence of the present system building methods is that the process from the user request to successful operation takes a long time. The elapsed time is a function of the size of the system, but several years is not unusual. The elapsed time can be reduced by carrying out some subphases in parallel; but this must be planned very carefully or it may result in inconsistencies and require more time rather than less.

A major problem in system building occurs right at the beginning in determining what the user wants. In fact, it is not even clear that this is the right way to state the objective. The user may not be able to articulate what he wants and usually is not the appropriate person to decide what he should have. The situation is further aggravated by the fact that the user usually is not accustomed to describing what he wants in sufficient detail to translate it into computer programs, a point that has been very well stated by Vaughn.[3]

There is no doubt that, ideally, it is very important to start with the "correct" requirements. Building a system to accommodate wrong requirements is a waste of time and effort. However, it is our view that in the present circumstances it is more important to develop methods to reduce the time to build systems *once the requirements are given.* The major reason for this is that it does not do much good to produce the "absolutely perfect" set of requirements if it will then take a long time (six months to several years) to produce the system that will accomplish the requirements. The absolutely perfect requirements are not constant; they change as the environment in which the organization exists changes. The organization itself changes, and the individual users change or learn to use the outputs from the computer-based system. We have therefore adopted as our basic objective the need to reduce the length of time from the point where requirements are first stated until the target system to accomplish the requirements is in operation. Obviously the major tool to accomplish this reduction must be the computer itself.

Once it is decided to use the computer in the system development process, the next step is to decide where to begin recording data in machine processible form. Here we try to apply the first principle of automation: record the input data in machine-readable form as close to the source as possible and thereafter process it with as little human intervention as possible. For the reasons mentioned above, we have decided to start at the point at which the requirements of the management system have been determined and the specification of individual inputs and outputs can begin. In the future we hope to extend our techniques to aid the process of determining what the requirements of the management system *should* be. There is no reason why any piece of data about requirements should not be recorded in machine-readable form the first time it appears in the system building process. (The proposed format for capturing the specification of requirements at this point and the software to process it are outlined later.)

Once it is decided to base the system building process on the use of the computer, there are other potential benefits than just the reduction in elapsed time. It should be possible to accommodate changes in requirements more easily both during the design process and during system operation. The computer can also be used as the basis for coordinating the activities of many analysts and to relieve them of many tedious and laborious clerical tasks which they now must do manually.

METHODS OF IMPROVEMENT OF SYSTEM BUILDING

The conclusion reached in the previous section is that the system building process itself should be automated, or at least computer-aided. Before describing our approach on how this might be done it is worthwhile examining some other ways to improve the process. The alternative methods may be grouped into four major categories: improve education and training of system builders,

---

[3] Vaughn, P. H., "Can COBOL Cope?" **Datamation**, Sept. 1, 1970, pp. 42–46.

provide aids (computer based and others) for the system builders, use application packages, and use generalized software.

System building, as a profession, is still in its infancy and most practitioners were trained in other fields. In the early days, practice was relatively simple and required little more than programming. Now, however, the practice is becoming more professional and educational programs for the "information engineer" are being developed. However, it is extremely unlikely that it will be possible to build the number of systems of the size and complexity desired by manual methods; there will not be enough people. It will be necessary to use the trained professionals more effectively by moving from "handcrafted" systems to "mass-produced" systems.

Many aids designed to facilitate individual tasks in system building have been proposed. Probably the most generally used are the general-purpose programming languages. Less widely used are programs for other aspects of the process such as flowcharters and system simulators (SCERT, CASE, etc.). Space does not permit a detailed analysis of these aids here; however, our conclusion from such analysis is that these aids tend to be useful in only one particular (and usually narrow) aspect of the whole system development cycle. Manual intervention and manual preparation of input is required at each stage. What is needed instead is a coherent system that covers all phases of the life cycle in which the output of one phase is automatically an input to the next.

Application packages have been available since the early days of computers. Their use has been limited primarily because the user needs are continuously changing and attempts to provide flexibility usually result in high processing cost. There is a spectrum of methods to build application packages so that they can be tailored for a specific set of requirements ranging from applications in which the user has no alternatives, to ones in which he has complete freedom. In the most completely specified packages the user can only enter data values. This approach tends to be satisfactory only where the problem is relatively small and very well defined. Some packages allow more freedom through the use of parameter values as well as data values. Another level of generality is reached by providing for a number of options which the user specifies by filling out a questionnaire or by completing a form. This method has been used to generate simulation programs[4] and is the basis of generating software for the IBM System/3. An even more general approach to application packages is represented by user-oriented languages. These give the user a relatively flexible method of specifying his problem but require less effort than would be required to write a program in a general-purpose language. To cover all user needs would require many different languages and maintenance of the associated software. Some standardization clearly is desirable.

Generalized software started from input/output subroutines, and packages such as sort, merge, report generators, etc., are now in common use. Generalized file maintenance packages, however, have only fairly recently evolved into "data base management systems." These systems differ from application packages in that they "generalized" in terms of operations inside the computerized system rather than in terms of view of the user from the outside. Sorting, for example, is a processing operation that is not dependent on the

---

[4]Ginsberg, A. S., H. M. Markowitz, and P. M. Oldfather, "Programming by Questionnaire," AFIPS Conference Proc. Vol. 30, 1967, SJCC, pp. 441–446 [CR 12764, 12149].

particular application. Data base management systems will undoubtedly achieve a major role in the next few years. They are attractive, despite their high processing cost, because they relieve the programmer of the need to program frequently used operations such as access methods for complicated data structures and variable-length items, records, and files.

All of the methods of improving system building described above have been used and will continue to be used in the future. What we are concerned with is the next major plateau. There has been a progression in which general-purpose programming languages have replaced assembly languages and general-purpose languages themselves have had to be augmented by data base management systems to provide the framework for the programmer to communicate with the machine. In turn, the limitations of data base management systems will be overcome through automation of the system-building process. The effectiveness of trained professionals can be amplified and the computer-based aids to system building integrated into a software factory that can produce user programs tailored to user requirements.

AUTOMATION OF THE SYSTEM-BUILDING PROCESS

The need to automate the whole system-building process, as contrasted with the development of aids for parts of the process, has been recognized. For example, this is the expressed goal of the CODASYL Systems Committee.[5] So far, however, the committee has been primarily concerned with data base management systems. A computer-aided approach, the TAG (Time Automated Grid) System, has been developed by IBM. A number of other systems have been proposed.[6] Many concepts from these systems have been incorporated into ISDOS.

ISDOS (Information System Design and Optimization System) is the name of a software package being developed by faculty, students, and research associates in the Department of Industrial Engineering at the University of Michigan. It consists of a number of major components which are shown in Fig. 1; this section gives a description, and purpose, of each component.

As mentioned earlier, ISDOS begins with the user requirements recorded in a machine-readable form. The problem definer (i.e., the analyst or the user) expresses the requirements according to a structure format called the Problem Statement Language. This language can be considered a generalization of those of Young, and Kant;[7] Information Algebra;[8] SYSTEMATICS;[9,10] TAG Input/Output

[5]The CODASYL Systems Committee states its objectives as: ". . . to strive to build up an expertise in, and to develop, advanced languages and techniques for data processing, with the aim of automating as much as possible of the process currently thought of as systems analysis, design, and implementation."

[6]For a discussion and comparison, see Teichroew, D., "A Survey of Languages for Stating Requirements for Computer Based Information Systems," ISDOS Working Paper No. 42.
[7]Young, J. W. and H. Kent, "Abstract Formulation of Data Processing Problems," J. of Ind. Engr., Nov.–Dec. 1958, pp. 471–479. Reprinted in Ideas for Management, Internat. Systems-Procedures Assoc., 1959.
[8]CODASYL Development Committee, "An Information Algebra-Phase I Report," Communications of the ACM, 5, 4, April 1962, pp. 190–204.
[9]Grindley, C. B. B., "SYSTEMATICS – A Non-Programming Language for Designing and Specifying Commercial Systems for Computers," Computer Journal, Vol. 9, August 1966, pp. 124–128.
[10]Grindley, C. B. B. and W. G. R. Stevens, "Principles of the Identification of Information," File Organization, IAG Occasional Publication, No. 3, Scolts and Zeitlinger N.V., Amsterdam, 1969, pp. 60–68.

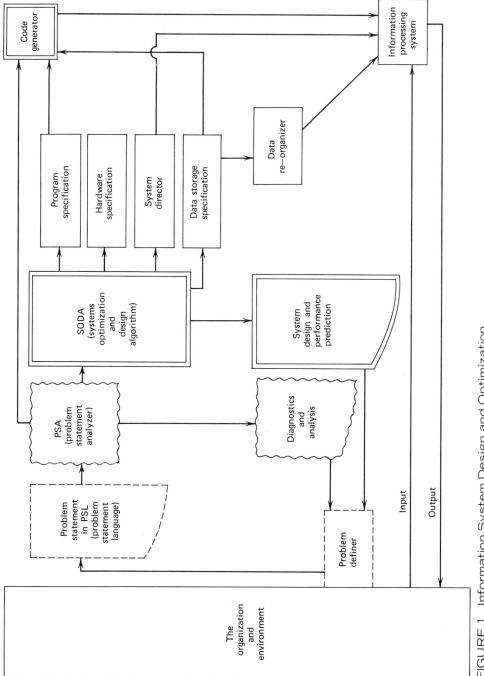

FIGURE 1 Information System Design and Optimization System (ISDOS).

Analysis Form; and ADS.[11] All of these languages are designed to allow the problem definer to document his needs at a level above that appropriate to the programmer; i.e., the problem definer can concentrate on *what* he wants without saying *how* these needs should be met.

It is very important to note that a problem statement language is not a general-purpose programming language or, for that matter, any programming language. A programming language is used by a programmer to communicate with a machine in the fifth phase of the system building process. A problem statement language, on the other hand, is used to communicate the needs of the user to the analyst and therefore is needed in the third phase. The problem statement language consequently must be designed to express what is of interest to the user: what outputs he wishes from the system, what data elements they contain, and what formulas are to be used to compute their values. Analogous information must be given for inputs. In addition, the user must be able to specify the parameters which determine the volume of inputs and outputs and the conditions (particularly those related to time) which govern the production of outputs and acceptance of inputs. The Problem Statement Language is designed to prevent the user from specifying processing procedures that should be selected in the fourth or fifth phase; for example, the user cannot use statements such as SORT and he cannot refer to physical files.

The Problem Statement Language has sufficient structure to permit a Problem Statement to be analyzed by a computer program called a Problem Statement Analyzer. This program is intended to serve as a central resource for all the various groups and individuals involved in the system building process as shown in Fig. 2.

Since the problem definer may be one of many, there must be provision for someone who oversees the problem definition process to be able to identify individual problem definitions and coordinate them; this is done by Problem Definition Management. One desirable feature of a system building process is to identify system-wide requirements so as to eliminate duplication of effort; this task is the responsibility of the System Definer. Also, since the problem definers should use common data, there has to be some standardization on their names and characteristics and definition by computations (these are referred to here as "functions"). One duty of the data administrator is to control this standardization. If statements made by the problem definer are not in agreement as seen by the system definer or data administrator, he must receive feedback on his "errors" and be asked to correct these.

All of these capabilities are being incorporated in the Problem Statement Analyzer, which accepts inputs in the Problem Statement Language, analyzes them for correct syntax, and produces, among other reports, a comprehensive data dictionary and a function dictionary that are helpful to the problem definer and the data administrator. It also performs static network analysis to ensure the completeness of the derived relationships, dynamic analysis to indicate the time-dependent relationships of the data, and an analysis of volume specifications. It also provides the System Definer with a structure of the problem statement as a whole. All these analyses are performed without regard to any computer implementation of the target information processing system. When these analyses indicate a complete and

---

[11]National Cash Register Company, Accurately Defined Systems, 1967.

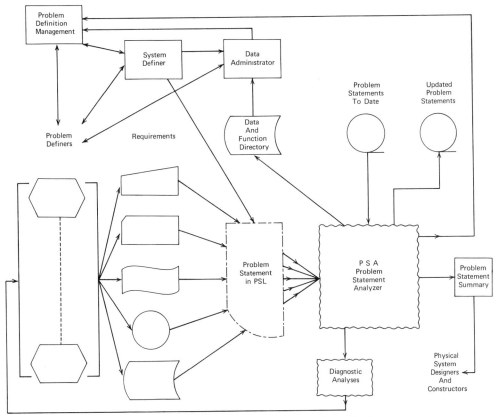

FIGURE 2 Information flows in problem statement analysis.

error-free statement of the problem, it is now available in two forms for use in the succeeding phases. One, the problem statement itself, becomes a permanent, machine-readable documentation of the requirements of the target system *as seen by the problem definer* (not as seen by the programmer). The second form is a coded statement for use by the physical systems design process and other modules of ISDOS.

In the conventional approach, the physical systems design phase (phase four) is concerned with accepting a consolidated statement of the requirements from the system analysts and outlining specifications for the actual construction of programs, files, the relevant schedules, etc. The number of alternatives available is usually so large

that the manual approach does not permit the examination of more than a handful of these. An objective of ISDOS is to formalize the physical design process along the lines pioneered by Langefors,[12] Grosz,[13] Turnburke,[14] Martin,[15] etc. The design problem is

[12]Langefors, B., "Theoretical Analysis of Information Systems," 2 Vol. Studentlitteratur, Lund, 1966. (Also available from National Computing Centre Ltd., Quay House, Quay Street, Manchester, England.)
[13]Grosz, M. H., "Systems Generation Output Decomposition Method," Standard Oil Company of New Jersey, July 1963.
[14]Turnburke, V. P, Jr., "Sequential Data Processing Design," IBM Systems Journal, March 1963.
[15]Martin, J., Design of Real-Time Computer Systems, Prentice-Hall, Englewood Cliffs, N.J., 1967.

formulated mathematically. Operations research methodology is used to develop methods to search over the range of alternatives. A multilevel approach, where the decision variables at one level become the constraints at the next level, is required. This makes it possible to evaluate various design strategies and aids the hardware acquisition group in the selection and justification of appropriate hardware. It also gives the performance officer (who is responsible for the efficient use of resources in computer operations) and the physical system designers a good indication of the expected performance of the system. In addition to the requirements as prepared by the Problem Statement Analyzer, a description of hardware characteristics is required. The outputs are specifications for program modules, storage structures, and scheduling procedures which are in a form suitable for processing by the next two ISDOS modules.

The Data Re-organizer accepts specifications for the desired storage structures from the physical systems design process, definition of data as summarized by the Problem Statement Analyzer, the specifications of the hardware to be used, and the data as it currently exists, and its storage structure. It then stores the data on the selected devices in the form specified. The Re-organizer also produces information for the data administrator and the performance officer. The other module, the Code Generator, accepts specifications from the physical design process and organizes the problem statements into programs recognizing the data interface as specified by the Data Re-organizer. The code produced may be either machine code, or statements in a higher level language (e.g., COBOL), or parameters to a software package. These two modules perform, automatically, the function of programming and file construction in the fifth phase of the system building process.

The final module of the ISDOS system is the Systems Director. It accepts the code generated, the timing specifications as determined by the physical design algorithm, and specifications from the Data Re-organizer and produces the target IPS. This IPS is now ready to accept inputs from the environment and produce the necessary outputs according to the requirements expressed in the problem statement.

The central concept which makes possible the automation of design and construction is the separation of user requirements from decisions on how these requirements should be implemented. This philosophy is incorporated in the design of the Problem Statement Language. From then on the problem statement can be manipulated by the Problem Statement Analyzer. The decisions which are made in the physical systems design are basically "grouping" decisions, which theoretically can be represented as combinatorial problems. In practice, of course, the number of combinations is very large; and therefore a major research task is to develop efficient algorithms.

ISDOS Development Plan

If a system such as the one outlined in the previous section were available it would go a long way towards improving the effectiveness of computer-based information systems. Since new requirements or modifications to existing requirements could be implemented at computer speeds, management would be able to get the information it asked for in a much shorter period of time. It would therefore be much less important to get the requirements right the first time since a change could be incorporated more easily than at present. The user would be closer to the requirement specifications since the language is closer to the one he is familiar with. Hardware could be used more effectively since design would be based on a formalized procedure using latest available parameters which specify volume of system inputs and outputs.

The system described is itself an information system; and the development of functional specifications, design, and construction is a substantial task which involves three major subtasks:

1. The specification of man-machine communication problems encountered by the analyst in acquiring and recording the requirements for the target system; in other words, in the design of the Problem Statement Analyzer we must ask, "What type of information, in what form, would most aid the analyst?"
2. The specification of the system development cycle with sufficient detail of subtasks to indicate what functions must be performed, and their interrelationships.
3. The development of algorithms using decision-making (operations research) methodology where possible, synthesizing wherever appropriate, the various "micro" decision models already available.

These tasks are being undertaken in the ISDOS Project. Basic engineering philosophy is followed: development of subsystems, evaluation, and validation in real life situations, and eventually, demonstration of the feasibility of the whole concept.

PART III
COST/EFFECTIVENESS ANALYSIS

Introduction to SECTION I

The Theory Behind Cost/Effectiveness Analysis

While this book concentrates on Phases I and II of the system development cycle, the delineation between Phases II and III is not so clear-cut that Phase III can be ignored. As pointed out by Teichroew, "Information needs interact with the characteristics of mechanisms (speed, cost, capabilities, etc.) that will be used to satisfy them; consequently, there must be iterative cycles between the analysis and design."

Cost/effectiveness analysis is an integral part of the iterative process between system analysis and design. Considerable sophistication has occurred since the simplistic, first generation approach to cost/effectiveness analysis. Computers were often justified on "intangible" savings.

This practice has all but disappeared. As Gregory and Van Horn say in their paper in this part, if some factors are treated as "unvalued" factors, the result may be a wrong decision. This section concentrates on approaches to determine the value of a system.

Section I of Part III deals with the theory of cost/effectiveness analysis. The material by Emery provides the introduction to this subject:

> "The organization responds in different ways to a technological advance. It can, as one alternative, choose to exploit the new technology primarily by lowering cost of producing information. This presumably is the motivation behind projects that merely convert an old system to the new technology, without making any basic changes in information outputs. If outputs remain essentially constant, so must value; the justification must therefore come solely from the lower cost of information.

INTRODUCTION

> Alternatively, the system can be redesigned in a more fundamental way that enhances information value. Benefits in this case might come from lower operating costs, improved services, or better decision-making information."

Emery discusses the factors in system design which affect the value of information, showing an approach to estimating "intangible" benefits.

Boyd and Krasnow provide a model for evaluating management information systems. Their model represents the gross characteristics of an information system within a dynamic model of the firm. The performance of the firm and, indirectly, that of the information system is measured in accordance with usual financial accounting procedure. The paper, therefore, shows an approach to measuring "intangible" aspects of a proposed information system in terms of its contribution to the dynamic control of the firm.

In the final paper in this section, Sharpe provides an analytical mechanism for determining cost/effectiveness of computer systems. This paper is taken from his book *The Economics of Computers*. The first part of that book explains concepts from microeconomic theory and their relevance for evaluating computer use. The remainder of the book presents application of microeconomic theory to various aspects of computing, e.g., the cost/effectiveness analysis of computer memory, pricing computer services, purchase versus rental of computers. The material selected for inclusion in this book concerns cost/effectiveness analysis of computer systems. Sharpe's material provides both theory and application, and therefore is the bridge to the next section—application of cost/effectiveness theory.

COST/BENEFIT ANALYSIS OF INFORMATION SYSTEMS

J. EMERY

VALUE AND COST AS A FUNCTION OF INFORMATION QUALITY

A system can be perfectly efficient (in the sense of meeting specifications at the lowest possible cost) and still be a very bad one. Efficiency is not enough; the desirability of a system depends on both the cost of meeting the specifications and the value derived from the information.

Thus, any attempt to define an optimum system must consider alternative specifications. Each specification defines information in terms of such characteristics as content, age, accuracy, and so on. For the time being, however, we can enormously simplify the discussion if we artificially think of information as having only a single characteristic, which we will call its *quality*.

In effect this assumes that it is possible to trade off all of the detailed characteristics of information into a single, overall index. The justification for this totally unrealistic assumption is that it permits us to discuss some important conceptual issues without the burden of unnecessary detail. Our immediate goal is to determine the relationship between the quality of information and its value and cost. This will in turn allow us to consider the optimum balance between value and cost.

Let us first consider value as a function of quality. As we have already seen, it is usually not possible to determine this relationship. Nevertheless, we can still discuss its general characteristics. The most important one is the declining incremental value of information as its quality increases.

The gross value of information continues to increase as quality goes up. Beyond a certain point, however, increased quality may add very little to value. For example, increasing the accuracy of invoices issued to customers

SOURCE: Emery, J., *Cost/Benefit Analysis of Information Systems,* The Society for Management Information Systems, 1971, pp. 16–46.

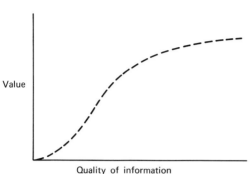

FIGURE 1 Value as a function of information quality. The value of information goes up as its quality increases. At high levels of quality further improvements yield relatively small incremental benefits.

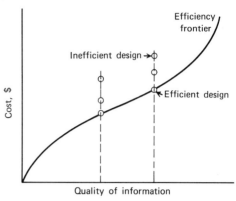

FIGURE 2 Cost as a function of information quality. Information is obtained from a specific system. Alternative systems vary in their efficiency, and so the cost of a given quality of information depends on the efficiency of the design used. The *efficiency frontier* represents the set of systems that provide each level of quality at the lowest cost. For a given level of efficiency, cost rises with increased quality.

from 98 to 99 percent reduces the number of errors by a half. This may be viewed as a highly worthwhile improvement. A further reduction by a half, to 99.5 percent, is of less value. Eventually the point will be reached at which continued improvement provides very little benefit. The same thing can be said of increases in level of detail, timeliness, or any other desirable characteristic of information. Figure 1 shows this general phenomenon.

We need a similar relationship between cost and quality. Each level of quality represents a different set of detailed specifications. For each set we are interested in finding the efficient system. The curve connecting the efficient points shows the tradeoff between cost and quality provided by current technology. In the terminology of the economist, this curve is the *efficiency frontier*. It is shown in Figure 2.

Of course, we do not actually prepare curves of this sort. At most we may look at a few alternative levels of quality—a "real-time" versus a batch processing system, for example. For each level of quality we may then evaluate a few alternative designs. The smooth curve shown in Figure 2 thus represents a considerable abstraction from reality.

It is difficult to defend any particular cost curve. Nevertheless, certain characteristics are probably fairly general. For example, over a considerable portion of the curve, economies of scale are common—that is, as quality goes up, costs go up less than proportionally. Eventually, however, quality increases to the point that costs start to rise very rapidly as the limits of current technology are approached.

BALANCE BETWEEN VALUE AND COST OF INFORMATION

Finding the Optimum System

Having discussed (abstract) curves relating value and cost to the quality of information, we are now in a position to consider the design of the optimal system. We should aim at finding the design that maximizes the *net* benefits—that is, the difference between gross

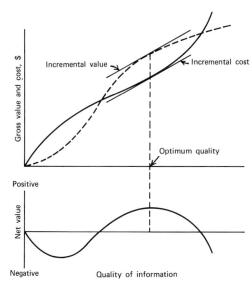

Positive / Net value / Negative / Quality of information

FIGURE 3 Determining the optimum system. The optimum level of quality occurs at the point at which net value (i.e., gross value minus cost) is maximum. This can also be viewed as the point at which incremental value equals incremental cost. Obviously, the optimum system does not provide all useful information; there will always remain unfulfilled information "requirements" that cost more to satisfy than they contribute in benefits.

benefits and cost. Equivalently, the optimum occurs at the point where incremental value just matches incremental cost. This is shown in Figure 3.

Since in practice such curves rarely exist, it is not possible to actually find the optimum design. Even if the curves were known, the optimum would tend to shift during the time span required to implement the system.

Nevertheless, an important and valid conclusion emerges from this simplified view of reality. The optimum system does not supply all useful information, since some information costs more than it is worth. Therefore, the specifications of systems requirements must simultaneously consider both cost and value of information.

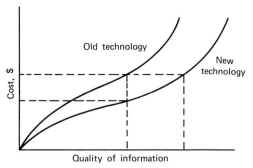

FIGURE 4 Lowering the cost curve with an advance in technology. As information processing technology advances, it becomes possible to obtain a given quality of information at lower cost than before. Alternatively, higher quality can be obtained at the same cost. The heavy line represents *dominant* systems that provide some combination of both higher quality and lower cost.

Effects of an Advance in Information Technology

The (gross) value of information does not depend on the technical means of obtaining it, but costs do. Suppose that an advance takes place in the technology of processing information, such as occurred, for example, between the early 1950's (when punched card technology prevailed) and the current computer era. The advance may come from either hardware or software improvements. The effect is to drop the current cost curve below the earlier one, as shown in Figure 4.

The organization can respond in different ways to a technological advance. It can, as one alternative, choose to exploit the new technology primarily by lowering the cost of producing information. This presumably is the motivation behind projects that merely convert an old system to the new technology, without making any basic changes in information outputs. If outputs remain essentially constant, so must value; the justification must therefore come solely from the lower cost of information.

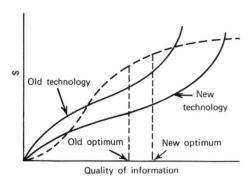

FIGURE 5 Effect of an advance in information processing technology on the optimum information quality. The optimum quality of information changes when an advance in technology lowers the cost of information processing. The new optimum is almost always at a higher level of quality. In this illustration the new optimum dominates the old (i.e., gives higher quality information at lower cost). In other cases, the new optimum could result in a higher cost than before but give benefits that more than offset the higher cost.

Alternatively, the system can be redesigned in a more fundamental way that enhances information value. Benefits in this case might come from lower operating costs, improved service, or better decision-making information. Figure 5 shows the alternatives available.

It is difficult to lay down hard and fast rules about the best strategy to follow. Clearly, however, an advance in information technology tends to shift the optimum design toward higher quality information. This is simply a manifestation of the general economic principle that a reduction in the price of a resource (relative to other resources) should normally lead to its greater use.

But even if the best long-run strategy is to upgrade the quality of information through a redesign of the system, attractive short-run benefits may also be possible through a relatively straightforward conversion of the existing system. The two approaches are not necessarily in conflict; they may proceed more or less concurrently. By the time the fundamental redesign is ready for implementation, an earlier short-term conversion may have already paid for itself handsomely through cost reductions. Unfortunately, too many organizations appear to pursue short-term savings at the exclusion of any long-term benefits.

Important Characteristics of an Information System that Govern its Value and Cost

In discussing the balance between value and cost, we found it convenient to use a composite characteristic of information called quality. In practice we cannot deal with information in this way; instead we must consider each of its individual characteristics. Although there are tradeoffs among the characteristics —between detail and timeliness, for example—it is useful to consider their separate effects on the overall value and cost of a system.

Allocation of Tasks between Man and Machine

An information system includes both human and automatic components. A critical characteristic of a system is the way in which these tasks are divided between human and computer. Certain tasks clearly belong to one or the other, but this is by no means always the case.

Complex decision-making that deals with ill-structured goals and relationships is typically best handled by man. So are tasks that occur rarely and do not involve major risks. It is exceedingly difficult (or impossible) for a computer to duplicate man's flexibility and ability to generalize, recognize complex patterns, and deal with unexpected or unusual situations. On the other hand, the computer enjoys an obvious edge over man in a number of respects—in speed, accuracy, volume of data, and the ability to draw inferences from complex models.

A system designer faces the job of allocating tasks between man and machine in the way that leads to the best overall performance. Problems of allocation arise at all levels in the system. At the operating level the designer must determine the extent to which clerical tasks should be replaced by the computer. Typical examples of such questions are:

- Should freight or passenger rates be calculated automatically within an airline information system?
- Should premiums be calculated automatically within an insurance system?
- Should detected errors be corrected automatically?
- Should a rare combination of circumstances be handled automatically (or as an "exception" dealt with by a clerk)?

Similar types of issues arise in connection with the design of decision-making systems:

- Should inventory order points and order quantities be calculated automatically?
- Should buy and sell orders be generated automatically in a trust management system?
- What thresholds should be set to require automatic decisions to be reviewed by a human?
- What tasks can the computer perform to aid human decision making?

It is difficult to provide many hard and fast generalizations about such questions. One generalization, however, is inescapable: the optimal system falls far short of complete automation. Insofar as possible, each task should be dealt with on its own merits, considering both the value and cost of performing it automatically.[1]

Content of the Data Base

The data base provides an organization with an image or analogue of itself and its environment. The more detailed the image, the greater its realism. This may improve the decisions that rely on the data base as a source of information, but it also increases costs.

The content of the data base depends on the data that enter the system and how long they are retained. High-volume data are usually captured in the form of transactions that feed some operational system. Sales data, for instance, are collected as part of order processing. Once immediate needs have been met —after an order has been shipped, for example—transaction data can be retained in detailed form in an accessible storage medium, or they can be retained only in aggregate form.[2] The level of aggregation and the length of retention are important system characteristics.

Let us examine this issue in more detail. Suppose we are designing an order processing system for a supermarket chain. Replenishment orders for each store are processed daily at a central location. The system generates

---

[1] The interdependencies among tasks make this approach difficult to apply in practice. For example, an on-line system for printing railroad passenger tickets would probably not be feasible if rates were calculated manually. We can broaden the definition of a "task" to include both printing and rate calculation, and then decide whether the combined task should be automated. When a task is defined too broadly, however, the designer may overlook some subtasks that could be better handled independently as a manual operation.

[2] Detailed transactions may also be retained in an inexpensive but relatively inaccessible form for archival purposes (such as microfilm). We are concerned here with data stored in a way that permits easy access for further processing.

shipping schedules and thus has access to data about the current day's shipments of each stocked item to each store (three cases of Campbell's tomato soup shipped to Store 53, say).

Now, it is highly unlikely that each individual replenishment order will be worth saving after it has been processed. The real issue is the level of aggregation to be retained. One alternative is to retain individual item data aggregated across all stores within a given week. Shipments to each individual store could be aggregated across all items.

Aggregation of this degree washes out information about the movement of a given item at a given store. Suppose the buying department wishes to analyze the sales of each product to determine if it generates enough gross margin to justify its use of shelf space. If only total figures are known, the decision must be based on an item's *average* sales per store. But an average can be very misleading because it hides all variation among the stores.

Sales of some items may vary greatly among stores. One supermarket firm found, for example, that virtually all of the movement of 25-pound sacks of flour occurred in a small number of rural stores serving women who were accustomed to baking their own bread. For these few stores it was essential to carry the item. However, if the decision to stock it were based solely on average sales per store, large sacks of flour would not be retained. Only a data base that retains item-store data would lead to the proper conclusion. See Figure 6.

Even if detailed data find little direct use, their retention may still be very desirable in order to allow aggregation in unanticipated ways. Suppose, for example, that an analyst performing a distribution study wishes to find the total weight of products shipped to each store. Unless item-store shipment data are maintained (as well as the unit weight of each product), it would be very difficult to retain reliable estimates of shipments. To be sure, this information could be accumulated as part of the periodic reporting scheme, but it is impossible to anticipate all possible aggregations of detailed data. The retention of relatively disaggregated data is the best way to overcome this difficulty. See Figure 7.

The incremental value of information diminishes as detail grows. This is so because it becomes increasingly likely that information needs can be satisfied without resorting to still finer detail. In the supermarket example, storage of item-store data by *week,* say, would probably meet most information needs; it would not be necessary to retain *daily* item-store shipments.

The incremental value of data declines with age as well as with detail. Retention of complete transaction data may be justified for a short time following a shipment to a store. After a few months, however, the details should be summarized into, say, item-store data by week; the transaction data can then be discarded (or stored in a low-cost medium for archival purposes). When older than a year, the data might be further summarized into sales by product groups (canned soups, for instance) and month. This process of increasing the degree of aggregation might continue for several years. Usually only highly aggregated data are retained permanently.

The size of the data base obviously grows as the level of detail and length of retention increase. In fact, it can grow explosively. For example, if sales data are maintained on a daily instead of a weekly basis, the size of the file may increase by a factor of six (assuming a six-day week). A similar factorial growth occurs as other dimensions of classification are added. Differentiating between

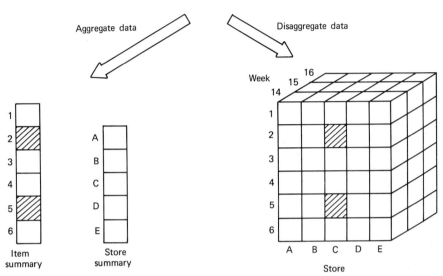

FIGURE 6 Retention of aggregate versus disaggregate data. Transaction data, such as the shipment transactions shown above, are rarely kept very long in complete detail; the question is, how much detail should be retained (and for how long). Maintaining only aggregate data—for example, total units shipped of each item and total dollar shipments to each store—greatly reduced storage requirements. Retaining item-store shipment by week, say, drastically increases the size of the file, but allows a much greater variety of information to be obtained.

credit and noncredit sales, for example, potentially doubles the size of the file. The length of retention also has an important effect on size. The data base can grow very rapidly if steps are not taken to cull out old data.

The cost of storing data naturally increases as the size of the data base grows. The cost of the storage medium increases, but usually less than proportionately (i.e., some economies of scale are exhibited, at least over a considerable range of volume).

Far more significant is the cost of retrieving desired information. This cost depends on the way in which data are organized and accessed. Retrieval may require a sequential scan of a portion of the data base or, alternatively, relatively direct access by means of indices or linked records. In either case cost of retrieval may increase more than proportionately as the size of the data base grows.

The decision regarding the proper level of detail and length of retention should be based on the expected value of stored data. The expected value of a given data element equals its value, if it is required, multiplied by the probability that it will be required. Thus, the storing of information may be justified on the

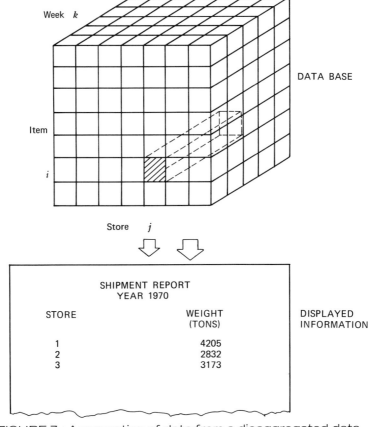

FIGURE 7 Aggregation of data from a disaggregated data base. A detailed data base allows great flexibility in calculating specific aggregations. For example, the total weight of product shipped to a store can be calculated if item-store shipment data are available. The item-store shipment data may, in turn, be aggregated from weekly shipment data. Thus,

total weight shipped to store $j = \sum_i w_i S_{ij} = \sum_i w_i \sum_k S_{ijk}$

where

w_i = unit weight of ith item
S_{ij} = total unit shipments of ith item to ith store
S_{ijk} = total unit shipments of ith item to jth store during kth week

grounds that (1) it will be extremely useful if asked for, even if this is fairly unlikely (such as a cancelled check); or (2) it is very likely to be needed and its value upon retrieval exceeds the cost of retaining it. In practice one must usually rely on subjective (and somewhat vague) estimates of value and probabilities, but the person making these judgments should at least have his objective clearly in mind.

We may summarize this discussion in the following way:

- Increasing the degree of detail and the length of retention of data may drastically increase the size of the data base.

- The value of the data base tends to increase with its size, but at a diminishing rate.
- The cost of maintaining and retrieving from the data base grows rapidly as its size increases.
- The optimum size occurs at the point where incremental value equals incremental cost. Typically this point falls considerably short of retaining (for very long, at least) complete transaction data.

Selectivity of Displayed Data

It is not enough merely to keep useful information in the data base; the information must be displayed if it is to be used for human decision-making. As the size of the data base grows, it becomes all the more critical to display only highly selective information. Ideally, the information displayed should always have some surprise content and lead to a better decision than would otherwise be made. In practice we can only approach this ideal.

Let us consider the total variety of ways information can potentially be obtained from the data base. This includes all of the individual data elements, as well as any arbitrary transformation performed on these elements. The transformations include simple aggregations, standard statistical analyses, preparation of graphical output, and even calculations using complex decision models.

Now, let us look at the information needs of a given user. The very large (in fact, infinite) set of potential information available from the data base can be partitioned in two independent ways: displayed versus nondisplayed information, and relevant versus irrelevant information.

Displayed information is that portion presented to the user. It may be displayed either in hard copy form or as a transient display on a CRT or some similar device.

The relevance of information is not as easy to define; it is a matter of degree. We will define information to be relevant if its value exceeds the incremental cost of using it—in short, if the user would benefit by having it displayed. This means that relevant information offers some surprise, leads to a decision that would otherwise not be taken, and improves payoff.

The total set of potential information can be broken down into four subsets:[3]

1. Relevant and displayed.
2. Irrelevant and not displayed.
3. Relevant and not displayed.
4. Irrelevant and displayed.

The designer of a system would naturally like to limit the information included in the third and fourth categories, since they represent "errors" of omission or commission. When relevant information is not displayed, decisions will not be as good as they otherwise would be. When irrelevant information is displayed, the user himself must select the useful information from the useless. The greater the proportion of irrelevant information, the greater the effort on the user's part to cull it out and the greater the risk that he will overlook valuable information. Like the gold in the ocean, relevant information too diluted with useless information ceases to have any value.

Errors in information selection are unavoidable. In order to display *all*

---

[3]The four-way breakdown is analogous to that faced in quality control, in which one is concerned with whether a product should be accepted as good or rejected as bad. The possibilities are: (1) satisfactory quality and accepted, (2) unsatisfactory and rejected, (3) unsatisfactory and accepted (the so-called "buyer's risk"), and (4) satisfactory and rejected (the "producer's risk"). This analogy is worth pointing out because the issues are much the same in the two contexts.

relevant information, and *only* relevant information, the system would have to determine what each user already knows and what his decision process is. This is clearly impossible, and so the choice of information to display must be a compromise between displaying too much or too little. The probability of displaying relevant information goes down as the degree of filtering increases, but this is accomplished by an increase in the probability of overlooking relevant information.

The optimum degree of selection depends on the relative penalties of the two types of errors. If failure to display critical information carries a very high penalty relative to the cost of displaying irrelevant data, then it is advantageous to err on the side of displaying too much. This is the implied motivation behind many existing information systems.

Both types of errors can be reduced simultaneously if sufficient resources are spent in implementing more effective selection.[4] It is primarily the cost of design effort and information processing that imposes an upper limit on the extent to which selection errors should be avoided.

Increased selectivity can be achieved through a variety of means. Some of the techniques have been used for many years, while others have become feasible only with relatively recent advances in information technology. Let us examine the more important techniques.

Appropriate Aggregation of Details. Virtually no one needs transaction data in complete detail (except for handling the transaction itself, of course); it is almost always necessary to aggregate the details before they can be used. The aggregation may be by product group,

---

[4]This is analogous to taking a larger sample in quality control applications in order to simultaneously reduce both the producer's and buyer's risks.

organizational unit, cost category, time period, or some similar dimension. The intent is to aggregate in a way that washes out irrelevant dimensions and preserves the relevant ones. For example, sales data used by the marketing vice president may be aggregated by major product groups and sales regions. Finer detail (sales of a specific product, say) or some alternative aggregation (sales classified by industry, perhaps) would probably not be particularly useful in assessing the performance of regional sales managers (although obviously such information might be entirely relevant for other purposes).

Simple aggregation, in which each data element is added to its appropriate category, implies that each element carries the same relative value. A summary sales report, for example, implicitly assumes that a dollar of sales for Product X is equivalent to a dollar of sales for Product Y if they both fall within the same product group. This is often valid, but it need not be. A production report that shows delivery performance may give quite misleading information if it merely provides a count of late jobs. A much more meaningful figure would be one that weights each late job by a measure of its lateness and importance (man-hours of labor applied, for instance).

Although aggregation is essential to reduce the volume of displayed information, it always carries some risk of washing out relevant information. A sales report may hide significant trends within a product group or sales territory, for example. Similarly, an inventory report that aggregates across all items may cancel out a serious imbalance in which some items are in critically short supply while an offsetting group has a large surplus. The remedy for this is to aggregate within finer (and hopefully more relevant) categories. For example, the inventory report can aggregate according to the current status of each item,

INVENTORY ANALYSIS

STATUS	NUMBER OF ITEMS	BALANCE ($000)	STANDARD ($000)	PERCENT OF STANDARD
IN CONTROL	9242	2530	2400	105
SHORT	1025	135	420	32
SURPLUS	779	832	205	406
TOTAL	11046	3497	3025	116

FIGURE 8 Aggregation of displayed information to reduce detail while preserving essential information. Aggregations are necessary to increase selectivity, but run the risk of washing out significant details. A complete aggregation of the above inventory information would give the misleading impression that inventory is in control (only 16 percent above standard). Aggregating the inventory items into three categories according to their current status gives a much more realistic picture at little increase in the amount of information displayed.

using the three categories "in control," "short," or "surplus." See Figure 8. Unfortunately, increasing the fineness of aggregation also increases the likelihood of displaying irrelevant information.

The risk of overlooking relevant information can be reduced by providing the user with backup details that explain each aggregation. For example, a report giving sales by product group and region might be supported with more detailed reports by sales office and product subgroups. Each detailed report may, in turn, be supported with still more detailed reports that show sales by individual salesmen and product. A detailed report should clearly show the relation between its data and the next higher aggregation. See Figure 9. Similar backup can be provided to support aggregations across cost categories, multiple time periods, and the like.

This hierarchical linking of reports provides an effective means of reducing unnecessary detail while still allowing the user to penetrate into the details when this appears to be warranted. An obvious requirement of a complete hierarchical linking of reports is the proper nesting of data elements in terms of data definitions, reporting periods, and classifications. In other words, a lower level data element must be identified uniquely with a higher level aggregation for a given classification scheme.[5]

Good Human Factors in the Design of Display Formats. Information must be perceived before it can have value for human decision-making. The effectiveness with which a user perceives information is largely governed by the way in which it is displayed. The interface between the system and user is one of the more critical design factors.

Some of the general principles of good display are as follows:

- Use standard report formats, headings, and definitions whenever possible. This permits a user to scan a display without having to interpret each item.

[5] Geographical boundaries provide a useful analogy. An example of proper nesting is the aggregation of United States counties to form states and the aggregation of states to form the United States. On the other hand, metropolitan areas do not nest within state boundaries. Proper nesting for a given classification does not preclude alternate nesting for other classifications. For example, the sales of a given item can be aggregated by geographical boundaries, product groupings, or industries.

MONTHLY SUMMARY SALES REPORT ($ MILLIONS)

		REGION					
		NORTH EAST	SOUTH EAST	CENTRAL	SOUTH WEST	WEST	TOTAL
PRODUCT GROUP	A	25	12	32	8	10	87
	B	17	9	35	4	12	77
	C	12	5	22	2	3	44
	D	(20)	7	15	15	18	75
	TOTAL	74	33	(104)	29	43	(283)

MONTHLY REGIONAL SALES REPORT ($ MILLIONS)
PRODUCT GROUP D   NORTHEAST REGION

		OFFICE			
		NY	BOS	PHIL	TOTAL
PRODUCT SUBGROUP	D1	4.2	1.5	2.1	7.8
	D2	1.8	.8	2.6	5.2
	D3	1.1	1.7	.3	3.1
	D4	3.0	.3	.8	4.1
	TOTAL	10.1	4.3	5.8	(20.2)

FIGURE 9 Hierarchic relationship among reports. It is desirable to provide detailed backup reports for all aggregate data. A figure on a high-level report should appear as a total on the next lower level report.

- Each item displayed should be labeled or have an obvious interpretation.
- Avoid unnecessary precision. Since an aggregation inevitably represents an approximation of reality, excess precision adds little value while it clutters up the display.
- Use graphical display when feasible. A graphical display reduces unneeded precision while often revealing relationships among variables much more perceptibly than a tabular display.
- Provide a basis for interpreting information. A given piece of information seldom has value by itself; it must be assessed relative to some standard or anticipated result. It is therefore important that a user be provided sufficient information to comprehend the significance or surprise content of new information. This can be done by displaying the new information (e.g., actual current results) in juxtaposition with the existing plan, standard, or past results.
- Provide links among separate displays. Each display should contain relatively little information in order not to swamp the user. It is therefore necessary to use multiple displays if much information is to be conveyed. The user should be able to relate one display to another. Hierarchical relationships among displays, as discussed in the previous section, provide one of the basic means of doing this.

Use of the Exception Principle. The exception principle is by no means a new idea. Its intent is to identify "exceptions" that require human attention. Information about exceptional conditions is displayed while all other information is filtered out. An exception is deemed to have occurred when actual results deviate from a standard by more than an established threshold value. See Figure 10.

Threshold values are ideally set at a level that correctly distinguishes between relevant and irrelevant information. This means that information about conditions outside control limits should lead to a new decision, such as altering an existing plan. The deviation from the plan may be either favorable or unfavorable. A favorable deviation opens new opportunities that would be lost if the current plan were not revised—an increase in sales, for example, may call for increased output. An unfavorable deviation may require a change in plan in order to minimize the cost of the deviation—a renegotiation of delivery schedules when difficulties are detected in an engineering development program, for instance.

The effectiveness of an exception reporting system obviously hinges on its ability to distinguish between relevant and irrelevant information. The ideal system cannot generally be achieved in practice, since this requires complete formalization of the decision process. Suppose, for example, that we wish to report inventory status according to the exception principle. In order to identify items of inventory requiring attention (to change the existing production schedule or the order point and order quantity, say), the system must be capable of comparing the existing plan with the new optimum based on the latest known conditions. Only when there exists a significant disparity between the two is it necessary to signal an exception.

Thus, the identification of a true exception requires (1) a prediction of the likely outcome if no change in plan is made, and (2) the penalty that this outcome would entail in comparison with the current optimum plan. Only if the

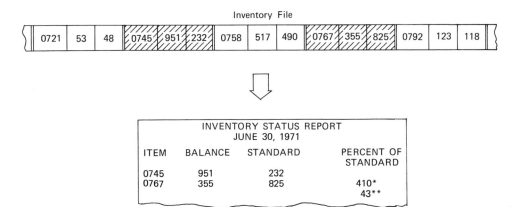

FIGURE 10 Exception reporting. An exception report displays only the conditions that fall outside of defined control limits. An inventory status report, for example, may show only the items that are over 200 or less than 50 percent of their standard. The system should readily handle changes in control limits in order to allow users to redefine their information requirements. The inventory standard used in the report may be set externally by management, or it could be computed periodically from an inventory model.

penalty exceeds the cost of changing the plan should the situation be labeled an exception. Such formalization rarely exists, and so it is necessary to strike a balance between the risk of displaying too much or too little information.

The control limits used to identify exceptions can be based on any (or all) of the variables used in a plan. In the case of an operating budget, for example, manufacturing overhead might be monitored in order to signal an alarm when costs become excessive. The deviation allowed each variable should be set according to its relative importance and its normal random fluctuations. Thus, a variation of 50 percent from standard may be allowed for a minor variable subject to considerable routine fluctuation, while a 5-percent variation in a well-behaved aggregate variable may be deemed worthy of management attention.

An exception can be defined in terms of trends as well as deviations. For example, an unfavorable deviation in raw material costs of, say, 8 percent may normally be considered unexceptional; but if it follows a month with a favorable variance of 6 percent, the unfavorable trend could be tagged as an exception.

Because of the imprecise nature of control limits, they should be viewed as parameters subject to change as users see fit. If too many exceptions are currently being displayed, a user may wish to broaden the limits in order to display only the most severe exceptions. If only a few exceptions occur, a user should be able to tighten limits in order to reveal the situations that may most benefit from his attention. The extreme case of zero limits should be allowed when an exhaustive display is desired, as well as "infinite" limits when no information is wanted.

A well-designed exception reporting scheme can add greatly to the value of an information system. To be sure, it carries some risk of overlooking important information. It is a profound mistake, however, to assume that this is a risk unique to formal exception reporting; a user may stand a much greater risk of overlooking significant information if it is immersed in a huge report containing mostly irrelevant data. Selection of relevant from irrelevant information must be performed either by the user himself or within the system. The capability of the computer to apply sophisticated selection criteria often gives it a tremendous advantage over the user in identifying likely candidates for closer inspection.

There are obvious costs of providing exception reporting services. The design and maintenance of a sophisticated system is by no means trivial. Users must be trained to use the system well. Some additional processing costs are usually incurred to filter out irrelevant data (although savings in display costs may in some cases more than offset this). However, when these costs are compared to the (largely hidden) costs of having users perform their own selection, the economies almost always favor formal exception reporting.

Use of ad hoc Inquiries. The typical periodic report is based on the anticipated recurring needs of a group of users. It therefore necessarily contains a great deal of information that is not relevant to a given user at a given point in time.[6]

[6] A specialized example of a periodic report is the standard telephone directory. It contains a very low density of useful information for any given subscriber. Nevertheless, economies favor such an approach because the cost of providing each subscriber with selected information (through a directory assistance operator or a tailor-made directory) would be prohibitive with existing technology. Besides, the cost of selection from the telephone book is relatively small (if one knows the name of the person whose number he is seeking) and is borne by the subscriber.

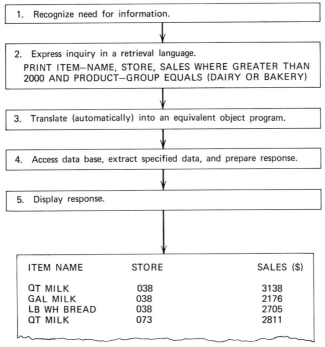

FIGURE 11 Processing of an ad hoc inquiry. A specific display is prepared in response to an ad hoc inquiry. The resulting information can therefore be highly selective.

An alternative to periodic reports is an ad hoc inquiry system that provides a response (within a reasonable time lag) to each specific request for information. See Figure 11. If information is supplied only on demand and is tailored to a specific user's needs, it stands a much higher probability of being useful.

The capabilities offered by inquiry systems vary widely. Some systems only allow data to be extracted from the data base, without any further manipulation. Of much greater use is a system that can perform appropriate aggregation of detailed data. Some systems also have built-in standard transformations of data, such as the calculation of an average, range, or standard deviation. A still more advanced capability is the ability to handle transformations defined by the user in the retrieval language of the system. Finally, in the most sophisticated systems the user can define transformations with any available language (such as FORTRAN, COBOL, or PL/1); these are properly termed data base management systems, since they serve as the basic interface with the data base. Olle (1970) discusses some of these issues.

The costs of handling ad hoc inquiries depend greatly on the particular system used, the size of the data base, and response time requirements. Costs include the effort required by the user to specify requested information, as well as the processing of the inquiries. A sophisticated system providing a powerful retrieval language and fast response may be quite expensive indeed. On the other hand, handling batch-processed inquiries of a standard nature may be quite inexpensive, particularly if the processing is combined with routine file updating.

It is sometimes feasible to maintain certain commonly used information, such as standard financial data, as a separate subset of the total data base. Relatively standard inquiries can then be handled by retrieving data from this subset. Its size may be very much smaller then the entire data base, permitting substantial economies.

Use of a Decision Model. A decision model can be viewed as a particularly effective information filter. The user of a model is presented with the output of the model, rather than its detailed inputs. Output normally is very much less voluminous than input, which drastically reduces the amount of information displayed.

The degree of filtering depends greatly on the type of model used. A decision model fully embedded within the information system, in which all input data are obtained automatically from the data base and all decisions are fed automatically into operations, provides the most extreme example. In this case only aggregate results need to be displayed for human monitoring.

This is obviously not the common situation. In most cases even "optimizing" models require human intervention in providing inputs and in reviewing and modifying outputs. Simulation models usually require active human participation in proposing alternatives and selecting the best one among them. Nevertheless, the amount of information presented to the user is usually much less than would be required if he made the decision unaided by a model.

A decision model can be combined effectively with exception reporting. The model can be run periodically to test if the latest conditions—e.g., sales, costs, etc.—should cause a revision in existing plans. If so, this fact can be displayed for human review; otherwise, nothing is reported (except perhaps aggregate performance).

Response Time

The response time of an information system is the time it takes to respond to a significant stimulus. The stimulus may be the occurrence of an event that is to be reflected in the data base, such as the arrival of a sales order from a customer. Alternatively, the stimulus may be a request for information already stored in the data base. Thus, in discussing response time we are concerned with two aspects of time lag: (1) the time it takes to update the data base, and (2) the time it takes to retrieve desired information from the data base.

It is useful to break down response time in this way because value and cost may depend heavily on which aspect is being considered. Some applications may benefit significantly from quick updating and retrieval, others may call for only quick retrieval, while still others may require neither quick update nor retrieval.

Value of Short Update Time. Let us first consider the value of information as a function of its recency. The timeliness of the data base depends on the update response time. In an environment that changes both rapidly and unpredictably, the data base must be updated quickly in order to keep a faithful representation of current reality. If decisions are highly sensitive to changes in the environment, substantial penalties may be suffered if the update time is not kept short relative to the rate of these changes.

Some applications clearly benefit from on-line updating. These are nearly always found at the operating level of the organization. It is only here that events are likely to take place frequently

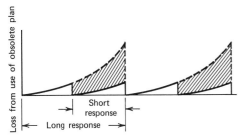

FIGURE 12 Loss due to use of old information. A plan, once made, begins to decay with time when unexpected events occur. Replanning brings the plan back to the new optimum based on current information. The more frequent the planning and current the information, the less the loss due to use of obsolete plans. The shaded area shows the incremental loss when response time is doubled; it therefore represents the incremental value of the shorter response time.

enough to justify very short update lags. Such "real-time" applications as air traffic control, industrial process control, and stock quotation services would scarcely be feasible without a data base kept current within a matter of seconds. Airline reservation systems begin to incur significant penalties (in the form of underbooking or overbooking, for example) as updating lags exceed an hour or so. High-volume production and inventory control systems similarly benefit from rapid file updating. All of these examples exhibit the same essential characteristic: the average time interval between (unpredictable) events is short, and so effective control demands a correspondingly short update lag. Failure to revise current plans in the face of unanticipated events may lead to significant loss. See Figure 12.

This is not a common characteristic of higher level decision processes. Individual events have little relevance at this level; all that usually matters is aggregate behavior. Aggregate variables, such as total sales and total capacity, exhibit significant change only slowly and often in a fairly predictable way. It is not necessary to provide rapid updating of the data base in order to track sluggish and well-behaved changes. Furthermore, the accuracy of predictions over the relatively long planning horizon typically associated with higher level decisions would not be significantly reduced by update lags of days or even a few weeks. We can therefore conclude that quick updating of the data base usually adds little value to higher level decision making. See Figure 13.

Value of Quick Retrieval Time. The value of quick retrieval is often quite independent of the age of information obtained. Thus, under some circumstances we may want fast access to relatively old information. A short retrieval time enhances value under two circumstances: (1) a decision (and the resulting action) must be made quickly, or (2) the decision process benefits from a series of accesses performed in "browsing" fashion.

The typical real-time system requires fast action (as well as fast file updating) in order to control a dynamically changing environment. Air traffic control, to use this example again, must deal with the problem of collision avoidance. A speedy decision is warranted when two aircraft on a collision path approach one another at Mach 2. The information on which the decision is based must therefore be both current and rapidly accessible. See Figure 13.

In less dynamic cases it may be important to make a fairly quick decision even if the information supplied is not particularly current. In an order entry application, for example, it may be quite valuable to respond immediately to a customer's inquiry concerning stock availability (while the customer waits on the telephone, say). In most cases there

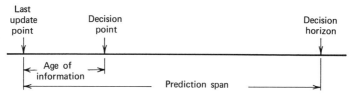

FIGURE 13 Effect of age of information on decisions. Decision-making—whether at the operational or strategic level—requires a prediction for each of the input variables used in the decision process (e.g., sales forcasts, inventory levels, aircraft positions, etc.). The prediction span for a given variable extends to the decision horizon from the point in time at which the data base was last updated. The quality of a decision depends in part on the accuracy of the predictions. Accuracy, in turn, depends on the length of the prediction span and the inherent variability of events within the span. A rapidly changing and unpredictable environment requires a short prediction span in order to maintain control; accordingly, information must be very current in order to provide suitable accuracy. Strategic decisions, on the other hand, tend to have long-term effects; and so the decision horizon is well into the future. In this case reducing the age of information adds little to predictability over the long span.

would be no great value in having an up-to-the-minute picture of inventory status; normally it would be sufficient to base the decision on status as of some earlier cut-off point[7] (e.g., at the end of the previous day). This allows file updating to be handled on an economical batch basis. Inquiries between file updates can be processed through an online system or a manual operation that uses a daily status report (the choice depending on speed and cost of processing an inquiry). Most on-line reservation systems are motivated much more by the requirement for fast confirmation than by the need to keep files current within a few seconds.

Data entry applications can also benefit substantially from quick access to the data base. An interactive system can perform validity checks while a clerk remains at a remote terminal. A detected error usually can be corrected immediately, thus avoiding the serious complications involved in off-line error correction (as well as speeding up the updating process). Data accepted as valid may then be used for on-line file updating (if a short update lag is required), or may simply be stored temporarily for later batch updating.

Although the above examples are representative of important quick response applications, it certainly is true that the bulk of decision-making within an organization cannot justify quick retrieval on the grounds that hasty action is called for. On the contrary, a delay in reaching a decision of days or weeks may not carry a serious penalty, especially at the strategic level. Nevertheless, even in these cases there may still be considerable value in providing information within a short response time.

Quick response tends to be especially valuable when one deals with an ill-structured problem. Typically one cannot specify in advance all required information. Instead the problem is best

---

[7]If inventory levels change rapidly, but fairly predictably, the system can adjust the inventory balance by the predicted withdrawals from the cut-off point up to the time of the decision. In any case, one can always increase the probability of having an item in stock by simply maintaining a larger safety stock.

approached through a sequential examination of responses. Each response may suggest new information that would shed additional light on the task at hand. This "browsing" process can continue until the problem solver feels that further probes are not justified.

A quick response allows more alternatives to be examined, which normally results in an improved decision. Even though there may be no great urgency in reaching a decision, there is always some upper limit on the time available. The number of sequential probes is therefore limited by response time.

If the response is fast enough, in the order of seconds or perhaps at most a minute,[8] the decision can be reached through the uninterrupted participation of the human problem solver. This allows him to retain a grasp of the problem that would otherwise be lost if he were forced to switch to some other activity while he waits for a response. Although the evidence is by no means overwhelming, it appears reasonably certain that interactive man-machine problem solving of this sort can be very effective in dealing with many types of complex tasks.

A capability of this sort may be particularly useful when a decision is reached through group cooperation. For example, setting a quarterly production schedule may require the participation of the managers of marketing, manufacturing, purchasing, and personnel. The value of a man-machine model may be fairly limited unless it allows the group as a whole to explore alternative schedules. This probably is feasible only if the model provides a response quick enough to make a decision in a relatively short meeting among all participants.

Cost of Short Response Times. Having discussed the value of information as a function of response time, we can now turn our attention to the matter of costs. Cost is governed largely by the frequency with which the data base is accessed for updating or retrieval.

Consider first the case of batch processing. Response time is clearly a function of the interval between successive batches. Suppose that this interval is I, and that there is a processing lag L between the cut-off point at the end of the batch cycle and the availability of output.

Update time depends on the timing of an event. If the event occurs just before the end of the cycle, update time is the processing lag L; if it occurs just after the cut-off point, update time is $I + L$. The average time, assuming that events are spaced uniformly over the processing cycle, is $I/2 + L$. This is shown in Figure 14. The same minimum, average, and maximum response times apply to retrieval, whether information is obtained from a periodic report or through an ad hoc inquiry.[9]

Thus, response time is a function of the processing cycle and processing lag. The processing lag can be reduced primarily through the use of more rapid means of collecting and transmitting data, such as an on-line data collection system tied directly to a central processor. The lag can also be reduced somewhat by reducing the average time a

---

[8]Humans become extraordinarily impatient in an interactive environment if they have to wait very long for a response.

[9]These times apply only to *status* information, which gives the value of a particular variable, such as inventory level, at the cut-off point. *Operating* information deals with the series of events that occur over a reporting interval, such as the orders shipped during the past month. Immediately before an operating report is produced, the latest information available covers events that occurred as long as $R + I + L$ time units earlier, where R is the reporting interval (usually, but not always, the same as I). The average age of operating information is $R/2 + I/2 + L$. Gregory and Van Horn (1963) discuss these matters in some detail (pp. 576–580).

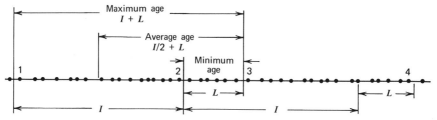

FIGURE 14 Response time as a function of the batch processing interval and processing lag. The time required to update a sequential file depends on the processing interval I and the processing lag L. All events that occur during a given interval are processed together in a batch. The first event within the interval (at time 1 in the figure) is not reflected in the file until $I + L$ time units later (at time 3). Information about the last event (at time 2) enters the file in only L units. The average, assuming a uniform distribution of events, is $I/2 + L$. The minimum, maximum, and average *retrieval* times are the same as the *update* times. The information retrieved from a periodic report or through batch processing of ad hoc inquiries is L time units old immediately upon completion of the processing cycle (at time 3); it is $I + L$ units old at the end of the cycle (at time 4). In some applications the minimum response time may be the most important design consideration (e.g., in the case of a financial statement reviewed by management only when it first becomes available); in others, the maximum (e.g., the handling of customer orders); and in still others, the average (e.g., inventory updating).

batch must wait in queue prior to processing. This is achieved by increasing surplus capacity to absorb fluctuations in demand. The actual run time taken to process a batch is rarely a significant portion of overall response time.

The most significant component of response time is typically the processing interval I. The number of runs is, of course, inversely proportional to this interval—e.g., twice as many runs are required if the interval is cut in half. Sequential processing runs tend to be input/output bound; this tendency increases all the more as the interval shrinks (since the number of transactions goes down when they are accumulated over a shorter interval). Therefore, processing time—and hence cost—is inversely related to response time.

As the processing interval shrinks, costs of conventional sequential processing begin to grow very rapidly. Eventually the point is reached where it becomes less expensive to use indexed sequential file organization, since this allows inactive portions of the file to be skipped. As the batch interval goes down still further, random processing—i.e., handling each transaction separately—eventually becomes the least expensive method. Figure 15 illustrates how cost varies as a function of response time.

Because of the dominant nature of input/output time for batch processing, it is usually advantageous to perform both file updating and retrieval during the same run. Retrieval may consist of handling ad hoc inquiries as well as preparing periodic reports. Since all records have to be accessed with each run, the only incremental cost of combining file updating and retrieval is the added size and complexity of the combined program.

Only active portions of the file need to be accessed with either indexed sequential or random file organization. Therefore, the number of active records becomes a principal determinant of cost (and not just the number of processing runs). It may be possible to substantially reduce the number of random accesses if updating is separated from retrieval. For example, all updating can be handled in batch fashion, while ad hoc inquiries are processed randomly. Batch updating cannot only lower processing costs, but it can also significantly reduce the problems of file security and reliability that always attend on-line file updating.

Accuracy

Information is accurate if the image it provides conforms closely with reality. Thus, an inventory report is accurate if the stated levels agree with actual inventory status. A sales forecast is accurate if it correctly estimates future sales.

The value of accuracy is derived from the improvements it brings in decisions or operating actions. Any decision or action is based on the information available. To the extent that this information does not accurately portray reality, outcomes will be less desirable than would otherwise have occurred.

The most stringent requirements for accuracy generally come from operational activities. A considerable loss may ensue from an error in a paycheck, purchase order, invoice, or bank balance. A good deal of effort is justified to avoid the unrecoverable losses, the confusion,

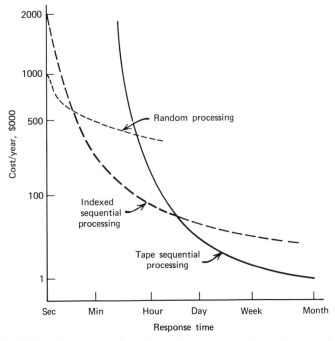

FIGURE 15 Cost as a function of response time. Economical batch processing with magnetic tape can be used when response time is not critical. As response time is reduced— and hence run frequency is increased—total input/output time grows rapidly. Eventually batch processing with an indexed sequential file becomes less expensive, since it allows skipping over inactive portions of the file and thus reduces input/output time. If response times are reduced still further, eventually random processing of a randomly organized file becomes the least expensive approach.

and the loss of goodwill usually associated with errors of this type.

Higher level decision making generally imposes relatively mild requirements for accuracy. Most decisions and payoffs are fairly insensitive to moderate errors, and so great accuracy adds little value.

Decision-making information that comes from the aggregation of transaction data usually is accurate by virtue of operational needs. Any errors that exist in the details tend to cancel out in the aggregation process.

Some data are collected specifically for decision-making purposes—for example, demographic data used in store location studies, consumer survey data used in marketing, business intelligence data used in competitive evaluations, and economic indicators used in forecasting. It usually is not necessary to subject these data to the same close control applied in collecting operational data.

Accuracy, like all desirable characteristics of information, has a cost. It is achieved by collecting error-free data that are then processed by suitable routines. Accuracy in data collection comes from error control procedures that both discourage creation of errors and provide the means for detecting and correcting any errors that enter the system. A great deal of sophistication and ingenuity can go into these procedures. Almost any desired degree of accuracy in data collection can be attained, but at an increasingly high cost as perfection is approached. Cost is incurred in the form of collecting and maintaining redundant data for error detection and correction purposes, as well as in the implementation and operation of the routines for performing error control functions.

Accuracy depends not only on the quality of data inputs, but also on the routines used in processing information. An obvious requirement for assured accuracy is that the routines be bug free. It virtually is impossible to eliminate all bugs from a large program, but careful (and expensive) testing and maintenance procedures can reduce the probability of an error to a manageable level.

A less obvious way to increase accuracy is to apply improved estimating procedures. Decision-making is concerned with predicted future conditions rather than historical accuracy. Sales data used in forecasting may be perfectly accurate; but if the estimating procedures are not suitable, the resulting errors may cause substantial penalties of stockouts or excessive inventory.

Any prediction has inherent errors, but the average size of error can be reduced through the use of appropriate estimating procedures. Accuracy is governed to some extent by the type of input data available for calculating predictions. Inputs include data about past transactions, past forecasts, economic indices, and the like. These inputs may be subjected to relatively simple procedures, such as exponential smoothing or other averaging techniques. On the other hand, accuracy may be improved significantly by using some of the more elaborate forecasting methods. Cost of such accuracy includes the maintenance of data inputs and the design and operation of the forecasting routines.

TECHNIQUES FOR MAKING COST/BENEFIT ANALYSES

We have seen that cost/benefit analysis of information systems faces some extremely complex issues. We are a long way from developing fully satisfactory approaches. Nevertheless, an organization that sinks vast sums of money into the development and operation of an information system cannot ignore the complexities; it must deal with them in the best way it can.

Analysis of Tangible Cost Reductions

Tangible costs and benefits are those that can be expressed in *monetary* terms. This clearly is possible in the case of projects aimed at clear-cut cost reductions in information processing. Elimination of clerical operations or lowering the cost of equipment rental (through greater efficiency or use of a later generation computer, say) are common examples of such projects. They are often accompanied by little or no change in information quality, essentially the same information is provided (hopefully) at lower cost. Under these circumstances the analysis need not concern itself with determining benefits, since they remain the same as before.

A cost reduction project thus can be viewed as a straightforward investment. As such, traditional methods of analysis can be applied. Certain expenditures must be made to implement the modified system. These are then followed (if all goes well) by reduced future operating expenditures. See Figure 16. One can analyze the investment in terms of net present value, internal rate of return, or payback period. Project selection can be dealt with in the same way used to set priorities on other forms of investment.

There lurks a hazard in this: it is too easy. Because cost reduction projects can be analyzed in traditional ways, the organization is often tempted to concentrate on them. This avoids the problem of having to assess benefits of improved information but runs the risk of misdirecting efforts away from projects that can make more fundamental improvements in organizational performance. This is not to say that worthwhile cost reductions should not be pursued, but the bias is often too heavily weighted in their favor.

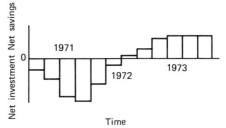

FIGURE 16 Analysis of tangible savings from an investment in an information system project. Tangible costs and benefits can be handled using traditional methods of investment analysis. The net cash flows that occur in different time periods can be translated into a single index by calculating the present value of the investment, its rate of return, or its payback period.

Analysis of Tangible Benefits

Really significant contributions normally come through enhanced information quality. Many of the benefits from improved quality may be perfectly tangible. The benefits are certainly tangible, for example, when faster customer billing reduces cash requirements. The translation into a monetary savings requires an estimate of both an annual rate of return and the total expected cash reduction. If the rate of return is estimated to be, say, 20 percent (based perhaps on the opportunity cost of internal investments), a $100,000 reduction in cash is worth $20,000 per year.

The principal difficulty in assessing benefits of this sort is the estimation of the effects of improved information. In the above example the analyst must estimate the relation between cash requirements and billing time. If it can be assumed that faster billing will not change the distribution of time lags between the receipt of an invoice by a customer and the receipt of his payment, then the relation can be estimated quite easily. For instance, a two-day reduction in billing, with average daily sales of

$50,000, will result in a $100,000 reduction in cash requirements.

The use of a formal decision model often greatly facilitates the estimation of benefits. Suppose, for example, that we would like to estimate the inventory reduction stemming from more frequent order processing. The reduction comes from lower inventory safety stocks made possible by the shorter reorder lead time (since more frequent order processing reduces the time from the breaking of an order point to the preparation of a replenishment order). The relation between processing cycle and inventory level is a very complex one, but it can be estimated easily if a suitable inventory model exists. Sensitivity studies of this sort can be used to consider the effects of altering any of the model's input data. Such studies should be an important part of implementing any decision model.

In the absence of a formal model, it becomes considerably more difficult to estimate the effects of changes in information quality. Formal models often do not exist, of course. In particular, they do not exist when the information project under consideration is the implementation of such a model.

One way to deal with this situation is to develop a "quick and dirty" model that gives a gross estimate of possible savings. For example, in estimating the benefits of more frequent order processing, we might analyze a typical inventory item using a standard inventory model. This will not give the same accuracy as an analysis of all items using a tailor-made model, but when used with discretion this approach gives a satisfactory first-order approximation. Often even a cursory study allows management to reject or accept clear-cut cases. Attention can then be focused on projects that appear to be borderline cases.

Sometimes a simple model can be used to provide boundary estimates of benefits—i.e., an upper or lower limit on possible benefits. An upper limit can be used to reject unworthy projects, while a lower limit can identify worthwhile projects. For example, in considering the implementation of an improved forecasting system one can establish an upper limit by assuming that the system will give *perfect* forecasts. In a similar fashion an upper limit on the improvement in production scheduling can be obtained by assuming 100 percent capacity utilization and no interference among jobs. A proposed system could be rejected out of hand if the estimated cost of implementation exceeds the upper limit on benefits. If this is not the case, more refined analysis can often lower the upper limit in order to bring it closer to the actual value.

Lower limits can be estimated in a variety of ways. One means is to put a value on only the most easily determined benefits. In analyzing a customer billing project, for example, estimated savings from a reduction in cash requirements gives a lower limit on total benefits. Actual benefits might exceed this limit by the (unknown) value of fewer errors in billing, better by-product information for market analysis, and the like. Another technique is to calculate benefits by using *worst case* estimates for probabilistic variables. If a project is acceptable using the most pessimistic estimates, actual benefits are highly likely to prove even more attractive. See Figure 17.

A similar approach is to estimate the break-even improvement necessary to balance the cost of implementing a project. Suppose, for example, that a firm spends $1 million per year on advertising in magazines and newspapers. The cost of obtaining and analyzing readership data is estimated to be, say, $10,000. Thus, a one-percent reduction in advertising expenditures (while holding exposure constant) would justify the cost

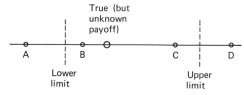

FIGURE 17 Use of upper and lower limits in analyzing benefits. Suppose that the upper and lower limits shown in the diagram have been estimated for a given project. The project must provide sufficient benefits to justify its estimated cost. If point A represents the minimum acceptable payoff, the project is worthwhile by virtue of the fact that the lower limit on benefits exceeds the acceptable payoff. Similarly, the project can be rejected if point D is the minimum acceptable payoff, since this value exceeds the upper limit on benefits. The acceptance of point B and the rejection of point C require further analysis to refine the estimated upper and lower limits to bring them closer to the true payoff.

of the analysis. Similarly, a supermarket firm might determine that the cost of maintaining detailed sales statistics could be balanced by an increase in sales of 0.3 percent. Given such an analysis, an experienced manager can often judge whether or not the likely improvement exceeds the break-even point.

Analysis of Intangible Benefits

The benefits discussed so far are tangible enough, even though they may be difficult to estimate. Some benefits, however, are especially difficult to translate into a monetary value. For example, it would be very difficult indeed for General Motors to put a dollar value on the customer goodwill brought about by a reduction from 10 percent to 5 percent in the probability that a dealer will have a stockout of a needed repair part. It would be equally difficult for the Southern Railway System to attach a dollar benefit to the improved service achieved by a one-day reduction in average delivery time of freight shipments.

Almost any benefit can be assigned a tangible value if sufficient effort is devoted to the task; the difference between a "tangible" and an "intangible" benefit thus lies in the difficulty of estimating monetary value. Even a benefit such as customer goodwill could be translated into a reasonable estimate of monetary value if the effort were justified. If this is not the case, however, we must deal with the problem in other ways.

At the outset it should be pointed out that difficulty in expressing a benefit in *monetary terms* does not imply that the benefit cannot be *quantified*. Failure to appreciate this fact has often resulted in an unnecessary lack of specificity in describing intangible benefits. Thus, in the above examples it was possible to quantify benefits (e.g., stockout probability reduced from .10 to .05), even if no dollar value was attributed to them. In some cases very little quantification may be possible (and so a narrative description must suffice), but this is the exception.

Under some circumstances the analysis of intangible benefits becomes fairly straightforward. This is the case when a proposed system can be justified on tangible grounds alone (i.e., through some combination of cost reductions and tangible benefits), while also contributing significant intangible benefits. The proposed system is said to *dominate* the existing one, since it is superior in terms of both its tangible and intangible characteristics.

Opportunities of this sort are not as uncommon as one might suppose. Existing systems are rarely as efficient as they could be, and therefore offer considerable potential for cost reductions. Such savings are made all the greater when technical advances permit new economies. Rather than concentrating solely on tangible effects, however, an organization usually finds it worthwhile to take some of its gains in intangible form.

Advanced "real-time" systems often provide a variety of benefits of this sort. For example, a comprehensive airlines reservation system may be justified on such tangible grounds as reduction in the salaries of reservation and ticketing clerks, higher seat bookings, and more efficient routing and scheduling of flights. Improved customer service may be as valuable as any of these, but it is probably not possible to assign a dollar estimate to it.

If a formal model exists, it may be possible to establish an explicit tradeoff between an intangible benefit and a tangible one. Suppose, for example, that an improved inventory model is implemented. Benefits can be taken in the form of lower inventory cost, lower stockout probability, or some combination of both. Even if a monetary value cannot be placed on fewer stockouts, its cost can be expressed in terms of foregone opportunity to reduce inventory costs. See Figure 18. An experienced manager presumably can resolve questions of this sort when he is presented with explicit tradeoff information.

USERS' ROLE IN COST/BENEFIT ANALYSIS

As information systems become more comprehensive and integrated, formal methods of analysis become increasingly difficult to apply. Joint costs and joint benefits often make it impossible to determine the payoff from any one subsystem. As the system begins to pervade day-to-day operations and higher level decision making, benefits become increasingly difficult to evaluate in monetary terms. Inevitably, then, we must rely on experienced judgment, as well as technical analysis, to make cost/benefit decisions.

Communication between User and Technical Staff

One of the complications of making a cost/benefit decision is that it requires close coordination between two disparate groups, the users and the technicians. A rational decision requires knowledge of both costs and benefits. Information acquires value only through use, and so it is the users who are in the best position to judge the value of information. On the other hand, users very often have only a hazy idea of the cost of information. As a result they may specify an information "requirement" that is very expensive to satisfy, or they may fail to ask for useful information that could be provided at very little extra cost. To get costs into the picture, the technical manager and his staff must play the vital role of (1) analyzing costs

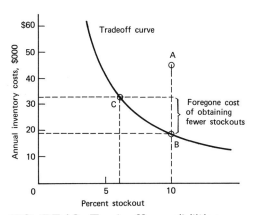

FIGURE 18 Tradeoff possibilities offered by increased efficiency. Changes in technology (or simply discovery of more efficient ways to exploit existing technology) allow new alternatives that provide tangible or intangible benefits. Such changes might permit, for example, the replacement of inventory system A by system B or C. All the benefits of system B are in the form of a tangible cost reduction. The selection of system C in preference to B implies that the intangible benefit of fewer stockouts is worth *at least* as much as the resulting reduction in cost savings.

Response time	Incremental benefit	Incremental cost	Net incremental value
10 seconds, 90% confidence	$100,000	$50,000	$50,000
10 seconds, 75% confidence	98,000	45,000	53,000
60 seconds, 75% confidence	75,000	30,000	45,000
2 hours	72,000	10,000	62,000
overnight	70,000	5,000	65,000

and benefits to the extent possible, and (2) providing users with tradeoff information so that they will know the cost implications of their specifications.

The knowledge necessary to strike a reasonable balance between the cost and value of information is thus split between users and technicians. Clearly, then, this knowledge should be shared in some way. The dialogue between users and technicians may take several forms. One important way is by means of an iterative revision of requirements.

This process can begin by having the users specify desired characteristics of information outputs. A user might, for example, request that the system provide a response time of ten seconds in handling a given type of inquiry.[10]

With this and other specifications as a starting point, the technicians can then design, in gross terms, a system that meets (most of) the stated requirements. The designers may find that some few of the requirements are wholly infeasible; in these cases an alternative should be proposed. Even when a specification is perfectly feasible, the designers should provide tradeoff information so that users have some indication of the cost consequences of their specifications. For example, users may be given the following tradeoffs:

Response time	Incremental annual cost
10 seconds, 90% confidence	$50,000
10 seconds, 75% confidence	45,000
60 seconds, 75% confidence	30,000
2 hours	10,000
overnight	5,000

The great value of such tradeoff information is that it gives users a basis for balancing benefits against costs. A user may very well choose to alter his initial specification in light of its cost implications; at least he should be presented with this option.

Let us continue with the example of the inquiry system. Suppose that benefits are estimated as shown at the top of the page. The *net* value of information can then be determined. Overnight service turns out to be the optimal response time in this example, and so the original specification should be changed.

Tradeoffs should be expressed in monetary terms whenever feasible. Ultimately, however, the final judgment usually involves considerable subjective evauation of intangible benefits. This

---

[10] Since the load on the system varies minute by minute, response time must be specified in probabilistic terms. The user may, for instance, specify that at least 90 percent of the transactions receive a response within ten seconds.

should not be viewed as a serious limitation—after all, managers are paid to exercise judgment. When presented with sufficient tradeoff information, a user is probably able to strike a satisfactory balance between cost and value.

An important question remains, however: Is the user motivated to choose what he considers to be the proper balance between cost and value? He may well not be if he receives benefits while not bearing the incremental cost of obtaining them. Question: In the above example if the user does not pay incremental costs, what response time is he likely to specify? A policy of charging users for incremental costs raises all sorts of costing problems, but it is very difficult otherwise to motivate users to make wise judgments.

Determination of Tradeoffs

The preparation of tradeoff information is obviously expensive. Each tradeoff point represents an alternative design in terms, say, of different capacities of main or auxiliary storage, number of I/O channels, or communication network configurations. The designers should therefore limit their consideration to a relatively few alternatives that stand some chance of being acceptable to users. For example, it serves no useful purpose for the designer to determine the cost of an order processing system that provides invoices of 95 percent accuracy if the marketing vice president insists on a minimum of 99 percent accuracy.

Some alternatives can be dismissed out of hand by having technical personnel work with users in preparing initial specifications. This will better insure that the alternatives considered will be limited to those that are reasonable from a technical point of view. An experienced designer can establish, without much analysis, fairly good boundaries on the range of feasible alternatives.

It is quite unlikely that users will be able to supply explicit estimates of value for alternative levels of information quality. A user may be willing to choose an alternative when presented with tradeoffs, and thereby establish an implicit estimate of relative values, but this is a considerably easier matter than giving explicit values. Nevertheless, the user should at least be able to provide *qualitative* tradeoff information in order that the designers can confine their attention to acceptable alternatives. In the case of the inquiry system, for example, users might be able to state in advance that overnight service is fully acceptable, interactive response is marginally useful, and any response time in between contributes very little extra benefit compared with overnight response. Designers then need only consider overnight batch processing and an interactive system.

Hierarchical Nature of
Cost/Benefit Analysis

Any cost/benefit analysis takes money to perform. The more money spent, the better the estimates that can be provided. In other words, one can reduce uncertainty about the economic payoff of a project by devoting resources to the task of analysis.

The organization thus faces a typical resource allocation problem in trying to decide how much to spend on cost/benefit studies. Like other information expenditures, resources used for analysis should be spent where the results have the greatest surprise content, will lead to the most significant modifications in decisions, and offer the greatest potential payoff from improved decision making. These conditions are not met if the payoff from a project is fairly obvious, if the level of benefits expected from a project cannot justify an elaborate study, or if the benefits are intangible enough so that trying to place a monetary value on them is not worth the effort.

Very often a project can be dismissed without a great deal of analysis. Although a superficial analysis may be subject to considerable uncertainty, the estimated cost/benefit performance for a given project may be unattractive enough that further refinement is not necessary. An experienced information specialist can be very effective in identifying these marginal projects.

The remaining projects require further study. Uncertainty is reduced as implementation proceeds through gross design, detailed design, and programming. Periodically during this process management should review the project to determine if it should be abandoned, accepted, or subjected to still further analysis. See Figure 19.

Thus, cost/benefit analysis should continue throughout the life of a project. During the early, uncertain stages of a project, the amount of money spent is relatively modest. The ante goes up significantly when it reaches the stages of detailed design and programming, but

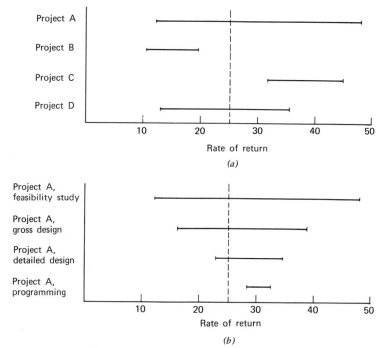

FIGURE 19 Uncertainty in estimating return from different projects. (a) shows the estimated return from different projects, based on preliminary feasibility studies. Such estimates are subject to a considerable range because they are made without detailed information about the proposed systems. Nevertheless, even a superficial analysis can identify projects that appear worth pursuing in greater detail (e.g., Project A). It can also screen out projects that clearly fail to meet the minimum acceptable return of, say, 25 percent (e.g., Project B). Uncertainty is reduced as the analysis proceeds through the various stages of implementation, as is shown for Project A in (b). More detailed analysis and design narrows the range of estimated payoff. The estimate may stay within the acceptable region, or it may fall below the limit and call for the project to be modified substantially or even abandoned.

by this point most of the uncertainty concerning payoff should have been eliminated.

Subjective Estimation of Benefits

Since it is usually necessary to rely at least partially on subjective evaluation of information benefits, it is important that users have a good conceptual grasp of the factors that contribute to value. The evidence is that these concepts are not at all universally understood. More than a few systems have been implemented in which there is an obvious imbalance between cost and value; either the system falls far short of providing the information quality that it should, or it supplies information at a cost drastically exceeding any possible value. Even a primitive understanding of the issues involved would have avoided many of the problems of this sort.

Surprise Content

A user, along with the technicians who aid him, should ask himself questions of the following sort:

- Does the contemplated information tell me something that I did not already know or strongly suspect?
- How frequently will I obtain a significant surprise?
- Is the information selective enough that I am likely to perceive significant facts?
- Would less detailed, accurate, timely, or reliable sources provide essentially the same surprise content?

Effect on Decisions

- Will any surprise information cause me to take an action that I would otherwise not take?
- Would my decisions be significantly altered by less accurate or timely information?

Effect on Performance

- What are the benefits from any change in action that results from receiving surprise information?
- Are the benefits sensitive to moderate deviations from "optimal" actions?
- Do benefits decrease significantly with delays in taking action?

REFERENCES

1. James C. Emery. 1969. *Organizational Planning and Control Systems.* Macmillan, New York. Chapter 4 discusses the economics of information, relying heavily on a theoretical model.
2. Robert H. Gregory and Richard L. Van Horn. 1963. *Automatic Data-Processing Systems,* second edition. Wadsworth Publishing Company, Belmont, California. A classical treatment of the technical and managerial issues of information systems. Chapters 14 and 15 deal with economic questions.
3. E. Gerald Hurst, Jr. 1969. "Analysis for Management Decisions." *Wharton Quarterly,* Winter, 1969. This paper presents some simple examples of Bayesian decision making and shows how the value of information can be calculated.
4. Börje Langefors. 1970. *Theoretical Analysis of Information Systems,* Volumes 1 and 2. Studentlitteratur Lund. Available fron Barnes and Nobel, New York. A valuable reference for anyone willing to wade through fairly heavy theory. Chapter 3 deals with the economics of information. Volume 2 is devoted mostly to questions of file organization.
5. Jacob Marschak. 1959. "Remarks on the Economics of Information." In *Contributions to Scientific Research in Management.* Graduate School of Business Administration, U.C.L.A., pp. 79–98. One of the fundamental theoretical contributions to an understanding of information economics.

6. Norman R. Nielsen. 1970. "The Allocation of Computer Resources—Is Pricing the Answer?" *Communications of ACM*, 13,8 (August 1970), pp. 467–474. An excellent general discussion of pricing for computer services. It contains useful references to other papers on the subject.
7. T. William Olle. 1970. "MIS: Data Bases." *Datamation*, November 1970. A survey article on data base management systems.
8. Howard Raiffa. 1968. *Decision Analysis —Introductory Lectures on Choices Under Uncertainty*. Addison-Wesley, Reading, Massachusetts. A very readable discussion of Bayesian decision making by one of the leading authorities in the field.
9. William F. Sharpe. 1969. *The Economics of Computers*. Columbia University Press, New York. The most complete discussion of economic issues connected with the use of computers. Chapters 2, 5, and 9 are particularly relevant to a discussion of cost/benefit analysis of information systems.

ECONOMIC EVALUATION OF MANAGEMENT INFORMATION SYSTEMS

D. F. BOYD
H. S. KRASNOW

The evaluation of data processing systems has traditionally rested upon the notion of cost displacement. This approach is a natural outgrowth of viewing such systems as essentially productive. However, significant economic benefits of many recent systems accrue from the so-called intangible benefits to management. Thus, the nature of current information systems suggests that they be viewed, for purposes of economic evaluation, in a broader context than that of a producing machine.

Here we view the contribution of an information system in maintaining control over a business system operating in a changing environment. This view implies a criterion of evaluation related to the dynamic performance of the firm. We hypothesize that better information will lead to better control which in turn will yield improved total performance. The control objective of the firm is to respond to the environmental demands in an economically efficient manner. The effectiveness of an information processing system in satisfying this objective may be evaluated by:

1. An accounting measurement of the financial performance of the firm over time in the face of changing demand (environment).
2. The accuracy, completeness, and timeliness with which that demand is satisfied.

These measures, being more complex, are more difficult to estimate than cost displacement and require an adequate model of the firm itself.

The objectives of the current study were, first, to define a method suitable for the economic evaluation of information systems when viewed in this manner; and second, to demonstrate its technical feasibility by applying it to a hypothetical firm.

SOURCE: Boyd, D. F., H. S. Krasnow, "Economic Evaluation of Management Information Systems," *IBM Systems Journal*, March, 1963, pp 2–23. © 1963 by the International Business Machines Corporation, and reprinted by permission of the *IBM Systems Journal*.

DESCRIPTION OF THE METHOD

The importance of the dynamic behavior of the firm to its own well being has been shown and it has been demonstrated that this behavior can be simulated.[1] Advanced information systems, which are often intimately and extensively involved in control, have also been successfully simulated. The problem, then, is to relate the mechanics of the information system to the dynamics of the business firm within a single model.

Physical System

The simple firm performs an economic function upon which its existence is based. (The modern corporation, of course, often performs many such functions.) A minimal set of activities is required in order to perform this function: we designate this set and its interrelationships as the *physical system*. In a manufacturing firm the elements of the physical system are the production processes and the resources which produce the end product. In a service firm, the physical system is composed of those activities and their associated resources which directly provide the customer with service.

Information Processing System

A total representation of the firm requires, in addition to the physical system, a second part referred to as the *information processing system*. The latter encompasses all activities of the firm whose direct or indirect function is to control the physical system (Figure 1). In a real firm there are, of course, activities which do not fall within either of these two categories (for example, janitorial services). These activities are of little interest for the purposes at hand, and appear only as fixed or variable cost elements within the accounting structure.

The information processing system is broader in concept than any existing data processing system, the latter serving as a component of the former. The information processing system can be represented by the following basic elements and their interrelationships:

Sensor. This type of element originates all data input to the information processing system. In included both manual and machine-generated input. It reports the occurrence of an event within the physical system (or perhaps within the environment).[2] A segment of a physical system is shown in Figure 2. Sensors record all possible events, the receipt of material into inventory, disbursements from inventory, and the receipt of requisitions (demand) for inventory.

Input Transmission. Sensed data are subject to delay and/or distortion during transmission. All delays associated with input are imagined to occur at this point (i.e., sensing alone is complete, accurate and instantaneous).

Image. The end result of data input and most conventional processing, whether machine or manual, is an image.

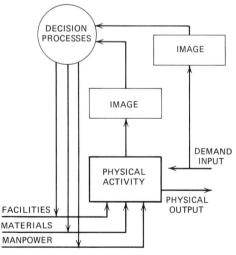

FIGURE 1 Elements of a dynamic model.

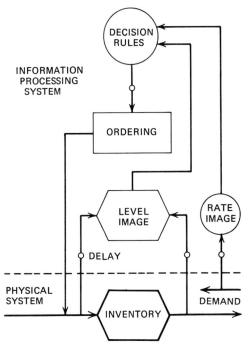

FIGURE 2 Segment of a dynamic model.

In Figure 2, the image of the true inventory is the inventory record. Images can be classified as levels (e.g., inventory) or rates (e.g., the arrival rate of inventory requisitions). If applied to continuous flow measurements, level images would be the time integral of one or more rate images. With appropriate sensors, images can be provided which describe any activity within the physical system. However, they are distorted as a result of input transmission delays and may be biased by the random or systematic loss of sensed data during transmission.

Decision Process. This is a crucial element of the information processing system. The term is used in the broadest possible sense to encompass all decision-making related to the control of the physical system. Decision processes can function with the aid of much or little information; with information which is accurate or distorted, timely or outdated. The information upon which the decision process depends (all of the information available to it) is contained in images. The decision process has no direct contact either with the physical system or the environment. In the example of Figure 2, the decision to order additional material for inventory utilizes images of the current requisitioning rate and inventory level.

Output Transmission. The result of a decision is a command which will ultimately produce some change in the activities of the physical system. A single time delay is associated with both the decision-making process and the transmission of its commands. In Figure 2, the command is in the form of an order for additional material. More generally, commands take the form of an adjustment to the resources committed within the physical system.

Environment

In addition to representing the firm in this manner (physical system-information processing system), a complete model requires explicit recognition of the interaction with its environment. In particular, it recognizes certain basic requirements (demands) which the environment places upon it and which it undertakes to satisfy. One basic measurement of the performance of the firm is the adequacy with which it satisfies these demands. The environment may also provide information inputs to the information processing system relevant to the future demand pattern. (It should be noted that for purposes of model building, the interface between the firm and its environment is somewhat arbitrary. The crucial distinction is between that which can and that which cannot be controlled by the firm. The former is classified within the physical system; the latter within the environment.)

Accounting Structure

Figure 3 suggests that the representation of the firm has two interfaces: one with its environment and one with the experimenter. This figure also suggests the experimenter may change the parameters governing the environment and the information processing and physical systems. In order to measure the results of these changes, he must make comprehensive observations regarding the performance of the simulated firm. The mechanism for accomplishing this observation has been designated the *accounting structure* because of the central role of financial accounting for performance evaluation. Cost is a critical element of performance and must be considered in any over-all evaluation. Conventional accounting procedures are introduced for this purpose. The complete accounting structure is capable of providing any desired data concerning the operation of the model, including data which are entirely independent of cost. No errors or time delays are introduced. In this sense it is perfect and provides an accurate and unbiased appraisal of the performance of the firm.

A SPECIFIC MODEL

We will now describe a specific model of a simple, hypothetical manufacturing firm.

Physical System

The physical system of the model shown in Figure 4 incorporates as much as possible of the dynamic complexity found in a typical manufacturing operation within a nominally simple model. Thus, a basic assumption is made that the general dynamic characteristics of a system can be adequately represented without the introduction of the large number of individual elements actually present. The components of the physical system are now described.

Two end-products are manufactured, designated as Products 1 and 2. Both products are assembled and shipped to customer order. Three finished parts

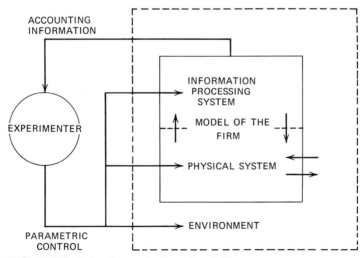

FIGURE 3 Interfaces in the simulation.

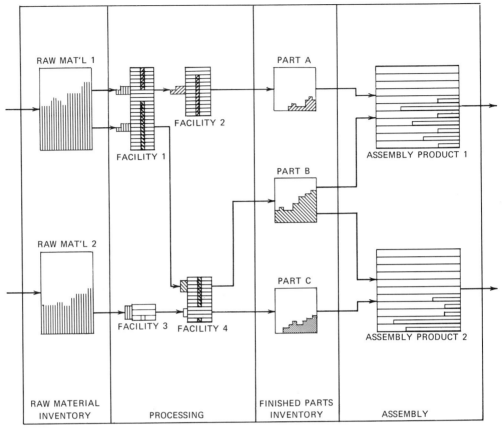

FIGURE 4 The physical system.

(Parts A, B, C) provide all of the components for the assembled products, in accordance with the Bills of Material shown in Table 1.

It can be seen that Part B is common to both products, introducing a conflict situation (with its related decision problems) of the type often found in practice.

The activities of the physical system are distributed over three stages of manufacturing: raw material procurement, parts processing (fabrication), and assembly and shipping. This introduces much of the dynamic complexity of the model, since overall response is dependent upon actions taken somewhat independently within each stage. Accurate control will require good planning to coordinate the activities within different stages. These activities are:

Raw Material Procurement. Inspection, receipt, and storage of raw material.

Processing. Requisitioning of raw material. Setup of a facility unit for processing a particular part. Processing a part on a facility unit (fabrication operation). Scrapping a part on a facility.

TABLE 1 Bills of material

	Units		
Part	A	B	C
Product 1	1	2	
Product 2		1	1

Movement of partially finished parts to next operation. Movement of finished parts into inventory. Storage of finished parts in inventory.

Assembly and Shipping. Requisitioning of finished parts required for assembly of an order. Movement of parts to assembly area. Assembly. Scrapping of parts during assembly. Requisitioning and movement of replacement parts. Shipment of completed orders.

The scale of an activity (e.g., time to perform, rate of occurrence, etc.) is either dependent upon other activities and therefore determined by the simulation (for example, number of parts in inventory), or it is a parameter of the physical system controllable by the experimenter (for example, time to assemble one unit of Product 1). In the latter case, the value may be specified determinately as a constant or a function, or stochastically as a random function.

The performance of an activity requires the commitment of one or more resources. Several activities have been structured so that they compete for the same resources, thereby creating typical conflict situations which can only be resolved by rational decisions. The resources available in the model are:

Processing Manpower. Men within the processing stage are entirely interchangeable, and may work on any valid operation, or remain idle.

Assembly Manpower. Men within the assembly stage may assemble orders for either product. However, no transfer of men between the assembly and processing stage is permitted.

Processing Facilities. Each facility within the processing stage possesses a discrete number of units of capacity. A processing operation commits one man and one unit of facility to the processing of one part. The facility units must be set up prior to processing; however, successive units of the same part may be processed on the same setup.

Material. The finished parts used in the assembly of the two products are fabricated from two raw materials. Two of the finished parts (Part A and Part B) compete for Raw Material 1.

Information Processing System

The prime objective in constructing the information processing system was to provide sufficient capability to permit effective dynamic control over the physical system. Within this context, the emphasis was placed upon building a conventional structure which could plausibly incorporate a range of data system types. Figure 5 is a schematic of the complete model depicting, among others, all of the major features of the information processing system.

Hierarchical aspects of an information processing system in the large firm are included. Decision-making occurs at various levels within the organization with considerable interaction between levels. Operational control, at the lowest level, responds to events on a fairly rapid time scale, in a highly constrained manner. At a higher level, tactical decisions are taken whose effect may be only indirect, leading to direct action at the operational level. These decisions are less frequent than those at the operational level, as well as more complex.

The physical system, as previously described, is also included in Figure 5. In the model, sensors are included at all points on the interfaces between the three stages of manufacture, and on the interface within the environment. The sensora are assumed to exert no direct influence on the physical system. It is believed that this generates a reasonable amount of data for this type of system. Additional sensors, placed within each

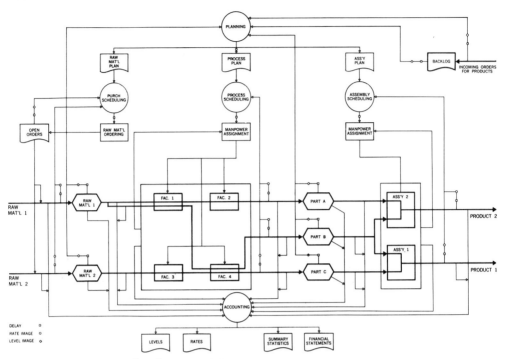

FIGURE 5 Schematic of the model.

stage (e.g., recording material movements between operations in processing), would suggest a rather highly advanced information system. Fewer sensors placed, say, only on the interface with the environment (e.g., recording orders and shipments) would probably not permit effective control over the physical system. The precise configuration shown in Figure 5 is arbitrary, and could be readily extended or curtailed. The sensors could be inserted at any point at which an event can occur.

Figure 5 also indicates delays associated with information transmission, the resulting images of the sensed data, and the decision processes which utilize these images.

Decision rules are themselves parameters of the information processing system, in the sense that they can be individually detached and replaced. However, only one set of decision rules has been utilized in the model thus far. These are designed to achieve reasonable control even under fairly poor information flow conditions. In practice, of course, the decision processes and the quality of the information flow are highly related. Improved flow may be ineffective if not accompanied by improvements in decision-making; conversely, major improvements in decision-making (e.g., utilization of mathematical techniques) may be impossible without parallel improvements in information flow.

The set of decision rules for the model relate to planning, purchasing, and manpower assignment. Descriptions follow:

Planning. This is the mechanism which permits the model to adjust to, and perhaps anticipate, systematic changes in customer demand. The crucial element in planning is projection of shipping requirements for the next two months, based upon the past pattern of orders and the current backlog of unstarted orders. Exponential smoothing is employed to generate the forecast of future orders, and the backlog is distributed to future requirements in an exponential manner. Once shipping requirements are established, they are used as the planning base at all three stages of manufacture. An assembly plan is produced from the shipping requirement by adjusting for assembly lead time. The processing plan and the raw material plans are generated from the assembly plan by the necessary parts explosions, adjustments for excess inventories, lead times, and scrap losses.

Purchasing. The raw material plan provides the basis for ordering raw material. Orders are placed periodically, at a time determined by the availability of a new plan. This time is later than the nominal date of the plan, due to the delay implicit in the planning process. (For example, the plan stating requirements for the months of January and February might not be available until the second week in January.) Before ordering, therefore, the plan must be updated for material received since the start of the month, and for any currently open orders. Allowance is also made for the possibility of receiving defective material. The actual order quantity is determined so as to cover requirements through an entire period (month) until the expected receipt of the next order.

Manpower Assignment. In the processing stage, the plan is used once each week to generate a scheduled load. The plan is first adjusted for parts produced since the first of the month, and is then extended in accordance with the work content (standard time) remaining in the month for each production operation. The available work force is then assigned to each operation (part to be processed on a facility) in proportion to the computed work loads and subject to the limitations set by facility capacities. Existing setups are not considered in arriving at this decision. The implementation of the decision will permit reassigned men to complete the operation on which they are currently engaged before moving to their new assignment. In the assembly stage, the assignment procedure (between products) is identical except that there are no facility constraints to be observed. Each stage makes assignments based on its own work force, with no exchanges permitted. The planning process is insensitive to local conditions prevailing "on the floor." As a result, it is possible for assignments to be made to operations for which material is perhaps temporarily unavailable. In such cases, it is desirable to consider reassigning the men to other idle facilities for which material may be available. The decision determines the number and location of idle men, and reassigns them in sequence to the remaining operations to the limit of facility capacity. In the assembly stage, this decision merely transfers idle men to the alternate product unless idleness is observed for both products.

As previously noted, the commands associated with the foregoing decision processes consist of purchase orders, which generate new material, and manpower assignments. All of the decisions are time triggered, although it would be equally straightforward to utilize event triggering. The lengths of the planning period (month) and the manpower assignment review period (week) are fully adjustable, as are all of the delays associated with decision-making and implementation.

Environment

The interactions between the firm and its environment are limited. They consist of the following items:

Customer Orders. An input to the physical system. The properties of an order are: it is for a single product; it specifies the quantity (number of units) required; it is held within the physical system until filled.

Product Shipments. An output of the physical system. No partial shipments are made. Orders are shipped as soon as completed.

Purchase Orders. An output of the information processing system. Each order is for a single raw material, specifying the quantity desired.

Receipt of Raw Material. An input to the physical system. The environment imposes a delay (lead time) upon the filling of purchase orders. At the end of this delay, material is entered into the physical system.

Experimenter

The nature of the interface between the model of the firm and the experimenter is indicated in Figure 3. Communicating the results of the simulation is the role of the accounting structure. It provides a wide variety of data needed for evaluation. Cost factors are a critical element of performance, and are incorporated in a fairly complete set of conventional financial statements. Direct data are also provided on all relevant features of the physical system (e.g., inventory levels, manpower utilization) and of the information processing system (e.g., shipping requirements, scheduled loads by operation in man-hours). Some of the data are provided as a function of time (i.e., periodically), some as a single aggregate measure for the entire simulation period.

The experimenter exerts control over the simulation by setting parameters for the physical system, the information processing system, and the environment. He is also free to independently set the cost elements (e.g., labor rates, material prices) of the accounting structure, which govern the absolute level of the financial results. The major controllable features of the model are summarized in Table 2. For stochastic variables the parameters are in the form of a probability distribution.

In addition to direct variation of system parameters, the experimenter may introduce more basic changes. Decision rules can be modified or entirely replaced without disturbing other parts of the model. It is also possible, though not quite as straightforward, to modify the structure of the physical system. For example, the flow of parts in the processing stage could be changed, or the material usage specifications could be altered.

DESCRIPTION OF THE SIMULATION RUNS

The experimental approach that is chosen depends entirely upon what one wishes to learn about the total system. It is possible to vary the parameters of the information processing system in order to evaluate the relative worth of a spectrum of data processing capabilities; or evaluate alternative decision processes. Alternatively one can vary the parameters of the physical system to suggest the range of industry characteristics for which a given information handling capability is worth while. As in all simulation work, a systematic approach is desirable. In particular, statistically designed experiments offer the best prospect of achieving soundly based conclusions at minimum cost in computer time.

We turn to the second purpose of this paper, which is to demonstrate the feasibility of the method for the economic evaluation of certain "intangible" benefits of improved information systems. For this purpose six simulation runs were selected.

Key Parameters

These runs were based on manipulating two aspects of the information processing system: first, the length of the plan-

TABLE 2 Parameters which can be controlled by the experimenter

Subsystem	Parameter	Stochastic
Physical	Setup times	Yes
	Processing and assembly times	Yes
	Material movement times	Yes
	Rejection rates	Yes
	Size of work forces	No
	Facility capacities	No
Information processing	Input transmission delays	Yes
	Command delays	Yes
	Decision parameters	
	Planning	
	Length of period	No
	Fcs't smoothing constant	No
	Backlog distribution constant	No
	Processing & ass'y lead times	No
	Inventory safety margins	No
	Planned manpower assignment	
	Standard times	No
	Purchasing	
	Scrap allowance	No
Environment	Purchase order lead time	Yes
	Customer order arrival rate	Yes
	Customer order quantity	Yes

ning period together with a related implementation delay; and second, the magnitude of information transmission delays.

The model contains a series of decision rule algorithms beginning with the generation of a sales forecast and continuing on through the detailed scheduling and assignment of materials and manpower. These algorithms are applied periodically, and new plans and schedules are generated based on the sensing of new demand information as well as "accomplishment-to-date" in the physical system. These algorithms closely parallel typical planning and scheduling sequences in a real manufacturing enterprise.

Thus, increasing the frequency of the planning cycle specifically implies the availability of information systems of increased capacity and sophistication.

Table 3 lists the characteristics of the three planning cycles used in the feasibility runs. The slow cycle corresponds to once-a-month, medium to every-two-weeks, and fast to once-a-week planning and scheduling. The implementation delay (output transmission delay) represents the time lag between the availability of the basic new planning information and actually putting the plan into effect.

The second aspect of the information processing system chosen for manipulation was that of information time lags (input transmission delays). The information processing system senses various aspects of the environment and physical system through more or less distorted images. A principal distorting influence is that of information delays. For example, it may be necessary to write today's purchase orders based on last week's inventory figures.

TABLE 3 Planning cycles

Characteristic	Slow	Medium	Fast
Length of period	1 month	2 weeks	1 week
Implementation delay	7 days	4 days	2 days

TABLE 4 Information delays

Information category	Slow	Fast
Incoming orders for products Open purchase orders for raw material Product shipments Raw material receipts	1 week	1 day
Part movements, finished parts to assembly	3 days	0
Raw material movements into process Finished part movements into inventory	2 days	0

Two sets of such delays were used in the feasibility runs as indicated in Table 4. In the slow set, incoming orders and shipping and receiving status are sensed through a one-week time lag and in-plant movements are assigned a two- or three-day delay as shown. In the fast set, the first category delays were reduced to one day and the inplant delays to zero. (The latter change implies some type of on-line production monitoring system.)

Three values of the planning cycle and two sets of information lags yield six combinations which were the basis for the feasibility runs. Three of these runs (as designated in Table 5) will be described in some detail.

Role of Demand Pattern

The activity which initiates the internal functioning of the simulation model is the stream of incoming orders for products. This demand pattern is also the most direct means for loading and testing the management control capabilities of the model. A prime function of the management is, in a broad sense, to respond in an effective way to the demand pattern. As noted above, the purpose of the feasibility runs was to determine whether significant differences in performance would result from changes in selected aspects of the information processing system. In order to

TABLE 5 Parameter combinations for simulation runs

		Planning cycle		
		Slow	Medium	Fast
Information delays	Slow	(1)	(2)	
	Fast		(3)	

amplify any such differences, a severe response requirement was placed on the model through the demand pattern. This was accomplished by imposing an abrupt change in the product demand mix.

Figure 6 is a graphical representation of the demand pattern used for all six of the feasibility runs. The initial level of demand for Product 1 is at the rate of 20 orders per week, and for Product 2, 95 orders per week. At the end of the first four weeks of simulated operation, Product 1 orders rise suddenly to 50 per week, while Product 2 orders drop suddenly to 15 per week. Demand remains at these levels for the balance of the 16-week period simulated.

Prior to starting each run, the model was initialized by providing a stock of raw materials and finished parts in the proportions required to supply the processing and assembly functions at the initial demand mix. The amount of the initial raw material stock was adjusted between runs so as to be compatible with the planning cycle used.

In addition, the forecasting algorithm was given "historical" demand levels which also reflected the initial demand mix.

The effect of these initializing values was to put the modeled enterprise approximately in the condition of having operated for an extended period at the initial demand mix and of having no expectation that this would change.

The nature of the management response problem presented can be anticipated by an examination of the demand pattern. When the abrupt change in demand mix occurs at the end of the 4th week, there are three major problems:

1. The nature of the change in demand must be assessed and extrapolated in the form of new sales forecasts.
2. Raw material orders must be initiated to rebalance the raw material

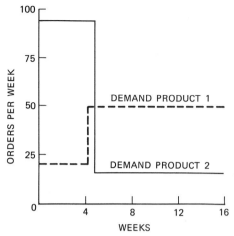

FIGURE 6   Demand pattern.

inventory to fit the new mix. Early action is especially critical here due to the substantial procurement lead times.
3. Processing manpower assignments must be shifted in order to supply the finished parts inventory with a new mix of finished parts for assembly.

It is apparent that the logistics of Product 1 will be much more critical than those of Product 2. At the time of the demand mix change, a 2-week supply of stocks supporting Product 1 is effectively reduced to only a 4-day supply whereas a 2-week supply for Product 2 is extended to a 12-week supply.

Simulation Output

The accounting structure of the model results in a very complete set of output data describing the behavior of the physical system during the course of the simulation. An extensive printout of virtually all pertinent physical data was produced at the end of each weekly reporting cycle, including listings of manpower distribution, queues at all facilities, inventory levels, backlogs of unfilled orders, product shipments, and raw material receipts.

As mentioned earlier, the total ac-

TABLE 6 Accounting parameter values

Category	Detail	Value
Product selling prices	Product 1 Product 2	$325 $215
Raw material costs	Raw material 1 Raw material 2	$ 20 $ 95

Direct labor standards	Wage rate
	Standard work content
	  Facility 1, Part A
	  Facility 1, Part B
	  Facility 2, Part A
	  Facility 3, Part C
	  Facility 4, Part B
	  Facility 4, Part C
		$2.50/hr.
		
		5 hrs.
		4 hrs.
		10 hrs.
		1 hr.
		2 hrs.
		1 hr.

Standard burden	Burden rate (on D.L.)	70%
Fixed cost assumptions	Depreciation charge Selling and admin. exp.	$ 4,000/mo. $20,000/mo.

counting structure contains a financial cost accounting framework which provides a set of standard costs for the evaluation of finished products and all raw material and in-process inventories. These standard costs are a function of input parameters specifying material costs, wage rates, labor standards, standard burden rate, etc. Table 6 is a tabulation of the parameter values used in the feasibility runs.

At each reporting cycle the pertinent physical rates and levels are sensed and extended by the appropriate actual and standard cost values to produce a set of conventional financial statements including a manufacturing expense statement, an income statement, a statement of cash flow, and an abbreviated balance sheet tabulating current assets. Table 7 illustrates the form of these statements.

RESULTS OF SIMULATION RUNS
Slow Planning, Slow Information

Perhaps the most direct indication of the response of the physical system to product demand is given by a comparison of the actual shipments of finished products with the demand pattern. Figure 7 gives this comparison.

In all the graphs of Figure 7, there is an initial rise from zero shipments which reflects the initializing phase of the run during which the assembly operation is loaded from the finished parts inventory. This process only affects shipments for the first two weeks.

In the case of Run 1 it will be noted that shipment of Product 1 responded rapidly to the demand step with shipment actually exceeding the new level by the 7th week.

This rapid initial response reflects the fact that assembly is "to-order." During the 11th and 12th weeks, however, Product 1 shipments dropped sharply. Shipment did not again match the demand rate until the 16th week.

The pattern of Product 2 shipments reflects the easier response problem posed by the downward step in demand.

Figure 8 displays two aspects of performance which summarize the relationships between the demand and shipping patterns, the backlog of unfilled orders and delivery time.

TABLE 7 Form of weekly financial statement

Manufacturing expense statement

Raw material purchases	$XXXX	
Direct labor expense	XXXX	
Indirect expense	XXXX	
Depreciation	XXXX	
Total expense		$XXXXX
Deduct inventory incr./decr.		
Change in raw material inventory	$XXXX	
Change in in-process inventory	XXXX	
Change in finished parts inventory	XXXX	
Change in assembly inventory	XXXX	
Net change in inventories		XXXXX
Cost of goods sold		$XXXXX

Income statement

Sales		$XXXXX
Deduct:		
Standard cost of goods sold	$XXXXX	
Manufacturing cost variance	XXX	
Cost of goods sold		XXXXX
Gross profit on sales		XXXX
Less selling and admin. expense		XXXX
Net profit/loss on operations		$XXXX

Cash flow $XXXX

Balance sheet

Cash		$XXXXX
Inventories	$XXXX	
In-process	XXXX	
Finished parts	XXXX	
Assembly	XXXX	
Total inventories		XXXXX
Total current assets		$XXXXX

The unfilled orders graph for Run 1 reflects the initial Product 1 shipping response to the demand step, with the backlog rising to about 50 units and being held approximately at that level through the 10th week. The abrupt rise in unfilled orders for Product 1 which begins at about the 11th week resulted from the shipping lag noted above. The Product 2 backlog pattern shows only small accumulations with complete elimination of unfilled orders in the final weeks.

For a firm of the type represented, perhaps the best single overall measurement of physical performance is that of delivery time, i.e., time from receipt of an order to shipment of the order. The lower portion of Figure 8 is in the form of histograms showing the distribution of delivery times for the entire 16-week simulated period. In Run 1, average delivery time for Product 1 was 12.1 days. The distribution, however, is a bimodal one. The left portion of the

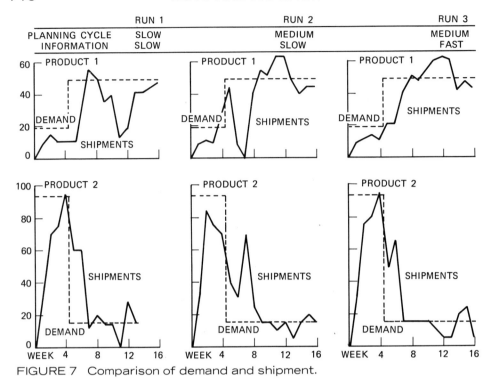

FIGURE 7 Comparison of demand and shipment.

histogram is representative of delivery performance before the 11th-week shipping lag. The right portion, with an average of about 18 days, represents performance for the latter part of the simulated period. As might be expected, delivery time for Product 2 was relatively much better, with an average of 4.5 days.

We can find the explanation for the 11th-week decline in Product 1 shipments by observing inventory behavior. Figure 9 is a week-by-week plot of inventory levels.

In Run 1, it will be noted that raw material outages developed during the 5th, 6th, and 7th weeks with corresponding dips in the in-process stock. Finished part stocks, however, were generally sufficient to support assembly and shipping.

A shipment of Raw Material 1 was received during the 7th week, but the quantity was not adequate to support the new demand level for parts A and B. During the 10th week the stock of Part B, the part common to both assemblies, was exhausted with the result that assembly was largely shut down during the 10th and 11th weeks. This was reflected in the poor shipping performance shown in Figure 7 for those weeks.

The material outages noted above were accompanied by substantial idleness of the work force during the corresponding intervals. As a result, manpower utilization for the run as a whole was only 77%.

In Run 1 it was not until the 11th week that adequate supplies of Raw Material 1 began to be received. Excessive quantities of Raw Material 2 continued to be received through the 9th week. One result may be seen in the soaring inventory of Part C.

Both of these phenomena are symptoms of delayed recognition of the magnitude of the change in the demand pattern and slow corrective action in raw material ordering. The secondary effects, as shown, were poor shipping performance and low average manpower utilization.

The run results discussed thus far represent only selected output values out of the total available from the program, but serve to illustrate the very comprehensive picture of physical behavior which is available from the model. In addition to the weekly values, two measures of physical performance were also illustrated: manpower utilization and delivery time. None of these data, however, provide a direct economic evaluation, which is our present objective. It remains for the financial accounting framework to provide this vital link.

Figure 10 summarizes financial results as tabulated in the weekly financial statements illustrated in Table 7. Weekly levels of income and expense are plotted and show the resulting profit or loss. The current assets graph pictures the weekly fluctuations in cash and inventories.

Cumulative financial performance in Run 1 for the 16-week period resulted in recording a net loss of $23,600. Current assets showed a net decrease of $7,600.

Medium Planning, Slow Information

It will be recalled from Table 5 that the only key parameter change between Runs 1 and 2 was in the planning cycle with a medium (two-week) cycle being substituted for the slow (one-month) cycle. The forecasting technique remained the same as did all the other decision rules. The demand pattern was identical for all runs.

Figure 7 permits a comparison of shipping performance with Run 1. An early dip in Product 1 shipments occurred in the 7th week but was accompanied by a high shipping rate for Product

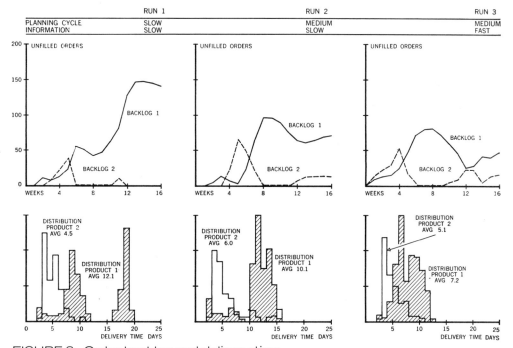

FIGURE 8 Order backlog and delivery time.

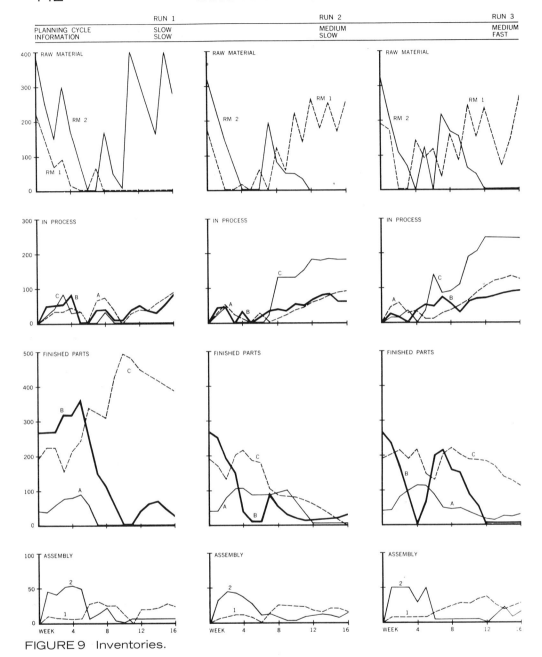

FIGURE 9 Inventories.

2. The backlog graph of Figure 8 indicates a general improvement over Run 1. Run 2 delivery time for Product 1 was reduced to 10.1 days.

In Figure 9, inventory behavior may be compared. The two-week raw material ordering pattern which accompanies the medium planning cycle in Run 2 is reflected in more frequent and smaller "saw teeth" in the raw material inventory graph. In the case of finished parts inventory, the over-shooting of Part C stock, which was noted in Run 1, is much less severe in Run 2.

In general, the improved responses of Run 2 shortened the period of readjustment and resulted in improved manpower utilization (82%), and better delivery performance.

These improvements in the physical performance of Run 2 are summarized and cast in an economic framework by the financial accounting output which appears in Figure 10. Run 2 performance for the 16-week period resulted in a net loss of only $3,900 and an increase in current assets of $12,100.

In comparing the financial outcomes between Runs 1 and 2, it is interesting to examine results in the 11th week. Run 1 recorded a net loss of $10,500 for the week whereas Run 2, having accomplished its rebalancing and material "turn-around," recorded a net profit of $4,800 with sustained earnings thereafter. It is revealing to note that during the same 11th week, Run 2 actually had a lower investment in inventories ($41,200) than Run 1 ($62,200).

At the 11th week, Run 1 inventories consisted largely of the unneeded inventory of Part C and unfinished materials,

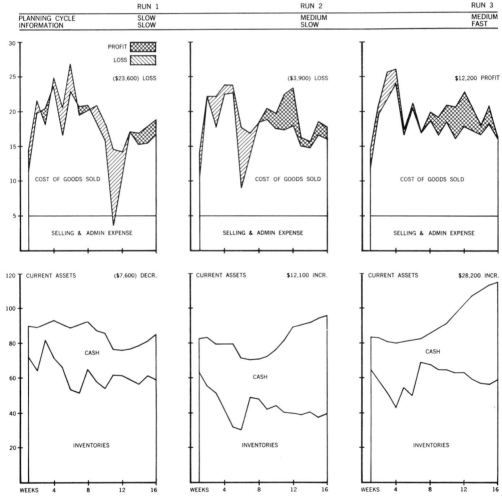

FIGURE 10 Financial results.

whereas, Run 2 had attained a reasonable balance of the right inventories throughout the physical system. It seems clear that Run 2 benefitted from better inventory management which in turn resulted in improved utilization of facilities and manpower.

The demand pattern was identical for both runs and thus presented the same hazards and opportunities. Run 2 management was no more "intelligent" (the decision rules were unchanged), but was simply made more effective through the improved response capability permitted by the shorter planning cycle. The value of this change from a one-month to a two-week planning cycle is of the order of $19,000 (reduction in loss from Run 1 to Run 2), in the model context for the period shown.

Medium Planning, Fast Information

In Run 3, the planning was kept at the medium (two-week) frequency as in Run 2, and the reduced information lags of the fast set were substituted for the slow lags.

A comparison of earnings between Runs 2 and 3 places a value of about $16,000 on this reduction in information delays.

The financial results for the runs described above, together with results for the other three runs, are summarized in Table 8. It will be seen that, within this very limited exploration, a continuous improvement in performance resulted from either an increase in planning frequency or a reduction in information delays.

Since there is stochastic "noise" in the model, statistical significance was tested by introducing a different random number sequence in repeat runs. The differences in economic performance, described above, were shown to be highly significant in the statistical sense.

SUMMARY AND CONCLUDING OBSERVATIONS

This paper has defined a method for evaluation of some major "intangible" aspects of an information processing system in terms of its contribution to the dynamic control of a firm as measured by the overall economic performance of the firm.

Application of the method has been demonstrated by a series of simulations carried out using a specific model of a hypothetical firm.

The feasibility of the method has been tested to the extent that selected parameter changes which are representative of "improved" information processing have been reflected in significant improvements in over-all economic performance of the modeled firm.

The extension of this method to useful economic evaluation of proposed systems in real firms will depend on how successfully the critical dynamics of the real enterprise can be described in model form. In addition, there is a need for

TABLE 8 Summary of profit or loss

		Planning cycle		
		Slow	Medium	Fast
Information delays	Slow	($23,600) Loss	($3,900) Loss	$11,500 Profit
	Fast	($1,100) Loss	$12,200 Profit	$24,000 Profit

fuller understanding of the effects of selective aggregation and/or scaling down of the multiple characteristics of the real firm, since some degree of abstraction will always be required to obtain models of manageable size.

Results such as those described in this paper, together with the current rapid rate of development in modeling and simulation techniques, serve to strengthen the authors' belief that the method described shows significant promise for eventual extension to useful evaluation of real information processing systems.

CITED REFERENCES AND FOOTNOTES

1. Forrester, Jay W., *Industrial Dynamics,* The MIT Press, 1961.
2. The model described in this paper is entirely discrete in nature. If one wished to describe continuous events within the physical system of a model (e.g., flow processes), sensors would report the *rate* of occurrence of such events.
3. Programming was accomplished with the use of the simulator described in a paper by G. Gordon, "A General Purpose Systems Simulator," *IBM Systems Journal,* Vol. 1, September, 1962.

THE COST AND EFFECTIVENESS OF COMPUTER SYSTEMS

W. SHARPE

A. COMPUTER SELECTION

1. The Problem

The correct approach to computer selection is as simple in theory as it is difficult to implement in practice. Assume that a selection must be made among M alternative computer configurations. Let there be N possible uses. For configuration i, devoted to use j, let

$$NV_{ij} \equiv TV_j - TC_{ij}$$

where TV_j = the total value of use j,
TC_{ij} = the total cost of configuration i devoted to use j, and
NV_{ij} = the net value obtained when computer i is devoted to use j.

SOURCE: Sharpe, W., "The Cost and Effectiveness of Computer Systems," *The Economics of Computers,* Columbia University Press, 1969, pp. 279–314.

For completeness, assume that these values are defined for all $M \times N$ combinations (in any case in which it is completely infeasible to perform some or all of the tasks included in use j with configuration i, TC_{ij} can be considered infinite, giving a net value, NV_{ij}, of minus infinity).

Configurations may be defined either narrowly (e.g., "IBM 360/50 with 6 tape drives") or broadly (e.g., "an RCA Spectra/70 System"), as may uses. A narrowly defined use would indicate precisely the jobs to be performed, the time each is to be submitted and completed, etc. Examples of very broad definitions would be "batch processing only" and "batch processing plus conversational computing." Obviously the broader the definition of configuration i and/or use j, the greater will be the analysis required to find the largest possible value of NV_{ij}. In any event, we assume that the required analyses have been performed and that each NV_{ij} represents such a value.

The computer selection problem is completely trivial once the set of net values has been obtained. The optimal configuration will be i^*, and its optimal use will be j^*, where

$$NV_{i^*j^*} \geq NV_{ij} \quad \text{for all } i \text{ and } j$$

In theory the optimal configuration and use cannot be obtained without explicit consideration of all possible alternatives (i.e., NV_{ij}'s). However, the choice can be made by using either of two stepwise procedures. One involves a selection among configurations on the basis of the maximal net value obtainable from each. Let

$$NV_i^{\max} \equiv \max_j (NV_{ij}).$$

Then select configuration i^*, where

$$NV_{i^*}^{\max} \geqq NV_i^{\max} \quad \text{for all } i$$

An alternative approach selects the best use on the basis of the maximal net value (or, equivalently, gross value less minimal cost) for each use. Let

$$NV_j^{\max} \equiv \max_i (NV_{ij})$$
$$= TV_j - \min_i (TC_{ij})$$

Then select use j^*, where

$$NV_{j^*}^{\max} \geqq NV_j^{\max} \quad \text{for all } j$$

Once the optimal computer configuration is known, the appropriate use is clearly the one giving the maximum net value. And once the optimal use is known, the appropriate configuration is clearly the one that will do the job(s) at lowest cost. But neither the optimal use nor the optimal configuration can, in theory, be determined without explicit consideration of all possible combinations.[1]

Users attempting to make explicit and quantitative analyses on which to base computer selection often evaluate alternative systems on the basis of the cost of performing a specified set of tasks. In our terms, given use j', select configuration i', where

$$NV_{i'j'} \geqq NV_{ij'} \quad \text{for all } i$$

[1] In other words, one must guard against procedures that may lead only to a local optimum instead of the global optimum.

or, equivalently,

$$TV_{j'} - TC_{i'j'} \geqq TV_{j'} - TC_{ij'} \quad \text{for all } i$$
$$TC_{i'j'} \leqq TC_{ij'} \quad \text{for all } i$$

If (by chance) the selected use is indeed optimal, this procedure will clearly give the optimal configuration. But if it is not, the result may be worse (i.e., give a lower value of NV_{ij}) than random selection of *both* a configuration and a use. In practice the selected tasks are often those performed by a currently installed system. Even if the tasks are optimal for that system, it is unlikely that they constitute the best use, given new types of configurations. Obviously, the greater the differences between currently available equipment and the equipment available when the present system was selected, the less satisfactory will be selection based on current use.

This discussion suggests that the objectivity of selection based on competitive bids in response to a set of "requirements" may be expensive, in the sense that it may result in a clearly suboptimal computer configuration. A less objective approach, in which each of several alternative configurations is rated on the basis of its overall value and cost if used in the best manner (i.e., best for the configuration in question), may give far better results. Of course the latter approach provides greater opportunity for malfeasance. If the interests of the person selecting a system diverge from those of the people to whom he is responsible, the problem becomes considerably more complex. An extreme example would include actual bribery by a manufacturer. However, more subtle but nonetheless damaging biases may affect the decision. If the maximum net value obtainable with computer A is 10% less than that obtainable with computer B, but computer A is more prestigious (e.g., costs more, has more impressive peripheral devices, or is made by a better-known manufacturer), the person charged with the task may be strongly tempted to select computer A. Although he may be able to accomplish this even under competitive bidding (by the appropriate definition of the required tasks), it may be more difficult than in the freer environment of a selection among alternative (and, in a sense, "incomparable") systems.

2. Competitive Bidding

Competitive bidding is widely used for procurement by federal government agencies. Since the prices of individual components are essentially fixed by the Federal Supply Schedule Price List, competitors bid against one another by offering configurations that meet a particular agency's requirements at the lowest possible cost (i.e., Federal Supply Schedule price). The Department of the Air Force, one of the largest users of computer equipment, has set up an agency charged solely with technical assistance in the selection procedure. The Electronic Data Processing Equipment Office, Electronic Systems Division, located at L. G. Hanscom Field, is

> the Air Force's centralized agency for the competitive evaluation and selection of commercially available computer systems for Air Force users worldwide. . . . [Its] job is to solicit proposals and to evaluate vendors' proposals, and to recommend a source from which the selected computer is to be acquired.[2]

The Air Force procedure involves a number of subjective evaluations and is in no sense based simply on minimizing cost for a specified level of performance. However, selection typically involves a request for proposal (RFP) stating a set of mandatory requirements; only vendors of configurations meeting these requirements are judged to be "responsive" and thus are considered further. Some of the implications of such a policy commanded considerable attention during 1967. The issue concerned an initial award to IBM of a contract for 135 computers for the Air Force Phase II Base Level Data Automation Standardization Program. The contract, involving a purchase cost of approximately $146 million, was reported to be the largest single order for computers ever placed.[3] It also turned out to be the most controversial.

The controversy centered on the fact that only the three alternative configurations submitted by IBM were judged "responsive" to the RFP; moreover, none of the other bidders (Honeywell, RCA, and Burroughs) were allowed to revise its proposal for re-evaluation. The problem involved the time required to process each of the two sets of benchmark problems. The RFP stated that each set must be completed within 200 hours of operational use time. Actual tests showed that only IBM's configurations met the requirement; Honeywell's equipment, for example, required 266.7 and 260.8 hours for the two prescribed workload levels.[4] But Honeywell's proposed equipment involved an initial cost approximately $65 to $70 million lower than that of IBM. Moreover, company spokesmen claimed that, had a revised proposal been allowed, Honeywell could have provided a configuration that would have met the mandatory requirement at cost only slightly above that originally proposed.[5]

The Air Force maintained that, considering maintenance costs, projected growth in workload levels, and similar

---

[2] Interview with Col. Sylvester P Steffes, reported in *Business Automation,* August, 1967, p. 31.
[3] *Business Automation,* August, 1967, p. 58.
[4] *Ibid.*
[5] The increase was reported to have been approximately $1.25 million (*ibid.*).

factors, the overall cost of the Honeywell proposal would not have differed from that of IBM by more than a "very few" million dollars.[6] Moreover, a process of iteration to obtain a responsive system was regarded by Air Force spokesmen as undesirable in such circumstances:

> We are talking about equipment that is available off-the-shelf. . . . We believe that it would be patently unfair to allow vendors to repair a proposal after live test demonstrations since the very purpose of these demonstrations is to prove that the system proposed meets the conditions of the request for proposal.[7]

Whatever the merits of the Air Force position as a general policy, the decision in this case was revoked. Honeywell filed an official protest with the Comptroller General (the head of the General Accounting Office). The resulting decision[8] was that "further written or oral discussions should be held with Honeywell as well as with other offerors. . . ." Although the overall source selection procedure used by the Air Force was held to be "reasonable," the selection of IBM in this case was considered unreasonable because of the failure to conduct further discussions with Honeywell well after the benchmark tests. The Air Force thus canceled the original award and reopened negotiations with the four bidders. The final contract was awarded to Burroughs, at a saving of $36 million compared to the original award, according to one source.[9]

This case provided a dramatic illustration of the conflict between efficiency and other goals such as equity, and objectivity. The imposition of any sort of rigid measure of performance and/or requirement for performance is almost certain to lead to a less-than-optimal result. If all parties are willing to assume that selection is in the hands of unbiased and highly knowledgeable individuals with the time, resources, and interest required to consider all relevant alternatives, a thoroughly subjective selection procedure is obviously preferable. In the real world, where these conditions are virtually never met, procedures such as that used by the Air Force, although less than optimal, may be far better than any realistic alternative.

3. Cost Minimization for Given Performance

Some writers have proposed that virtually all subjective elements be removed from the computer selection process and that the goal be to select the cheapest configuration capable of meeting a clearly specified set of requirements. The most explicit statement of such an approach is that given by Norman Schneidewind,[10] who advocates a mathematical programming formulation. The decision variables would be the numbers of various types of devices, such as tape drives, printers, and processors; the constraints would indicate the elapsed time within which each of several jobs must be run; and the objective would be to minimize cost. Schneidewind shows that an analyst with thorough knowledge of both equipment and the tasks to be performed can in some cases formulate the selection process as an integer linear programming problem. However, even in simple cases it is a far from trivial exercise to prepare coefficients that capture all the intricate interrelationships involved. The

[6]*EDP Industry and Market Report,* May 31, 1967, p. 2.
[7]*Ibid.*
[8]Comptroller-General Decision B161483.
[9]*EDP Industry and Market Report,* Dec. 29, 1967, p. 2.

[10]Norman Schneidewind, "Analytic Model for the Design and Selection of Electronic Digital Computers," doctoral dissertation, University of Southern California, January, 1966.

4. Scoring Systems

Frequently those charged with computer selection attempt to combine objectivity with the consideration of apparently nonquantifiable factors. Relevant considerations are enumerated and assigned weights. Then each competing system is subjectively rated (e.g., given a score from 1 to 10) with respect to each attribute by one or more judges. The scores for each system are averaged (using the assigned weights), and the best system is selected on the basis of the overall scores.

Cost is seldom included as one of the factors in a scoring system. A common approach is to consider only configurations of comparable cost, selecting the one with the best overall score. Often the cost level chosen is that of the currently installed system, on the (often implicit) grounds that (1) no more money can be obtained for computing and (2) the optimal amount is at least this great (and probably greater).

One study of several equal-cost systems[11] considered 123 separate items, organized into the following seven major divisions:

Division	Number of Items	Weight
Hardware	38	0.27
Supervisor	18	.27
Data management	8	.08
Language processors	31	.16
General programming support	4	.02
Conversion considerations	8	.12
Vendor reliability and support	16	.08

[11]Performed at the RAND Corporation in 1966.

Another study,[12] designed to choose among competing families of equipment (with detailed configurations to be selected later), utilized a stepwise procedure to arrive at a final set of relevant weights. As a first step, a set of high-level goals was defined and weighted:

Goal	Weight
1. Increase employee productivity	0.20
2. Improve the availability, relevance, and timeliness of information used by administrators at all levels	.25
3. Reduce current and future corporate operating costs	.20
4. Improve the company's responsiveness	.25
5. Maximize the capacity to cope with change	.10
	1.00

Next a set of six characteristics was defined, and a matrix relating characteristics to goals specified in such a manner that all column sums were equal to 1:

Multiplication of this matrix by the vector of goal weights provided the following set of characteristic weights:

Characteristic	Weight
1. Low data-processing costs	0.19
2. Interchangeability	.22
3. Capability to exploit technological advances	.22
4. Adaptability	.11
5. Low risk	.12
6. Good support from supplier	.14

Next a matrix relating each of forty-one attributes (rows) to each of the six char-

[12]Performed at North American Aviation; see Alan C. Bromley, "Choosing a Set of Computers," *Datamation,* August, 1965, pp. 37–40.

| | Goal | | | | |
Characteristic	1	2	3	4	5
1. Low data-processing costs	0.20	0.20	0.30	0.10	0.15
2. Interchangeability	.10	.25	.25	.25	.30
3. Capability to exploit technological advances	.25	.30	.10	.30	.05
4. Adaptability	.10	.05	.10	.05	.40
5. Low risk	.15	.10	.15	.10	.05
6. Good support from supplier	.20	.10	.10	.20	.05

acteristics (columns) was defined, again with each column sum equal to 1. Multiplication by the vector of characteristic weights gave a set of attribute weights. Then a matrix relating each of the four competing systems (rows) to each of the forty-one attributes (columns) was defined, with each column sum equal to 1. Finally, this matrix was multiplied by the vector of attribute weights to obtain the weight (score) for each of the four systems.

Weighting schemes must be used with considerable care. It is interesting to note that, as part of the latter study, sensitivity analyses were performed to investigate the impact of different assumptions regarding the appropriate weights. According to the author, "We were especially concerned with the sensitivity of the end score to changes in goal ratings . . . [but] we found that supplier scores were almost completely insensitive to even severe changes in goal weights."[13] Such a result may be cause for concern, not complacency, but analyses of this type are certainly desirable.

The usual weighting scheme assumes that the user's objective function is cardinal and linear. Letting S_i be the score for factor i and W_i its weight, the overall score is given by

$$S^* \equiv \sum_{i=1}^{N} W_i S_i \quad \text{with} \quad \sum_{i=1}^{N} W_i = 1$$

[13]Bromley, op. cit., p. 40.

Such a function is inconsistent with the usual assumptions of economic theory, since it asserts that the marginal rate of substitution of factor i for factor j is independent of the amounts (scores) of the two factors—that is, an equally desirable system can be obtained by substituting factor j for factor i at a rate equal to W_i/W_j. This is illustrated in Fig. 1 for a case involving only two factors, with weights $W_1 = \frac{2}{3}$ and $W_2 = \frac{1}{3}$. The indifference curves for $S^* = 2.5$ and $S^* = 5$, as shown, are linear. This implies, for example, that computer A will receive the same score as computer B, even though the former's superior hardware performance may never be available because of the complete lack of "vendor support" (i.e., $S_2 = 0$).

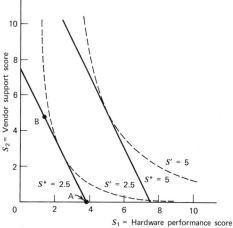

FIGURE 1 Isoquants based on two scoring schemes.

Economic theory usually assumes that indifference (iso-objective) curves are convex to the origin. A simple modification of the typical weighting scheme provides a function with such characteristics. Let

$$S' \equiv S_1^{W_i} \cdot S_2^{W_2} \cdot \ldots \cdot S_N^{W_N}$$

or, equivalently,

$$\ln S' = \sum_{i=1}^{N} W_i \ln S_i \quad \text{with} \quad \sum_{i=1}^{N} W_i = 1$$

S' is simply the weighted geometric average of the factor scores, while S^* is the weighted arithmetic average. In each case the weights are assigned to sum to 1. But note that S' will take on a value of 0 whenever any factor score is 0. Moreover, it does display the characteristics expected of such functions, as shown in Fig. 1 by the curves for $S' = 2.5$ and $S' = 5$ based on the original weights ($W_1 = \frac{2}{3}$ and $W_2 = \frac{1}{3}$).

Nothing that has been said here implies that linear weightings are necessarily inappropriate, especially for "well-balanced" systems. Note, for example, that when all factors are given the same score, $S' = S^*$ and the curves are tangent. In practice the two measures are likely to be very close, as shown by the values obtained for five computer systems evaluated in one study:[14]

System	S^*	S'
A	4.44	4.27
B	5.51	5.42
C	5.64	5.57
D	6.27	6.19
E	6.44	6.40

[14] The RAND study referred to in footnote 11. The score for each of the major factors was computed by taking a weighted arithmetic average of the scores assigned to the relevant subcategories. Thus S' is not the geometric mean of all 132 scores. In fact, the geometric mean is likely to be inappropriate for a detailed breakdown of factors, since a score of zero on one or more relatively minor items may not really be disastrous.

Economic theory cannot provide a "correct" form for an objective function for this (or any other) purpose. However, a linear function is not likely to prove applicable over a wide range of alternatives. In general, the function should reflect a willingness to give up less and less of A to obtain a unit of B as the amount of B is increased and the amount of A decreased. The geometric mean is one function meeting this criterion,[15] although it is only one of many that do.

Before leaving the subject of weighted scores, the treatment of cost deserves attention. It is perfectly consistent with economic theory to use a weighting scheme to measure performance, considering only equal-cost systems or, better yet, considering alternative levels of cost, with the final solution based on the best performance (score) obtainable for each cost. In either case no assumption about the relative importance of performance vis-à-vis cost is implicit in the procedure. However, some have

[15] Let $S' = S_1^{W_1} \cdot S_2^{W_2} \cdot \ldots \cdot S_A^{W_A} \cdot S_B^{W_B} \cdot \ldots \cdot S_N^{W_N}$. For given values of all S_i except S_A and S_B.

$$S' = K S_A^{W_A} S_B^{W_B}$$

and

$$S_A = \left(\frac{S' S_B^{-W_B}}{K}\right)^{1/W_A}$$

$$= \left(\frac{S'}{K}\right)^{1/W_A} (S_B)^{-W_B/W_A}$$

$$= K'(S_B^{-W_B/W_A})$$

where

$$K' = \left(\frac{S'}{K}\right)^{1/W_A}$$

Now, for constant S',

$$\frac{dS_A}{dS_B} = \left(\frac{-W_B K'}{W_A}\right)(S_B^{-W_B/W_A - 1})$$

$$= -\left[\frac{C}{S_B^{1+(W_B/W_A)}}\right]$$

where C is a positive constant, since W_A, W_B, S', and K are all positive.

The formula thus has the desired characteristic: as S_B increases, dS_A/dS_B becomes less and less negative.

advocated that cost be included directly in the overall score. This clearly involves a more heroic set of specifications.

Consider a case in which cost is the Nth factor and its score is determined as follows:

$$S_N = \frac{K}{TC}$$

where K is a constant chosen so that $0 \leq S_N \leq$ for all systems, and TC is the total cost of the system. Assume that a linear scoring scheme is to be used. Then

$$S^* = \sum_{i=1}^{N} S_i W_i$$
$$= \sum_{i=1}^{N-1} S_i W_i + S_N W_N$$

Now, define a measure of performance based on the scores for all factors other than cost with all weights rescaled to sum to 1:

$$P^* \equiv \sum_{i=1}^{N-1} \left(\frac{W_i}{1-W_N}\right) S_i$$

Obviously,

$$S^* = \left[P^* + \frac{W_N K}{1-W_N}\left(\frac{1}{TC}\right)\right][1-W_N]$$

As shown in Fig. 2 (for a case in which $W_N = 0.2$ and $K = 10$), this type of scoring system assumes that the greater the total cost of a computer system, the greater is the additional expense that should be incurred to obtain a given increase in performance. This assumption is hardly likely to be consistent with the user's true objective function. Note that the weight assigned to the cost factor (W_N) will change the positions and slopes of curves such as those shown in Fig. 2, but not their general shape.

The effect of including cost in a geometric-average scoring system depends more heavily on the weight assigned. As before, let

$$S_N = \frac{K}{TC}$$

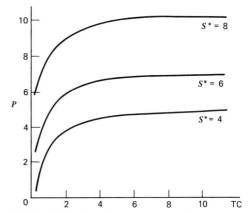

FIGURE 2 Performance and cost combinations with equal overall scores using a system in which cost is included.

The overall score S' will be

$$S' = S_1^{W_1} S_2^{W_2} \cdots S_N^{W_N}$$

Let P' be the measure of performance, with weights rescaled to sum to 1:

$$P' \equiv S_1^{W_1/1-W_N} \cdot S_2^{W_2/1-W_N}$$
$$\cdots \cdot S_{N-1}^{W_{N-1}/1-W_N}$$

Then

$$S' = K^{W_N} \left(\frac{P'}{TC^{W_N/1-W_N}}\right)^{1-W_N}$$

For given S':

$$\frac{P'}{TC^{W_N/1-W_N}} = K$$

where K is a constant. This formula, which defines an iso-objective curve, shows the importance of the weight assigned to the cost factor. If $W_N < 0.5$, the curves become flatter as TC increases. If $W_N = 0.5$, the curves are all linear through the origin: maximizing S' is equivalent to maximizing the performance/cost ratio (P'/TC). Finally, if $W_N > 0.5$, the curves become steeper as TC increases.

This discussion suggests that, if cost is to be included in a scoring system, the geometric average is to be preferred, since it can be made to have reasonable characteristics by selecting a value of $W_N \geqq 0.5$. However, the assumptions required are still substantial and, perhaps most important, far from obvious to the casual observer (and possibly to the eventual decision-maker). Assuming that an appropriate measure of performance can be obtained by weighting factor scores, it is far better to find the system giving maximum performance for each of several levels of cost and then choose the preferred cost and performance level explicitly.

2. Simulation Methods

One of the key tasks in computer selection is to estimate the manner in which each of several configurations will behave with one or more workloads. Some techniques use relatively simple formulas to obtain a single measure of performance; they will be discussed in Section B. Here we briefly consider methods designed to obtain relatively detailed estimates of performance, usually characterized by many measures, such as elapsed time for a task, percentage of idle time, percentage of time compute-bound, percentage of time input-output-bound, and probability of response time $\geqq 3$ seconds.

Perhaps the most popular system of this type is SCERT (Systems and Computer Evaluation and Review Technique), developed by Comress, Inc., and offered as a commercial service. The system includes a substantial file of information on computer components (e.g., timings, rental costs, and purchase prices). Instead of detailed simulation, SCERT uses "table-look-up and a series of empirically determined equations to estimate a computer systems behavior under a given job mix."[16] It is designed to be used for many purposes. According to a Comress spokesman, for hardware selection it serves to facilitate the choice of "that particular configuration which will process the defined workload in acceptable time-frames and which achieves the best cost/performance ratio."[17] Figure 3 summarizes the system.

Other approaches utilize true simulation: tasks are created and then processed by various units in the proper sequence, and detailed statistics gathered on the overall operation. Usually a number of values are drawn randomly from prespecified probability distributions. Such simulations often are designed primarily to help select a preferred operating system or scheduling algorithm, or simply to predict the behavior if a given system under as-yet unencountered loads. However, they have been used to evaluate alternative configurations as well.

Three general methods have been employed. Some investigators utilize special languages or routines in conjunction with a standard algebraic language.[18] Others use a general-purpose simulation language.[19] And at least two groups have developed special languages designed specifically for simulating computer system operation.[20] Whatever the method used, studies of this type attempt to obtain many highly detailed estimates of performance; rarely is an effort made to obtain a single overall measure of "effectiveness."

[17] F. C. Ihrer, "Computer Performance Projected through Simulation," *Computers and Automation,* April, 1967, p. 27.
[18] For example, the SDC system based on JOVIAL, Neilsen's system based on FORTRAN-IV, and Scherr's CTSS system based on MAD. See Huesmann and Goldberg, *op. cit.*
[19] For example, GPSS, SIMSCRIPT, and SIMTRAN. See Heusmann and Goldberg, *op. cit.*
[20] IBM's CSS (Computer Systems Simulator) and Lockheed's LOMUSS II (Lockheed Multipurpose Simulation System). See Huesmann and Goldberg, *op. cit.*

[16] L. R. Huesmann and R. P. Goldberg, "Evaluating Computer Systems through Simulation," *Computer Journal,* August, 1967, p. 150.

6. Computer Selection in Practice

A survey of 69 installations drawn randomly from the readers of *Datamation* was made in 1966 to determine the way in which computer selection was performed in practice.[21] Five major techniques were described; the percentage using each is as follows:

1. Evaluation of benchmark problems   60.9%
2. Published hardware and software evaluation reports   63.8
3. Programming and executing test problems   52.2
4. Computer simulation   15.9
5. Mathematical modeling   7.2

Note that the sum exceeds 100%, since some readers reported two or more techniques.

Respondents were also asked to rank each of eight selection criteria in order of significance, with the most important given a rank of 1 and the least important a rank of 8. The average ranks were as follows:

Item	Average Rank
1. Hardware performance	2.63
2. Software performance	2.69
3. Cost	4.10
4. Support provided by manufacturer	4.15
5. Compatability with present hardware and software	4.54
6. Potential for growth (modularity)	4.63
7. Delivery date	6.40
8. Availability of application programs	6.85

[21]Norman F. Schneidewind, "The Practice of Computer Selection," *Datamation*, February, 1967, pp. 22-25.

Not surprisingly, installations with large complements of equipment made greater-than-average use of more sophisticated techniques. However, the relative importance attributed to the selection criteria varied little among major groupings of users. The author of the article describing the survey found the emphasis on "objective" (hardware and software) criteria relative to the other criteria surprising: "This result is the most significant one of the survey. It was anticipated that subjective criteria would play a greater role."[22]

When considering these results (or, for that matter, the results of any survey of this type), it is useful to be skeptical. As the author states, the finding in regard to the importance of objective criteria

> is based on the assumption that the rankings provided by the respondents are truly indicative of the weight given the various criteria in the actual selection of a computer. It is possible that some users do not want to admit that a selection is made on other than a rational basis.[23]

Note also that the importance of each of the selection criteria was evaluated on the basis of ordinal rankings—no method was provided for a respondent to indicate, for example, that the first four criteria differed little in importance but that each was a great deal more important than criteria 5-8 taken together. Averages of such rankings are particularly deceptive, the more so because they appear to be cardinal measures.

B. MEASURES OF COMPUTER EFFECTIVENESS

1. Measuring Effectiveness

For some purposes any attempt to obtain a single cardinal measure of computer

[22]*Ibid.*, p. 24.
[23]*Ibid.*

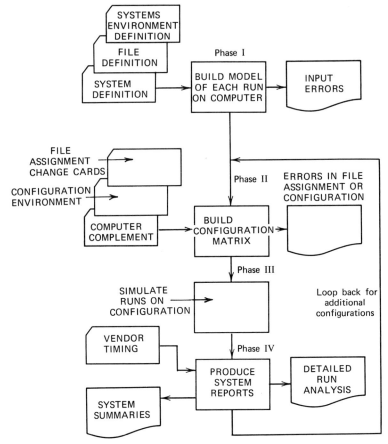

FIGURE 3 The SCERT system. (Source: "Evaluating Computer Systems through Simulation," L. R. Huesmann and R. P. Goldberg, *Computer Journal*, August 1967, p. 151.)

effectiveness ("performance," "throughput") would be ludicrous. But for other purposes it may be most sensible. Among the questions that may be answered reasonably well with such a measure are the following:

a. What has been the rate of technological progress (i.e., improvement in cost/effectiveness) for (1) computer systems and (2) particular components?
b. Are there economies of scale in computing, and, if so, of what magnitude?
c. To what extent is the technological progress achieved by one manufacturer dispersed among all manufacturers?
d. What is the relative (not absolute) effectiveness of computer 1 compared with that of computer 2?

The value of such measures for computer selection is subject to considerable dispute. It is obvious that the simpler the measure, the less "complete," "realistic," and "correct" it will be. However, it is usually also true that simpler measures are less expensive (and time-consuming) to use. It may thus be best to use a relatively simple measure after all. Only in a world in which information and analysis are free goods can it be stated categorically that the most realistic and complete method is the best.

A number of terms have been used to denote computer effectiveness. *Response* usually refers to the capacity of a system to react to some type of request; it is typically measured by the average or maximum time required for a response. Terms such as *throughput, performance,* and *capacity* usually deal with the system's capability in a steady-state operation. The measure may be the number of hours required to perform some specified set of tasks or the number of such sets of tasks that can be performed in a specified time period. The goal is usually to measure performance for a "typical" set of tasks. The importance of selecting an appropriate set cannot be minimized. Market forces should ensure that no computer dominates another, that is, provides better performance per dollar for every type of job. Any given system should perform some type of task more cheaply, or at least as cheaply, as any other system; if not, no sales will be made until its price is lowered. But no market mechanism guarantees uniformity of cost/effectiveness among systems for any single task. For some types of analysis it may be convenient to deal with "the" effectiveness of a system, but in general one must consider effectiveness for task A, effectiveness for task B, etc.

In this section we consider some important measures of effectiveness that have been used in the past. Several were designed for studies of technological change and/or economies of scale; we defer an extended discussion of such studies, which are covered in subsequent sections.

2. Simple Formulas

Some investigators have used extremely simple formulas for measuring the effectiveness of at least a portion of a computer system. The reciprocal of the time required to perform some rudimentary operation has been proposed as a measure of central processor effectiveness. Bourne and Ford found that the use of a single attribute, such as add time or internal clock time, as a measure implied that the effectiveness per dollar cost of 1960 computers was only slightly higher than that of computers of the early 1950's.[24] Since it is generally believed that major improvements took place during this period, such results suggest that these measures are not very satisfactory. Somewhat more useful results were obtained by Hillegass, who measured central processor effectiveness by the reciprocal of the time needed to add two numbers and store the result.[25] The record shows substantial improvement in effectiveness per dollar cost in the mid-1960's, with post-1964 computers giving almost three to four times the ratio obtained with pre-1964 equipment.

Since so-called central processors often include both processing units and high-speed storage (although not as commonly as at one time), several investigators have attempted to include estimates of the capabilities of both

---

[24]Charles P. Bourne and Donald F. Ford, "The Historical Development and Predicted State-of-the-Art of the General Purpose Digital Computer," *Proceedings of the Western Joint Computer Conference* (May 3–5, 1960), pp. 1–21.

[25]John R. Hillegass, "Hardware Evaluation," *DPMA Proceedings,* Vol. VIII, 1965, pp. 391–392; the measure used is "the time to access the contents of storage locations A and B, add them together, and store the results in location C. This eliminates the usual bias in favor of single-address computers when add times are quoted. Furthermore, all operations are at least five decimal digits in length to eliminate bias in favor of computers with very short word-lengths."

units in a single measure of performance. Schneidewind[26] and, later, Skattum[27] used a simple measure of this type:

$$E^S_{\text{cpu}} \equiv M \cdot N_c$$

where E^S_{cpu} = effectiveness,
M = high-speed memory storage capacity (in thousands of characters),
N_c = cycles per second = $1/t_{\text{cycle}}$, and
t_{cycle} = time required to read a word from memory and regenerate it (if required).

The number of storage cycles per second (N_c) may measure processor speed imperfectly, but it has the virtue of relative ease of measurement. Multiplication of M by N_c, while essentially an arbitrary choice, at least provides an index with expected properties; in particular, each curve connecting equal-effectiveness combinations of M and N_c is convex to the origin.

Schneidewind and Skattum used even simpler measures for the performance of other components.

For tape drives:

$E^S_{\text{tape}} \equiv$ maximum transfer rate (in thousands of characters per second);

For line printers:

$E^S_{\text{printer}} \equiv$ maximum number of lines printed per minute;

For card readers:

$E^S_{\text{reader}} \equiv$ maximum number of cards read per minute; and

For card punches:[28]

$E^S_{\text{punch}} \equiv$ maximum number of cards punched per minute.

A more complicated formula, proposed by Gruenberger,[29] attempts to take into account a computer's speed in arithmetic processing and other factors:

$$E^G = \frac{M(N_a + N_m)}{L}$$

where M = high-speed memory storage capacity (in bits),
N_a = the number of additions per second = $1/t_a$,
t_a = the time required to perform an addition (in seconds),
N_m = the number of multiplications per second = $1/t_m$,
t_m = the time required to perform a multiplication (in seconds), and
L = the instruction length (in bits).

The inclusion of the instruction length may seem unusual; according to Gruenberger, "L attempts to measure inefficiencies due, for example, to decimal capability."[30] None of the elements is specified completely enough to be measured directly. For example, do t_a and t_m refer to fixed-point or floating-point, decimal or binary, operations? If instructions are of variable length, how is L to be measured? As Gruenberger indicates, "None of the . . . factors is wholly ob-

---

[26] Norman F. Schneidewind, "Analytic Model for the Design and Selection of Electronic Digital Computing Systems," *op. cit.*, pp. 204, 205.

[27] Stein Skattum, "Changes in Performance of Components for Computer Systems" (unpublished). This paper was written as a term project for a seminar given by the author at the University of Washington in 1967.

[28] Schneidewind did not consider card punches; the definition is that given by Skattum. In Skattum's study, combination units (reader-punches) were considered to be two units, each costing half the total cost.

[29] Fred Gruenberger, "Are Small, Free-standing Computers Here to Stay?" *Datamation*, April, 1966, pp. 67–68.

[30] *Ibid.*

jective ... and some are extremely difficult even to estimate for some machines."[31]

3. Instruction Mixes

Solomon[32] has proposed the following technique for comparing two processors with similar sets of instructions. Let C_j be the cost per unit time (e.g., microsecond) of processor j, and T_{ij} the time (e.g., in microseconds) required to execute instruction i on processor j;[33] then the cost of executing instruction i on processor j is

$$C_{ij}^s = T_{ij} \cdot C_j$$

Processor j^* can obviously be said to cost less per unit of effectiveness than processor j if

$$C_{ij^*}^s \leq C_{ij}^s \quad \text{for all } i$$

and

$$C_{ij^*}^s < C_{ij}^s \quad \text{for at least one } i$$

Unfortunately, such cases are rare. Typically one processor will be better (i.e., give a lower value of C_{ij}^s) for some instructions and poorer (i.e., give a higher value of C_{ij}^s) for others. In such instances some weighting scheme must be invoked. Let W_i be the weight assigned to instruction i, and let there be N instructions in all. For convenience, assume that

$$\sum_{i=1}^{N} W_i = 1,$$

so that W_i can be interpreted directly as the relative importance (or frequency) of instruction i.

[31] Ibid.
[32] Martin B. Solomon, Jr., "Economies of Scale and the IBM System/360," *Communications of the ACM*, June, 1966, pp. 435–440.
[33] In some advanced systems, T_{ij} may not be a constant—the time may depend on other activities taking place concurrently. Such complications are ignored here.

Processor j^* can be said to cost less per unit of effectiveness than processor j if

$$\sum_{i=1}^{N} W_i C_{ij^*}^s < \sum_{i=1}^{N} W_i C_{ij}^s$$

or, equivalently,

$$\sum_{i=1}^{N} W_i T_{ij^*} C_{j^*} < \sum_{i=1}^{N} W_i T_{ij} C_j$$

This can be rewritten as

$$\frac{C_{j^*}}{E_{j^*}^{IM}} < \frac{C_j}{E_j^{IM}}$$

where

$$E_j^{IM} = \frac{1}{T_j^{IM}} \quad \text{and} \quad T_j^{IM} = \sum_{i=1}^{N} W_i T_{ij}$$

T_j^{IM} can be interpreted as the time required by processor j to execute a "typical" instruction, while its reciprocal (E_j^{IM}) measures the effectiveness of the processor in terms of the number of "typical" instructions performed per unit time. The superscript indicates that the measures are based on an "instruction mix," defined by the weights W_i.

Obviously E^{IM} is appropriate only for measuring the effectiveness of a central processor. Moreover, its usefulness depends critically on the selection of relevant weights. In practice, instructions, are normally grouped into relatively broad classes for this purpose; the more diverse the central processors to be considered, the broader are the classes (and, perhaps, the less relevant the results). Two approaches have been taken to obtain weights. The first uses the actual frequencies of execution for a "typical" mix of tasks, based on dynamic traces taken during the operation of an actual system. The second approach uses estimates of the relative frequencies that would be encountered if particular codes were executed.

TABLE 1. Weights for a Scientific Mix and a Commercial Mix*

Instruction Category†	Scientific Weight	Commercial Weight
1. Fixed add (subtract) and compare instructions	0.10	0.25
2. Floating add (subtract) instructions	.10	0
3. Multiply instructions	.06	.01
4. Divide instructions	.02	0
5. Other manipulation and logic instructions	.72	.74
	1.00	1.00

*Source: Kenneth E. Knight, "A Study of Technological Innovation—The Evolution of Digital Computers," doctoral dissertation, Carnegie Institute of Technology, November, 1963, pp. IV-5, IV-6, IV-7.
†Category descriptions:
1. "These instructions are the fixed additions, subtractions and compare operations performed. We may obtain the fixed add time for each system from the computing literature."
2. "The floating add time is given in the computing literature for machines with built-in floating-point arithmetic. For other machines the figure can be approximated by multiplying the fixed-point add time by 10 . . . (the mean value for six computing systems considered)."
3. "We have included only one multiply category since the operating times for these two operations on systems capable of both floating- and fixed-point arithmetic are approximately equal. The multiplication time is a characteristic available in the computing literature."
4. "The fixed- and floating-point operations were combined . . . the divide time represents a characteristic of each system published in the computing literature."
5. "This category combines a large number of branch, shift, logic and load-register instructions. . . . For computers with parallel arithmetic, the time . . . is the shortest of . . . add time or . . . 2 [times] the memory access time for one word. . . . For computers with serial arithmetic, the . . . time equals the shortest of (1) add time or (2) [the time required to access an instruction, slightly modified]."

Table 1 shows two sets of weights obtained by Knight from dynamic traces. The "scientific" weights are based on approximately 15 million operations of an IBM 704 and an IBM 7090 performed on a set of more than 100 problems. The "commercial" weights are based on approximately 1 million operations of an IBM 705 performed on a set of nine programs (two inventory control, three general accounting, one billing, one payroll, and two production planning). Another set of weights, obtained by Arbuckle using a dynamic trace, is shown in Table 2; according to the author, it "represents a composite of a number of scientific and engineering applications."[34] Although Arbuckle's mix is not directly comparable with Knight's scientific mix, the two appear to be reasonably consistent.

Table 3 shows weights based on three programs analyzed by Solomon:

> The first is highly scientific, a matrix multiplication problem; the second [a floating square root program] is also scientific but utilizes arithmetic capabilities less heavily; the third . . . is perhaps more closely related to data processing (and compiling) applications. It is a field scan of a card for control options.[35]

For purposes of comparison, the weights are also summarized by major instruction category.

[34] R. A. Arbuckle, "Computer Analysis and Thruput Evaluation," *Computers and Automation*, January, 1966, p. 13.

[35] Solomon, *op. cit.*, pp. 437, 438.

TABLE 2. Weights for a Scientific Instruction Mix*

Instruction Category	Weight
Floating-point add/subtract	0.095
Floating-point multiply	.056
Floating-point divide	.020
Load/store	.285
Indexing	.225
Conditional branch	.132
Miscellaneous	.187
	1.000

*Source: R. A. Arbuckle, "Computer Analysis and Thruput Evaluation," *Computers and Automation,* January, 1966, p. 13.

The differences among the three sets of weights given by Solomon suggest that the selection of an instruction mix may greatly influence the results of any comparison. And the contrast between Solomon's detailed instruction weights and the much broader classes used by Knight and Arbuckle suggests the dangers associated with using any single set of weights when comparing systems with radically different instruction sets. Calingaert provides an example of the problem:

> The members [of a group of experienced system engineers] were asked to specify the time in microseconds on System/360 Model 40 for the compare class of instructions, given only the fact that the original mix was based on the 7090.... The ten answers ranged from 11.88 to 30.66 with a mean of 21.5 and standard deviation of 7.0.[36]

4. Kernel Timing Estimates

One way to deal with differences among processors is to compare the times (and costs) required to perform a specified task, called a *kernel,* assuming efficient coding for each machine analyzed. According to Calingaert, a kernel is "the central processor coding required to execute a task of the order of magnitude of calculating a social security tax, or inverting a matrix, or evaluating a polynomial."[37] An attempt is generally made to have the problem "coded with equal levels of sophistication by experienced programmers in assembly language."[38]

The three programs that Solomon used are typical kernels, and he identified them as such. However, since he wished to compare only processors with the same instruction set, each problem was coded only once. In this special case, the kernel approach degenerates to an instruction-mix comparison.

Table 4 describes seven kernels used to compare an IBM 360/67 with an IBM 7094-I. As shown in Table 5, the power of the 360/67 relative to that of the 7094-I varies considerably among the seven kernels; the appropriate overall ratio depends, of course, on the relative importance of each kernel. According to the study, "Estimation of computing center workload indicates that it may be represented by the distributions [shown in Table 5] between compiling and object code execution."[39] By using these weights, the ratio of the performance of the 360/67 processor to that of the 7094-I was estimated to be 3.991 for compilation and 3.157 for execution. No weights were given for combining the two ratios into a single result.

Note that some set of weights is required if a single figure of merit is to be obtained from timing estimates for

---

[36] Peter Calingaert, "System Performance Evaluation: Survey and Appraisal," *Communications of the ACM,* January, 1967, p. 15.

[37] *Ibid.*

[38] *IBM System/360 Model 67 Time-sharing System, Technical Summary,* Aug. 18, 1965, p. E-1.

[39] *Ibid.,* p. E-5.

TABLE 3. Instruction Weights for Three Programs*

Instruction	Operation†	Weight		
		Matrix Multiplication	Floating Square Root	Field Scan
Fixed-point 32-bit operations				
A RX	C(storage) + C(reg) → reg			0.0015
AR RR	C(reg 1) + C(reg 2) → reg 2	0.1559		.1773
L RX	C(storage) → reg	.1753	0.0634	
LM RS	{C(storage) → reg} 4 times	.0002		.0015
LR RR	C(reg 1) → reg 2	.0368		.0443
LTR RR	{ C(reg 1) → reg 2 / set condition code }			.0443
ST RX	C(reg) → storage			.0015
Floating-point 32-bit operations				
AE RX	C(storage) + C(reg) → reg	.0421		
AER RR	C(reg 1) + C(reg 2) → reg 2	.1559	.1745	
DER RR	C(reg 1)/C(reg 2) → reg 1		.1429	
HER RR	C(reg 2)/2 → reg 1		.1587	
LER RR	C(reg 1) → reg 2		.1429	
ME RX	C(storage)·C(reg) → reg	.1559		
STE RX	C(reg) → storage	.0597	.0159	
Logical operations				
CLC SS	{ C(storage 1) : C(storage 2) (4 bytes) }			.1773
CLR RR	C(reg 1) : C(reg 2)			.0443
LA RX	{ C(storage) → reg (24 bits) }	.0002		
STC RX	{ C(reg) → storage (8 bits) }			.0044

several kernels. Needless to say, the collection and use of such weights involve problems similar to those associated with instruction mix comparisons. Moreover, both methods assume suboptimization at some level. Consider matrix multiplication and BCD arithmetic. The cost of the latter, in terms of the amount of the former sacrificed, is clearly lower for the 360/67 than for the 7094-I. Truly optimal use of the 360/67 would almost certainly involve more BCD arithmetic relative to matrix multiplication than would optimal use of the 7094-I. Any single set of weights must thus represent suboptimal use of one (or both) systems. Note, however, that the weighted-kernel approach at least allows optimal use of each system's instruction set; the suboptimization thus occurs at a higher level than in an instruction-mix comparison.

The importance of selecting appropriate weights has been emphasized

TABLE 3. (continued)

Instruction	Operation†	Weight		
		Matrix Multiplication	Floating Square Root	Field Scan
Branching				
BALR RR	PSW → reg	.0002	.1429	.0015
BC RX	Branch on condition to address in register			.2792
BCR RR	Branch on condition to address in register modified	.0002		
BCT RX	[C(reg) − 1] → reg Branch if C(reg) = 0		.1429	
BCTR RR	[C(reg) − 1] → reg			.0443
BXH RS	Branch on index high	.2174		.1328
EX RX	Modify instruction and execute			.0443
Status switching				
SVC RR	Supervisor call	.0002	.0159	.0015
		1.0000	1.0000	1.0000
Summary by Major Instruction Category				
	Fixed-point operations	0.3682	0.0634	0.2704
	Floating-point operations	.4136	.6349	0
	Logical operations	.0002	0	.2260
	Branching	.2178	.2858	.5021
	Status switching	.0002	.0159	.0015
		1.0000	1.0000	1.0000

* Sources: Weights are based on frequencies given in Martin B. Solomon, "Economies of Scale and the IBM System/360," *Communications of the ACM*, June, 1966, pp. 435–440. Instruction descriptions and classifications are based on *IBM System/360 Principles of Operation*, IBM Form A22-5821-1.
†C(x) stands for the contents of x.

Reg, reg 1, and reg 2 signify (arbitrary) registers.
Storage, storage 1, and storage 2 signify (arbitrary) locations in storage.
PSW represents the program status word.
a:b indicates that a is compared to b, and the condition code set on the basis of the result.

by Calingaert: "In one study comparing the performance of one CPU relative to another, different kernels yielded performance ratios as high as 9.5 and as low as 3.3. I am aware of no rational technique for weighting kernels."[40] Although the situation may not be quite that hopeless, results based solely on kernel timing estimates clearly must be used only after careful analysis.

[40]Calingaert, *op. cit.*, pp. 15, 16.

5. Benchmark Problem Times

One of the major drawbacks of the kernel approach is its concentration on processor performance. The point is often made that, to evaluate an entire computer system, much more must be taken into account—in particular, non-overlapped input-output operations. Knight has proposed a general formula for accomplishing this; it is discussed in Section B-6. Here we deal briefly

TABLE 4. Seven Kernels*

1. Matrix multiplication

This is a matrix multiplication subroutine. Two 10 × 10 matrices were generated with single-precision floating-point elements. The matrix multiplication subroutine was then entered, and, using the standard formula below, the product was generated:

$$C_{i,j} = \sum_{K=1}^{n} a_{iK} b_{Kj} \quad \begin{cases} i = 1, 2, 3, \ldots, n \\ j = 1, 2, 3, \ldots, n \end{cases}$$

2. Square root approximation

This kernel is indicative of the type of functional subroutine used often in a scientific program. In this case,

$$X = N^{1/2}$$

is computed to the accuracy of the floating-point word or to 10 approximations, using the formula

$$X_{n+1} = \frac{1}{2}\left[X_n + \frac{N}{X_n}\right]$$

It is assumed that N is in storage; the result is left in storage at X.

For the first approximation, $X_1 = N$ is used. No test for negative or zero X is required. For timing purposes, it was assumed that 10 iterations are performed.

3. Field manipulation

Control card scans, which this kernel represents, are similar to source statement scans found in FORTRAN and COBOL; consequently, this kernel is somewhat representative of both control card scans and source statement scans. Here, a variable field is scanned, starting in column 16 and ending with either the first blank or column 72, whichever comes first. The field that is scanned will have options delimited by commas (or a comma and a blank). Each option, 1–6 characters in length, is matched against an option dictionary of 8 items; an indicator is set if a match is found. For timing purposes, 30 columns were scanned in which 5 options (separated by 4 commas) are found.

4. Editing

A common problem in commercial programs is to edit a field of decimal digits—suppressing or leaving leading zeros, inserting commas and a decimal point, etc. In this kernel, a field of 10 decimal digits is edited in the following manner: leading zeros in the field are suppressed; a decimal point is inserted between the second and third digits from the right; commas are inserted between the fifth and sixth digits and between the eighth and ninth digits (but, in each case, only if the high-order digit is nonzero); and a dollar sign if "floated," i.e., it precedes and is in juxtaposition to the first significant digit or the decimal point, whichever comes first. A field of 10 zeros should appear as $.00. A minus sign is carried in machine notation; and, if it appears in the original number, it should appear after the edited field. For timing purposes, the number 0007777512 was edited. (It would appear as $77,775.23 after editing.)

5. Field comparison

This kernel is often found in programs when a decision is to be made on the basis of whether one number is greater than, equal to, or less than another number. A field consisting of N consecutive characters is compared with another field of N consecutive characters. An indication of whether the first field is less than, equal to, or greater than the second field should be made so that it can be interrogated later (this indication is normally made automatically by the machine). For timing purposes, two fields of 10 digits were compared.

TABLE 4. (*continued*)

6. BCD arithmetic

This kernel shows an execution time for a typical decimal addition if both addend and augend must be preserved. One field is moved to a work area, and the other field is added to it in the work area. For timing purposes, two fields of 10 decimal digits were added after moving the first field to a work area.

7. Character manipulation

This kernel represents a typical data movement. A source field of *N* bytes of alphanumeric information is moved (and left justified) into a target field at least 2 bytes longer than the source field. The timings were made for a 12-character source field and a 16-character source field and a 16-character target field.

*Source: *IBM System/360 Model 67 Time-sharing System, Technical Summary,* Aug. 18, 1965, pp. E–1, E–2, E–3.

with an alternative approach: the estimation of the total time required to complete certain "benchmark" tasks.

Perhaps the most extensive set of estimates of this type is that prepared by Auerbach Info, Inc., for inclusion in the company's *Standard EDP Reports*.[41] Six major benchmark problems are utilized; however, the definition of a given problem may include one or more parameters, giving rise to a range of subproblems. Thus the standard file-updating problem is defined in terms of the average number of detail records per master record (among other things), and estimated times are given for values of this ratio from 0 to 1.0.[42]

The six benchmark problems used by Auerbach are as follows:[43]

- Updating sequential files.
- Updating files on random-access storage.
- Sorting.

- Matrix inversion.
- Evaluation of complex equations.
- Statistical computations.

Times are estimated, not obtained directly. The following quotation outlines the general approach:

To help insure objective comparisons, the standard problems are rigidly specified in terms of available input data, computations to be performed, and results to be produced. On the other hand, factors such as master file arrangement and detailed coding methods are left flexible to permit maximum utilization of the distinctive capabilities of each computer.

To assure realistic comparisons between competitive systems, the equipment configurations, as well as the problems, must be standardized. For example, one configuration includes six magnetic tape units on a single channel and an on-line card reader, card punch and printer.

The execution time for each standard problem on each standard configuration is determined by computing all input-output times and central processor times, and then combining them with due regard for the system's capabilities for simultaneous operations. The problems are coded and timed in detail, and submitted to the

---

[41]*Auerbach Standard EDP Reports,* Auerbach Info, Inc., Philadelphia, Pa.; subscription rates (1967): $900 for one year. $695 per year thereafter

[42]John R. Hillegass, "Standardized Benchmark Problems Measure Computer Performance," *Computers and Automation,* January, 1966, pp. 16–19.

[43]J. B. Totaro, "Real-time Processing Power: A Standardized Evaluation," *Computers and Automation,* April, 1967, and Hillegass, "Hardware Evaluation," *DPMA Proceedings,* Vol. VIII, 1965, p. 405.

TABLE 5. Relative Power: The IBM 360/67 versus the IBM 7094-I*

Kernel	Relative Power† 360/67:7094-I	Weight Compilation	Execution
Matrix multiplication	2.29	0	0.30
Square root approximation	3.15	0	.35
Field manipulation	2.37	0.35	.07
Editing	5.09	.20	.05
Field comparison	4.00	.20	.11
BCD arithmetic	7.10	.10	0
Character manipulation	4.22	.15	.12
		1.00	1.00

*Source: *IBM System/360 Model 67 Time-sharing System, Technical Summary*, Aug. 18, 1965, pp. E–4, E–5.

†Relative power = $\dfrac{\text{processor time for 7094-I}}{\text{processor time for 360/67}}$

computer manufacturers for checking to help assure their validity. The results are presented in the form of graphs that show the computer system's performance over a wide range of problem parameters and equipment configurations.[44]

Since many installations rely heavily on higher-level programming languages, evaluation of computer hardware alone may not suffice. To assess the capabilities of both hardware and software, estimates of the times required to compile and execute benchmark programs written in appropriate problem-oriented languages may be used. Such estimates are extremely difficult to obtain without actual runs on equipment that is at least similar to that being evaluated. However, even such a seemingly straightforward approach is likely to prove difficult in practice, as shown by the results of one study.[45]

Seven benchmark problems coded in FORTRAN were prepared; Table 6 summarizes their characteristics. Each was compiled and executed on the "old" computer system. The goal was to compile and execute each program on each of four new systems under consideration. However, this proved impossible. One program (number 6), could not be compiled on one of the computer systems. Execution times for another (number 3) proved incomparable because the execution path was dependent on the sequence of pseudo-random numbers generated and each system generated a different sequence. Execution times for yet another (number 4) could not be compared because "one manufacturer ran the problem in a multi-programmed mode and obtained an elapsed processor time of nearly zero. Another simulated tapes on a magnetic drum. Another used much smaller physical records." Finally, the last program (number 7) could not be executed on two of the four systems "because of problems in random number generation."

Even the times that were obtained proved in most instances to be estimates. The figures given for the first system were obtained by doubling the actual times required on a faster system from the same family. Those for the second system were derived by multiplying the

[44]Hillegass, *op. cit.*, pp. 405, 406.
[45]Performed at the RAND Corporation in 1966.

actual times required on a slower system by 0.6. In the case of the third system, actual times were adjusted to reflect improvements expected from extensive system modifications (e.g., a new loader and replacement of a disk system with a drum); these adjustments were substantial: compilation times were reduced to one-sixth and execution times to one-third of the actual amount. Of the four system considered, actual times could be used without modification for only one.

In spite of all these problems, some comparisons were possible. For example, Table 7 shows the ratio of the time required to perform each of several tasks on the "old" system to that required on one of the new systems: Note the variation. How should these results be summarized? The ratio of total time required to compile all seven programs on the old system versus a new one is approximately 18 to 1. But the ratio of the time required to execute the four that could be executed is only 4.6 to 1. Since compilation is more time-consuming in this instance than is execution, the ratio of total time is far above the mean of the compile and execute ratios (approximately 14.6 instead of 11.3). However, all these ratios fall

TABLE 6. FORTRAN Benchmark Problems

Description	Size (cards)	No. of Subprograms	Input (cards)	Output (pages)
1. Evaluates a set of formulas to study blood and oxygen transfer between a pregnant ewe and her fetus.	85	3	36	14
2. Computes performance characteristics of rocket vehicles in simulated trajectories. 34 of the 68 subprograms were null; all but 4 of the remaining contained an identical set of 57 specification cards.	3868	68	96	5
3. Evaluates a mass-accretion hypothesis on the evolution of the solar system. Contains a relatively large number of CALL and IF statements. 11 subprograms contain 6 or fewer statements.	1188	25	5	84
4. Writes and rewinds two utility units n times.	26	1	1	1
5. Given the number of fragments and total weight of a fragmented object, applies Mott's law to compute the distribution of fragments by weight.	100	1	11	5
6. Simulates adaptive routing techniques for a distributed communications network. Contains essentially no floating-point arithmetic.	728	5	332	3
7. Computes the trajectories of two missiles in a simulated interception.	3208	44	723	99

TABLE 7. Comparison of System Times

Problem	Compile or Execute	Time Required on Old System (minutes)	Time Required on New System (minutes)	Ratio of Old to New
1	compile	0.717	0.017	42.18
2	compile	22.5	1.357	16.58
3	compile	6.683	0.177	37.76
4	compile	0.183	0.007	26.14
5	compile	0.483	0.02	24.15
6	compile	1.783	0.16	11.14
7	compile	13.05	0.75	17.40
1	execute	0.317	0.077	4.12
2	execute	0.85	0.043	19.77
5	execute	0.183	0.003	61.00
6	execute	2.733	0.767	3.56
Total compile time:		45.399	2.488	18.25
Total execute time:		4.083	0.890	4.59
Total time:		49.482	3.378	14.65
Average Ratio Compile times:				25.05
Execute times:				22.11
Total time:				23.98

below the figures obtained if the ratios for individual tasks are averaged. As shown in Table 7, the latter values all exceed 20 to 1. Clearly the problem of selecting appropriate weights is as difficult and important in this case as it is in any other.

One final problem deserves mention. Even if accurate benchmark problem times can be obtained, in general they may not be considered additive. If T_i is the time required to perform task i alone, how long will it take to perform N different tasks? All that can be said with certainty is that the total time will lie within the following range:

$$\max_i (T_i) \leq \text{total time} \leq \sum_{i=1}^{N} T_i$$

Clearly the possibility of substantial overlapping through multiprogramming and/or multiprocessing makes even more difficult the already impossible task of specifying for each system a set of benchmark problem weights that will give an overall indication of its effectiveness if used optimally.

6. Knight's Formula

We conclude this section with a description of the formula used by Knight to measure the "computing power" of an entire system:[46]

Computing power = memory factor
× operations per second

Considering first the latter term, we have

$$\text{Operations per second} \equiv \frac{10^{12}}{t_c + t_{I/O}}$$

where t_c = the time (in microseconds) required to perform one million opera-

[46] Kenneth E. Knight, "A Study of Technological Innovation—The Evolution of Digital Computers," doctoral dissertation, Carnegie Institute of Technology, November, 1963, pp. IV-1 through IV-16 and A-2 through A-5.

tions, and $t_{I/O}$ = the nonoverlapped input-output time (in microseconds) necessary to perform one million operations.

The computing time (t_c) is based on the weights given in Table 1. Knight measures two kinds of computing power—commercial and scientific; the weights obtained from the appropriate mix are thus used to compute t_c.

The estimation of nonoverlapped input-output time is rather complex. It is based on the channel width, transfer rate, and start, stop, and rewind times for both primary and secondary input-output devices, plus estimates of the extent of possible overlaps and the utilization of primary and secondary input-output systems. Several of the required coefficients are specified by Knight, often with one value for commercial computation and another for scientific.

The other component in computing power is defined as follows:

$$\text{Memory factor} \equiv \frac{[(L-7)N(WF)]^P}{K}$$

where K = a constant,
L = word length (in bits),
N = the total number of words in high-speed memory,
$WF = \begin{cases} 1 \text{ for fixed word length memory,} \\ 2 \text{ for variable word length memory,} \end{cases}$
$P = \begin{cases} 0.5 \text{ for scientific computation,} \\ 0.333 \text{ for commercial computation.} \end{cases}$

This formula is based primarily on opinions:

> A total of 43 engineers, programmers and other knowledgeable people were contacted and asked to evaluate the influence of computing memory upon performance.[47]

> Authorities estimate that variable word length memories are twice as valuable as fixed word length . . . with an equivalent bit capacity.[48]

> We also found that if word length is very short, the system encounters difficulties in carrying out many scientific and commercial calculations. For this reason we decided, upon the advice of the experts, to subtract seven binary digits from the actual word length, thus serving to penalize the short words.[49]

From the opinions of the experts the following approximations were made: (1) for scientific problems the computing power increases as the square root of the bit value of memory; (2) for commercial problems the computing power increases as the cube root of the bit value of memory.[50]

Knight's approach is certainly subject to criticism. However, it has advantages: it is relatively straightforward and can be applied without excessive effort. Perhaps most important, Knight has used it to obtain estimates of both the commercial and scientific computing power of more than 300 systems. No other measure has been applied consistently to such a wide range of computers.

C. ECONOMIES OF SCALE IN COMPUTING

According to economic theory, average cost will be inversely related to output volume (given rate of output) and directly related to rate (given volume). Moreover, the strength of the volume effect is purported to decrease with volume and that of the rate effect to increase with rate. This implies a U-shaped average cost curve for proportional changes in rate and volume.

[47] *Ibid.*, p. IV-12.
[48] *Ibid.*, p. IV-13.
[49] *Ibid.*
[50] *Ibid.*

Consider computers of different sizes. A "larger" system can produce computation at a faster rate; over any given period it can also produce a larger volume. For simplicity, consider a period of one month, with the manufacturer's rental charge (including maintenance) as total cost. Then rate (computation per month) equals volume (total computation during the month), and any system can be represented by a point on a (presumably U-shaped) average cost curve.

On the assumption that many system designs are available, and that each gives a point along a U-shaped average cost curve, which systems will be placed in production? One might expect that only those giving the minimum attainable cost per unit of effectiveness would be produced, the market being limited to machines that were neither too large nor too small, but "just right." However, this would occur only under very special circumstances. For many users a small machine may in fact be cheaper overall than a larger one, even though the latter can give a lower cost per unit of effectiveness if utilized to capacity. A larger system used only to perform tasks that could be completed with a smaller computer will clearly give a higher cost per unit of computation, since (by assumption) the two provide equal effectiveness and the larger has a greater cost. If computer sharing were cost-free, of course, any system could be used to capacity and a part-time user would pay only a proportional share of the total cost of the equipment, thus obtaining computation at the machine's optimal cost per unit of effectiveness. But sharing is not free: there are overhead costs, communications costs, and political problems (e.g., who gets top priority?). Thus small systems with nonoptimal cost/effectiveness are likely to be found on the market.

The case for larger-than-optimal systems is not as strong. Assume that the optimal system costs C^* and gives a total effectiveness (computation) of E^*. A system twice as large (i.e., giving $2E^*$ units of effectiveness) will cost more than twice as much. But two optimally sized computers will give $2E^*$ units of effectiveness at a cost of precisely $2C^*$. Why, then, would anyone buy one giant system instead of two or more optimal systems? Presumably because the former can do things that the latter cannot— things not adequately reflected in the measure of effectiveness. For example, assume that the giant computer processes jobs twice as fast as the optimal machine. Obviously any processor-bound job that must be performed sequentially can be completed in half the time with the giant machine. For certain applications (e.g., real-time control of a complex missile system) rapid response may be worth the higher cost.

In summary, economic theory implies a U-shaped average cost curve (although it may be very flat over a wide range), but only a portion of such a curve may actually be observed. There are reasons to expect that for computers much of the downward-sloping portion of the curve, and perhaps some of the upward-sloping portion, may be observed, although the question is essentially an empirical one.

Introduction to SECTION II

Application of Cost/Effectiveness Analysis Techniques

As shown in the previous section, it is no longer appropriate to classify value of computer use as tangible and intangible. Each area of anticipated results has a probability associated with it. Emery made this point: "In discussing the balance between value and cost, we found it convenient to use a composite characteristic called quality. In practice we cannot deal with information in this way; instead we must consider each of its individual characteristics."

Gregory and Van Horn, in the opening paper of this section, concur with Emery:

> When systems are being discussed, the comment is often heard that a proposed system is worthwhile because it will provide *better* information for operations and for management decisions. But 'better' is seldom defined in measurable terms. Improvements are merely treated as plus factors, or intangibles, and no value is assigned. Although intangibles are often used to reinforce a decision to adopt a new system, if they are treated as unvalued factors, the result may be a wrong decision. A change not warranted when the value of intangibles is omitted might be warranted if their value is counted.

Gregory and Van Horn discuss the quality, quantity, and timeliness of information in respect to its relevance to the decision-making process. They use similar logic in identifying all costs associated with the system:

> The concept of the value of information is basic to a study of data processing and information production. Value is usually defined as that property of a thing which makes it esteemed, desirable, or useful, or the

degree to which this property is possessed. . . . There is greater potential gain in attending to areas that were neglected in the past when satisfactory processing methods were not available. The fact that the consequences are a final link in determining the value of information must be kept in mind. The effect of a manager's knowing or not knowing some piece of information and the action that follows such knowledge are important determinants of its value.

The costs of operating an information system depend on many factors, the most important of which are discussed here: quality, quantity, timeliness, and capacity. Some other factors are touched on briefly: flexibility, communications, processing schemes and rate of transition.

The authors include examples of measuring cost and value. The first of two papers by Fried provides an organized approach to cost/effectiveness analysis, showing a worksheet and the necessary entries.

Costs are illustrated over a five-year period, for both the existing system and the proposed system. Fried ties the cost analysis to the master plan of the organization, showing the effect of varying levels of business.

Benefits are estimated and evaluated from the standpoint of cash flow and payback analysis. The probabilistic aspects of benefits are included by the assignment of weights by management responsible for the activity being computerized.

In his second paper, Fried specifies techniques for estimating the cost of system development and implementation. While his first paper discussed overall cost/effectiveness comparison, this one concentrates on the process of estimating data processing costs.

This section provides the application of the theory in the prior section, demonstrating that cost/effectiveness analysis is an integral part of the system development cycle.

VALUE AND COST OF INFORMATION

ROBERT H. GREGORY
RICHARD L. VAN HORN

VALUE OF INFORMATION

The value of information and the cost of obtaining and processing data to produce information deserve study, for final design of a system is usually governed by economic considerations. The concept of the value of information is basic to a study of data processing and information production. *Value* is usually defined as that property of a thing which makes it esteemed, desirable, or useful, or the degree to which this property is possessed. Theories of the value of information fall into three categories: intangibles, cost outlay, and managerial

SOURCE: *Automatic Data-Processing Systems: Principles and Procedures,* Second Edition, by Robert H. Gregory and Richard L. Van Horn. © 1963 by Wadsworth Publishing Company, Inc., Belmont, California 94002. Reprinted by permission of the publisher

use. The first two are only mentioned here; the third is discussed in detail.

Intangibles

When systems are being discussed, the comment is often heard that a proposed system is "worthwhile because it will provide *better* information for operations and for management decisions." But "better" is seldom defined in measurable terms. Improvements are merely treated as plus factors, or intangibles, and no value is assigned. Although intangibles are often used to reinforce a decision to adopt a new system, if they are treated as unvalued factors, the result may be a wrong decision. A change not warranted when the value of intangibles is omitted might be warranted if their value is counted. This point is considered later under the managerial use of information.

Cost Outlay

Many systems analysts, and business managers, too, adhere to the theory that the value of information is equal to the *cost outlay* for obtaining it. When considering changes for an existing system, analysts may insist that the same information, or even more, should be acquired without any increase in cost. They are, in effect, accustomed to the existing outlay for processing data and are satisfied with the results obtained. An extremely simple change, such as a large decrease in processing costs, poses an interesting test of the cost-outlay theory of value. One choice is to save the reduction in cost and spend less than before for processing. Another choice is to spend as much as before and obtain better quality, larger quantity, or more timely information. But the cost-outlay theory of value cannot answer the simple question of whether to save the cost reduction or spend the savings to produce more information.

Managerial Use

A more useful concept is that the value of information should be studied in terms of its effect on an organization's operating performance or the revenue obtained. Assume that all factors influencing operating performance can be held constant. If some report or portion of a report is dropped or changed, the resulting decrease in performance would be an indicator of the value of information supplied by that report. If no decrease occurred, the report might be considered valueless.

Actually, the effect of a single report on an overall result is difficult to measure. Within a large organization many departments may not sell their products in the market but merely transfer them to other departments at arbitrary values. Changes in revenue associated with any particular report are difficult to estimate. Nevertheless, it is worth examining the value of information in four of its aspects: quality, quantity, timeliness, and relevance to management's ability to take action.

Quality. The quality of information is judged by the degree of correspondence between the report about a situation and the actual situation. Information of high quality is valuable for several reasons, two of which are that it reduces the range of uncertainty about what action to take and that it makes management more willing to take prompt and vigorous action. If a report about a given situation lacks the minimum degree of accuracy, a user is likely either to demand another report or to simply wait for it and take no action in the interim. At the other extreme, there is a limit to the degree of accuracy useful for decision-making purposes because management rules are usually broad enough that they are not sensitive to a minor drop in quality. Operating personnel, however, may need higher-quality reports, since they deal with a higher level of detail. Higher quality is warranted whenever lower quality is likely to cause the user to make a less desirable decision.

> The inventory replenishment rule applied to item X is to reorder when the quantity on hand falls to 1000 units. If 2000 units are actually on hand, the decision to reorder is the same whether 1001 or 1,000,000 are reported to be on hand.

In such a case, the decision is not affected by the degraded quality of the information, even though the reported quantity of units on hand ranges from minus 50 percent to plus hundreds above the actual quantity. On the other hand, if the actual quantity were 1010, an inaccuracy of even 1 percent could lead to an incorrect decision. Nevertheless, when carried to the extreme, the apparent quality of information may be increased beyond the point where it aids decision-making; such superfluous increases may be ignored or discarded.

One organization set up inventory control records with six-digit fields for the size and weight of each item. Size was supposed to be stated to the nearest .01 cubic foot and weight to the nearest 0.1 pounds. The organization found that it was not practicable to measure the size of an item more closely than about 10 percent. Commercial grade scales would not weigh heavy items more accurately than about one pound in 500.

The specified quality of data for size and weight—0.01 cubic foot and 0.1 pound—could not be achieved in practice and was therefore fictitious. For the purposes intended—warehousing and shipping—less accurate volumes and weights were adequate.

Quantity. The quantity of information available to management depends on the predictability of events and the degree of detail worth having about each event. Information value is related to the difficulty of predicting what is going to happen next and the potential action that can be taken to improve the situation. If it is possible to predict events with certainty, there is no need to have an elaborate data-processing system. Predictable events are similar to the cowboy movies in which the good guys always beat the bad guys: there is no need to watch the movie to know the outcome nor any need, some would say, to make the movie at all. Similar comments apply to processing data and preparing reports when the outcome is fairly certain. A simple example illustrates the idea of a high degree of predictability and the consequent reduced usefulness of pertinent reports:

> Commercial banks use essentially identical procedures to update all depositors' accounts despite the fact that a large fraction of depositors rarely have an overdraft and, if so, readily pay it. Different processing and reporting schemes might well be used for classes of depositors based on past and probable future experience with them. For most depositors, it probably is not worthwhile, from a management and operating viewpoint, to update their accounts every day.

Unpredictability—as reflected in the number of possible alternatives and variations—has an important bearing on the value of information and the design and implementation of business data systems. A system to handle the number of hours worked on individual jobs is more elaborate than a system to keep track of the number of people on the payroll, for variations in the number of hours worked is much greater than in the number of people employed. A job-order production shop demands a more complex data-processing system than a factory operating an assembly line. And highly standardized factory operations are adequately served by relatively simple data-processing systems. Careful observation of the alternatives and variations that may arise helps determine how much detail is useful for an information control system.

Timeliness. Information may be timely in the sense of being available at a suitable time or being well-timed. Timeliness, or the age of information, has three components: interval, delay, and reporting period. *Interval* is the period of time—minutes, days, months, etc.—between the preparation of successive reports or answers to inquiries. *Delay* is the length of time between the cut-off point—the time when no more transactions are accepted for inclusion in the particular report—and the distribution of reports to users. The delay covers the time required to process data and to prepare and distribute reports or to answer questions. *Reporting period* is the length of time that an operating report covers and may be longer or shorter than the interval between the preparation of reports. That is, the time periods covered by successive reports may be continuous, overlapping, or have skips.

At intervals of one month, a filling station owner prepares an income-expense report for a reporting period of one month.

Each month a data-processing service bureau prepares its operating report of revenues and expenses for the current month and the year to date.

A Christmas tree retailer prepares, each December 26, a cash receipts and payments report for the period December 1 to December 24.

For moving or running totals or averages—for example, shipments during the most recent thirteen periods of four weeks each—the reporting period consists of a series of four-week periods extending over 52 weeks, but the reports are prepared at four-week intervals. In short, the reporting periods may be entirely different from the interval used.

Managers often think they must have immediate or up-to-the-minute information in order to operate effectively. *Immediately* literally means pertaining to the present time or moment. The interval, delay, and reporting period would have to be extremely short or even zero in order to get immediate information. The practical results of immediate information would be immediate issuance of a continuous stream of reports covering only a short reporting period. Since a flood of reports without any time coverage is hard for the reader to assimilate and use, "up-to-the-minute information" more often means reports issued with a short delay but with the interval long enough to keep the number of reports manageable and the reporting period long enough to make their contents meaningful.

For *status or point-of-time information,* the minimum age of the newest information ever available is equal to the processing delay involved. The *delay* is the number of units of time—seconds, hours, days, etc.—that it takes to prepare and distribute reports to users. If it takes ten units of time to produce a report of inventory as of December 31, then it is first available ten units of time later. That is to say, the information is ten units of time old when it becomes available. If reported as of the end of each month, the next time new information will become available is January 31 plus ten units of time. Therefore, the age of information pertaining to a point in time ranges from a *minimum* of the delay involved in preparing the report to a *maximum* of the interval plus the reporting delay. For on-line, real-time systems, it is possible to make inquiries whenever desired to find the most recent status and to keep the maximum age near the minimum. But attempting to keep the maximum age extremely close to the minimum may keep the system fully loaded with inquiries and replies.

If information is used as soon as it becomes available, the processing delay is the primary determinant of age. The time that a manager spends reading reports to make decisions counts as part of the processing delay for operational purposes. Reports may come so thick and fast that the receiver cannot make full use of one before he receives other reports, or even the next issue of the same report. He must then decide whether to plow through all the reports and try to get on a current basis or to skip some and start afresh. Therefore, the timing of reports must, of course, be matched with the user's ability to use them.

The North Country sourdough's approach to reading newspapers is interesting in this regard. The sourdough may get all the newspapers for a whole year at one time and spend the next year reading them at the rate of one a day. The news of the world unfolds for him day by day just as it does for everyone else, except that it is a year or so late.

The sourdough could, of course, discard all but the most recent issue or, preferably, order only the most recent copy of the paper. The businessman can seldom afford the luxury of steadily progressing through a huge backlog of reports. He must skip to the current situation and return to prior reports only to obtain background for later situations and to get answers for long-range questions.

Operating information about events occurring throughout a period of time — widgets manufactured, hours worked — covers the period of time during which the events happened. Therefore, the age of operating information depends on the length of the operating period, since some of the events happened early in the period, some near the middle, and some near the end. Information about operations throughout a period of time has an inherent age of one-half a period merely because the events involved are spread throughout the reporting period. The processing delay involved in producing reports must be added to this inherent time lag of one-half a period to give the minimum age of information before it is available for use. As was pointed out earlier, the reporting period can be the same as or different from the interval between reports.

The maximum age of operating information is reached one report preparation interval later just before new information becomes available. The age of operating information is one-half a reporting period more than the corresponding age of status information because operating information has an inherent age of one-half a reporting period, no matter how long or short the period is. That is, the age of operating information is more dependent on the length of the reporting period and the interval than is the case for status information; delay is less important as a determinant of age.

General rules for the optimum length of the interval, delay, and reporting periods are easy to state, although they may be difficult to apply in practice. The optimum length interval is determined by two factors: (1) the amount of variation away from the projected operating result that is likely to occur during the interval used, which corresponds to the sampling period; and (2) the cost of sampling, processing, and reporting. Highly variable processes deserve short-interval reporting, even though costs of data processing increase, so that variations from plans can be detected before they grow excessive. Short-interval reporting does, however, have a drawback in that it increases the number of reports that must be prepared and used or merely scanned and discarded. During a year, the use of monthly intervals will result in 12 reports, weekly intervals in 52, and hourly intervals in 2000 or more.

The optimum length delay is determined, on the one hand, by the amount of variation away from the situation existing at the time of cut-off that is likely to occur before reports are ready for use. On the other hand, the optimum length delay depends on the cost of a higher capacity processing-reporting system to furnish outputs with an extremely short delay. A shorter delay in reporting is beneficial because a quick fix on a situation is more useful than one obtained after a long lag in which important variations may have occurred.

Short-delay reporting has the advantage of enabling management to take action earlier. The value gained from earlier action is equal to the amount of improvement in operations for the time saved by the shortened delay. For most business-management situations, decision-making and delays in reporting are measured in days, and any consequential improvement requires reducing the delay by a half day, a day, or more. But many operating-control situations need reporting delays measured in hours, minutes, or seconds. In such cases, a small reduction in delay, measured in hours, minutes, or seconds, can sometimes yield large relative improvements in the degree of control over operations.

There is no clear-cut relationship between the length of the reporting period and the value of results. Extremely short reporting periods may permit unusual or superficial events to outweigh the real developments. Conversely, long reporting periods submerge unusual events in longer-term totals or averages. Some intermediate-length reporting period permits optimum sampling to obtain information that represents what is actually happening in the particular situation.

Two examples show the relationship between interval, delay, and reporting period and help draw the discussion together.

> Each department manager in a store may demand reports by 10 o'clock each day showing inventory for the previous day, daily sales for the past week, weekly sales for a month, and cumulative sales for the season to date. Details are wanted for item, style, size, color, manufacturer, etc. Reports at daily intervals and one business hour of delay emphasize yesterday's activities but also cover a longer period to put each day in perspective—each report has four reporting periods: day, week, month, and season to date.

Fire alarm messages are so simple that merely pulling the handle of an alarm box transmits the message, and the firemen roar out of the station whether the call is real or false. A fire call gets a standard response with short delay and with little regard for the interval since the last call, although, if alarms occur too frequently, the whole system will be overtaxed; and they cannot be answered except with an increased delay or the aid of additional equipment.

Relevance and Consequences. In order to be useful, information must be relevant to a manager's problems—or, more carefully stated, to areas needing improvement. A manager is able to make either better or more decisions when he does not have to spend his time sorting relevant facts from irrelevant ones. If available information is not related to the decision to be made, it has little if any value. The problem of information relevance is intimately related to the managerial organization pattern. The quantities and location of truck tires in inventory may be valuable information to a motor pool commander but useless to a tank company commander.

Another measure of the value of information is the consequence of having it. One useful yardstick is the benefit obtained from the information over a period of time. An inventory-control procedure that saves a penny per part is valuable if millions of parts are involved.

In some cases, a manager cannot act on information because the decision required is made at a different level of management. Or past decisions may limit his present ability to act because "sunk costs" carry over to the future.

In such cases, the consequences of more information are trivial.

The possible range of consequences should be considered. If operations are already 85 percent efficient, perfect information can increase efficiency by only 15 percent. Information is more valuable in areas that are greatly in need of improvement. However, despite this fact, some areas of business data processing that are easy to understand and change are favorite targets for improvement. Since these areas have had so much attention already, it is unlikely that further attempts at improvement will have much payoff. There is greater potential gain in attending to areas that were neglected in the past when satisfactory processing methods were not available. The fact that the consequences are a final link in determining the value of information must be kept in mind. The effect of a manager's knowing or not knowing some piece of information and the action that follows such knowledge are important determinants of its value.

COST OF DATA AND INFORMATION

Cost is the price paid for obtaining and processing data to produce reports. This section discusses methods of cost assignment for processing data and producing information and the factors affecting such costs.

Cost Measurement Schemes

There are two widely used methods for assigning cost to data: average cost and marginal cost. Long used for factory cost accounting, these methods are equally applicable to the costs of processing.

Average Cost. Some costs can be identified with a particular project or job for obtaining and processing data. These costs correspond to factory production costs of direct material and direct labor. Other costs of processing data cannot be easily identified with a particular project but must be distributed over all the work done. Such costs correspond to indirect factory costs—equipment depreciation, supervision, space, supplies —and are distributed by overhead rates that supplement the direct costs.

Data-processing costs for large operations may be identified readily with such major areas as purchasing, inventory control, production control, and payroll. But a detailed costing by jobs within each major area may be difficult to obtain because data-processing operations in these areas are interrelated. One set of data may be used for several different purposes, and data from different sources may be combined for one purpose. If costs vary with the amount of work done, job cost accounting can give useful answers for guiding management. Direct costs may be closely related to the volume of work; indirect costs may be essentially fixed and change little or not at all with changes in production volume. The average cost scheme is often used to distribute costs so that the total costs are completely distributed among all the users. The arithmetical niceties of the average cost scheme appear to give it stature beyond its actual merits for decision-making. Highly automatic operations, whether in a factory or office, have a high fraction of fixed costs that vary little with changes in the volume of data handled. In such cases, job cost accounting loses much of its meaning, and another method for assigning costs is more useful.

Marginal Cost. Marginal cost, popularly called "out-of-pocket cost," is the amount that costs change as volume changes. Despite large changes in the volume of data handled or reports produced, the operating costs of an automatic data-processing system are essentially constant and marginal cost is small. Marginal costing charges a job with only the additional costs incurred because of that job. For example, a report might be prepared at a small marginal cost by starting with data already collected for other purposes and be completed within the basic operating schedule for both equipment and people. The marginal cost concept is often used when considering the installation of a data-processing system. People talk of starting equipment on important "bread-and-butter" applications to absorb the total cost of the system. They start other applications that benefit by not having to absorb any costs. No charge is made against additional applications because the marginal cost is zero.

The use of either the average cost or the marginal cost scheme has certain consequences when changes occur in the volume of data handled. For an automatic system with large fixed costs, the average cost per unit is high when volume is low and such costs may further discourage use of the equipment. The opposite is also true. Average costing may lead to either too little or too much work. Actually, it is probably wise to encourage use of an idle system and discourage use of an overloaded system. Marginal costing is sensitive to the system load. If equipment is idle, marginal cost is small and encourages use. If equipment is fully loaded, marginal cost is high and discourages use. Marginal cost is the full increase in costs for equipment, personnel, and supplies, and is high for the first application and the application that requires more equipment or that causes second-shift operation. To overcome this feature of marginal costs, basic system costs may be lumped together and not allocated to individual jobs. Actually, decisions either to tolerate idle capacity of equipment or to use it to full capacity are often made implicitly on the basis of marginal costs.

Factors Determining Cost

The costs of operating a data-information system depend on many factors, the most important of which are discussed here: quality, quantity, timeliness, and capacity. Some other factors are touched on briefly: flexibility, communications, processing schemes, and rate of transition.

Quality. The nature of the quality of data and information were covered earlier in this chapter when discussing the value of information. The point was made that costs increase rapidly as the degree of accuracy of information is pushed toward the limit of perfection. The degree of quality maintained in a system is related to the costs of achieving any particular degree and the benefits obtainable from having it. An organization usually keeps track of the total amount of accounts receivable with more accuracy than it does the value of inventory. Costs involved in determining inventory value, or even the identification of items and the number of units on hand, may force the organization to stop appreciably short of the ultimate degree of accuracy. The retail inventory scheme, for example, uses the ratio of cost to selling price for merchandise purchased in order to convert the inventory, valued at selling price, back to cost. The use of ratios is an indication that the loss of accuracy in finding the cost of inventory might be more than balanced by the expense reduction from not keeping detailed records.

Results may differ from true values because of people's mistakes, errors in the calculation plan, or malfunctions in equipment. Inaccuracies can arise at any stage from data origination to report preparation. People may misunderstand or misread original data, or make mistakes in operating typewriters and keypunches when preparing data for input. Instruction routines may have errors in logic that make all results erroneous. Even when most cases are handled correctly, others are unanticipated and may give erroneous results. Equipment malfunctions on either a repetitive or intermittent basis.

Many schemes based on partial or complete duplication of operations are used to increase the accuracy of results, and data origination is commonly verified by repetition. A parity-bit associated with each character helps guard against the accidental gain or loss of a bit that would change a character. Instruction routines are checked for logic and test-checked with simulated or real data to debug them. Duplicate circuitry, programmed checks, or even both, are used to detect malfunctions in automatic processors. Double-entry, balancing, and proof schemes are commonly used to ensure accuracy in manual or mechanical accounting systems; more elaborate plans are used with automatic processing systems.

These plans for increasing accuracy (decreasing the difference between results and the true value) also increase the cost of processing data. Additional precautions are required to increase the margin of accuracy from 0.99 to 0.999, 0.9999, and so forth. The precautions required grow rapidly as the margin of accuracy approaches 1.0. The important point here is that increasing the accuracy of results, measured in terms of either their precision or their reliability, adds to the cost of originating and processing data. Much of the cost increase arises from the trouble involved in organizing and policing data origination and transcription methods.

Quantity. The quantity of reports, files, and data handled has an important effect on the costs of systems operations. For example, an increase in the number of transactions of a particular type may cause a strictly proportional increase in the costs of originating raw data. Some costs may remain essentially constant, provided there is any idle capacity. An increased volume of data may be transmittable over leased lines at almost no cost.

Important economies may be available in other ways, such as by processing a larger volume of transactions by means of the same programs so that only small additional programming and set-up costs are involved. Cost increases that are less than proportional to volume increases are obtainable in most systems using large amounts of equipment. The fixed costs of the system buy a large basic capability; increased capacity is obtained at a small outlay merely by more intensive use of equipment.

Some costs, on the other hand, probably increase more rapidly than larger volume alone would indicate. More intensive and more elaborate controls are required to achieve a certain degree of quality when the number of transactions increases. More care is required to ensure that data are originated for all transactions, that none are lost in the communication network, and that the right programs and files are used in processing transactions. The quantity of data in each record in the files, as reflected in the number of data elements, appears to have more than a proportional effect on processing costs. A larger number of data elements in the record design requires the preparation of a more elaborate program to process the file and to prepare the myriad analyses that are possible from the larger quantity and variety of data. Also, a larger quantity of input data is required to set up the initial record and to keep it current thereafter. The mere fact that a record has more data elements in it means there is a higher incidence of change: a wider variety of input data must be obtained, and obtained more frequently, in order to keep an elaborate record current and correct. Also, a more elaborate record offers better potential for analysis and is likely to be retained longer and analyzed more carefully than a short record. Consider the implications of just one aspect of a utility record—the amount of electricity consumed and the quantity of the detail kept in the record for each customer.

- Consumption for the most recent month.
- Consumption for the most recent month and the minimum and maximum amounts for any month in the current calendar year.
- Consumption for the most recent month and for each of the twelve preceding months.

The effort involved in establishing the file is appreciably greater for the second and third cases. In fact, when first adopting the latter record design, it is impossible, at any reasonable cost, to set up the record, since a year's history is not available. Only the new meter reading is necessary each month to calculate consumption for the most recent month, but appreciably more processing is involved in keeping track of the minimum and maximum consumptions for the year to date or consumption by months for the preceding year. This progression in the work involved in file updating arises solely because of record design and content, for only one input each month is actually needed to update the record. From this simple case involving only one part of a record it is possible to gain some idea of how a large and intricate record is likely to increase data-gathering and processing costs.

Timeliness. Three classes of operations are worth considering here. Some operations involved in processing data are carried on throughout a reporting interval without regard for the length of the interval. Data origination and some processing operations may be continuous. Other operations, such as file maintenance and processing, may be variable; but they must be done at least once before reports are prepared.

The use of short *report preparation intervals* involves additional processing cycles during any time period. For example, a high fraction of the total cost of processing magnetic-tape files is incurred merely by passing tape through the processor. The tape read-write time may be essentially the same for low or high activity of the records. In such tape-limited processing—file updating can be done during the time required to read and write the files—costs are more closely related to the number of file-processing cycles than to the number of transactions handled. The cost of summarizing files and preparing reports is related to the number of reports prepared, and some costs of processing —updating files and preparing reports— may double if the reporting interval is cut in half. The ultimate, as the interval between reports approaches zero, is for costs to become infinite.

Going in the other direction, as the report preparation interval becomes longer, the costs of processing data are also likely to increase. Extremely long intervals require holding more data in active storage until reports are prepared and a new period starts. The preparation of reports serves the useful but little noticed purpose of permitting a purge of active files and adoption of a new starting point. The cost implications of extremely short intervals have practical importance for systems utilizing magnetic-tape files.

Systems analysts in one company reported that, over a long period, weekly processing of a policy holder's file was about four times as expensive as monthly processing. Similarly, a manufacturing company reported that daily inventory processing would cost about four times as much as weekly processing.

To the extent that tape file processing is limited by internal operating speeds, processing at shorter intervals may be obtained at little cost, since processor running time is not increased in proportion to the frequency of updating. This discussion of the effect of changing the length of the processing interval presumes that the processing technique remains constant. Another method of file storage would, of course, have a different set of cost curves, although probably with similar characteristics. It appears that the cost of random-access file processing on addressable bulk-storage disks or drums would be at a minimum for some report preparation interval, but would increase if the interval was either shorter or longer. But costs would increase less rapidly for processing disk or drum files than for tape files at short intervals because random-access files get at the next active record in less time. On the other hand, a long interval for file updating and reporting would, it seems, greatly increase processing costs because of the large number of disks or drums required to hold the larger volume of files. Protracted intervals mean that data must be carried forward for a long time before reports are prepared and files can be purged. A curve with a minimum point in the center and higher points on either side is said to be "U" shaped. The objective in systems design is to avoid ending up in either of the high cost areas.

Report preparation may cost less for random-access equipment than for magnetic-tape files because the data transfer rate from some types of disks and drums is faster than for sequential data on magnetic tapes. This is especially true for limited volumes of data to be selected from a large file. In short, data-origination costs may be the same for systems using either magnetic-tape or random-access storage; short-interval processing may be less for a random-access system; and report preparation may be less for a random-access system, especially for small volumes of reports and short intervals.

For any particular system using certain methods—manual, electromechanical, or electronic—there is some optimum *processing delay* that gives the lowest operating costs. System and equipment capacity are used at a high fraction of capacity throughout the interval and neither an overload nor idleness occurs. Shortening the processing delay, for a particular type of system and equipments, increases costs. Shorter delays increase costs because additional capacity is required, scheduling is more difficult, and average usage is likely to be lower. As the delay is decreased toward zero, costs skyrocket because no system, even a "blue-sky" one, is capable of producing results with zero delay. Moving in the other direction, longer delays are also likely to increase processing costs. The system may bog down because it must store and deal with a great quantity of data before processing can be completed and the files purged or downgraded to inactive storage.

These general comments on the cost of processing and length of delay are not restricted to any particular system and type of equipment, but apply to all. Changes in a system or its equipment merely alter the cost-delay relationship but do not destroy it. It appears that a "U"-shaped cost curve is generally true for the costs of processing delay for any processing scheme.

Capacity. The capacity of a system must be large enough to handle peak loads. For a steady work load, the use of either faster equipment or more equipment can reduce delay. But work loads are uneven; some tradeoff must be made between increased system capacity and longer processing delays. The fixed costs of a system for equipment, space, basic personnel, and programming are likely to be determined chiefly by the maximum capacity of the system. Operating costs for supplies, second-shift rental, and others, are more sensitive to total volume. When the peak load is far larger than the average processing load, either more capacity must be provided to handle the peak, or the elapsed time required for processing will grow. Such added capacity may then be idle until the next peak load occurs.

Staggering or overlapping intervals so that peak loads occur at different times may smooth out loads on the system. Cycled work loads can increase the use of processing facilities, but this approach does not directly attack the basic problem of short delays and the high cost of adequate capacity for peak-load processing. Eventually, a "one-hoss shay" effect occurs because a number of different-length intervals end at the same time to cause a huge work load. In effect, the staggered scheduling system falls apart, and a new schedule has to be adopted to ease the work load. As the acceptable delay is permitted to increase, the maximum capacity required drops off from the peak-load quantity, and levels off just above the average load. Obviously, a system must have enough capacity to handle present applications; and additional capacity may have to be obtained in advance because available equipment comes in only a few sizes. The choice among central processors is restricted to a small number of sizes, although much more freedom exists to add relatively small units of peripheral equipment later when needed.

Other Factors. Several other factors affect the cost of a data-processing system. Some that are merely touched on here are flexibility, communication methods, processing scheme, and rate of transition.

Flexibility costs money. A system limited to one or a few specific applications that do not change has a minimum cost. At the other extreme, a system may be flexible enough to deal with any application. In such a case, more capacity, systems analysis, and programming capability are required to handle applications. *Communication methods* used may range from regular or airmail to wire and radio transmission. Each, of course, has different cost functions that vary with the channel capacity and volume of data transmitted. *Processing schemes* may be standardized or selective. The exception principle, internal decision, adaptive, and variable processing plans described earlier in this chapter are examples of selective processing schemes. Analysis and programming costs for a selective processing scheme are higher than for standardized processing. After a selective processing scheme is set up, however, its operating costs may be low because attention is focused on situations where it is needed.

The *rate of transition* from the old system to the new also affects costs. Rapid changes from one system to another result in confusion and lost motion. Some people, in fact, hold that important changes in a system cannot be made efficiently more frequently than once every three to five years by a large organization.

RELATIONSHIP BETWEEN VALUE AND COST

From an economic viewpoint, information is a factor in production similar to manpower, equipment, and material. Obtaining better information — measured in terms of quality, quantity, and timeliness — may permit larger savings in other factors.

Railroads have recently introduced automatic systems for reporting train location and, by using sidings for passing, are able to get most of the benefits of double tracks yet save the investment and maintenance costs. If high-speed communication and control systems had been available a hundred years ago, probably few double-track roads would have been built.

Many companies buy credit reports to get more information about a prospect's rating in order to balance the risk of loss against the profit from accepting an order. The organization supplying credit reports deals solely in gathering data and interpreting them into credit ratings for sale to any purchaser. The report buyer weighs the costs of information against the increased profits from accepting orders from customers whose credit can be appraised.

The general rule for deciding how much data and information to obtain corresponds to the rules applicable to the other factors of production. Continue to use more of a factor until the cost of the next unit is just equal to the benefits obtained from using it. While applying this rule to data-information systems is difficult, it is probably no more so than applying it to other production factors.

PROBLEMS IN DEVELOPING GENERAL PRINCIPLES

The foregoing discussion of value and cost of information has been lengthy because few general principles or rules exist to serve as guides for designing data-processing systems. Truly general rules could, if they existed, be stated very briefly. The causes of this lack of general principles deserve some comment.

The first reason for a lack of general principles is that operating data-information systems are *difficult to describe* in simple terms. This is true whether the system uses manual, punched card, or electronic methods. Analysis is expensive and time-consuming. Systems do not remain in a "steady state" but continue to change even during analysis. In fact, the rate of change may outrun the analysts who must keep abreast of the old system while designing a new one. Any attempt merely to analyze a system will change it and, where people are involved, change it appreciably. Simply asking a responsive person, "What do you do?" or "Why do you do it?" is likely to lead to some change.

Second, the operating *environment* and the *problems* to be handled differ from one case to the next. Reports, files, and data inputs are different, even though equipment and operations are similar; and such differences limit the value of comparing operations in different environments. Comparison of a proposed system with one that does not change or with an extrapolation of the present system may be valuable for drawing valid conclusions about the effect of new proposals. Laboratory models or controlled experiments are sometimes useful for showing the effect of certain changes. But it is difficult to use a business *per se* as a laboratory for testing new ideas about data processing because of the unsettling effect of experimentation and the risk that untested initial operations may fail and discredit the whole system effort. Hardheaded managers insist upon having solutions that work when first installed.

A third reason for the lack of any general rules is that the use of electronic equipment and advanced *techniques* is still relatively new. Experience now available covers only a small number of years and a limited number of complete cycles from initial installation to introduction of still newer equipment or discard of the system and return to the old. Imagine the problem of an insurance actuary asked to predict average life expectancy from life histories of many children and a few adults, but no deceased persons. The massive introduction of electronic equipment by many organizations in a few years has foreclosed many of the benefits from learning from the experience and mistakes of others. More knowledge would have been generally available from a slower rate of introduction. Important changes in both equipment and the systems built around the equipment occur frequently, and they will continue to occur indefinitely.

Fourth, data-processing systems are *complex*. Many alternate configurations of equipment are available to solve a particular problem. The system built around the equipment in order to handle the origination of data and distribution of results reaches throughout all parts of a business organization. The combination of equipment configurations and highly complex management information systems produces a set of possible data-processing systems that is too numerous to evaluate properly.

These four factors—difficulties of description, lack of laboratory conditions, newness of electronic equipment, and complexity of equipment and systems—impede the development of exact rules for systems design and operation. A formula would be useful for determining system and equipment needs and ways of using them. Lacking a formula, one approach to solving the problem, an approach useful in any field, is to search for some of the basic ideas involved for an understanding of the concepts or general principles. This approach is especially important in such a new and developing field as automatic data processing. General principles must be developed before useful, specific rules can be devised. A search for general principles reveals new facets of a subject, puts various features into perspective, and discloses new relationships.

SUMMARY

At the *syntactic* level, data are collections of symbols or characters arranged in some orderly way—for example, documents, files, or reports—to serve as the vehicle for information. Information is the meaning derived from data and represents the *semantic* level—the relationship between a symbol and the actual object or condition that is symbolized. The impact of the objects or conditions on the receiver and the action that he can take represent the *pragmatic* level of information.

Complete, detailed reports may show the variations between actual and forecasted results; this is a first step toward reporting by the exception principle, which can be extended to omit items that do not have significant variations. Using a normal range is one way to set limits for reporting exceptional values. Items within the normal range can be handled by one decision rule, whereas items falling outside require different action. Normal ranges for screening items that are not controlled can be used at many stages of report preparation.

Expected values may be based on the actual or average amount experienced in the past or upon a forecast of future results. Variations from budgeted or projected values can be measured in the same units as the data or in relative terms—percentages, or variations. The exception-principle scheme is designed to improve the content of *action* reports by increasing their impact on the receiver. Items reported infrequently can be dropped from action reports; complete reports can be prepared for *reference* purposes.

A processing plan for making internal decisions applies management decision rules during the main stream of data processing. Managers can still review the results of applying the rules before using the results from processing, and analyze in detail any situations out of control. A manual intervention plan provides for equipment to follow rules where suitable, but situations not covered by rules are turned over to people in order to combine the best abilities of man and machine. More elaborate and sophisticated systems that can better adapt to their environment are in the offing.

The *value* of information is often treated as an intangible not amenable to analysis. Another approach is to treat the value of information as being just equal to the cost of processing. A better approach to studying the value of information is to examine its managerial implications. The crucial question is, "What does the information contribute to managerial decisions and overall operations?" Factors that affect the value of information are quality, quantity, timeliness, and relevance. The timeliness of information depends on the length of the interval, the reporting period, and the processing delay. The minimum, average, and maximum ages of information suitable for each situation must be considered during systems design.

Information needs to be relevant to the problems that the receiver of a report can handle. The consequences arising from knowing something depend on how much change in operations that knowledge will lead to in actual practice. Operations that are predictable require little or no information for effective control, whereas dynamic operations require elaborate control.

The concepts of *cost*, both average and marginal, are pertinent to the volume of data handled and to decisions for changing procedures. Factors that have an important bearing on the cost of processing data are quality, quantity, timeliness, and relevance. Short intervals and frequent reports go together; they are two sides of the same coin. The costs of reporting probably double each time the interval is cut in half. On the other hand, the value of reports first increases and then may actually decrease as intervals are made shorter. In extremely short intervals, unusual events may outweigh and mask the underlying events. The delay—the length of time before a report about a single event or a series of events is available—can be shortened to get up-to-date

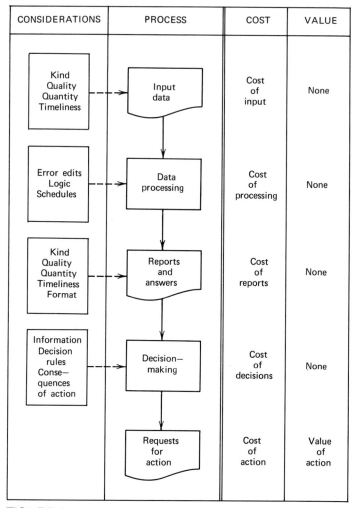

FIGURE 1   Value and cost in a data processing system.

reports. Costs may increase rapidly if large facilities are used to meet peak demands for quick processing because such facilities are underutilized most of the time. The increase in the value of results learned sooner—whether an hour, minutes, or seconds earlier—depends on the circumstances of the situation.

Some of the basic ideas in this chapter are summarized in Figure 1. The *process* column shows the operation of an information system starting from input data and extending to the action flowing from decisions. The *considerations* show the factors important at each stage of processing. The nature of input data is fundamental to the design and operation of the whole system. All four factors are considered for selection of input data. The data-processing system has some measure of control over kind, quality, quantity, and timeliness, at least at the syntactic level. The makeup of reports is determined in terms of their information content. A decision maker gets certain quantities of reports with some specified degree of kind, quality, quantity, timeliness, and format. The primary considerations for a report user are the meaning and impact of reports — their semantic aspect. Following these, he must consider the practical consequences of his action.

The *cost* and *value* emphasize the point that costs are incurred at every stage of processing while benefits result only from managerial action utilizing system output. The *raison d'être* of an information system is to improve the operation of the organization. Even good decisions are valueless unless they lead to action.

For several reasons, few general principles or rules exist for guidance in designing data-processing systems: analysis is expensive and time-consuming, systems are essentially unique to an environment, insufficient experience with new equipment is available to permit long-run conclusions, and data-processing and information-production systems are complex because they pervade an entire organization.

HOW TO ANALYZE COMPUTER PROJECT COSTS

LOUIS FRIED

The management processes involved in the design and implementation of a computer-oriented business system can be thought of as being identical to those involved in the process of new product design and development. If the new system is viewed as a tool to enhance the performance of the organization, then the justification should be performed on the basis of a cost and payback analysis similar to that used for any other capital investment. This will include costs of savings that will accrue to the new system and the calculation of the length of time the new system will take to pay for itself (payback).

This article will look at the general procedure that can be followed for the cost and payback analysis of almost any computer project. The example used is a personal trust accounting system for a trust department servicing over 4,000 accounts. This trust accounting system was to perform the following activities:

- Accounting for all assets held in trust including securities, real property, businesses, and other assets.
- Posting pending and final securities transactions to appropriate accounts and distributing dividends.
- Pricing securities each month and preparing statements for clients.
- Preparing client tax returns.
- Preparing client billing.
- Recording trust officer time charged by account.
- Analyzing and reporting account performance.
- Controlling disbursements to trust beneficiaries.

Accounts range in size from $50,000 to $1,500,000 and the total assets controlled are close to two billion dollars. Before the new system, the trust accounting was handled by a unit record installation with files containing over 500,000 cards. An investigation indi-

SOURCE: Fried, L., "How to Analyze Computer Project Costs," *Computer Decisions,* Aug., 1971, pp. 22–26.

cated that development costs in banks which had programmed their own systems had run from $500,000 to $1,000,000 and more. It was therefore reasonable to assume that development of this system would involve a similar expenditure.

Because of the scope, complexity, and potential cost of the new system, a detailed preliminary study was undertaken to select the alternative courses of action and determine their costs. This study occupied over one-man year of systems analyst time as well as the time of personnel in the user department. The company performs this type of cost, benefit, and payback analysis for all proposed projects costing over $25,000.

CHECKING THE ALTERNATIVES

After identifying the problem, several alternative solutions were examined. Initially these alternative solutions ranged from "blue-sky" approaches to minor changes in the existing routines. Each alternative solution was first analyzed in a cursory fashion to identify and eliminate those which were obviously unattractive or not feasible.

After a rough design of the alternative solutions, each design was analyzed to determine its impact on personnel, available resources, computer time, organization performance, and established organization goals. Tentative operating costs were established and compared with the actual costs of the present operating methods (which represented one possible alternative). No decision, or a decision not to change, is actually the selection of the present method as an alternative.

One element of the operating cost of alternatives must be the amortization of the implementation and conversion costs of the new system over the anticipated life of the system. Implementation and conversion plans had to be developed for each of those alternatives that passed an initial test of feasibility on the basis of operating cost and the impact on the organization.

The operating costs for the computer portion of the system were estimated in a variety of ways. For software packages, present users were consulted and computer times were projected for our transaction volume on the basis of their experience. For functions not performed by the packages, we estimated the transaction volume and the consequent input-output time necessary on the computer for the number of programs involved. The computer times developed were priced at our internal billing rate for the computer.

With the aid of line management, we determined the number of clerical staff required. Clerical costs were based on actual salaries paid at the present time.

Cost estimates were also prepared for all the recognized elements such as systems analysis and file conversion, involved in implementing the system and converting to it. In this process some alternative solutions were eliminated for some of the following reasons:

- The technical capability for the proposed solution did not exist or was extremely new and untried.
- The environment of the organization was not suitable to the proposed solution or it required more resources than the company was willing to commit.
- The time needed to implement the system eliminated some more complex or time-consuming alternatives.
- Some alternatives were politically or psychologically unpalatable.

These activities resulted in a list of requirements or specifications characteristic of the proposed solutions available.

The specifications were rated as to degree of importance and priority for installation. This accomplished, implementation methods for alternatives were considered.

The two most common alternatives are to design and program the system in-house or to obtain a packaged system — the classic "make or buy" decision. During the past three years the growth of the commercial software package market has been such that any firm seeking a solution to a common problem can frequently find that solution in a packaged form. In fact, the firm may have the choice of several alternative packages from which to select a "best fit" to its existing organization. There are many factors, including reduced cost and shorter implementation time, that encourage the purchase of a package. On the other hand, the unique needs of the organization might tend to encourage the development of a new system in-house.

After technically feasible make or buy alternatives had been identified, it was then necessary to compare them with each other. This involved field work in which the analyst and members of line management visited companies that had installed their own systems and companies that had installed packages that were under consideration. This developed into an extremely valuable phase of the investigation since it provided actual case histories on which to base our estimated costs of design and implementation for each of the alternatives. After applying what was learned in the field studies to our own situation, it was then necessary to compare the estimated costs of each alternative in such a manner as to permit the organization's management to make an intelligent, informed business decision. This required examining the comparative impact of the alternatives on company resources, profit, cash-flow, and benefits attained. It also required the presentation of these considerations to management in a concise, understandable manner. For this application it was decided to buy a software package. All the estimates, to be given later, are for the "buy" case. The use of a software package turned out to be considerably less expensive than similar in-house designs.

This presentation described the basic concepts of the existing system and its major problems. It outlined the concepts of the proposed alternative, its major objectives, and the primary benefits to be derived from its approval. A comparison of the operating cost of the present system with the estimated operating cost of the proposed alternative was illustrated, and the cost of implementation and conversion for the alternative was clearly indicated. The effect of the present system and the proposed alternative on the cash flow of the company was described. A payback analysis for the alternative was presented. The impact of the alternative on company employees, organization structure, and business practices was clearly set forth. Finally, the carefully considered recommendations of the group making the presentation were delineated. These recommendations identified the alternative selected as being the most desirable and described the specific steps that should be taken to implement the recommendation.

The Implementation Cost Worksheet (right) was prepared for the analysis of all elements of implementation cost for each alternative. System implementation costs were divided into major task groups which included:

- System design and programming if the alternative was to make the system.
- System modifications if the alternative was to buy the system (in this

Implementation cost work sheet

Project tasks	Cost elements								
	Systems analysis	Programming	Key-punch	Computer	Forms & supplies	Outside contract services	User adminis-tration	User person-nel	Project manage-ment	Total
Systems design and programming	$13,500	$16,000		$5,200			$80			$34,780
System modifications			$1,600	1,250		$8,200		$11,860		22,910
Preparation for conversion	9,750	750			$1,000			840		12,340
Clerical and operating procedures	2,675	2,050	9,660	16,075			200	3,350		34,010
File conversion	1,875						240	1,560		3,675
Training			900	5,250			540	1,400		8,190
Pilot and parallel operation									$14,800	14,800
Other project tasks	$27,800	$18,800	$12,160	$27,775	$1,000	$8,200	$1,060	$19,010	$14,800	$130,605
Subtotal		38,000								38,000
Capital expenditures	$27,800	$56,800	$12,160	$27,775	$1,000	$8,200	$1060	$19,010	$14,800	$168,605

Implementation cost: Summary of out-of-pocket costs

	First quarter	Second quarter	Third quarter	Fourth quarter	Fifth quarter	Total
Optimistic	$31,545	$27,670	$19,865	$32,945	$21,165	$133,190
Most likely	36,281	31,820	22,844	37,887	24,340	153,172
Pessimistic	41,723	36,593	26,270	43,570	27,991	176,147

Implementation cost: Summary of total costs

	First quarter	Second quarter	Third quarter	Fourth quarter	Fifth quarter	Total
Optimistic	$39,921	$33,682	$31,346	$36,346	$27,410	$168,705
Most likely	45,909	38,734	36,044	43,897	30,120	194,704
Pessimistic	52,795	44,544	41,450	52,582	34,050	225,421

case it was decided that purchasing a software package was most economical).
- Preparation for conversion which included review and clean-up of the present files whether manual or automated.
- Preparation and publication of clerical and operating procedures.
- Conversion of manual files to machine readable form or conversion and reformatting of existing computerized files.
- Training of the personnel who would utilize the system.
- Pilot operation of a small segment of the data in the new system and/or parallel operation of the new system with the old.
- Miscellaneous project tasks.
- Capital expenditures for equipment or packaged software.

A subtotal was taken at the end of those items which are usually charged to expense in the year in which they occur. The final total included capital expenditures which were to be amortized over five years.

Personnel costs such as those associated with system analysis, programming, user personnel, and project management were based on actual salaries paid plus overhead. Keypunch and computer costs were based on established internal billing rates. Since it was decided to buy a software package, there are no entries in the Implementation Cost worksheet for system programming. Programming to adapt to the package is included under "system modifications."

Once all costs had been accumulated it was necessary to indicate their occurrence over time by the use of a PERT chart or Gantt chart. A development schedule was prepared for the proposed system; and as the costs for each task were analyzed, they were estimated on an optimistic, most likely, and pessimistic basis. While the Implementation Cost worksheet shows the optimistic total cost only, separate worksheets were developed for the most likely and pessimistic estimates as well as for the out-of-pocket costs. The different levels of estimated costs could also have been approximated by the application of contingency factors to the various cost elements in the initial worksheet.

These various levels of costs were not consolidated into an average for presentation to management. Instead they were presented in a manner that clearly showed management the extent of the risk involved in the proposed project. The anticipated variance from the most likely estimates was from 13 percent under to 15 percent over.

A second worksheet (right) was used to analyze annual operating costs. This sheet was prepared for the anticipated life of the system or for five years of operation, whichever is shorter. The example illustrates five years' costs only. The elements of the annual cost of operation are:
- Computer costs, generally based on time against an hourly rate or representing the total cost of equipment dedicated to the application.
- Supplies, including cards, magnetic tape, forms, and similar computer or general office items.
- Operating personnel, generally user organization members and administrative management.
- Maintenance cost of the system, including systems and programming personnel, computer time for test and assembly of programs, and other expenses.
- Overhead, including floor space, electricity, telephone service, and labor related costs.
- Amortization of system implementation cost, including capital expenditures.

Operating cost worksheet

Elements of annual cost	Present system		Alternative system					
			Optimistic		Most likely		Pessimistic	
	Employees	Amount	Employees	Amount	Employees	Amount	Employees	Amount
Data processing	14.5	$254,600	2.5	$30,000	3	$36,000	4	$48,000
Supplies & forms		13,300		10,700		11,900		12,800
Operating personnel	94	583,425	76	482,175	85	527,550	90	561,825
Overhead		252,817		208,942		228,602		243,457
Maintenance cost:								
Personnel	0.2	4,800	0.3	7,200	0.3	7,200	0.3	7,200
Computer time		17,000		10,000		11,700		12,300
Other								
Amortization of special equipment or software				7,600		7,600		7,600
Sub-total	108.7	$1,125,942	78.8	$756,617	88.3	$830,552	94.3	$893,682
Amortization of implementation cost				26,638		23,034		35,229
Final total	108.7	$1,125,942	78.8	$783,255	88.3	$853,586	94.3	$928,911

Operating cost summary

Year	Present system		Alternative system					
			Optimistic		Most likely		Pessimistic	
	Employees	Amount	Employees	Amount	Employees	Amount	Employees	Amount
1	108.7	$1,125,942	78.8	$783,255	88.3	$853,586	94.3	$928,911
2	119.5	1,238,536	81.9	822,418	91.8	904,801	99.9	1,003,224
3	131.4	1,363,389	85.2	863,539	95.5	959,089	105.9	1,083,482
4	144.5	1,498,628	88.6	906,715	99.3	1,016,635	112.3	1,170,160
5	158.9	1,648,491	92.2	1,045,336	103.3	1,077,633	119.1	1,263,773
Total	158.9	$6,873,986	92.2	$4,421,263	103.3	$4,811,744	119.1	$5,449,550
			Capital expenditure		$88,000			

A subtotal was taken at the end of those elements which represented cash costs (or costs that affect the profit and loss statement) with a final total that included the "non-cash" costs of implementation that have been charged to expense when they occurred.

Only employee count and dollar amount were shown for the present system. These two measures were also used for the proposed alternative; but within the alternative, an optimistic, most likely, and pessimistic figure was shown. The alternative costs varied from the most likely figure by 8.2 percent under to 8.8 percent over.

These annual worksheets were consolidated into a five-year operating cost chart for presentation to management. The annual costs reflected the impact of increased levels of business expected in future years. The totals indicate that a savings in operating cost of 30 percent of the cost of the present system is most likely.

EVALUATING THE BENEFITS

The benefits of any proposed alternative can be either tangible or intangible. The tangible benefits can be calculated and put in tabular form (page following).

Cash flow analysis

Year	Present system	Alternative system			
		Implementation	Operation	Earnings	Total
1	$1,125,942	$164,584	$1,125,942	–	$1,290,526
2	1,128,536	30,120	921,676	$(6,000)	945,796
3	1,362,389	–	891,998	(28,000)	863,998
4	1,498,628	–	945,517	(33,000)	912,517
5	1,648,491	–	1,002,248	(37,000)	965,248
Total	$6,763,986	$194,704	$4,887,381	$(104,000)	$4,978,085

Payback analysis

		Alternative system		
Year	Present system	Total cost	Saving (loss)	Saving (loss)
1	$1,125,942	$1,290,526	$(164,584)	$(164,584)
2	$1,238,536	$945,796	$292,740	$128,156

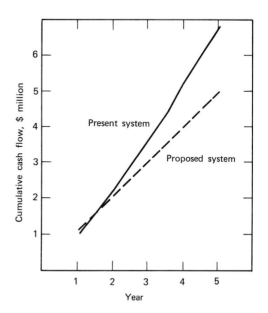

Charts are shown for one alternative. It may be necessary to present several cases to management; but, in this case, only the present system and its most desirable alternative were presented.

The cash-flow analysis shows the present system and the proposed alternative for a five-year period. The projected operating costs of the present system are brought forward from the worksheet previously prepared. For the proposed alternative the implementation costs are also brought forward from the worksheet. Operating costs included the cost of operating the present system until it would be replaced by the proposed alternative system (1.25 years in this example) and the operating cost of that system brought forward from the worksheet. Earnings directly attributable to this system resulted from the anticipated sale of new services and were entered as negative figures (enclosed in parentheses) with the total being a net amount. This chart used the "most likely" figures from previous charts, but it could have been repeated for each of the three levels of cost estimated. On this basis a cumulative savings of $1,895,901 was projected for the five-year period.

The final exercise was a payback analysis. Only two years need to be considered to arrive at a positive figure in the cumulative savings column (actually, one year and one month). The costs of the present system were brought forward from the annual totals for the alternative in the cash-flow analysis. The savings (loss) column was computed for each year by subtracting the total cost of the proposed alternative from the cost of the present system. The payback period was determined by the point at which the cumulative savings (loss) column becomes a positive figure.

There are several areas of intangible benefits that should be presented to management decision makers. While these were not used in our example, they could also include the following. The morale of the user organization staff may improve (with resultant increased output) as a result of the attention paid to employees in the area during design and implementation of a new system. As the famous Western Electric (Hawthorne) studies indicated, increased output frequency results from increased management attention without regard to actual improved working conditions.

Improved statistical information may add to business earnings by improving decisions and performance. The ability to control operations such as production is heavily dependent upon the speed and accuracy of processing data. A cybernetic feedback loop can only be effective if the required data arrive in time to properly control the process being measured. The implementation of computer oriented systems may serve to improve the firm's image or by analysis of sales and market research data to more properly direct the thrust of the company's marketing efforts.

Although the foregoing items are classified as intangible benefits, it is not unreasonable to ask user management to place a dollar value on some of these benefits. These dollar values can be estimated on the basis of such things as improved operational capability, decreased product rejection rate, increased marketing potential, and similar items. Estimates of dollar value so created can be presented to management by inclusion in the earnings column of a cash-flow chart and in a payback analysis chart. Such charts should be clearly labeled to indicate that they included the estimates of value placed on intangible benefits.

ESTIMATING THE COST OF SYSTEM IMPLEMENTATION

LOUIS FRIED

Estimating systems development and programming is a difficult and time-consuming task not generally recognized as a major overhead cost of the systems and programming functions. (In fact, estimating represents a significant drain of available manpower time.) Furthermore, estimates establish a standard against which performance may be measured . . . often without validity.

The requirement for estimates forces the edp executive to consider not only the project itself, but the attitude of management toward that project. If management views the project with favor, it would like an estimate that demonstrates justification for the project and it will excuse or ignore overruns. If it is doubtful or negative about the project, they would like the estimate to be realistic or even high, diminishing the justification for the system and providing an excuse for refusal.

Finally, the edp executive is required to provide estimates before sufficient data has been collected to determine the scope of the project. As a result, poor estimates on major projects are the rule, rather than the exception, in industry today.

EARLY ESTIMATE NEEDED

The reason that the "customer" (the line executive) within the company requires an estimate as early as possible in the investigation of a possible application is obvious. He must make the economic decision as to whether or not the application is worth doing. The line executive also must decide if he can afford the application and establish a budget or plan to pay the implementation cost.

Unfortunately, management often feels that designing and implementing a business system are similar to designing and building a product. There are, of course, some parallels (Table 1); but there are some notable exceptions (Table II). It is difficult for management

---

SOURCE: Fried, L., "Estimating the Cost of System Implementation," *Data Processing Magazine,* March-April, 1969, pp. 77–91.

TABLE 1 Comparison Between Manufacturing Functions and Systems Implementation

MANUFACTURING FUNCTIONS	SYSTEMS IMPLEMENTATION
1. Customer requirement established	1. Problem recognized
2. Customer specifications drawn and Request for Quotation released	2. Problem definition, system survey
3. Applications engineering study	3. System synthesis
4. Bid or quotation	4. System proposal
5. Product engineering	5. System specification
6. Manufacturing engineering	6. Program definition
7. Production	7. Programming, manual writing, etc.
8. Quality Control	8. Systems testing
9. Prototype test or first article qualification	9. Parallel operation
10. Delivery	10. Implementation

TABLE 2 Dissimilarities Between Manufacturing Functions and Systems Implementation

MANUFACTURING FUNCTIONS	SYSTEMS IMPLEMENTATION
1. Production standards available	1. Production standards often not applicable
2. Performance a factor of group average effort	2. Performance a factor of background and speed individual aptitude
3. Operations clearly defined	3. Operations require creative skills
4. Specifications known from customer	4. Specifications to be developed as part of project
5. Product to meet limited flexibility requirements	5. System to provide maximum flexibility
6. Limited coordination needed	6. Constant coordination and approval required

to understand the size of the estimate, the amount of work necessary to arrive at an estimate, and any overrun of cost by the project.

It is almost inconceivable that the complexities of a system frequently exceed those of most products. In fact, there is often little similarity between systems bearing identical names (such as Payroll or Accounts Payable) that exist in different companies.

This accounts for the notable lack of success or use of computer manufacturers' application "packages." Very few organizations are willing to change to fit a system . . . *nor should they.* The effective system must be designed to fit the organization.

The unique elements of an application constitute one of the major constraints on the predictability of systems and programming cost.

In addition to these elements, changes in the scope or approach to solutions *after* the original estimate is made, may invalidate that estimate.

Such changes almost inevitably occur since the feasibility study cannot possibly recognize or forsee all the factors involved in the system.

ESTIMATING REQUIRES ANSWERS

These problems relating to estimating require that several questions be answered.

1. What are the major elements of systems and programming?
2. Are systems analysis and programming equally subject to estimation?
3. What methods of estimating are available?
4. Are the available methods realistic?
5. Is accurate estimating possible?
6. Is there a formula for estimating?
7. How should management view estimates?

Most published studies of estimating have had limited benefit due to the failure to clearly define the elements of work within each major function. It should be recognized that each organization varies in the particular elements that may be assigned to the analyst or the programmer. For example, which function is responsible for preparing record layouts, keypunch instructions, or control instructions.

Since a comparative study of estimating is presented later in this article, it is necessary to define the elements on which the study is based.

ELEMENTS OF SYSTEMS ANALYSIS

Systems analysis is divided into eight primary elements.

Problem Definition: investigating the problem or current system in sufficient detail to present alternative suggestions for solution or improvement. The problem definition phase of the systems study should also result in determining the scope and objectives of the solutions.

Systems Survey: gathering, recording, and analyzing the facts relating to current operation of the area being studied. This phase should provide the detailed confirmation of the problem definition.

Synthesis: creating the methods for meeting the objectives delineated in the problem definition.

Systems Development: defining the EDP system, programs, controls, forms, and I/O requirements. This includes coordinating all definitions with potential system users for approval and documenting the system.

Analyst-Programmer Consultation: guiding the programmers working on the project by answering questions related to the definition and resolving any discrepancies in the definition.

Systems Testing: creating or obtaining test data that can be used to simulate all possible data conditions in the system; observing and approving the final results of a simulated run of the whole system.

Procedure or Manual Preparation: writing and publishing the manual or procedures necessary to operate the system.

Implementation: assisting in supervision of pilot or parallel operation, coordinating conversion to the new system, and training of personnel in the operation of the new system.

The documentation performed in the Systems Development phase usually contains the following items:

- Systems Abstract—a verbal description of the system.
- System Flowchart—a charted representation of the system in its organizational context.
- Program Definitions—verbal descriptions of each computer program, its input, and necessary processes.
- Card and Record Layouts—illustrations of data fields and their characteristics within each specific record.
- Input Document Samples
- Output Report Layouts or Samples
- Keypunch Instructions
- Control Procedures
- Paperwork Flowcharts

ELEMENTS OF PROGRAMMING

The four primary elements of programming are:

Program Design: often referred to as "preparing to code," this function includes study of the program definition, logic flowcharting, file organization, etc., which are required before coding can start.

Coding: the actual writing of the program.

Check-out: assembling, testing, and debugging a program until it is proven operational. This function is not concluded until the system is implemented and accepted by the "customer."

Documentation: the completion of those documents necessary to provide a record of the intent and design of the program and to operate the program. This documentation consists of both program and operating documentation.

1. Program Documentation:

- Logic Flowchart—a symbolic representation of the program's logic.
- Assembly Listing—a list of the coding and its machine language translation.
- Sample Program Output—a report sample or a printout of a file created by the program.
- Static Test Data—test data prepared by the programmer to meet the specified conditions of the program definition.
- Source Program—the program deck of tape as written in the coding language.

2. Operating Documentation:

- Object Program—the machine language program ready for operation.
- Parameter Cards—control cards for the program that may be required by the Operating System.
- Operating Instructions—describing how the computer is set up to run the program. These instructions also define any unusual conditions for which the program may halt.
- Off-line Processing Procedures—describing operations not performed on the computer.
- System Operating Flowchart—symbolically representing the relationships among all programs in the system.

ESTIMATING SYSTEMS ANALYSIS

Table 1 draws parallels between systems implementation and the production process. The inference made is that, as many line managers insist, the cost of systems implementation can be estimated in a manner as reliable as the cost of a product.

This inference is correct *insofar as the parallels drawn are true.* For example, if a standard number of labor hours per unit is used to estimate product cost and no comparable standard exists for systems and programming units, then no direct comparison is valid.

It is important to note that parallels between the production process and the systems elements of problem definition, systems survey, and synthesis (the Systems Specification Phase) do not exist. These elements are the ones that establish the parameters, specifications, and scope of the application.

The closest parallel that can be drawn to this phase of systems development is that work done by the *customer* in establishing specifications for the product he is ordering. It may also include some elements of the product specification work done by the "applications engineer."

In any event, that effort necessary to define the product sufficiently for the prospective producer to prepare a fixed-price bid for its manufacture is generally *not included in the direct cost of the product itself*. The Department of Defense has recognized this fact and established the so-called "Product Definition Phase" as a cost-plus development phase preceding the actual fixed price contract for production.

When applied to the systems and programming activity, this concept is so important that Charles P. Lecht says:

> One of the most catastrophic mistakes which a computer programming manager can make in estimating is to assume the existence of a project upon receipt of the External Functional Specifications ... Performance of the Analysis Phase must occur before any accurate estimating can even begin to be achieved.[1]

Lecht identifies the Analysis Phase as occurring between receipt of the external Functional Specification (developed by the "customer") and the Programming Functional Specification from which programs will be written.

Dick H. Brandon calls for a problem analysis phase to determine the detailed specifications of a solution. He suggests that the "problem Analyst" be a systems analyst expert in the particular application and possibly a member of the using department.[2]

Professor John Dearden maintains that the entire task of "systems specification" should be done by analysts in the user departments with the functions of data processing implementation (EDP system design only) and programming performed by a centralized group in the same organization as the EDP equipment.[3]

ESTIMATES NOT ACCURATE

The experience of these authors and of many systems or EDP executives illustrates the fact that the cost of that portion of the systems function concerned with problem definition, systems survey, and synthesis *cannot be estimated with any degree of accuracy*. Any estimate prepared for these elements will generally be based on the previous experience of those doing the estimating. It cannot, however, be related to any consistent set of standards.

The cost of these elements varies with:

a. The nature, scope, and complexity of the system
b. The experience of the analyst in that application
c. The level of the user organization from which the effort is being directed (higher level reduces cost)
d. The exposure of user personnel to EDP systems and methods
e. The coordination required to establish specifications (higher level decision-making can reduce the cost of coordination)

One of the chief reasons for management dissatisfactions with the EDP organization is due to cost overruns and schedule slippage. These conditions often arise from the inclusion of the Systems Specification Phase in the systems estimate.

To help solve this problem, several firms today have established independent System Requirements Groups in the "customer" organization. These groups define the general system and its inputs and outputs, i.e., complete the Systems Specification Phase.

Estimates for the balance of the systems and programming work are then prepared by the EDP group. Generally, however, the EDP organization has failed to communicate to line management the fact that the Systems Specification Phase cannot be estimated and must be considered a part of the organization's overhead in the same manner as a feasibility study.

CLOSER ESTIMATION POSSIBLE

The remaining elements of the systems function comprise two phases:

a. The Systems Definition Phase includes the elements of systems development, analyst/programmer consultation, and procedure or manual preparation.
b. The Systems Implementation Phase includes the elements of systems testing and implementation.

Reasonably accurate estimates may be prepared for the Systems Definition Phase based on the number of programs to be defined, the complexity of the programs, and the number of interfaces with other systems. Statistical bases for estimating the system development element can be established in much the same manner as for programming (described later). The analyst/programmer consultation element is constrained to a duration equivalent to that of programming and debugging the system. It can be reasonably estimated after the project schedule is developed.

The procedure or manual preparation element can also be estimated based on statistical evidence for time per page. There is some element of risk in this estimate arising from the need to coordinate procedures. Any element of the systems task that requires coordination with user personnel is subject to variations as described above in the Systems Specification Phase. If, however, the specifications have been clearly stated and coordinated, this risk in the Systems Definition Phase is sharply reduced.

The need to coordinate design proposals and obtain management approvals can act as a major inflationary factor to both elapsed and actual time used in the Systems Specification Phase. Ideally, the system design is independent of the invididuals within the user organization, the individual being, within realistic limits, a replaceable component of the system.

In practice, however, the system reflects the characteristics of the analyst who designed it and the management who approved each segment of the design. The completed design is invested with the concepts and prejudices of the line manager participating. If this management changes during the development of the system, the entire design may have to be redeveloped and re-coordinated.

Similar excessive development costs can arise from changed management thinking due to altering business needs. Inflated coordination costs can also result from the need to coordinate with too many line personnel when centralized project direction is not assigned by management.

ESTIMATING PROGRAMMING

While it appears that accurate estimating of many elements of the systems function is not practical, the estimating of programming based on a clearly defined programming specification from the systems analyst is definitely feasible.

There is, however, a wide range of methods from the extremely simple (with little cost) to the highly complex, very costly estimating process. Each of these may be valid for use depending upon:

a. the size of the project being estimated
b. the need for accuracy in the estimate
c. the size of the programming group

Obviously, a great deal of cost should not be incurred in preparing an estimate for a relatively small project. Also, if only an approximation is required by management, an extensive study is wasteful.

It is not quite so obvious that the size of the programming group can be a factor in choice of the estimating method. Statistical data related to average programmer performance can be reliably applied to major projects spread over a large group of programmers. Average performance data can be misleading, however, when applied to either a small project assigned to a single programmer or a large project assigned to a small group of programmers. In these cases the previous performance of the *individuals* is far more pertinent than averages.

For this reason, while some generalized guidelines can be established for large project estimating in large programming organizations, these guidelines cannot be applied in the majority of smaller organizations. As a result, a primary constraint in estimating programming cost is that generalized methods for estimating programming can be defined, but generalized average guidelines or *prefabricated* estimates do not exist.

ESTIMATES EMPIRICALLY BASED

All estimates are based, in some way, on experience. This can be in the form of personal experience of the estimator, historical data collected relating to similar projects, data collected relating to previous work by the same group or individual, or data acquired from reference sources.

The most reliable experience upon which estimates can be based is that most closely paralleling the project to be estimated. Therefore, the ideal estimate would be based on having previously completed the same project with the same staff under the same conditions (such as language, hardware, definition, etc.). This ideal is obviously not practical, so experience must be developed to reflect data on those elements which are common from one project to the next.

In developing an estimating method applicable to extremely large projects, E. A. Nelson identifies 97 possible variables that can affect the cost of programming.[4] Such an analysis is not often economically feasible so that most variables must usually be assumed to be constants. Estimates can then consider a few major variable factors within the areas of hardware, software, personnel, procedures, and the complexity of the specific project.

Hardware Variables

Is there sufficient memory and peripheral devices to accommodate the program easily? Reducing program size to fit the machine is time-consuming work.

Software Variables

1. Is the program able to utilize available macroinstructions, subroutines, or utility programs? "Canned" sort, merge, or other utility programs can be used to save work.
2. The programming language used can influence the cost. Some studies have indicated that on average business applications a problem-oriented language (such as COBOL) saves time over a machine-oriented language (such as Assembler Language or Autocoder). For problems requiring complex processing and calculations or maximum efficiency a machine-oriented language will probably be best.[5]
3. Is the assembler or compiler reliable? New software issued by most computer manufacturers can be expected to have "bugs." Some extra effort will be required to debug both the program and the vendor software if a recently released compiler is to be used.
4. Diagnostic aids vary between compilers and between manufacturers. The better versions can considerably reduce program debugging time.

Personnel Variables

1. Is the programmer experienced with the organization and the application? This type of experience can save time in communication and reduce errors in interpretation of commonly used terminology for the organization or application.
2. Is the programmer experienced with the programming language and the equipment being used?
3. Is the programmer experienced with the standards, conventions, and documentation requirements of the organization?
4. Is the programmer fast, average, or slow? Is he a senior or junior man?

Program Complexity Variables

1. If the program stands alone and does not have to relate to other programs, less effort is required for implementation.
2. Is creative logic planning required? Innovative thinking requires more time than routine program development.
3. Programming complexity can be judged by estimating the number of source instructions that must be coded, by the number of files processed by the program, and by the number of record layouts processed in the program. Increases in all these variables require increases in programming time.

NEED HISTORICAL DATA

The key to more accurate estimating is in perfection of the historical information to establish a base for estimates and to eliminate variables. For example, if statistical data have been collected for some time on a group using COBOL, then estimates for future programming in COBOL can eliminate those variables having to do with the language used, the compiler used, and the diagnostic aids. These elements have become constant for the group.

The history of one group cannot easily be transferred for use by another group. This condition exists due not only to the above variables, but also to the differences in the definitions of the programming function. This can be illustrated by the results of a one-year

TABLE 3 Comparison of One-Year Study With External Information

	DELANEY[7]	RCA[8]	BRANDON[9]			TEST GROUP[10]	
			Simple	Average	Complex	Autocoder	Assembler
Program Design	25%	30%	11%	27.5%	30%	11.9%	7.3%
Coding	31%	20%	11%	24.0%	27%	30.0%	29.4%
Checkout	31%	45%	50%	34.5%	33%	39.9%	46.9%
Documentation	13%	5%	28%	14.0%	10%	18.2%	16.4%

study of a programming group compared with available external information in Table 3.[6]

The table indicates that variances exist not only between groups but between the same group using different languages, operating systems, and computers. But meaningful historical information can be developed for the group.

There have been several attempts to design estimating methods that are independent of those variables related to one or more of the major variable factors. The personnel variable is most generally discarded in an attempt to generalize the method for application in many installations.

Dick Brandon has presented several formulae for estimating, based on assigning a fixed number of man-days for levels of complexity, program size, and the major elements of the programming function (logic, coding, documentation, etc.). These formulae are presented as being valid for specific equipment, memory size, programming language, average program size, and a specified documentation standard.[11] Brandon also indicates the value of building performance history to provide a base for future estimating.

ESTABLISH GUIDELINE

RCA has established a guideline for estimating programming manpower requirements.[12] This method considers the variables of program size, program type (edit, update, etc.), programmer experience, language, and elements of the programming function. The working formula for computing elapsed time estimates is:

$$T = \frac{I}{22PR}$$

where T = time in months
I = total number of instructions to be coded
P = number of programmers
R = number of instructions/day/man
22 = constant (work days per month)

The R factor is determined on the basis of a table of production rates for language and programmer experience.

A table of average program sizes (number of machine instructions) is provided for the different program types.

The danger in estimating by expected lines of coding is that individual programs vary considerably in complexity. For example, an update program can vary in complexity by a factor of five or more depending on the nature of the system and files. Not only do instructions vary from program to program, but also from programmer to programmer. An experienced programmer may save as high as 50 percent of the number of coded instructions necessary.

The number of instructions is also affected by the standards (for editing, label checking, etc.) of the installation, and by the macroinstructions available for use in the program.

Probably the most popular method of estimating is the SOP method (seat of the pants method). This method consists of the estimator using his past experience and his knowledge of the program requirements to estimate the man-hours and elapsed time required. Because the estimator is generally familiar with the installation, its standards, its personnel, its language, and often with the system requirements, the SOP method can be relatively accurate.

METHOD TOO SUBJECTIVE

The SOP method suffers from a personal bias. Quite often those in charge of estimating have attained a senior or supervisory position due to their own superior talents and capabilities in programming. As a result, they may tend to project their own rate of work into the estimate. When this occurs, SOP estimates may be far short of the actual effort since the work may be done by average, rather than superior, programmers.

It is fairly obvious that no panacea to the programming estimating problem is available in the form of a single formula, method, scale, or standard. Instead, several methods must be used in a manner appropriate to the situation. The extensive and costly method delineated in E. A. Nelson's "Handbook" is certainly appropriate to large-scale contracting, but rarely appropriate to the internal operation of an organization.[13]

For the estimating normally required in intraorganizational work, three methods are proposed depending on the size and scope of the applications considered. These are:

1. The SOP method for small projects or minor changes to existing systems
2. The SOP/PERT method for major projects
3. The Statistical method for major projects

The SOP method of estimating is simply based on the knowledge of the estimator relating to the complexity of the project, the past performance of available personnel, and the operating environment of the installation. The SOP method consists of an experienced, educated guess.

SELF-ESTIMATION AVERAGED

The SOP/PERT method, on the other hand, lends reliability to an estimate by combining the experience of several people in a statistical approach. The program specifications are presented to several programmers. Each programmer prepares three estimates for every program—optimistic, most likely, and pessimistic. The programmer is expected to estimate the time it would take him to write the program. This method makes use of the natural bias of the estimator.

Estimates are tabulated and averaged for a final optimistic (O), most likely (M), and pessimistic (P) estimate. The final three figures may be presented in that manner or they may be combined in a weighted average using the formula:

$$E = \frac{O + 2M + P}{4}$$

This method may be used not only for estimating actual time on the job, but for estimating elapsed time. The resulting elapsed time estimates can be utilized in a PERT network for project control. Again, this method makes use of the intimate subjective experience of the estimators with all the variables of the situation.

The Statistical method is a relatively objective method of estimating. It is based on the compilation of programmer performance and project performance data. The following reports contribute to this information.

1. Percent of time for each programmer and for all programmers by

element of effort (i.e., preparing to code, coding, documentation, testing, debugging). This report is without regard to specific program but should be prepared separately for new project work versus maintenance work.
2. Average time and percent of time by element of effort within program type. Program type differentiates between the following four general classifications.
   a. Type A. Utility sort or merge.
   b. Type B.
      (1) Report printing from consecutive file with multiple total levels.
      (2) Extract from one file, reformat selected records and create one or more output files.
      (3) Card to disk or card to tape with editing (without batch balancing).
      (4) Pass file, accumulate totals by categories working through table lock-up.
   c. Type C.
      (1) Update with limited processing and editing.
      (2) Card or disk or tape with editing and batch balancing against predetermined control.
      (3) Match two files (detail against master file), create new output file, limited processing. Print unmatched records.
   d. Type D. Same as Type B with expanded processing and/or multiple files in or out.
   A more detailed listing by element within program type should be prepared to detect any unusual single programs which might distort the average (in either direction).
3. Further reports may be utilized to provide additional information for improved decisions. One might be by programmer within type of report, to give some indication of who should be assigned specific projects.

HIGHLY RELIABLE

Reports such as these may be used with some high degree of confidence providing major variables remain relatively constant, If, however, there is a change in programming language, operating system, computer, documentation standards, or a large number of personnel at one time, new data must be collected for some time before the reports will again be reliable.

Experience in one installation indicates that estimates based on statistics derived from second generation programming were an average of 50 percent low when the job was performed for third generation equipment (using a third generation language and operating system).

The reports described above that are oriented toward programmers are intended to aid in selecting the team for the project. Estimates prepared without selecting the programming team have a higher risk factor unless the estimate is made for a large project including many programmers.

The program type reports are intended to form the primary basis for preparing estimates. From these reports a range of time can be constructed for each type. For example, a recent study performed with a group utilizing a third generation, tape/disk computer, assembler language, and a tape/disk operating system showed the following ranges:[14]

PROJECT HOURS

Type A	4 to 6
Type B	35 to 55
Type C	85 to 115
Type D	120 and Up

The complexity of the program determines which part of the range is applicable. It may be completely outside the range or types indicated. If this is the case the program may be a combination of two types or may have to be estimated by the SOP method supported by the available statistics.

CONCLUSION

Management must know the estimated cost of system development and implementation in order to make proper economic decisions. Experience and evidence, however, indicate that the Specification Phase of system development cannot be reliably estimated. The cost of the balance of the system definition can be estimated with some reliability and the programming can be estimated upon completion of the program specifications.

To achieve maximum management satisfaction with system implementation efforts, the following procedure is recommended.

1. Establish system and reporting requirements and specifications through the use of an activity phase completely independent of the system definition phase. This can be done by establishing a system specification group within the "customer" organization or by devoting systems analyst personnel from the EDP systems group to such an effort *before an estimate is required*.
2. Once a System Specification has been completed *in writing*, a reliable estimate of the systems work required for implementation can be prepared. This can be accompanied by a tentative estimate of programming effort; but since the programs have not yet been designed, no reliable estimate can be made.
3. Upon completion of program specifications, a reliable estimate of the programming cost can be prepared based on methods outlined in this article.

Two rules are basic to successful measurement of performance against estimates. First, reliable estimates cannot be prepared without written specifications of the work to be performed. Second, changes in the scope or specifications must be adequately reflected by changes to the original estimate.

[1] Lecht, Charles P., *Management of Computer Programming Projects,* American Management Association, New York, N.Y., 1967, p. 50.
[2] Brandon, Dick H., *Management Standards for Data Processing,* Van Nostrand Co., Princeton, N.J., 1968, pp. 34-36.
[3] Dearden, John, *Computers in Business Management,* Dow-Jones, Irwin, Inc., Homewood, Ill., 1966, pp. 164-166.
[4] Nelson, E. A., *Management Handbook For the Estimation of Computer Programming Costs,* SDC, Santa Monica, Calif., Commerce Clearinghouse No. AD648 750, pp. 119-131.
[5] Software Sciences Corp., *Dimensions in Data Processing,* SSC, New York, N.Y., 1968, p. 6.
[6] Study of statistical reports of programming group during 1967-68 at Bourns, Inc.
[7] Delaney, William A., Predicting the Costs of Computer Programs, *Data Processing Magazine,* October 1966. The percentages presented by Mr. Delaney included 20 percent analysis, which was removed for purposes of this comparison.
[8] RCA, *Manpower Management Techniques,* RCA internal publication, 1967. RCA assigns much of the documentation to systems or clerical personnel.
[9] Brandon, Dick H., *op. cit.,* pp. 249-298.
[10] Jeffries, Stanley, Bourns, Inc. internal reports. The analysis of group activity using Autocoder was performed in September 1967. The analysis using RCA assembler language for the Spectra 70 was performed in July 1968 after six months of using the new language. Personnel remained relatively constant during this period as did documentation requirements.
[11] Brandon, Dick H., *op. cit.,* pp. 249-298.
[12] RCA, *op. cit.*
[13] Nelson, E. A., *op. cit.*
[14] Jeffries, Stanley, internal report, *op. cit.,* August 1968.